Industrial Applications of Nanoceramics

Industrial Applications of Nanoceramics

Edited by

SHADPOUR MALLAKPOUR
Department of Chemistry, Isfahan University of Technology, Isfahan, Iran

CHAUDHERY MUSTANSAR HUSSAIN
Department of Chemistry and Environmental Sciences, New Jersey Institute of Technology, Newark, NJ, United States

ELSEVIER

Elsevier
Radarweg 29, PO Box 211, 1000 AE Amsterdam, Netherlands
125 London Wall, London EC2Y 5AS, United Kingdom
50 Hampshire Street, 5th Floor, Cambridge, MA 02139, United States

ISBN: 978-0-323-88654-3

For Information on all Elsevier publications
visit our website at https://www.elsevier.com/books-and-journals

Publisher: Matthew Deans
Acquisitions Editor: Ana Claudia A. Garcia
Editorial Project Manager: Isabella C. Silva
Production Project Manager: Fizza Fathima
Cover Designer: Greg Harris

Typeset by MPS Limited, Chennai, India

Contents

List of contributors

Sameera Khathoon A.
Post Graduate and Research Department of Zoology, J. B. A. S. College for Women, Chennai, Tamil Nadu, India

Tanvi Agarwal
Department of Chemistry, Sri Venkateswara College, University of Delhi, Delhi, India

Md. Ahmaruzzaman
Department of Chemistry, National Institute of Technology Silchar, Silchar, Assam, India

S. Ajith
Department of Mechanical Engineering, Stella Mary's College of Engineering, Aruthenganvilai, Tamil Nadu, India

Khairunnisa Amreen
MEMS, Microfluidics and Nanoelectronics Lab, Department of Electrical and Electronics Engineering, Birla Institute of Technology and Science, Hyderabad, Telangana, India

V. Arumugaprabu
Department of Mechanical Engineering, Kalasalingam Academy of Research and Education, Krishnankoil, Tamil Nadu, India

Kantha D. Arunachalam
Faculty of Sciences, Marwadi University, Rajkot, Gujarat, India

Th. Babita Devi
Department of Chemistry, National Institute of Technology Silchar, Silchar, Assam, India

S. Kishore Babu
Department of Physics and Nanotechnology, SRM Institute of Science and Technology, Kattankulathur, Tamil Nadu, India

Arikta Baul
Department of Chemistry, Sri Venkateswara College, University of Delhi, Delhi, India

Bhooma Bhagat
Department of Chemistry, School of Technology, Pandit Deendayal Energy University, Gandhinagar, Gujarat, India

Bhaskar Chandra Das
Department of Chemical Engineering, University of Calcutta, Kolkata, West Bengal, India

Shashi Chawla
Department of Chemistry, Amity Institute of Applied Sciences, Amity University, Noida, Uttar Pradesh, India

Gayathri D.
Research Department of Zoology, National College, Trichy, Tamil Nadu, India

Rusham Das
Department of Pharmacy, School of Chemical Sciences and Pharmacy, Central University of Rajasthan, Ajmer, Rajasthan, India

Kalyanjyoti Deori
Department of Chemistry, Dibrugarh University, Dibrugarh, Assam, India

Uttakantha Dixit
Department of Mechanical Engineering, School of Technology, Pandit Deendayal Energy University, Gandhinagar, Gujarat, India

J. John Donald Raj
Department of Physics and Nanotechnology, SRM Institute of Science and Technology, Kattankulathur, Tamil Nadu, India

Mohammed Junaid Hussain Dowlath
Department of Anatomy, SRM Medical College Hospital and Research Centre, SRM Institute of Science and Technology, Kattankulathur, Tamil Nadu, India

Darul Raiyaan G.I.
Department of Biotechnology, College of Science and Humanities, SRM Institute of Science and Technology, Kattankulathur, Tamil Nadu, India

Tanya Garain
Department of Chemistry, Amity Institute of Applied Sciences, Amity University, Noida, Uttar Pradesh, India

Sanket Goel
MEMS, Microfluidics and Nanoelectronics Lab, Department of Electrical and Electronics Engineering, Birla Institute of Technology and Science, Hyderabad, Telangana, India

Priyanka Gogoi
Department of Chemistry, Dibrugarh University, Dibrugarh, Assam, India

Amit K. Goyal
Department of Pharmacy, School of Chemical Sciences and Pharmacy, Central University of Rajasthan, Ajmer, Rajasthan, India

Anjali Goyal
Department of Environmental Engineering, Delhi Technological University, Delhi, India

Kartika Goyal
Department of Chemistry, Sri Venkateswara College, University of Delhi, Delhi, India

Ramesh K. Guduru
Department of Mechanical Engineering, School of Technology, Pandit Deendayal Energy University, Gandhinagar, Gujarat, India

Shikha Gulati
Department of Chemistry, Sri Venkateswara College, University of Delhi, Delhi, India

Lovleen Gupta
Department of Environmental Engineering, Delhi Technological University, Delhi, India

Somoprova Halder
Department of Chemical Engineering, University of Calcutta, Kolkata, West Bengal, India

Chaudhery Mustansar Hussain
Department of Chemistry and Environmental Sciences, New Jersey Institute of Technology, Newark, NJ, United States

Arpan Kar
Department of Chemical Engineering, University of Calcutta, Kolkata, West Bengal, India

Sathish Kumar Karuppannan
School of Materials Science and Engineering, Gwangju Institute of Science and Technology, Gwangju, South Korea

Richa Kasana
Department of Chemistry, Amity Institute of Applied Sciences, Amity University, Noida, Uttar Pradesh, India

Areeba Khalid
Department of Environmental Engineering, Delhi Technological University, Delhi, India

Aruna Jyothi Kora
National Centre for Compositional Characterisation of Materials (NCCCM), Bhabha Atomic Research Centre (BARC), ECIL PO, Hyderabad, Telangana, India; Homi Bhabha National Institute (HBNI), Mumbai, Maharashtra, India

Gyanendra Kumar
Department of Chemistry, University of Delhi, Delhi, India

Sanjay Kumar
Department of Chemistry, Sri Venkateswara College, University of Delhi, Delhi, India

Shakila L.
Research Department of Zoology, Ethiraj College for Women, Chennai, Tamil Nadu, India

S. Sofiya Lawrence Mary
Department of Chemistry, Bharathidasan Government College for Women (Autonomous) (Affiliated to Pondicherry University, Pondicherry), Puducherry, India

Harish Neela Lingam B.
Department of Chemistry, Sri Venkateswara College, University of Delhi, Delhi, India

Sowmik Maji
Department of Chemical Engineering, University of Calcutta, Kolkata, West Bengal, India

Dhanraj T. Masram
Department of Chemistry, University of Delhi, Delhi, India

Kalisadan Mukherjee
Department of Chemistry, School of Technology, Pandit Deendayal Energy University, Gandhinagar, Gujarat, India

K.S. Prakash
Department of Chemistry, Bharathidasan Government College for Women (Autonomous) (Affiliated to Pondicherry University, Pondicherry), Puducherry, India

Prateek Rai
Department of Chemistry, Amity Institute of Applied Sciences, Amity University, Noida, Uttar Pradesh, India

Shilpa Rajkumar
Department of Physics and Nanotechnology, SRM Institute of Science and Technology, Kattankulathur, Tamil Nadu, India

Pragati Rathore
Department of Chemistry, Amity Institute of Applied Sciences, Amity University, Noida, Uttar Pradesh, India

Saptarshi Roy
Department of Chemistry, National Institute of Technology Silchar, Silchar, Assam, India

Subhasis Roy
Department of Chemical Engineering, University of Calcutta, Kolkata, West Bengal, India

Mohamed Khalith S.B.
Department of Chemistry, Sadakathullah Appa College, Tirunelveli, Tamil Nadu, India

R. Sarika
Department of Chemistry, Fatima College (Affiliated to Madurai Kamaraj University, Madurai), Madurai, Tamil Nadu, India

Diganta Sarma
Department of Chemistry, Dibrugarh University, Dibrugarh, Assam, India

V. Selvam
Department of Chemistry, The Madurai Diraviyam Thayumanavar Hindu College (Affiliated to Manonmaniam Sundaranar University, Tirunelveli), Tirunelveli, Tamil Nadu, India

Nandini Sharma
Department of Biological Science, Sri Venkateswara College, University of Delhi, Delhi, India

Uday Shashikumar
Department of Chemistry, Amity Institute of Applied Sciences, Amity University, Noida, Uttar Pradesh, India

Deepankara Vrushabhadas Shastri
Department of Physics and Nanotechnology, SRM Institute of Science and Technology, Kattankulathur, Tamil Nadu, India

D. Silambarasan
Department of Physics, The Madurai Diravsiyam Thayumanavar Hindu College (Affiliated to Manonmaniam Sundaranar University, Tirunelveli), Tirunelveli, Tamil Nadu, India

Gaurav Yadav
Department of Chemistry, National Institute of Technology Silchar, Silchar, Assam, India

CHAPTER 1

Synthesis and characterization of carbon-based ceramic nanocomposites

Gyanendra Kumar and Dhanraj T. Masram
Department of Chemistry, University of Delhi, Delhi, India

1.1 Introduction

Ceramics are nonmetallic inorganic solids that are used in various materials due to their outstanding physical and chemical properties. They have excellent features and possess high melting points, hardness, high mechanical strength, good temperature ability, excellent thermal stability, low density, and low electrical properties [1]. These features allow their use across a range of applications such as electronics, biomedical engineering, aerospace, chemical, and machinery industries [2]. Ceramic powder is made of ceramic particles and additives that make its use easier during fabrication of components. These mixtures are used through various manufacturing techniques such as slip casting, dry pressing, injection molding, tape casting, gel casting, and others to form the required simple shape and size of the composite material [3,4]. Traditional ceramic forming techniques have been well-studied but they have several disadvantages like high costs, long processing times, and the impossibility of producing pieces with extremely complex shapes. Furthermore, mechanical postprocessing work is required to obtain a sintered ceramic part with good surface quality and accuracy. This postprocess work is costly and time-consuming because of the natural high strength and brittleness of ceramic composites [5]. Ceramic matrix composites (CMC) are considered suitable for harder and more mechanical improvement of materials by including the use of low-dimensional materials like graphene and graphene oxide (GO) nanocomposites, nanofibers, nanotubes, or nanoplatelets [6]. The purpose of this strategy is to reduce ceramics' inherent fragility which has limited their use for structural materials. As a result, the majority of studies have focused on improving crack toughness. However, nanocomposites are one of the most demanding research areas due to new science and applications that have gradually been established in the last two decades [7–10]. It is very rapidly growing area because of versatile applications almost everywhere. There are various inorganic membrane materials have been developed with to improve chemical, mechanical, and thermal stabilities. However, graphene has been frequently used as a reinforcing phase for materials due to its excellent mechanical

Industrial Applications of Nanoceramics
DOI: https://doi.org/10.1016/B978-0-323-88654-3.00001-9

properties and special physical characteristics, such as theoretical surface area 2630 m^2 g and monolayer thickness of 0.335 nm [11,12]. Apart from the wonderful intrinsic mechanical and physical properties, the special thermal efficiency was provided by graphene and graphene-reinforced ceramic (GRC) [13]. Graphene's thermal conductivity is around 5000 W/m/K [14,15]. Thus the thermal application of GRC has been greatly encouraged owing to the excellent thermal conductivity of graphene [16,17]. The insertion of graphene-based nanostructures in ceramic matrices has allowed the production of new composites with excellent structural and functional properties. Graphene composites increase the toughness and wear resistance of monolithic ceramics or transform the electrical response from electrical insulators with Si_3N_4 ceramics [18,19]. Silicon nitride (Si_3N_4) is a high-temperature tolerant ceramic (up to 1500°C) and is often known to be the most reliable structural ceramic due to the development of the interlocking microstructure of R-Si_3N_4 reinforced with β-Si_3N_4 rod-like grains [20]. Current developments in graphene and GO have prepared the path for the development of ceramic nanocomposites [21,22]. A major problem in graphene-based ceramic composites is usually the inability of graphene to homogeneously disperse in the composite material; the agglomeration of graphene in the water and/or slurry matrix makes it extremely difficult to fully exploit its characteristics in the composites [23,24]. To solve this problem, graphene is chemically oxidized to develop GO. Further, GO reduces the mechanical and electrical properties of graphene by modifying a proportion of the sp^2 to sp hybridization [25]. Since the functional groups attached to the graphene nanosheet eliminate carbon as the key source of interactions between graphene and the surrounding matrix, the electrical properties of the composites are significantly affected [26,27]. Evidence of increased attention to graphene as being a suitable step for mechanical strengthening is given in Fig. 1.1. The Web of Science (WOS) and Scopus databases were used to search the literature [28], and it was discovered that the level of GRC research increased rapidly after 2012, although GRC has been explored since the discovery of graphene in 2004.

Carbon nanotubes (CNTs) have excellent mechanical properties, thermal durability, and chemical inertness, which are widely used for ceramic reinforcement [29]. CNTs are ideal candidates to be integrated into the ceramic matrix due to their superior Young's modulus, or tensile strength, which allows them to resolve the fragility described previously [30]. Recently, CMCs doped with multiwalled carbon nanotubes (MWCNTs) have been shown to have improved mechanical properties at room temperature, such as crack toughness and wear resistance [14,15]. The mechanical strengthening of CMCs by the addition of carbon allotropes is based on three diverse advantages [29]. First, the reinforcing step is often uniformly distributed at the nanometer scale. Second, stable suspensions of CNT dispersions are difficult to achieve due to excessive van der Waals interactions, which contribute to the forming of bundles and entanglements [28]. Furthermore, mechanical degradation occurs as a result of low

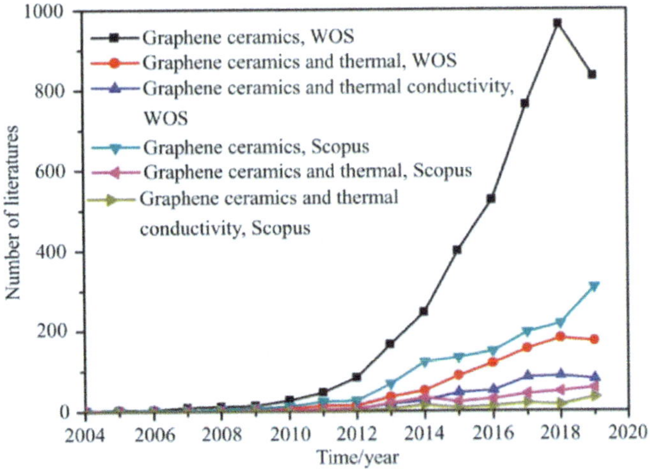

Figure 1.1 Trends in the investigation of graphene-reinforced ceramic (GRC) thermal properties. *GRC*, Graphene-reinforced ceramic. Source: *Reproduced with permission from F. Chen, K. Yan, J. Sun, J. Hong, Y. Zhu, Z. Huang,, From the research state of the thermal properties of graphene reinforced ceramics to the future of computer simulation, Ceram. Int. 46 (11) (2020) 18428–18445. https://doi.org/10.1016/j.ceramint.2020.04.265.*

CNT dispersion within the ceramic matrix. Third, the presence of nondispersed CNTs, which form aggregates in CNT-reinforced CMCs, result in residual pores and defects and lower mechanical resistance. Moreover, the strengthening phases are typically situated near the grain boundaries [31], but this location significantly restricts the potential improvements in fracture and wear properties since the toughening processes of the composite, such as crack-bridging and CNTs pulling back, can be reduced or not directly appear [32,33]. Finally, the failure of reinforcements has been attributed to weak bonds between carbon allotropes and the ceramic matrix. The sol–gel process has been considered for ceramic fabrication and more recently for the fabrication of alumina-based CMCs with CNTs [34,35]. The aim is to achieve good CNT dispersion by using liquid precursors, controlling the pH, and using high-power ultrasound to aid in the gelation process [36]. Furthermore, this method may facilitate intragranular placement of the CNTs, which are present during the very first stages of the forming of ceramic grains. However, the sol–gel approach can improve bond formation between the CNTs and the ceramic matrix by using functionalized CNTs and metallic hydroxides or alkoxides as precursors [37]. Consequently, the manufacturing problems with the inclusion of carbon allotropes into CMCs lead to inaccurate results and disparities in the reinforcement effect. However, a very large distribution of reported mechanical results can be found and characterization protocols are used for the toughness measurement. There are two most important methods for the measurement of fracture toughness: single-edge notched beam (SENB) and Vickers indentation fracture

(VIF). They require very different sample sizes, geometries, and models for the interpretation of the experimental results [38,39]. After all, the same elastic modulus is usually used in CNT compositions, which can lead to oversimplification [40]. Therefore, close interactions between alumina and CNTs are desirable with strong interfacial bonding [41,42].

1.2 Synthesis of ceramic-based composites

1.2.1 Mixture of preparation

GRC manufacturing typically consists of two steps: mixture preparation and sintering, which is shown in Fig. 1.2. Monolayer graphene [43], bilayer graphene, few-layer graphene (FLG), and multilayer graphene (MLG) are all different types of graphene nanoplatelets (GNPs). However, the agglomeration effect is easily induced in graphene due to the effects of van der Waals forces and conductivity [44]. The grain wrapping effect and heat transfer network will be difficult to construct once graphene agglomerates in

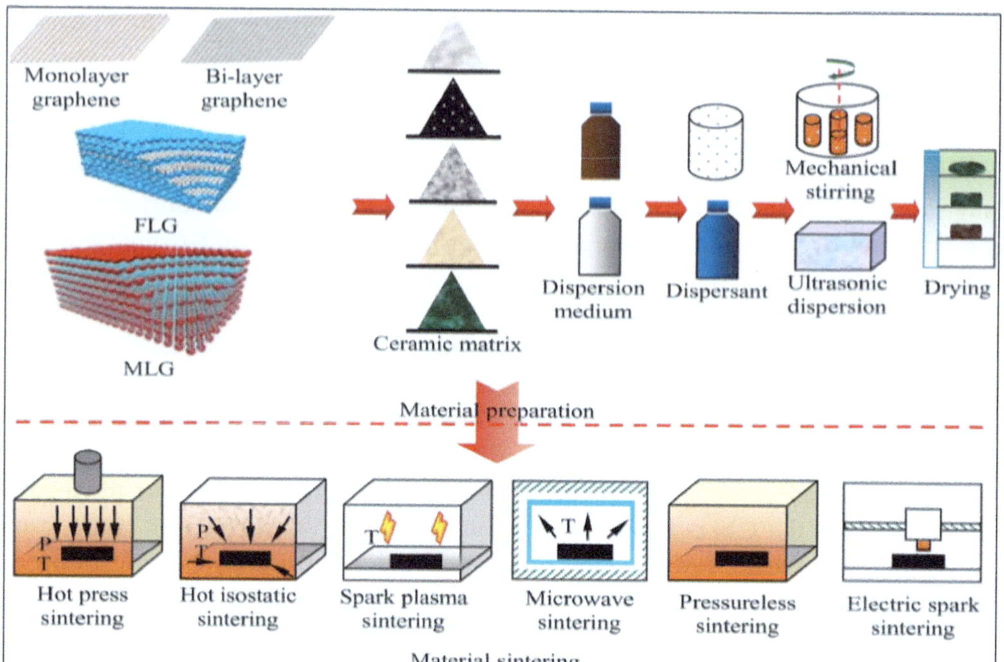

Figure 1.2 The manufacturing process of graphene reinforced ceramic (GRC). *GRC*, Graphene reinforced ceramic. Source: *Reproduced with permission from F. Chen, K. Yan, J. Sun, J. Hong, Y. Zhu, Z. Huang,, From the research state of the thermal properties of graphene reinforced ceramics to the future of computer simulation, Ceram. Int. 46 (11) (2020) 18428–18445. https://doi.org/10.1016/j.ceramint.2020.04.265.*

GRC, resulting in a weakening of GRC's thermal performance. As a result, overcoming the agglomeration effect has become an important aspect of ceramic material modification. In this case, there are effects of *N*-methyl-2-pyrrolidone (NMP) and alcohol solution of polyvinyl pyrrolidone (A-PVP) on the homogenization of graphene dispersion [45]. Until now, the most widely used method in graphene dispersion has been coupling the dispersion solvent with mechanical stirring and ultrasonic dispersion [46,47].

1.2.2 Sintering methods

Sintering occurs in mineral deposits naturally or as part of a production process for metals, ceramics, plastics, and other materials. Consequently, determining and selecting a suitable sintering method is important for improving GRC thermal properties. Fig. 1.2 shows the most widely used sintering methods for synthesizing GRC. There are various methods for the synthesis of GRC by spark plasma sintering (SPS) [44], hot press sintering (HP) [48,49], hot isostatic sintering (HIP) [50], microwave sintering (MWS), and pressureless sintering (PS) [46,47]. In addition, various methods are used for the clarification and reorganization of graphite nanoplates powder (GNP) into large amounts, as described [51] in Table 1.1.

1.2.3 Synthesis of multiwalled carbon nanotubes composites by sol–gel method

The sol–gel approach was used to develop Al_2O_3 matrix composites with MWCNTs [52]. To achieve optimal dispersion (Fig. 1.3), high-power ultrasound (15 minutes, 50 W) was used to disperse the MWCNTs into the sol. When the MWCNT bundles could no longer be separated optically, NH_3(aq) (pH = 11.6) was applied to the dispersion, and rapid homogeneous gelation (1–2 minutes) occurred, reducing the possibility of MWCNT aggregation after gelation. The gels were dried for 2 days at 40°C, milled in an agate mortar, and sieved (212 μm). To facilitate dihydroxylation of the aluminum hydroxides and eliminate moisture and other potentially undesirable traces present in the as-prepared gels, Boehmite-MWCNT composite powders were calcined for 1 hour (heating ramp: 10°C/min) at 600°C in an inert argon atmosphere (to avoid MWCNTs burnout). Calcination at 600°C produced Al_2O_3, as predicted by the thermal transformation series of aluminum hydroxides into metastable alumina polymorphs [53]. These powders were sintered using a Dr. Sinter Lab Inc. device (Model 515 S, Kanagawa, Japan, pulsed high DC current, 0–1500 A) at 1300°C for 5 min with 75 MPa of constant uniaxial pressure (heating ramp: 100°C/min, cooling ramp: 50°C/min) [54]. These parameters have already been used to successfully sinter alumina [55].

Table 1.1 Process techniques for graphite nanoplates (GNP), reinforced metal matrix composites (MMC), and ceramic matrix composites (CMC).

Synthetic process	Applicability metal/ceramic, powder consolidation	Advantages	Applications
Ball milling	Metal or ceramic, powder mixing	High shear forces are used to exfoliate and eliminate agglomeration of GNPs. The nanoparticles must be uniformly distributed between the metal and ceramic matrices.	GNPs can be damaged by high shear forces, and their lateral dimensions may shrink, resulting in less effective reinforcement.
Wet chemistry/ultrasonication	Metal or ceramic, powder mixing	Nanofillers disperse well in the matrix material. Surfactants can be used effectively to improve dispersion.	Cavitation-induced agitation may not be enough to break up severe agglomeration. Moisture is introduced into the powder, which can inhibit sinterability and obtain powders unsuitable for spraying.
Ultrasonication + ball milling	Metal or ceramic, powder mixing	The same level of dispersion is not achieved by ultrasonication or ball milling alone.	This method is ineffective for processing large amounts of powder because it requires a large number of liquid media. Ball milling and ultrasonication have limitations that have been reduced but not eliminated: moisture is still present, GNPs are damaged, and large amounts of media are required.
Molecular-level synthesis	Ceramics/powder mixing	Nanofillers have excellent dispersion. Filler and matrix bonding is superior. Can embed GNPs in ceramic particles,	As methods become more expensive, their complexity increases. Scalability is a problem because desired reactions must

		promoting toughening and strengthening.	be induced and controlled. The use of liquid media in spraying methods may also eliminate the use of powders (AM, thermal spray)
Mechanical alloying	Metals; powder mixing	The route is commercially viable. Outstanding dispersion is achieved by using high milling times.	High milling times can provoke GNP structure to degrade and defects to increase.
Spark plasma sintering (SPS)	Ceramics or metals; consolidation/forming	Rapid consolidation. High densification. Low grain growth no significant damage to GNPs. May induce graphitization of GNPs. By following this RCEP protocol (rectifying, cutting, embedding, and polishing).	Primarily effective as surface modification technique. Currently limited to pieces on the order of few centimeters. Results vary based on the conductivity of a material. Batch process.
Hot(isostatic) Pressing (HIP/HP)	Ceramics; consolidation	High densification. Suitable for production of large ceramic parts.	Long sintering times (can exceed 1 h). Higher grain growth than SPS.
Pressureless Sintering	Ceramics; consolidation	High densification. Direct sintering and densification of complex-shaped parts.	Require initial (and expensive) CIP step to consolidate powders.
Plasma spraying	Metal or ceramic coatings	Superior adhesion to the substrate and other particles. Ability to process nanocomposites with minimal damage to the nanofiller.	Higher grain growth than SPS/HIP. Relatively low densification. Powder with good flowability is needed.
Additive manufacturing	Metals or ceramics; consolidation/forming	Direct ability to form highly complex shapes. Metal and composites manufacturing technique at the industrial level.	Powder with good flowability is needed. Dense ceramics in their early stages of processing. Densification is poor.

(*Continued*)

Table 1.1 (Continued)

Synthetic process	Applicability metal/ ceramic, powder consolidation	Advantages	Applications
Epitaxial growth	Ceramics; consolidation	GNP and matrix have a good cohesion interface. GNPs in the matrix have excellent uniformity. Saves time and resources by avoiding the challenging stage of composite powder processing.	In situ graphene growth makes it difficult to determine and measure the amount of graphene produced. The amount of graphene produced is limited to a few matrix materials with favorable orientation interactions.
Cold spraying	Metal or ceramic coatings	Densification is high. Technique at the industrial level Metastable and oxidation-sensitive materials can be treated at low temperatures. In metals, high velocities might cause severe plastic deformation.	Powder with good flowability is needed. Many ceramics are difficult to process due to the low temperature. In ceramics, high impact forces can cause microcracking.

Reproduced with permission. from Source: Reproduced with permission from A. Nieto, A. Bisht, D. Lahiri, C. Zhang, A. Agarwal, Graphene reinforced metal and ceramic matrix composites: a review, Int. Mater. Rev. (2016) 241–302. https://doi.org/10.1080/09506608.2016.1219481.

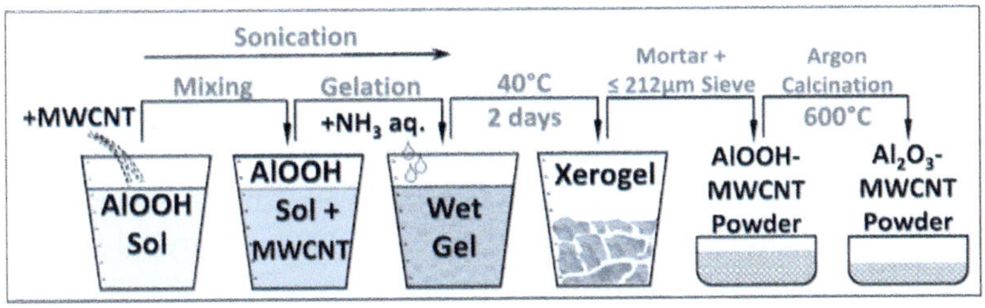

Figure 1.3 The synthesis procedure of the precursor powder is summarized in this sketch. Source: *Reproduced with permission from P. Rivero-Antúnez, R. Cano-Crespo, L. Esquivias, N. de la Rosa-Fox, C. Zamora-Ledezma, A. Domínguez-Rodríguez, V. Morales-Flórez, Mechanical characterization of sol–gel alumina-based ceramics with intragranular reinforcement of multiwalled carbon nanotubes, Ceram. Int. 46 (11) (2020) 19723–19730. https://doi.org/10.1016/j.ceramint.2020.04.285.*

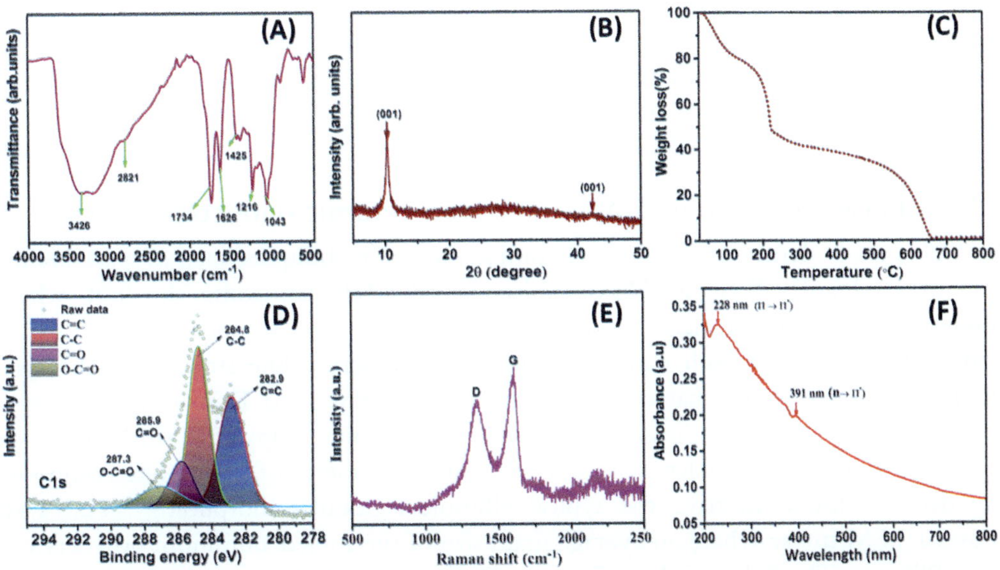

Figure 1.4 Pure spectra of GO: (A) FTIR; (B) XRD; (C) TGA; (D) XPS; (E) Raman; (F) UV. *FTIR*, Fourier-transform infrared spectroscopy; *GO*, graphene oxide; *TGA*, thermogravimetric analysis; *UV*, ultraviolet–visible spectroscopy; *XPS*, X-Ray Photoelectron Spectroscopy; XRD, X-Ray diffraction analysis. Source: *Reproduced with permission from G. Kumar, N.K. Mogha, D.T. Masram, Zr-based metal-organic framework/reduced graphene oxide composites for catalytic synthesis of 2,3-dihydroquinazolin-4(1H)-one derivatives, ACS Appl. Nano Mater. 4 (3) (2021) 2682–2693. https://doi.org/10.1021/acsanm.0c03322.*

1.3 Characterization of pristine graphene oxide

The Fourier-transform infrared spectroscopy (FTIR), spectrum of pristine GO exhibits high characteristic bands at 3395, 2924, 1729, 1619, 1407, 1211, and 1035 cm^{-1} due to the presence of functional groups such as hydroxyl, phenolic, carbonyl, carboxylic, and epoxy. Fig. 1.4A shows in the FTIR spectrum of pristine GO [56]. The X-Ray diffraction analysis (XRD),range of GO, which shows distinct peaks at $2q = 10.3°$, which corresponds to the plane <0 0 1> of GO, is depicted in Fig. 1.4B [57]. Three weight losses were seen in the thermogravimetric analysis (TGA),spectra of the GO as shown in Fig. 1.4C. The elimination of water that could be physically adsorbed on the surface is the first weight loss measure over the temperature range of 25°C–110°C. The second weight loss occurs at 110°C–219°C due to the loss of oxygenated functional groups. The third weight loss, which occurs between 219°C and 650°C, could be due to a loss of GO sheet [56]. In the C1s X-Ray Photoelectron Spectroscopy (XPS), spectra of GO Fig. 1.4D, the peaks with binding energies of 282.9, 284.8, 285.9, and 287.3 eV correspond to C=C, C–C, C=O, and O–C=O, respectively [58]. In Fig. 1.4E, the Raman spectrum of pure GO revealed distinct bands at 1350 and 1609 cm^{-1}, which correspond to the D and G bands of GO, respectively [59]. The ultraviolet–visible spectroscopy (UV), Fig. 1.4F of GO exhibits a high absorption peak at 228 nm and another peak at around 391 nm, which correspond to the $\pi-\pi^*$ transitions of the graphitic C–C bonds and to the $n-\pi^*$ transitions of the C=O bonds, respectively [56].

1.3.1 Characterization of graphene-based ceramic nanocomposites

In 2020 a study demonstrated a novel combination of techniques, such as spark plasma sintering and hot rolling, to fabricate bimodal Al-GNP nanocomposite [60], which possessed an excellent amalgamation of strength and ductility. Bimodal Al-GNP nanocomposites comprised both nanocrystalline and microcrystalline structures. The synthesized materials were characterized using SEM, TEM, and Raman spectroscopy. Fig. 1.5A shows an field emission scanning electron microscopy (FE-SEM) image of the as-synthesized Al powders with a mean particle diameter of 9 μm. The composite powders are found to have the typical ellipsoidal-globular morphology. Fig. 1.5B shows a TEM image where an average grain size of the milled aluminum powders was 100–200 nm. The Selected Area Electron Diffraction (SAED) pattern is showing the crystallinity of the high-energy ball-milled aluminum powders. A TEM image of the functionalized GNPs is given in Fig. 1.5C which shows the crimple and folds layered structure of graphene sheets. The stacks of graphene sheets are visible in the TEM picture in the inset of Fig. 1.5C, which was found to have 0.34 nm of lattice spacing. The Raman spectra of functionalized GNPs are shown in Fig. 1.5D. where D, G, and 2D have three peaks at 1360, 1577, and 2744 cm^{-1}, respectively. The disordered defect structure of graphene and graphitic carbon causes the D peak. The G band is associated with ordered graphene structures, while the 2D band (broad second order) is centered at 2695 cm^{-1} for a single graphene sheet. The 2D peak was shifted and observed due to multiple layers of GNP. The observed ID/IG ratio (0.17) is lower and showed a high sp^2 quantity with respect to sp^3.

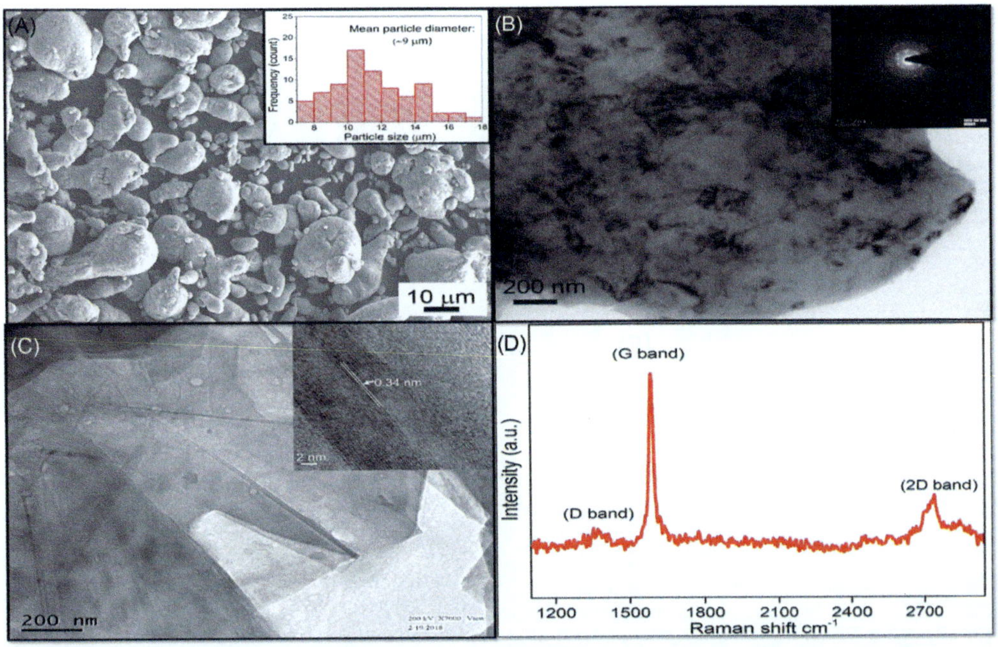

Figure 1.5 (A) FE-SEM image of microcrystalline Al powders with size distribution in the inset; (B) TEM images of ball-milled Al powder; (C) physiochemically functionalized GNPs; and (D) Raman spectrum of the functionalized GNPs. *FE-SEM*, Field emission scanning electron microscopy; *GNPs*, graphene nanoplatelets; *TEM*, transmission electron microscopy. Source: *Reproduced with permission from A. Bhadauria, L.K. Singh, S.K. Nayak, T. Laha, Tensile deformation behavior and strengthening mechanism in graphene nanoplatelet reinforced bimodal grained aluminum nanocomposite synthesized by spark plasma sintering and hot rolling, Mater. Charact. 168 (2020). https://doi.org/10.1016/j.matchar.2020.110568.*

In 2019 ceramic composites Al_2O_3-Ti(C, N) reinforced with graphene nanoplatelets (GNPs) were fabricated by hot-pressing sintering (HP) [45]. The microstructure, mechanical properties, thermal conductivity, and reinforcing mechanisms of composites are a function of the GNP material. *N*-Methyl-2-pyrrolidone (NMP) was used as a dispersing medium and an alcohol solution of polyvinyl pyrrolidone was used as a reference to distribute GNPs homogeneously in a ceramic matrix. According to the results, NMP is an efficient dispersing medium for powder mixing. As shown in Fig. 1.6, the dispersion of GNPs with different contents in Al2O3-Ti(C, N) matrix ceramic composites was investigated using the microstructure of fractured surface. GNPs with a content of 0.4 wt.% (Fig. 1.6B and E) are well dispersed in the ceramic matrix, while GNPs with a content of 0.2 wt.% (Fig. 1.6A and D) are not visible due to their lower content. As shown in Fig. 1.6C and F, when the GNP content reached 0.6 wt.%, agglomeration of GNPs occurred, resulting in the formation of a continuous layered structure. Furthermore, the layers of agglomerated GNPs in Fig. 1.6F are

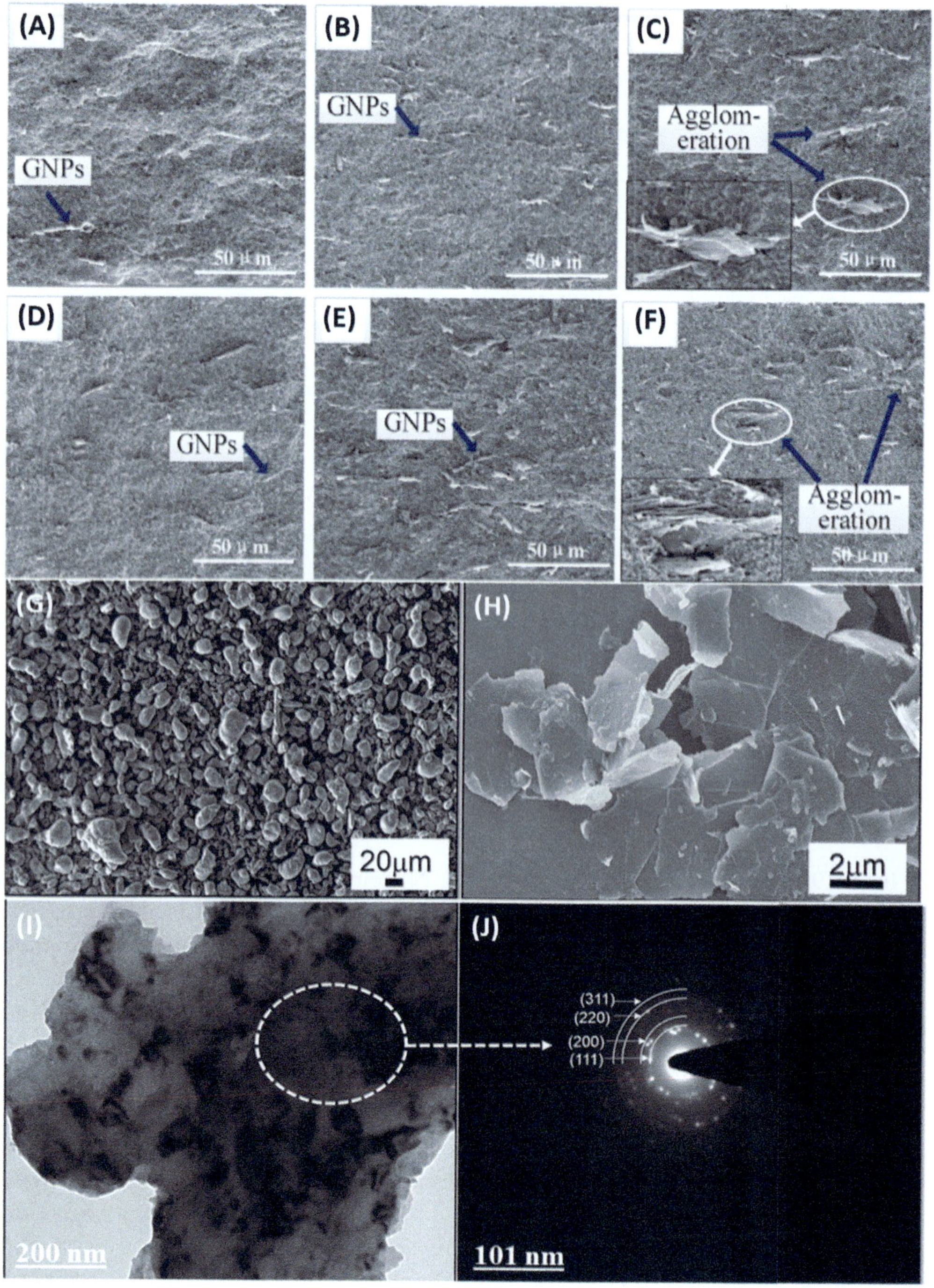
(A)
GNPs
50 μm
(B)
GNPs
50 μm
(C)
Agglom-
eration
50 μm
(D)
GNPs
50 μm
(E)
GNPs
50 μm
(F)
Agglom-
eration
50 μm
(G)
20μm
(H)
2μm
(I)
200 nm
(J)
(311)
(220)
(200)
(111)
101 nm

(Continued)

greater than those in Fig. 1.6C, implying that NMP is better for GNP dispersion than A-PVP. Prolonged ball-milling (with NMP as media) can achieve uniform dispersion of GNPs in the ceramic matrix, which improves the mechanical properties of the composites, especially fracture toughness.

In 2019 spark plasma sintering was used to develop Al-0.5 wt.% GNP nanocomposites for both microcrystalline and nanocrystalline matrices [61]. The effects of nanocrystalline Al matrix and GNP reinforcement on strength have been extensively investigated. FE-SEM micrographs of as-procured microcrystalline Al powders and functionalized GNPs are shown in Fig. 1.6G and H. The ellipsoidal-globular morphology of the Al powders is shown in Fig. 1.6G. The layered structure of graphene sheets with crimples and folds is depicted in Fig. 1.6H. The TEM picture and SAED pattern of the high-energy milled Al powders are shown in Fig. 1.6I and J. The average grain size of the powder was found to be 100 nm, as seen in the TEM picture, and the ring pattern in the corresponding SAED pattern supports the development of nanocrystalline grains in milled Al powders.

1.4 Conclusion and future outlook

The formation of a ceramic matrix enhances electrical conductivity which can be increased with the addition of GO-based composites by providing efficient pathways for electron transport. The phenomenon of high-temperature sensor and low thermal conductivity of bulk GO composites is essential for their applications as thermal insulators and energy sources. High-thermal-performance ceramics are becoming especially important in thermal energy management, microelectronics, machinery, aerospace, and other fields. Ceramics are made up of metallic and nonmetallic parts that are generally ionically or covalently linked because their bonds lack of free electrons. Ceramics are poor electrical and thermal conductors. The GRC has attracted a lot of attention in recent times, and it has been recognized as a potential way to improve the

◀ **Figure 1.6** Dispersion of GNPs with different content in ceramic specimens: (A) Al_2O_3–TiC–0.20 wt.% GNPs (ATG0.20), (B) ATG0.40, and (C) ATG0.60, NMP is served as dispersant. (D) ATG0.20, (E) ATG0.40, and (F) ATG0.60, A-PVP is served as the dispersant, ATG represents Al_2O_3-TiC-GNPs and the number after ATG indicates the content of GNPs. and FE-SEM images of (G) as-received Al powders and (H) functionalized GNPs, and (I) TEM image of high energy ball-milled Al powders and (J) corresponding SAED pattern depicting nanocrystalline nature of milled Al powders. *GNPs*, Graphene nanoplatelets; *NMP*, N-methyl-2-pyrrolidone; *TEM*, transmission electron microscopy; *SAED*, Selected Area Electron Diffraction. Source: *Reproduced with permission from E. Cui, J. Zhao, X. Wang, Determination of microstructure and mechanical properties of graphene reinforced Al2O3-Ti(C, N) ceramic composites, Ceram. Int. 45 (16) (2019) 20593–20599. https://doi.org/10.1016/j.ceramint.2019.07.041.*

thermal performance of traditional ceramics. As a result, this analysis has summarized the most recent developments in GRC preparation, microstructure, and thermal properties. On the thermal efficiency of GRC, the effects of sintering techniques, sintering parameters, and graphene/ceramic matrix parameters were highlighted. In the evaluation of GRC thermal properties, SPS and HP methods are the most widely used sintering techniques. SPS and HP methods have been used to investigate the effect of sintering temperature on graphene integrity and GRC thermal efficiency. However, the research into the consequences of heating rate, dwell time, sintering pressure, sintering additives, and gas flow, among other factors, on graphene integrity and GRC thermal efficiency is insufficient at present. Also, there is a lack of literature on the related mechanisms. As a consequence, these crucial points need to be further examined. The graphene/ceramic parameters have the greatest impact on GRC's thermal efficiency. The impact of graphene material on graphene integrity and GRC thermal properties has been thoroughly investigated and the mechanisms, synthesis, and characterizations involved have been extensively discussed. On the other hand, there have been few studies of the effects of graphene layers, graphene forms, graphene average particle sizes, and ceramic matrix parameters on the thermal properties of GRC. In the near future, nanoceramic coatings may play a larger role in the coating of automobile engines. Nanoceramic coating methods would be more suitable for industrial use. Graphene-based ceramic composites have proven their commercial influence and are expected to have a brighter future in many applications, despite the challenges and research gaps. Significant attempts have been made to increase graphene production scale. The application of graphene-ceramic composites will likely accelerate within the marketplace if the overarching issues with graphene can be solved appropriately and research continues to increase.

References

[1] W.D. Kingery, H.K. Bowen, D.R. Uhlmann, R. Frieser, Introduction to ceramics, J. Electrochem. Soc. (1977) 152C. Available from: https://doi.org/10.1149/1.2133296. -152C.

[2] C.B. Carter, M.G. Norton, Ceramic Materials: Science and Engineering, Springer New York, United States, 2013, pp. 1–766. Available from: http://doi.org/10.1007/978-1-4614-3523-5.10.1007/978-1-4614-3523-5.

[3] D.W. Richerson, W.E. Lee, Modern ceramic engineering: properties, Process. Use Des. (2018) 1–791.

[4] V.P. Meshalkin, A.V. Belyakov, Methods used for the compaction and molding of ceramic matrix composites reinforced with carbon nanotubes, Processes (2020) 1004. Available from: https://doi.org/10.3390/pr8081004.

[5] L. Liu, K. Morita, T.S. Suzuki, B.N. Kim, Effect of the heating rate on the spark-plasma-sintering (SPS) of transparent Y2O3 ceramics: microstructural evolution, mechanical and optical properties, Ceram. 4 (2021) 49–56. Available from: https://doi.org/10.3390/ceramics4010006.

[6] J.N. Coleman, U. Khan, W.J. Blau, Y.K. Gun'ko, Small but strong: a review of the mechanical properties of carbon nanotube-polymer composites, Carbon. 44 (9) (2006) 1624–1652. Available from: https://doi.org/10.1016/j.carbon.2006.02.038.

[7] P. Marchetti, M.F. Jimenez Solomon, G. Szekely, A.G. Livingston, Molecular separation with organic solvent nanofiltration: a critical review, Chem. Rev. 114 (21) (2014) 10735–10806. Available from: https://doi.org/10.1021/cr500006j, http://pubs.acs.org/journal/chreay.

[8] J. Kujawa, S. Al-Gharabli, W. Kujawski, K. Knozowska, Molecular grafting of fluorinated and non-fluorinated alkylsiloxanes on various ceramic membrane surfaces for the removal of volatile organic compounds applying vacuum membrane distillation, ACS Appl. Mater. Interfaces 9 (7) (2017) 6571–6590. Available from: https://doi.org/10.1021/acsami.6b14835, http://pubs.acs.org/journal/aamick.

[9] N. Rangnekar, N. Mittal, B. Elyassi, J. Caro, M. Tsapatsis, Zeolite membranes – a review and comparison with MOFs, Chem. Soc. Rev. 44 (20) (2015) 7128–7154. Available from: https://doi.org/10.1039/c5cs00292c, http://pubs.rsc.org/en/journals/journal/cs.

[10] S.M. Samaei, S. Gato-Trinidad, A. Altaee, The application of pressure-driven ceramic membrane technology for the treatment of industrial wastewaters – a review, Sep. Purif. Technol. 200 (2018) 198–220. Available from: https://doi.org/10.1016/j.seppur.2018.02.041, http://www.journals.elsevier.com/separation-and-purification-technology/.

[11] S. Yadav, A.P.S. Raman, H. Meena, A.G. Goswami, Bhawna, V. Kumar, P. Jain, G. Kumar, M. Sagar, D.K. Rana, I. Bahadur, P. Singh, An update on graphene oxide: applications and toxicity, ACS Omega 7 (40) (2022) 35387–35445. Available from: https://doi.org/10.1021/acsomega.2c03171.

[12] X. Du, I. Skachko, A. Barker, E.Y. Andrei, Approaching ballistic transport in suspended graphene, Nat. Nanotechnol. 3 (8) (2008) 491–495. Available from: https://doi.org/10.1038/nnano.2008.199, http://www.nature.com/nnano/index.html.

[13] E. Pop, V. Varshney, A.K. Roy, Thermal Properties of Graphene: Fundamentals and Applications, MRS Bulletin. 1273.

[14] C. Lee, X. Wei, J.W. Kysar, J. Hone, Measurement of the elastic properties and intrinsic strength of monolayer graphene, Science. 321 (5887) (2008) 385–388. Available from: https://doi.org/10.1126/science.1157996.

[15] D. Yoon, Y.W. Son, H. Cheong, Negative thermal expansion coefficient of graphene measured by Raman spectroscopy, Nano Lett. 11 (8) (2011) 3227–3231. Available from: https://doi.org/10.1021/nl201488g.

[16] A.A. Balandin, Thermal properties of graphene and nanostructured carbon materials, Nat. Mater. 10 (8) (2011) 569–581. Available from: https://doi.org/10.1038/nmat3064, http://www.nature.com/nmat/.

[17] A.A. Balandin, S. Ghosh, W. Bao, I. Calizo, D. Teweldebrhan, F. Miao, et al., Superior thermal conductivity of single-layer graphene, Nano Lett. 8 (3) (2008) 902–907. Available from: https://doi.org/10.1021/nl0731872.

[18] C. Ramírez, S.M. Vega-Diaz, A. Morelos-Gómez, F.M. Figueiredo, M. Terrones, M.I. Osendi, et al., Synthesis of conducting graphene/Si3N4 composites by spark plasma sintering, Carbon. 57 (2013) 425–432. Available from: https://doi.org/10.1016/j.carbon.2013.02.015.

[19] J. Sun, J. Zhao, F. Gong, Z. Li, X. Ni, Design, fabrication and characterization of multi-layer graphene reinforced nanostructured functionally graded cemented carbides, J. Alloy. Compounds. 750 (2018) 972–979. Available from: https://doi.org/10.1016/j.jallcom.2018.04.108, https://www.journals.elsevier.com/journal-of-alloys-and-compounds.

[20] F.L. Riley, Silicon nitride and related materials, J. Am. Ceram. Soc. 83 (2) (2000) 245–265. Available from: https://doi.org/10.1111/j.1151-2916.2000.tb01182.x.

[21] X. Li, A.H. Korayem, C. Li, Y. Liu, H. He, J.G. Sanjayan, et al., Incorporation of graphene oxide and silica fume into cement paste: a study of dispersion and compressive strength, Constr. Build. Mater. 123 (2016) 327–335. Available from: https://doi.org/10.1016/j.conbuildmat.2016.07.022.

[22] Z. Pan, L. He, L. Qiu, A.H. Korayem, G. Li, J.W. Zhu, et al., Mechanical properties and microstructure of a graphene oxide-cement composite, Cem. Concr. Compos. 58 (2015) 140–147. Available from: https://doi.org/10.1016/j.cemconcomp.2015.02.001, http://www.sciencedirect.com/science/journal/09589465.

[23] A.H. Korayem, N. Tourani, M. Zakertabrizi, A.M. Sabziparvar, W.H. Duan, A review of dispersion of nanoparticles in cementitious matrices: nanoparticle geometry perspective, Constr. Build. Mater. 153 (2017) 346–357. Available from: https://doi.org/10.1016/j.conbuildmat.2017.06.164.

[24] A. Gholampour, M. Valizadeh Kiamahalleh, D.N.H. Tran, T. Ozbakkaloglu, D. Losic, From graphene oxide to reduced graphene oxide: impact on the physiochemical and mechanical properties of graphene-cement composites, ACS Appl. Mater. Interfaces 9 (49) (2017) 43275–43286. Available from: https://doi.org/10.1021/acsami.7b16736, http://pubs.acs.org/journal/aamick.
[25] X. Fan, Z. Shen, A.Q. Liu, J.L. Kuo, Band gap opening of graphene by doping small boron nitride domains, Nanoscale. 4 (6) (2012) 2157–2165. Available from: https://doi.org/10.1039/c2nr11728b, http://www.rsc.org/publishing/journals/NR/Index.asp.
[26] D. Hou, Z. Lu, X. Li, H. Ma, Z. Li, Reactive molecular dynamics and experimental study of graphene-cement composites: structure, dynamics and reinforcement mechanisms, Carbon. 115 (2017) 188–208. Available from: https://doi.org/10.1016/j.carbon.2017.01.013, http://www.journals.elsevier.com/carbon/.
[27] K.I. Bolotin, K.J. Sikes, J. Hone, H.L. Stormer, P. Kim, Temperature-dependent transport in suspended graphene, Phys. Rev. Lett. 101 (9) (2008). http://oai.aps.org/oai?verb = GetRecord&Identifier = oai:aps.org:PhysRevLett.101.096802&metadataPrefix = oai_apsmeta_2.10.1103/PhysRevLett.101.096802 United States.
[28] F. Chen, K. Yan, J. Sun, J. Hong, Y. Zhu, Z. Huang, From the research state of the thermal properties of graphene reinforced ceramics to the future of computer simulation, Ceramics International 46 (11) (2020) 18428–18445. Available from: https://www.journals.elsevier.com/ceramics-international0.10.1016/j.ceramint.2020.04.265.
[29] C.B. Mo, S.I. Cha, K.T. Kim, K.H. Lee, S.H. Hong, Fabrication of carbon nanotube reinforced alumina matrix nanocomposite by sol–gel process, Mater. Sci. Eng. A. 395 (1-2) (2005) 124–128. Available from: https://doi.org/10.1016/j.msea.2004.12.031.
[30] I. Ahmad, H. Cao, H. Chen, H. Zhao, A. Kennedy, Y.Q. Zhu, Carbon nanotube toughened aluminum oxide nanocomposite, J. Eur. Ceram. Soc. 30 (4) (2010) 865–873. Available from: https://doi.org/10.1016/j.jeurceramsoc.2009.09.032.
[31] G.D. Zhan, J.D. Kuntz, J. Wan, A.K. Mukherjee, Single-wall carbon nanotubes as attractive toughening agents in alumina-based nanocomposites, Nat. Materials. 2 (1) (2003) 38–42. Available from: https://doi.org/10.1038/nmat793, http://www.nature.com/nmat/.
[32] L. Esquivias, P. Rivero-Antúnez, C. Zamora-Ledezma, A. Domínguez-Rodríguez, V. Morales-Flórez, Intragranular carbon nanotubes in alumina-based composites for reinforced ceramics, J. Sol–gel Sci. Technol. 90 (1) (2019) 162–171. Available from: https://doi.org/10.1007/s10971-018-4834-4.
[33] M.K. Satam, L. Gurnani, S. Vishwanathe, A. Mukhopadhyay, Development of carbon nanotube reinforced bulk polycrystalline ceramics with intragranular carbon nanotube reinforcement, J. Am. Ceram. Society. 99 (9) (2016) 2905–2908. Available from: https://doi.org/10.1111/jace.14425, http://www.blackwellpublishing.com/aims.asp?ref = 0002-7820.
[34] B.E. Yoldas, Alumina sol preparation from alkoxides, Am. Ceram. Soc. Bull. 54 (3) (1975) 289–290.
[35] C. Barrera-Solano, L. Esquivias, G.L. Messing, Effect of preparation conditions on phase formation, densification, and microstructure evolution in La-β-Al2O3/Al2O3 composites, J. Am. Ceram. Soc. 82 (5) (1999) 1318–1324. Available from: http://onlinelibrary.wiley.com/journal/10.1111/(ISSN)1551-2916.10.1111/j.1151-2916.1999.tb01914.x.
[36] M. Piñero, M.D.M. Mesa-Díaz, D. de los Santos, M.V. Reyes-Peces, J.A. Díaz-Fraile, N. de la Rosa-Fox, et al., Reinforced silica-carbon nanotube monolithic aerogels synthesised by rapid controlled gelation, J. Sol–gel Sci. Technol. 86 (2) (2018) 391–399. Available from: https://doi.org/10.1007/s10971-018-4645-7.
[37] S.R. Inbaraj, R.M. Francis, N.V. Jaya, A. Kumar, Processing and properties of sol gel derived alumina-carbon nano tube composites, Ceram. Int. 38 (5) (2012) 4065–4074. Available from: https://doi.org/10.1016/j.ceramint.2012.01.064.
[38] H. Miyazaki, Y. Yoshizawa, A reinvestigation of the validity of the indentation fracture (IF) method as applied to ceramics, J. Eur. Ceram. Society. 37 (15) (2017) 4437–4441. Available from: https://doi.org/10.1016/j.jeurceramsoc.2017.07.027, http://www.journals.elsevier.com/journal-of-the-european-ceramic-society/.
[39] H. Miyazaki, Y.I. Yoshizawa, Correlation of the indentation fracture resistance measured using high-resolution optics and the fracture toughness obtained by the single edge-notched beam (SEPB) method for typical structural ceramics with various microstructures, Ceram. Int. 42 (6) (2016) 7873–7876. Available from: https://doi.org/10.1016/j.ceramint.2016.01.116.

[40] J. Sun, L. Gao, M. Iwasa, T. Nakayama, K. Niihara, Failure investigation of carbon nanotube/3Y-TZP nanocomposites, Ceram. Int. 31 (8) (2005) 1131–1134. Available from: https://doi.org/10.1016/j.ceramint.2004.11.010.

[41] T.A. Saleh, V.K. Gupta, Characterization of the chemical bonding between Al_2O_3 and nanotube in MWCNT/Al_2O_3 nanocomposite, Curr. Nanosci. 8 (5) (2012) 739–743. Available from: https://doi.org/10.2174/157341312802884418.

[42] J. Yi, W. Xue, T. Wang, Z. Xie, Mechanical and electrical properties of chemically modified MWCNTs/3Y-TZP composites, Ceram. Int. 41 (7) (2015) 9157–9162. Available from: https://doi.org/10.1016/j.ceramint.2015.03.008.

[43] J. Lee, H.R. Chae, Y.J. Won, K. Lee, C.H. Lee, H.H. Lee, et al., Graphene oxide nanoplatelets composite membrane with hydrophilic and antifouling properties for wastewater treatment, J. Membr. Sci. 448 (2013) 223–230. Available from: https://doi.org/10.1016/j.memsci.2013.08.017.

[44] X. Wang, J. Zhao, E. Cui, S. Song, H. Liu, W. Song, Microstructure, mechanical properties and toughening mechanisms of graphene reinforced Al 2 O 3 -WC-TiC composite ceramic tool material, Ceram. Int. 45 (8) (2019) 10321–10329. Available from: https://doi.org/10.1016/j.ceramint.2019.02.087.

[45] E. Cui, J. Zhao, X. Wang, Determination of microstructure and mechanical properties of graphene reinforced Al2O3-Ti(C, N) ceramic composites, Ceram. Int. 45 (16) (2019) 20593–20599. Available from: https://doi.org/10.1016/j.ceramint.2019.07.041, https://www.journals.elsevier.com/ceramics-international.

[46] B. Lee, M.Y. Koo, S.H. Jin, K.T. Kim, S.H. Hong, Simultaneous strengthening and toughening of reduced graphene oxide/alumina composites fabricated by molecular-level mixing process, Carbon. 78 (2014) 212–219. Available from: https://doi.org/10.1016/j.carbon.2014.06.074, http://www.journals.elsevier.com/carbon/.

[47] J. Liu, H. Yan, K. Jiang, Mechanical properties of graphene platelet-reinforced alumina ceramic composites, Ceram. Int. 39 (6) (2013) 6215–6221. Available from: https://doi.org/10.1016/j.ceramint.2013.01.041.

[48] K. Wang, Y. Wang, Z. Fan, J. Yan, T. Wei, Preparation of graphene nanosheet/alumina composites by spark plasma sintering, Mater. Res. Bull. 46 (2) (2011) 315–318. Available from: https://doi.org/10.1016/j.materresbull.2010.11.005.

[49] P. Rutkowski, P. Klimczyk, L. Jaworska, L. Stobierski, A. Dubiel, Thermal properties of pressure sintered alumina-graphene composites, J. Therm. Anal. Calorim. 122 (1) (2015) 105–114. Available from: https://doi.org/10.1007/s10973-015-4694-x, http://www.springer.com/sgw/cda/frontpage/0,11855,1-40109-70-35752391-0,00.html.

[50] S.J. Yan, S.L. Dai, X.Y. Zhang, C. Yang, Q.H. Hong, J.Z. Chen, et al., Investigating aluminum alloy reinforced by graphene nanoflakes, Mater. Sci. Eng. A. 612 (2014) 440–444. Available from: https://doi.org/10.1016/j.msea.2014.06.077, http://www.elsevier.com.

[51] A. Nieto, A. Bisht, D. Lahiri, C. Zhang, A. Agarwal, Graphene reinforced metal and ceramic matrix composites: a review, Int. Mater. Rev. (2016) 241–302. Available from: https://doi.org/10.1080/09506608.2016.1219481.

[52] P. Rivero-Antúnez, R. Cano-Crespo, L. Esquivias, N.D.L. Rosa-Fox, C. Zamora-Ledezma, A. Domínguez-Rodríguez, et al., Mechanical characterization of sol–gel alumina-based ceramics with intragranular reinforcement of multiwalled carbon nanotubes, Ceram. Int. 46 (11) (2020) 19723–19730. Available from: https://doi.org/10.1016/j.ceramint.2020.04.285.

[53] P.S. Santos, H.S. Santos, et al., Standard transition aluminas. Electron microscopy studies, Mater. Res. 104–114 (2000). Available from: https://doi.org/10.1590/S1516-14392000000400003.

[54] K.Q. Dang, M. Kawahara, S. Takei, M. Nanko, Effects of pulsed current waveforms on sample temperature and sintering behavior in PECS of alumina, Funtai Oyobi Fummatsu Yakin/Journal Jpn. Soc. Powder Powder Metall. 56 (12) (2009) 780–787. Available from: https://doi.org/10.2497/jjspm.56.780.

[55] R. Cano-Crespo, B. Malmal Moshtaghioun, D. Gómez-García, A. Domínguez-Rodríguez, R. Moreno, High-temperature creep of carbon nanofiber-reinforced and graphene oxide-reinforced alumina composites sintered by spark plasma sintering, Ceram. Int. 43 (9) (2017) 7136–7141. Available from: https://doi.org/10.1016/j.ceramint.2017.02.146.

[56] G. Kumar, N.K. Mogha, M. Kumar, et al., NiO nanocomposites/rGO as a heterogeneous catalyst for imidazole scaffolds with applications in inhibiting the DNA binding activity, Dalton Trans. 49 (6) (2020) 1963–1974. Available from: https://doi.org/10.1039/c9dt04416g, http://pubs.rsc.org/en/journals/journal/dt.
[57] N.K. Mogha, K. Chaudhary, G. Kumar, D.T. Masram, Fur-imine-functionalized graphene oxide-immobilized copper oxide nanoparticle catalyst for the synthesis of xanthene derivatives, ACS Omega 3 (11) (2018) 16377–16385. Available from: https://doi.org/10.1021/acsomega.8b01781.
[58] G. Kumar, N.K. Mogha, D.T. Masram, Zr-based metal-organic framework/reduced graphene oxide composites for catalytic synthesis of 2,3-dihydroquinazolin-4(1*H*)-one derivatives, ACS Appl. Nano Mater. 4 (3) (2021) 2682–2693. Available from: https://doi.org/10.1021/acsanm.0c03322, https://pubs.acs.org/journal/aanmf6.
[59] K. Chaudhary, K. Kumar, P. Venkatesu, D.T. Masram, In-depth understanding of a nano-bio interface between lysozyme and Au NP-immobilized N-doped reduced graphene oxide 2-D scaffolds, Nanoscale Advances. 2 (5) (2020) 2146–2159. Available from: https://doi.org/10.1039/d0na00155d, http://pubs.rsc.org/en/journals/journalissues/na?_ga2.190536939.1555337663.1552312502-1364180372.1550481316#!issueidna001002&typecurrent&issnonline2516-0230.
[60] A. Bhadauria, L.K. Singh, S.K. Nayak, T. Laha, Tensile deformation behavior and strengthening mechanism in graphene nanoplatelet reinforced bimodal grained aluminum nanocomposite synthesized by spark plasma sintering and hot rolling, Mater. Charact. 168 (2020). Available from: https://doi.org/10.1016/j.matchar.2020.110568, https://www.journals.elsevier.com/materials-characterization.
[61] A. Bhadauria, L.K. Singh, T. Laha, Combined strengthening effect of nanocrystalline matrix and graphene nanoplatelet reinforcement on the mechanical properties of spark plasma sintered aluminum based nanocomposites, Mater. Sci. Eng. A. 749 (2019) 14–26. Available from: http://www.elsevier.com.10.1016/j.msea.2019.02.007.

CHAPTER 2

Nanoceramics: fabrication, properties, and applications

Richa Kasana, Pragati Rathore, Uday Shashikumar and Shashi Chawla
Department of Chemistry, Amity Institute of Applied Sciences, Amity University, Noida, Uttar Pradesh, India

2.1 Introduction

Nanoceramics are ceramic materials made up of nanometer-sized grains/crystallites with at least one dimension less than 100 nm. Nanoceramics have excellent mechanical properties, such as greater strength, excellent toughness, and high fatigue resistance [1,2]. The low-stress resistance, low fracture toughness, and poor ductility/machinability of most nanoceramics, however, limit their practical application [3]. The nanoscale creates new mechanical, magnetic, optical, and other properties [4]. The ductility and superplasticity properties of nanoceramics have been discovered and are cost-effective.

The needed characteristics of precursor nanodimensional powders are nanometer size, spherical, uniform, do not form a cluster, and have a high purity level. Due to high packing density, nanopowders are very reactive in solid state during the sintering process, resulting in nanocomposites [5].

Nanosized particles put at grain boundaries of a matrix increase mechanical parameters such as fracture durability and integrity. Nanoceramic composites can have microsized grid grains and nanosized grid grains. Nanoceramic composites have 200% higher strength retention at elevated temperatures and higher creep characteristics than pure ceramics [6].

The excessive use of fossil fuels has led to a deteriorating environment and climate change. The search for low-cost, high-efficiency energy materials is the right solution. Potential uses of nanoceramic materials are in heterogeneous catalysis, solid oxide fuel cells, photoluminescence, dielectrics, and superconductors [7]. Furthermore, nanoceramics are currently widely used in varied fields including electronics, biomedicine, pharmaceuticals, coating materials, nuclear energy, and so on, which are discussed in this chapter [1,8].

2.2 Synthesis of nanoceramics

2.2.1 Sol−gel technique for the synthesis

Particles that are nanosized metal oxides are prepared using this technique. Sol is generated by utilizing inorganic or metal−organic salts. The sol is then hydrolyzed, and

Industrial Applications of Nanoceramics
DOI: https://doi.org/10.1016/B978-0-323-88654-3.00003-2

condensation processes are used to turn it into a gel. The gel is a 3-D structure with many interconnected pores [9]. Finally, any surplus liquid is removed from the gel, which becomes dehydrated, shrinks, and eventually gets converted to the required product. This method allows complicated structures to be formed from a gel phase with a uniform chemical composition without the need for high processing temperatures.

During the synthesis of cationic nanoceramics materials, careful control of hydrolysis and condensation reactions is essential to reduce segregation problems. For synthesis of nanopowders of metal oxide and ceramics, the sol–gel technique is preferred because of its advantages over other techniques [10,11]. The drawbacks of this method are its excessive cost and moisture sensitivity.

2.2.2 Precipitation from solution

A salt precursor is dissolved in water, base solutions are added, and a precipitate of metal hydroxide is obtained in water. After filtering and washing, the hydroxide is calcined to obtain the oxide powder.

Gel precipitation (GPT), precipitation (PPT), and washed precipitation (WPT) can be used [12] for the processing of powder of alumina/ZrO_2 composite; here aqueous ammonia is the precipitating agent. The precipitate is washed with heated water and alcohol before drying in WPT. The precipitate is separated from the liquid before drying in the PPT process. The larger and harder agglomerates are obtained in the GPT approach, whereas smaller and softer agglomerations are obtained in the WPT route. As a result, WPT produces compact agglomerates that sinter well at lower temperatures (Fig. 2.1).

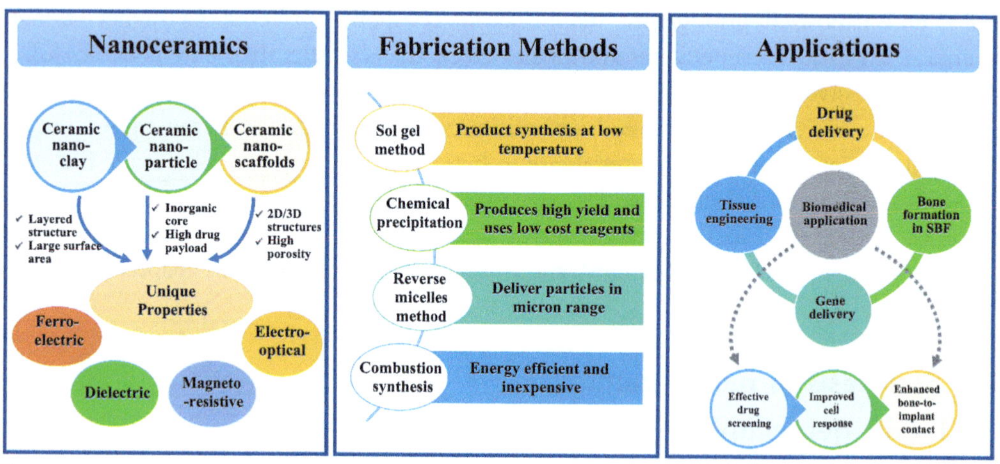

Figure 2.1 An overview of properties, synthesis, and applications of nanoceramics.

2.3 Fabrication techniques

The inclusion of particles of varying sizes provides a higher packaging density. Due to the coarseness of bigger grains, it is difficult to manage microstructural changes during sintering. It also affects densification. Furthermore, when the degree of agglomeration increases, heterogeneous packing occurs, resulting in sintering rate discrepancies and heterogeneous microarrangements. Large defects and lower sintered composite strength can result from a high degree of variability [5]. As a result, fine-sized, freely dispersed powders for both the matrix and reinforcing components are used to enable effective mixing and appropriate particle dispersion in the final powder (Fig. 2.2). The crystalline fraction of the untreated powder can also be used to influence densification and microstructural development. The advantages and disadvantages of various techniques of fabrication can be analyzed.

2.4 Properties of nanoceramics

In contrast to metals, ceramics have their own unique physical, mechanical, and chemical properties. The size of ceramic material has a significant impact on its properties.

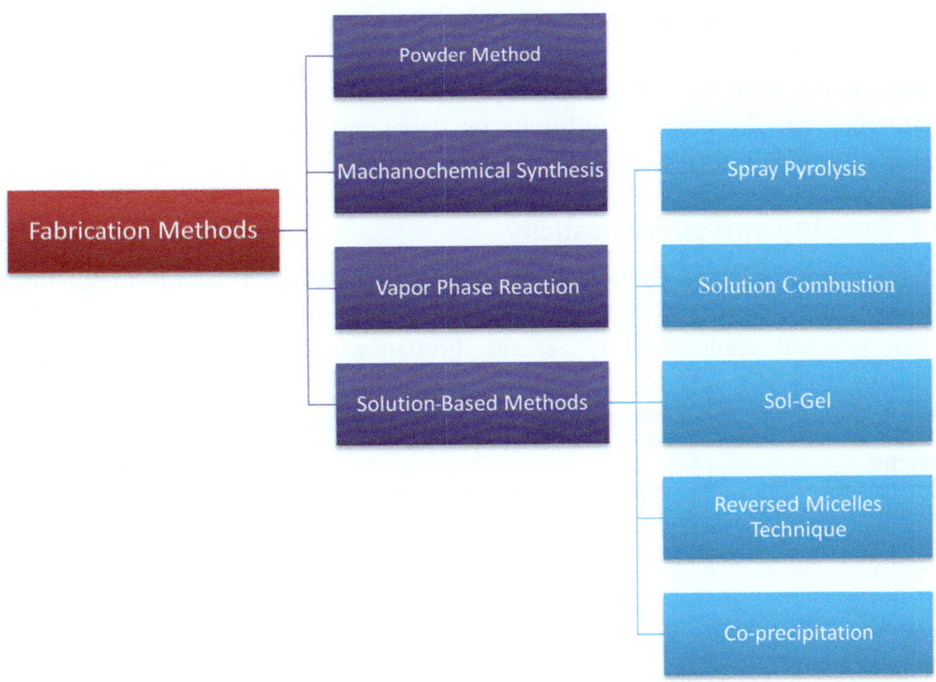

Figure 2.2 Types of fabrication procedures.

2.4.1 Chemical properties

The reactivity [13] of ceramics depends on sites where defects exist. The possibility of getting ceramic powders with excessive porosity [14] makes them suitable for packages. Heterogeneous catalysis and biomedical applications are extremely popular.

2.4.2 Mechanical properties

Ceramics have a high resistance to stress and deformation, which is due to the particle size. Silicon carbides and nitrides, for example, can retain energy at temperatures as high as 1400°C. Some possess metallic strength. Many physical and mechanical properties are affected by particle size [15]. They show useful and scalable mechanical properties like toughness with high Young's modulus.

2.4.3 Sintering properties

Firstly, powders made from individual ceramics are compacted to process nanoceramics from nanophase powders into a crude shape (known as a green body). This compacted powder is then warmed at higher temperatures [16]. Dissemination out of pores (to grain limits) leads to shrinkage of the sample and thus densification occurs. This cycle is known as pressureless sintering. Luckily, nanophase powders were found to be compacted easily. The parameters that need to be maintained during the application of this procedure are shown in Table 2.1.

The sample must be sintered at the lowest temperature feasible for an adequate time to eliminate the porosity and build up stable and well defined grain boundaries. The sintering would finally enhance the effective hardness of the synthesized ceramic.

2.4.4 Ductility and superplasticity

These properties refer to the ability of some polycrystalline materials to go through extensive tensile distortion without necking or breaking. The greatest specialized obstruction in practical applications is ceramic brittleness [17]. Brittle ceramics can be

Table 2.1 The parameters for induction heated sintering process.

Parameter	Applied value
Total power capacity	15 kW
Vacuum level	1×10^{-3} Torr
Resistance heating frequency	50 kHz
Heating rate	100°C–1200°C/min
The output of total power	0%–100%
Applied pressure	10–300 MPa
Cooling rate	500°C/min
Duration time	~10 min

superplastically distorted at modest temperatures. When made from nanoparticles and afterward reheated at higher temperatures high-temperature strengthening is obtained. Brittleness [18] is decreased in ceramics. Ductility [19] and superplasticity tend to be improved.

2.5 Application of nanoceramics

Because of the advantages of the nanoscale, such as the high surface-to-volume ratio, permittivity, porosity, piezo properties, and more, nanoceramics have a diverse and expandable practical scale. Table 2.2 gives a brief overview of the core properties of the nanoceramics for fields of applications and the scalability in certain areas.

2.5.1 Biomedical application

Medical implants made of nanoceramics [53] are gaining popularity because of their excellent properties, like strength, hydrophilicity, and biological compatibility. The present focus is to produce new nanoceramics that are useful for biomedical applications [54]. Cancer treatment, bacterial and viral infection treatment [55], oxygen delivery to damaged tissue, tissue regeneration, drug delivery [56], and gene delivery are some examples.

2.5.2 Simulated body fluid

Surface manipulation and nanoscale coatings are important in the development of substances that are body-interactive and aid in tissue rejuvenation and physiology restoration. Hydroxyapatite layers [56] are produced in simulated body fluid (SBF) on organic or inorganic substrates. Biomaterials show exceptionally fine crystals of carbonate ions that contain apatite after immersion in SBF. Bone cells [57] differentiate and proliferate on this apatite layer, displaying high bioactivity, and the layer facilitates the formation of new bone (as shown in Fig. 2.3).

Under normal conditions, SBF is supersaturated in calcium and phosphate in comparison to apatite. Therefore, if there is an effective functional group in a material for the nucleation of apatite upon its surface, apatite will form spontaneously. The rate of nucleation can be increased if there is an excessive amount of Ca^{2+}, Ti-OH, or the presence of similar groups. This approach may be used to assess an artificial material's bioactivity by looking at its surface's capacity to generate apatite.

2.5.3 Tissue engineering

In tissue engineering, [58] nanoceramics combine with living cells and bioactive scaffolds to deliver essential cells to the sites that are damaged in the body. Stem cells are incorporated into bioceramics. They can reproduce for a longer duration. Adult stem cells can be found in differentiated tissues as pluripotent cells. They regenerate and

Table 2.2 Applications, core qualities that aid them, and the scalable applicative points of nanoceramics.

	Application sector	Types of nanoceramics employed	Unique characteristic	large scale usage	References
Eco-conservation	Pollutant adsorbent for wastewater management	Al_2O_3, SiO_2, ZrO_2, and $Ca_3(PO_4)_2$ γ-Al_2O_3 [79], [80]	Mechanical and chemical resistance, High surface: volume, high liquid permeability with adjustable pore sizes at low pressure.[81]	Removal of bacteria and viruses Removal of dyes Removal of inorganic ions Removal of organic debris Nanofiltration	[20–22] [23,24] [25–27] [28] [2]
Medicine and health	Screening equipment, spectacles, bone tissue repair, guided medication administration, implants, radiation treatment, prosthetics, endoscopy, and orthopedics.	Al_2O_3, SiO_2, ZrO_2, [Al_2O_3 + ZrO_2]-ATZ and ZTA, $Ca_3(PO_4)_2$, TiNbN, Zwitterionic nanoceramics,	Nontoxic, Compatible, elevated bioactivity	Osseous/articulation clinical devices for the hip, shoulder, and more Osteoblast proliferation and Bone construction aids Hip and knee Arthroplasty Knee Prostheses Dentistry	[29–33] [29,34–36] [37,38] [39] [40–43]
Defense and space	Protective ballistics, nuclear fuel pellets, tiles for space shuttles, jet blades, and missile nose cones [79], [108]	Al_2O_3, Si_3N_4, SiC, and AlN, TiO_2	Excellent durability and endurance, exceptional weathering resilience, dimensional consistency, resistance to abrasion, minimal toxic effects, and thermal resilience. [109]	Ballistic helmets Optoelectrical-photoelectrochemical anticorrosion system	[44] [45]
Electronics	Lasers, capacitors, IC, modulators, and microphones.	$BaTiO_3$, lead zirconate titanate films,	High and low (versatile) dielectric coefficient, piezoelectric, permittivity [112], [113]	Capacitors, transformers, and microphones	[7,46–49]
Pharmaceuticals	Drug developmental biochemical research	Carboxymethyl-β-cyclodextrin + Fe_3O_4/SiO_2	Enantioselective environment-stereo selectivity, electric and spin polarization properties physical interactions, and assembly frameworks [118]	Chiral drug separation	[50,51]
Spintronic research	Sensors, detectors, and optoelectronic devices.	[La^{3+} + $BaTiO_3$]	excellent heat stability, strong dielectric constant, and reduced dielectric loss	Pyroelectric, chemical, and biosensors, IR-detectors	[52]

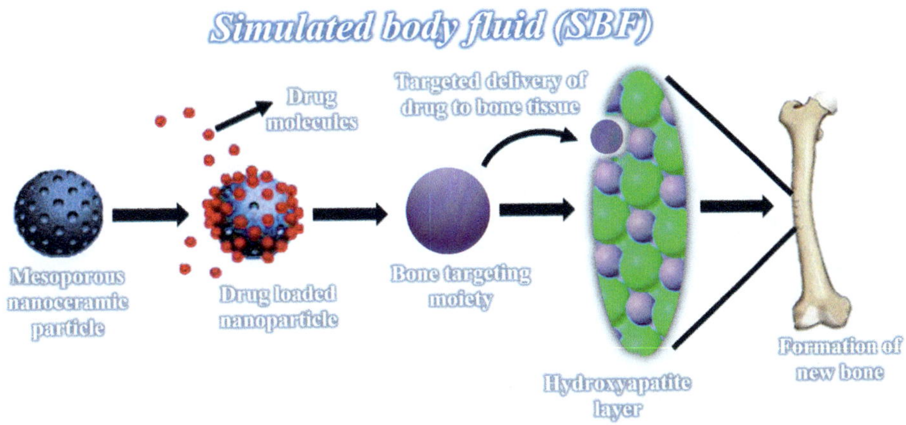

Figure 2.3 Formation of bone on the layer of apatite in SBF. *SBF*, Simulated body fluid.

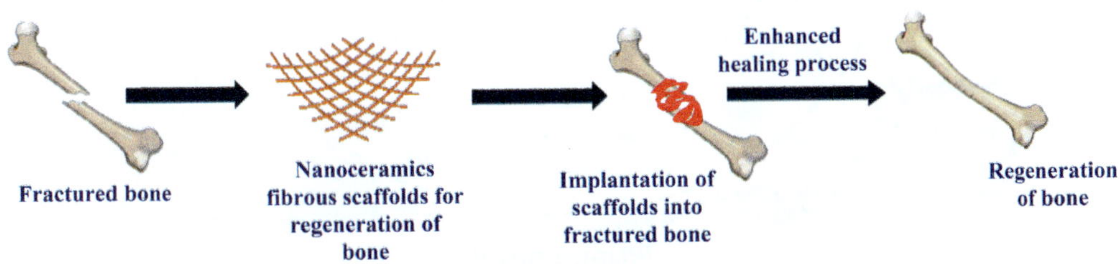

Figure 2.4 Tissue engineering mediated by nanoceramics for bone regeneration.

specialize to produce a certain cell type of tissue from which they emerged [59]. Nanoscale bioactive three-dimensional scaffolds can be useful for the treatment of gaps in long-bone shafts (Fig. 2.4) [60].

Due to their remarkable bioactivity and propensity to attach to hard tissue, Si-based bioactive nanoceramic composites have been developed as appealing materials for bone and tooth regeneration. Different pore diameters are critical for the creation of new tissue in bone tissue regeneration. Micropores operate as mineralization nucleation sites, mesopores as nutrition and developmental component reservoirs, while macrospores cause cell infiltration.

Because of its better potential to induce bone tissue regeneration while also acting as a drug delivery carrier, biologically active glass is rapidly being studied for use in the manufacture of bone-repair scaffolds. By altering the surface topography or the morphology of scaffolds and transplants, clay-based nanoceramics are being employed as coatings to increase the biological effect and incorporation (Fig. 2.5) [61].

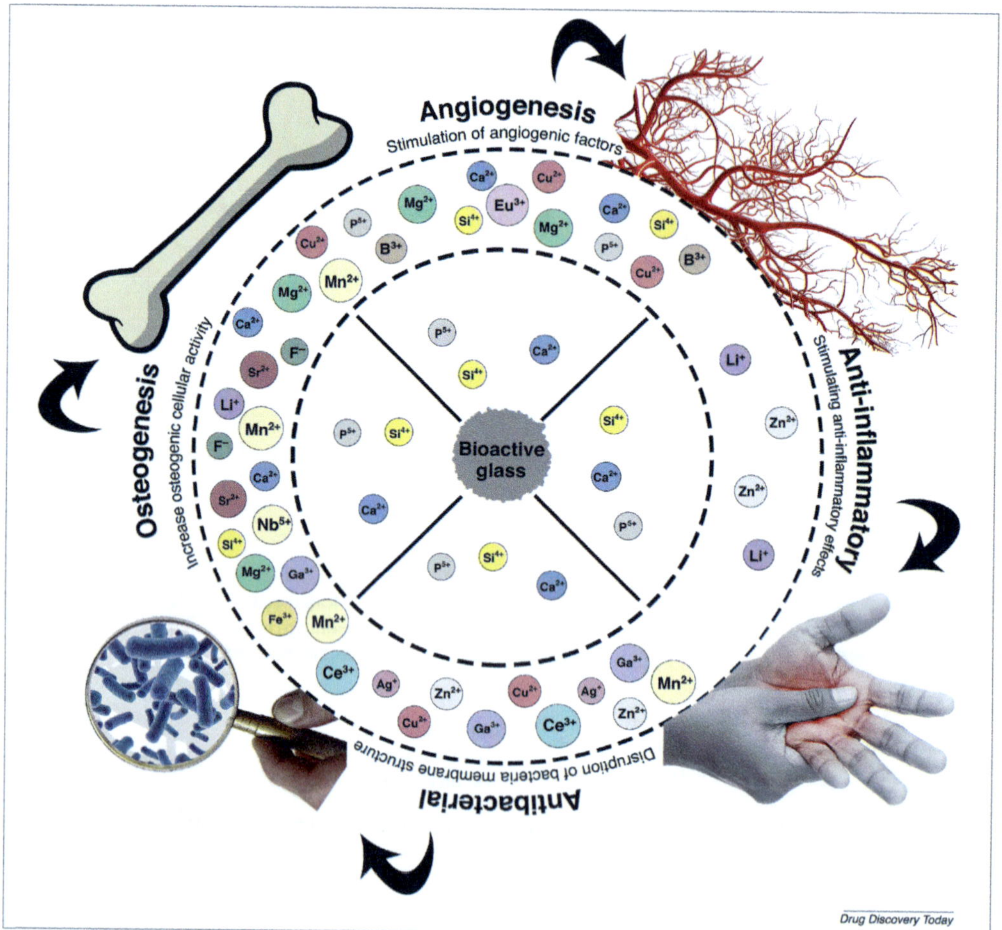

Figure 2.5 Representation of applications of nanoceramic (bioactive glass) in the therapeutic and biomedical field. Source: *From S. Kargozar, F. Baino, S. Hamzehlou, R. G. Hill, M. Mozafari, Bioactive glasses entering the mainstream, Drug Discovery Today, 23 (10) (2018) 1700–1704. https://doi.org/10.1016/j.drudis.2018.05.027.*

2.5.4 Drug delivery

Ceramic nanoparticles are formed of substances that lay between metals and nonmetals in terms of properties. They have poor electrical and thermal conductance, as well as a strong elastic modulus, stiffness, and resistance to corrosion [62]. Nanoceramics have also been made using a combination of these materials to improve biocompatibility, drug release mechanisms, therapeutic impact, and other features. Mesoporous silica nanoceramics are one of the most potent pharmaceutical administration methods in anticancer medication. The large surface area, configurable pore diameter, customizable particle size and shape, and exceptional biocompatibility are just a few of the advantages of this material over others. Mesoporous silica nanoparticles can deliver both hydrophobic and hydrophilic medicinal molecules with precision. They have also been used for therapeutic and diagnostic applications, such as visualizing the delivery location. Functionalizing these particles with chemicals like folic acid and VEGF-121, or lactose amination of the surface provides an effective targeting capability. Combining silica with thermally responsive polymers has also resulted in pH-sensitive, temperature, and magnetic field-sensitive drug delivery systems [62,63]. Table 2.3 shows the application of silica nanoceramics in anticancer drug delivery.

Table 2.3 Silica nanoceramics used in anticancer drug delivery.

Nanoceramic product	Challenges	Inference	References
Doxorubicin, a mesoporous silica-coated layer by layer	DOX has harmful effects due to uncontrolled and untargeted delivery	This nanosystem was designed to respond to acidic environments in a certain way (endosomal compartments of cancer cells). It has great stimuli-responsive drug release capability, excellent biostability, and minimal premature drug leakage at pH 7.4. Experiments in vivo revealed an extremely efficient tumor growth inhibition rate.	[64]
Charge reversal, pH-responsive polymer-coated doxorubicin-loaded with mesoporous silica nanoceramic	It has toxicity due to nontargeted delivery	Nanoceramics were made with 3-aminopropyl-triethoxysilane. DOX was effectively transported and released to the nucleus of HeLa cells using nanoceramics.	[57]
Lactosaminated mesoporous SiO_2 nanoceramics filled with docetaxel for asialoglycoprotein receptor targeting.	Systemic adverse effects of drug	HepG2 and SMMC7721 cells were effectively inhibited by Lactosaminated mesoporous silica nanoceramics loaded with docetaxel.	[65]
Cisplatin-loaded mesoporous silica nanoparticles with an avidin cap	Systemic harmful effects of cisplatin due to nontargeted delivery	Release of cisplatin via MMP9, which resulted in apoptotic cell death restricted to the lung tumor locations. MMP9 responsive nanoceramics also enabled excellent cisplatin and proteasome inhibitor combination drug delivery.	[66]

(*Continued*)

Table 2.3 (Continued)

Nanoceramic product	Challenges	Inference	References
Mesoporous silica nanoceramics loaded with conjugated-sunitinib	Insufficient medication administration limits therapeutic efficacy.	Due to effective targeting, Sunitinib was supplied to U87MG tumor cells in larger amounts.	[67]
Cisplatin, mesoporous hollow silica nanospheres	Application is limited to traditional dosage form	Cisplatin, folic acid, and rhodamine isothiocyanate were coupled to the external interface, which was acid-functionalized. Encapsulated superparamagnetic $CoFe_2O_4$ nanoparticles and a hydrophobic anticancer agent Nanospheres showed improved cytotoxicity, as well as MRI contrast characteristics, as compared to individual drugs.	[68]

2.5.5 Nanoceramics as disinfectant agents for water and wastewater purification

Waterborne bacterial infections are linked to various illness outbreaks recorded around the world because of contaminated water, creating a substantial hazard to the health-care system. Solar treatment, UV treatment, catalytic oxidation, heat disinfection, and advanced oxidation processes have all been investigated as alternatives to chemical disinfection of wastewater. Despite their apparent effectiveness, each of these drinking water treatment technologies has drawbacks that limit or prevent their usage in specific situations. Certain negatively strained bacteria, for example, may need longer treatment for complete decontamination by solar water disinfection method; due to bacterial regeneration documented after a day of the treatment because of initiation of the DNA repair process [69,70].

Magnetic nanostructures with well-controlled dimensions and components have received interest from researchers due to their successful implementation in versatile procedures as a secure alternative for the potential threat of bacterial infections. Due to their structural features, quasicrystal and ceramic nanoparticles have shown promising properties among the produced nanoparticles. Nano-porous Co_2O_3/Cu_2O_3 with alumina and silica was found to be the most efficient composite in removing Gram-positive and Gram-negative bacteria from water in recent research, showing that it is a harmless, environment-friendly, and cheap alternate disinfectant for water supply and

Table 2.4 The nanomaterials and their corrosion resistance capabilities.

Material	Nanomaterial coating	Coating thickness	Substrate	Corrosion resistance	References
Zirconia oxides	Thin films of ZrO_2	50–350 nm	Preoxidized 304L stainless steel	Best when the surface is oxidized before coating	[78]
		500 nm	Alumina-silica refractory material	Corrosion loss decreases by 18%	[79]
		0.5–0.8 μm	316L stainless steel	Better with temperature aid 400°C–650°C	[80]
Titanium oxides		TiO_2 NP	550 nm	Greatest for coating thickness 464 nm	[81]
		TiO_2 thin films	370 nm	Greatest with The n-modified surface of tio_2	[82]
Aluminum Oxides	Al_2O_3 thin film	200 nm	X40crmov5–1 steel	Greatest with deposition temperature 300°C	[83]
		50 nm	Carbon steel	Unable to protect the steel	[84]
		10, 20, 50 nm	Copper	10 nm is superior to 50 nm	[85]
Nickel alloys	SiO_2–Ni–P	29 μm	API-5L X65 steel substrates	Optimal conc. At 2 wt.% SiC	[86]
	SiC–Ni–P	19–21 μm	St37 tool steel substrate	4 wt.%	[87]

wastewater treatment. As a result, highly porous magnetic nanomaterials could be regarded as a new nanotechnique for disinfecting wastewaters. The doped Co_3O_4 nanoceramics materials have the potential to be utilized in sensing devices as well [20].

2.5.6 Corrosion-resistant coatings

Nanoceramic coatings are beneficial for solving corrosion problems in every electrode and material body [71,72]. Types of nanoceramic [73] coating include thermal spraying, plasma spraying [74], sputter coating, and electrochemical coating [24]. The ceramic films range from 50 nm to a few μm in thickness. Elevated-temperature resistible structural materials are Si_3N_4, diamond-like coatings, silicon carbide [75], and boron nitride [76]. They have magnificent thermomechanical properties [72]. The weathering potential of coated substrates is positive, but the corrosion current density decreases. The rate of corrosion slows down. Some materials and their abilities in corrosion resistance are shown in Table 2.4.

2.6 Conclusion and prospects

The development and vivid research of nanoceramics have increased ceramic material performance in various applications utilizing their chemical, mechanical, electrical, and other characteristics. These nanoceramics synthesized and fabricated in film-like forms have allowed tremendous progress in simplifying the daily life of producers and customers of various products for different sectors.

Ceramic processing has expanded on an industrial scale in fields such as medicine, pollution control, automotive durability and shielding, the aircraft industry, or improving therapeutic practices not only as an alternative commodity but as a superior need. Nanoceramic coatings have critical features that could lead to a revolution in the upcoming implantable devices for bone regeneration. It has been demonstrated that the implantation need not have to be balanced to the internal milieu of the patient's body where bioactivity is sought for bone implants without the use of an expensive graft and thus revolutionizing the medical field. Due to the widespread adaptability, this field has caught the attention of the whole science establishment and guarantees a lot of technology in the coming years. As a result, scaling for these nanoceramics might be a challenging task.

References

[1] I.A. Ovid'ko, Micromechanics of fracturing in nanoceramics, Philos. Trans. R. Soc. A: Math. Phys. Eng. Sci. 373 (2038) (2015). Available from: https://doi.org/10.1098/rsta.2014.0129, http://rsta.royalsocietypublishing.org/content/roypta/373/2038/20140129.full.pdf.

[2] O. Van Der Biest, Nanoceramics: issues and opportunities, Int. J. Appl. Ceram. Technol. 10 (4) (2013) 565–576. Available from: https://doi.org/10.1111/ijac.12074.

[3] C. Voisin, S. Guillemet-Fritsch, P. Dufour, C. Tenailleau, H. Han, J.C. Nino, Influence of oxygen substoichiometry on the dielectric properties of BaTiO3-δ nanoceramics obtained by spark plasma sintering, Int. J. Appl. Ceram. Technol. 10 (2013) E122−E133. Available from: https://doi.org/10.1111/ijac.12058.

[4] S. Laasri, M. Taha, A. Laghzizil, E.K. Hlil, J. Chevalier, The affect of densification and dehydroxylation on the mechanical properties of stoichiometric hydroxyapatite bioceramics, Mater. Res. Bull. 45 (10) (2010) 1433−1437. Available from: https://doi.org/10.1016/j.materresbull.2010.06.040, http://www.sciencedirect.com/science/journal/00255408.

[5] Uday Shashikumar, et al., Nanoceramics: Fabrication, properties and its applications towards the energy sector, FUEL (2023). Available from: https://doi.org/10.1016/j.fuel.2022.126829.

[6] S. Maitra, J. Roy, Nanoceramic matrix composites: types, processing, and applications, Second Edition, Advances in Ceramic Matrix Composites, 27−48, Elsevier Inc, India, 2018. Available from: http://www.sciencedirect.com/science/book/9780081021668.10.1016/B978-0-08-102166-8.00003-7.

[7] M. Ubaidullah, J. Ahmed, A.M. Al-Enizi, A. Tyagi, S.F. Shaikh, N. Tarannum, et al., Metal organic precursor derived $Ba_{1-x}Ca_xZrO_3$ ($0.05 \leq x \leq 0.20$) nanoceramics for excellent capacitor applications, J. King Saud. Univ. − Sci. 32 (3) (2020) 1937−1943. Available from: https://doi.org/10.1016/j.jksus.2020.02.009, http://www.sciencedirect.com/science/journal/10183647.

[8] D.G. Calatayud, T. Jardiel, M.S. Bernardo, V. Mirabello, H. Ge, R.L. Arrowsmith, et al., Hybrid hierarchical heterostructures of nanoceramic phosphors as imaging agents for multiplexing and living cancer cells translocation, ACS Appl. Bio Mater. 4 (5) (2021) 4105−4118. Available from: https://doi.org/10.1021/acsabm.0c01417, http://pubs.acs.org/journal/aabmcb.

[9] S. Mandizadeh, F. Soofivand, S. Bagheri, M. Salavati-Niasari, SrCrxFeO nanoceramics as an effective catalyst for desulfurization of liquid fuels: green sol−gel synthesis, characterization, magnetic and optical properties, PLoS ONE 12 (5) (2017). Available from: https://doi.org/10.1371/journal.pone.0162891, http://journals.plos.org/plosone/article/file?id = 10.1371/journal.pone.0162891&type = printable.

[10] S. Sakhala, S. Tatavarthy, Synthesising Silica Nanoparticles using Sol−Gel Technique and Studying Various Factors Affecting It. (Academia.edu).

[11] A. Kumar, N. Yadav, M. Bhatt, N.K. Mishra, P. Chaudhary, R. Singh. Sol-Gel derived nanomaterials and it's applications: a review. Res. J. Chem. Sci. 5 (12) (2015) 98−105. www.isca.in.

[12] R.P. Rana, S.K. Pratihar, S. Bhattacharyya, Powder processing and densification behaviour of alumina-high zirconia nanocomposites using chloride precursors, J. Mater. Process. Technol. 190 (1−3) (2007) 350−357. Available from: https://doi.org/10.1016/j.jmatprotec.2007.02.009.

[13] M. Lledos, V. Mirabello, S. Sarpaki, H. Ge, H.J. Smugowski, L. Carroll, et al., Synthesis, Radiolabelling and in vitro imaging of multifunctional nanoceramics, ChemNanoMat. 4 (4) (2018) 361−372. Available from: https://doi.org/10.1002/cnma.201700378, http://onlinelibrary.wiley.com/journal/10.1002/(ISSN)2199-692X.

[14] B. Charbonnier, C. Laurent, G. Blanc, O. Valfort, D. Marchat, Porous bioceramics produced by impregnation of 3D-printed wax mold: ceramic architectural control and process limitations, Adv. Eng. Mater 18 (10) (2016) 1728−1737. Available from: https://doi.org/10.1002/adem.201600308, http://onlinelibrary.wiley.com/journal/10.1002/(ISSN)1527-2648.

[15] J.A. Wollmershauser, B.N. Feigelson, E.P. Gorzkowski, C.T. Ellis, R. Goswami, S.B. Qadri, et al., An extended hardness limit in bulk nanoceramics, Acta Mater. 69 (2014) 9−16. Available from: https://doi.org/10.1016/j.actamat.2014.01.030.

[16] K. Abdelrazek, Advanced Sintering of Nano-Ceramic Materials, InTech, 2012. 10.5772/38287.

[17] F. Baino, G. Novajra, C. Vitale-Brovarone, Bioceramics and scaffolds: a winning combination for tissue engineering, Front. Bioeng. Biotechnol. 3 (2015). Available from: https://doi.org/10.3389/fbioe.2015.00202, http://journal.frontiersin.org/journal/bioengineering-and-biotechnology#archive.

[18] R.E. Smallman, A.H.W. Ngan, Creep, Fatigue and Fracture, Elsevier BV, 2014, pp. 571−616. Available from: 10.1016/b978-0-08-098204-5.00015-8.

[19] J. Bauer, C. Crook, A. Guell Izard, Z.C. Eckel, N. Ruvalcaba, T.A. Schaedler, et al., Additive manufacturing of ductile, ultrastrong polymer-derived nanoceramics, Matter. 1 (6) (2019) 1547−1556. Available from: https://doi.org/10.1016/j.matt.2019.09.009, http://www.cell.com/matter.

[20] A.B. Abou Hammad, A.M. El Nahwary, B.A. Hemdan, A.L.K. Abia, Nanoceramics and novel functionalized silicate-based magnetic nanocomposites as substitutional disinfectants for water and wastewater purification, Environ. Sci. Pollut. Res. 27 (21) (2020) 26668–26680. Available from: https://doi.org/10.1007/s11356-020-09073-9, https://link.springer.com/journal/11356.

[21] S. Ciston, R.M. Lueptow, K.A. Gray, Controlling biofilm growth using reactive ceramic ultrafiltration membranes, J. Membr. Sci. 342 (1–2) (2009) 263–268. Available from: https://doi.org/10.1016/j.memsci.2009.06.049.

[22] M.M. Fidalgo De Cortalezzi, M.V. Gallardo, F. Yrazu, G.J. Gentile, O. Opezzo, R. Pizarro, et al., Virus removal by iron oxide ceramic membranes, J. Environ. Chem. Eng. 2 (3) (2014) 1831–1840. Available from: https://doi.org/10.1016/j.jece.2014.08.006.

[23] G.M.K. Tolba, A.M. Bastaweesy, E.A. Ashour, W. Abdelmoez, K.A. Khalil, N.A.M. Barakat, Effective and highly recyclable ceramic membrane based on amorphous nanosilica for dye removal from the aqueous solutions, Arab. J. Chem. 9 (2) (2016) 287–296. Available from: https://doi.org/10.1016/j.arabjc.2015.05.009. http://colleges.ksu.edu.sa/Arabic%20Colleges/CollegeOfScience/ChemicalDept/AJC/default.aspx (ScienceDirect http://www.sciencedirect.com/science/journal/18785352).

[24] X. Cheng, N. Li, M. Zhu, L. Zhang, Y. Deng, C. Deng, Positively charged microporous ceramic membrane for the removal of Titan Yellow through electrostatic adsorption, J. Environ. Sci. 44 (2016) s. Available from: https://doi.org/10.1016/j.jes.2015.11.018.

[25] S. Sklari, A. Pagana, L. Nalbandian, V. Zaspalis, Ceramic membrane materials and process for the removal of as(iii)/as(v) ions from water, J. Water Process. Eng. 5 (2015) 42–47. Available from: https://doi.org/10.1016/j.jwpe.2014.12.010, http://www.journals.elsevier.com/journal-of-water-process-engineering/.

[26] C.P. Athanasekou, S.K. Papageorgiou, V. Kaselouri, F.K. Katsaros, N.K. Kakizis, A.A. Sapalidis, et al., Development of hybrid alginate/ceramic membranes for Cd^{2+} removal, Microporous Mesoporous Mater. 120 (1–2) (2009) 154–164. Available from: https://doi.org/10.1016/j.micromeso.2008.07.048.

[27] K.F. Lam, H. Kassab, M. Pera-Titus, K.L. Yeung, B. Albela, L. Bonneviot, MCM-41 "LUS": alumina tubular membranes for metal separation in aqueous solution, J. Phys. Chem. C. 115 (1) (2011) 176–187. Available from: https://doi.org/10.1021/jp1065874.

[28] K. Szymański, A.W. Morawski, S. Mozia, Humic acids removal in a photocatalytic membrane reactor with a ceramic UF membrane, Chem. Eng. J 305 (2016) 19–27. Available from: https://doi.org/10.1016/j.cej.2015.10.024, http://www.elsevier.com/inca/publications/store/6/0/1/2/7/3/index.htt.

[29] B.J. McEntire, B.S. Bal, M.N. Rahaman, J. Chevalier, G. Pezzotti, Ceramics and ceramic coatings in orthopaedics, J. Eur. Ceram. Soc. 35 (16) (2015) 4327–4369. Available from: https://doi.org/10.1016/j.jeurceramsoc.2015.07.034, http://www.journals.elsevier.com/journal-of-the-european-ceramic-society/.

[30] S.V. Dorozhkin, Bioceramics of calcium orthophosphates, Biomaterials. 31 (7) (2010) 1465–1485. Available from: https://doi.org/10.1016/j.biomaterials.2009.11.050.

[31] I. Izquierdo-Barba, M. Colilla, M. Vallet-Regí, Zwitterionic ceramics for biomedical applications, Acta Biomater. 40 (2016) 201–211. Available from: https://doi.org/10.1016/j.actbio.2016.02.027, http://www.journals.elsevier.com/acta-biomaterialia.

[32] S.V. Dorozhkin, Multiphasic calcium orthophosphate (CaPO4) bioceramics and their biomedical applications, Ceram. Int. 42 (6) (2016) 6529–6554. Available from: https://doi.org/10.1016/j.ceramint.2016.01.062.

[33] M. Vallet-Regí, D. Arcos, Bioceramics for drug delivery, Acta Mater 61 (3) (2013) 890–911. Available from: https://doi.org/10.1016/j.actamat.2012.10.039.

[34] S.N. Sethu, S. Namashivayam, S. Devendran, S. Nagarajan, W.B. Tsai, S. Narashiman, et al., Nanoceramics on osteoblast proliferation and differentiation in bone tissue engineering, Int. J. Biol. Macromol. 98 (2017) 67–74. Available from: https://doi.org/10.1016/j.ijbiomac.2017.01.089.

[35] L. Tan, X. Yu, P. Wan, K. Yang, Biodegradable materials for bone repairs: a review, J. Mater. Sci. Technol. 29 (6) (2013) 503–513. Available from: https://doi.org/10.1016/j.jmst.2013.03.002.

[36] E. Vunain, S.B. Mishra, A.K. Mishra, B.B. Mamba, Nanoceramics: Fundamentals and advanced perspectives, Sol–gel Based Nanoceramic Materials: Preparation, Properties and Applications, Springer International Publishing, South Africa, 2016, pp. 1–20. Available from: http://doi.org/10.1007/978-3-319-49512-5.10.1007/978-3-319-49512-5_1.

[37] C. Zietz, D. Kluess, P. Bergschmidt, M. Haenle, W. Mittelmeier, R. Bader, Tribological aspects of ceramics in total hip and knee arthroplasty, Semin. Arthroplasty 22 (4) (2011) 258–263. Available from: https://doi.org/10.1053/j.sart.2011.10.006.

[38] J.J. Yoo, P.W. Yoon, Y.K. Lee, K.H. Koo, K.S. Yoon, H.J. Kim, Revision total hip arthroplasty using an alumina-on-alumina bearing surface in patients with osteolysis, J. Arthroplasty 28 (1) (2013) 132–138. Available from: https://doi.org/10.1016/j.arth.2012.04.030.

[39] B. Denkena, J. Köhler, A. Turger, P. Helmecke, T. Correa, C. Hurschler, Manufacturing conditioned wear of all-ceramic knee prostheses, Procedia CIRP 5 (2013) 179–184 http://www.sciencedirect.com/science/journal/221282715.

[40] Z. Pang, A. Chughtai, I. Sailer, Y. Zhang, A fractographic study of clinically retrieved zirconia-ceramic and metal-ceramic fixed dental prostheses, Dental Mater. 31 (10) (2015) 1198–1206. Available from: https://doi.org/10.1016/j.dental.2015.07.003, https://www.journals.elsevier.com/dental-materials.

[41] C.Â.M. Volpato, L.G.D. Garbelotto, M.C. Fredel, F. Bondioli, Application of zirconia in dentistry: biological, mechanical and optical considerations, Advances in ceramics-electric and magnetic ceramics, Bioceramics, Ceram. and Environ. 25 (2011).

[42] I.K. Konstantinidis, S. Jacoby, M. Rädel, K. Böning, Prospective evaluation of zirconia based tooth- and implant-supported fixed dental prostheses: 3-Year results, J. Dent. 43 (1) (2015) 87–93. Available from: https://doi.org/10.1016/j.jdent.2014.10.011, http://www.elsevier.com/locate/jdent.

[43] D. Fabris, J.C.M. Souza, F.S. Silva, M. Fredel, J. Mesquita-Guimarães, Y. Zhang, et al., The bending stress distribution in bilayered and graded zirconia-based dental ceramics, Ceram. Int. 42 (9) (2016) 11025–11031. Available from: https://doi.org/10.1016/j.ceramint.2016.03.245.

[44] S.G. Kulkarni, X.L. Gao, S.E. Horner, J.Q. Zheng, N.V. David, Ballistic helmets – their design, materials, and performance against traumatic brain injury, Composite Struct. 101 (2013) 313–331. Available from: https://doi.org/10.1016/j.compstruct.2013.02.014.

[45] A.A. Daniyan, L.E. Umoru, O.S.I. Fayomi, A.P.I. Popoola, Structural evolution, optoelectrical and corrosion properties of electrodeposited WO3 integration on Zn-TiO_2 electrolyte for defence super application, Defence Technol. 14 (5) (2018) 396–402. Available from: https://doi.org/10.1016/j.dt.2018.06.015, https://www.journals.elsevier.com/defence-technology.

[46] P. Sharma, P. Kumar, R.S. Kundu, J.K. Juneja, N. Ahlawat, R. Punia, Structural and dielectric properties of substituted barium titanate ceramics for capacitor applications, Ceram. Int. 41 (10) (2015) 13425–13432. Available from: https://doi.org/10.1016/j.ceramint.2015.07.131.

[47] S.R. Bharadwaj, S. Varma, B.N. Wani, Electroceramics for fuel cells, batteries and sensors, Functional Materials, Elsevier Inc, India, 2012, pp. 639–674. Available from: http://www.sciencedirect.com/science/book/9780123851420. 10.1016/B978-0-12–385142-0.00016–7.

[48] A.K. Mishra, Mishra, A.K. ed. *Sol-gel based nanoceramic materials: preparation, properties and applications* (pp. 1-20). Cham, Switzerland:, Springer International Publishing, 2017https://doi.org/10.1007/978-3-319-49512-5.

[49] T. Zheng, J. Wu, Quenched bismuth ferrite-barium titanate lead-free piezoelectric ceramics, J. Alloy. Compd. 676 (2016) 505–512. Available from: https://doi.org/10.1016/j.jallcom.2016.03.205.

[50] Y. Fu, T. Huang, B. Chen, J. Shen, X. Duan, J. Zhang, et al., Enantioselective resolution of chiral drugs using BSA functionalized magnetic nanoparticles, Sep. Purif. Technol. 107 (2013) 11–18. Available from: https://doi.org/10.1016/j.seppur.2013.01.007.

[51] J. Fan, N.A. Kotov, Chiral nanoceramics, Adv. Mater. 32 (41) (2020). Available from: https://doi.org/10.1002/adma.201906738, http://onlinelibrary.wiley.com/journal/10.1002/(ISSN)1521-4095.

[52] B.H. Devmunde, S.B. Somwanshi, P.B. Kharat, M.B. Solunke, Rare earth ion (La3 +) doped $BaTiO_3$ Perovskite nanoceramics for spintronic applications, Journal of Physics: Conference Series 1644 (2020) 012055 http://iopscience.iop.org/journal/1742-65961644.

[53] Y. Liu, J. Luo, X. Chen, W. Liu, T. Chen, Cell membrane coating technology: a promising strategy for biomedical applications, Nano-Micro Letters 11 (1) (2019). Available from: https://doi.org/10.1007/s40820-019-0330-9, http://www.springer.com/engineering/journal/40820.
[54] W. Paul, C.P. Sharma, Bioceramics, towards nano-enabled drug delivery: a mini review, Trends Biomater. Artif. Organs 19 (1) (2005) 7–11. Available from: http://www.sbaoi.org/pdf/vol-19%281%29/19-1-7-11.pdf.India.
[55] P. Choudhary, S.K. Das, Bio-reduced graphene oxide as a nanoscale antimicrobial coating for medical devices, ACS Omega 4 (1) (2019) 387–397. Available from: https://doi.org/10.1021/acsomega.8b02787, http://pubs.acs.org/journal/acsodf.
[56] S.V. Dorozhkin, Nanodimensional and nanocrystalline apatites and other calcium orthophosphates in biomedical engineering, biology and medicine, Materials 2 (4) (2009) 1975–2045. Available from: https://doi.org/10.3390/ma2041975, http://www.mdpi.com/1996-1944/2/4/1975/pdf.
[57] P. Zhang, T. Wu, J.L. Kong, In situ monitoring of intracellular controlled drug release from mesoporous silica nanoparticles coated with pH-responsive charge-reversal polymer, ACS Appl. Mater. Interfaces 6 (20) (2014) 17446–17453. Available from: https://doi.org/10.1021/am5059519, http://pubs.acs.org/journal/aamick.
[58] L. Bačáková, E. Filová, F. Rypáček, V. Švorčík, V. Starý, Cell adhesion on artificial materials for tissue engineering, Physiol. Res. 53 (1) (2004) S35–S45.
[59] Z. Singh, Nanoceramics in bone tissue engineering: the future lies ahead, Trends J. Sci. Res. 3 (3) (2018) 120–123. Available from: https://doi.org/10.31586/nanomaterials.0303.03.
[60] M. Mastrogiacomo, S. Scaglione, R. Martinetti, L. Dolcini, F. Beltrame, R. Cancedda, et al., Role of scaffold internal structure on in vivo bone formation in macroporous calcium phosphate bioceramics, Biomaterials. 27 (17) (2006) 3230–3237. Available from: https://doi.org/10.1016/j.biomaterials.2006.01.031.
[61] S. Kargozar, F. Baino, S. Hamzehlou, R.G. Hill, M. Mozafari, Bioactive glasses entering the mainstream, Drug Discovery Today 23 (10) (2018) 1700–1704. Available from: https://doi.org/10.1016/j.drudis.2018.05.027, http://www.elsevier.com/locate/drugdiscov.
[62] M. Manzano, M. Vallet-Regí, Mesoporous silica nanoparticles for drug delivery, Adv. Funct. Mater. 30 (2) (2020). Available from: https://doi.org/10.1002/adfm.201902634, http://onlinelibrary.wiley.com/journal/10.1002/(ISSN)1616-3028.
[63] Y. Wang, Q. Zhao, N. Han, L. Bai, J. Li, J. Liu, et al., Mesoporous silica nanoparticles in drug delivery and biomedical applications, Nanomed.: Nanotechnol. Biol. Med. 11 (2) (2015) 313–327. Available from: https://doi.org/10.1016/j.nano.2014.09.014, https://www.journals.elsevier.com/nanomedicine-nanotechnology-biology-and-medicine.
[64] Q.L. Li, Y. Sun, Y.L. Sun, J. Wen, Y. Zhou, Q.M. Bing, et al., Mesoporous silica nanoparticles coated by layer-by-layer self-assembly using cucurbit[7]uril for in vitro and in vivo anticancer drug release, Chem. Mater. 26 (22) (2014) 6418–6431. Available from: https://doi.org/10.1021/cm503304p, http://pubs.acs.org/journal/cmatex.
[65] G. Quan, X. Pan, Z. Wang, Q. Wu, G. Li, L. Dian, et al., Lactosaminated mesoporous silica nanoparticles for asialoglycoprotein receptor targeted anticancer drug delivery, J. Nanobiotechnol 13 (1) (2015). Available from: https://doi.org/10.1186/s12951-015-0068-6, http://www.jnanobiotechnology.com/start.asp.
[66] S.H. Van Rijt, D.A. Bölükbas, C. Argyo, S. Datz, M. Lindner, O. Eickelberg, et al., Protease-mediated release of chemotherapeutics from mesoporous silica nanoparticles to ex vivo human and mouse lung tumors, ACS Nano. 9 (3) (2015) 2377–2389. Available from: https://doi.org/10.1021/nn5070343, http://pubs.acs.org/journal/ancac3.
[67] S. Goel, F. Chen, H. Hong, H.F. Valdovinos, R. Hernandez, S. Shi, et al., VEGF121-conjugated mesoporous silica nanoparticle: a tumor targeted drug delivery system, ACS Appl. Mater. Interfaces 6 (23) (2014) 21677–21685. Available from: https://doi.org/10.1021/am506849p, http://pubs.acs.org/journal/aamick.

[68] S. Mohapatra, S.R. Rout, R. Narayan, T.K. Maiti, Multifunctional mesoporous hollow silica nanocapsules for targeted co-delivery of cisplatin-pemetrexed and MR imaging, Dalton Trans 43 (42) (2014) 15841–15850. Available from: https://doi.org/10.1039/c4dt02144d, http://www.rsc.org/Publishing/Journals.

[69] A. Saravanan, P. Senthil, S. Kumar, et al., Photocatalytic disinfection of micro-organisms: mechanisms and applications, Environ. Technol. Innov. 24 (2021) 101909. Available from: https://doi.org/10.1016/j.eti.2021.101909.

[70] A.M. El Nahrawy, B.A. Hemdan, A.B. Abou Hammad, A.L.K. Abia, A.M. Bakr, Microstructure and antimicrobial properties of bioactive cobalt co-doped copper aluminosilicate nanocrystallines, Silicon. 12 (10) (2020) 2317–2327. Available from: https://doi.org/10.1007/s12633-019-00326-y, http://www.springer.com/chemistry/inorganic/journal/12633?cm_mmc = AD-_-Journal-_-PSE10941_V1-_-12633.

[71] A. Popoola, O.E. Olorunniwo, O.O. Ige, Corrosion Resistance through the Application of Anti-Corrosion Coatings, InTech, 2014. Available from: 10.5772/57420.

[72] D.H. Abdeen, M. El Hachach, M. Koc, M.A. Atieh, A review on the corrosion behaviour of nanocoatings on metallic substrates, Materials 12 (2) (2019). Available from: https://doi.org/10.3390/ma12020210, https://www.mdpi.com/1996-1944/12/2/210/pdf.

[73] V.S. Saji, J. Thomas, Nanomaterials for corrosion control, Curr. Sci. 92 (1) (2007) 51–55. Available from: http://www.ias.ac.in/currsci/jan102007/51.pdf.India.

[74] S. Sathish, M. Geetha, S.T. Aruna, N. Balaji, K.S. Rajam, R. Asokamani, Sliding wear behavior of plasma sprayed nanoceramic coatings for biomedical applications, Wear. 271 (5-6) (2011) 934–941. Available from: https://doi.org/10.1016/j.wear.2011.03.023.

[75] B. Díaz, E. Härkönen, V. Maurice, J. Światowska, A. Seyeux, M. Ritala et al. Failure mechanism of thin Al2O3 coatings grown by atomic layer deposition for corrosion protection of carbon steel, Electrochim. Acta 56 (2011) 9609–9618.

[76] E. Charalampopoulou, K. Lambrinou, T. Van der Donck, B. Paladino, F. Di Fonzo, C. Azina, et al., Early stages of dissolution corrosion in 316L and DIN 1.4970 austenitic stainless steels with and without anticorrosion coatings in static liquid lead-bismuth eutectic (LBE) at 500 °C, Mater. Charact. 178 (2021). Available from: https://doi.org/10.1016/j.matchar.2021.111234, https://www.journals.elsevier.com/materials-characterization.

[77] ADA307697, n.d.

[78] M. Atik, P. de Lima Neto, L.A. Avaca, M.A. Aegerter, Sol–gel thin films for corrosion protection, Ceram. Int. 21 (6) (1995) 403–406. Available from: https://doi.org/10.1016/0272-8842(95)94466-N.

[79] Q. Li, X. Zhong, J. Hu, W. Kang, Preparation and corrosion resistance studies of zirconia coating on fluorinated AZ91D magnesium alloy, Prog. Org. Coat. 63 (2) (2008) 222–227. Available from: https://doi.org/10.1016/j.porgcoat.2008.06.004.

[80] A. Keyvani, M. Saremi, M. Heydarzadeh Sohi, Microstructural stability of zirconia-alumina composite coatings during hot corrosion test at 1050°C, J. Alloy. Compd. 506 (1) (2010) 103–108. Available from: https://doi.org/10.1016/j.jallcom.2010.06.110.

[81] S.S. Fouad, B. Parditka, A.E. Bekheet, H.E. Atyia, Z. Erdélyi, ALD of TiO2/ZnO mutilayers towards the understanding of optical properties and polarizability, Optics Laser Technol. 140 (2021). Available from: https://doi.org/10.1016/j.optlastec.2021.107035, https://www.journals.elsevier.com/optics-and-laser-technology.

[82] C.X. Shan, X. Hou, K.L. Choy, Corrosion resistance of TiO2 films grown on stainless steel by atomic layer deposition, Surf. Coat. Technol. 202 (11) (2008) 2399–2402. Available from: https://doi.org/10.1016/j.surfcoat.2007.08.066.

[83] S. Mirhashemihaghighi, J. Światowska, V. Maurice, A. Seyeux, S. Zanna, E. Salmi, et al., Corrosion protection of aluminium by ultra-thin atomic layer deposited alumina coatings, Corros. Sci. 106 (2016) 16–24. Available from: https://doi.org/10.1016/j.corsci.2016.01.021.

[84] L.Q. Zhu, Y.H. Liu, H.L. Zhang, H. Xiao, L.Q. Guo, Atomic layer deposited Al 2 O 3 films for anti-reflectance and surface passivation applications, Appl. Surface Sci. 288 (2014) 430–434. Available from: https://doi.org/10.1016/j.apsusc.2013.10.051, http://www.journals.elsevier.com/applied-surface-science/.

[85] P. Boryło, K. Lukaszkowicz, M. Szindler, J. Kubacki, K. Balin, M. Basiaga, et al., Structure and properties of Al2O3 thin films deposited by ALD process, Vacuum. 131 (2016) 319–326. Available from: https://doi.org/10.1016/j.vacuum.2016.07.013.
[86] J. Vazquez-Arenas, M. Pritzker, Steady-state model for anomalous Co-Ni electrodeposition in sulfate solutions, Electrochim. Acta 66 (2012) 139–150. Available from: https://doi.org/10.1016/j.electacta.2012.01.067.
[87] L. Shi, C. Sun, P. Gao, F. Zhou, W. Liu, Mechanical properties and wear and corrosion resistance of electrodeposited Ni-Co/SiC nanocomposite coating, Appl. Surface Sci. 252 (10) (2006) 3591–3599. Available from: https://doi.org/10.1016/j.apsusc.2005.05.035, http://www.journals.elsevier.com/applied-surface-science/.

CHAPTER 3

Industrial applications of nanoceramics: from lab to real-time utilization in the food and bioprocessing industry

Gayathri D.[1], Darul Raiyaan G. I.[2], Shakila L.[3] and Kantha D. Arunachalam[4]
[1]Research Department of Zoology, National College, Trichy, Tamil Nadu, India
[2]Department of Biotechnology, College of Science and Humanities, SRM Institute of Science and Technology, Kattankulathur, Tamil Nadu, India
[3]Research Department of Zoology, Ethiraj College for Women, Chennai, Tamil Nadu, India
[4]Faculty of Sciences, Marwadi University, Rajkot, Gujarat, India

3.1 Introduction

Global population growth and the ever-increasing demand for quality and safe food have required innovations within the food, food biocatalyst, and bioprocessing industries in a sustainable and environment-friendly manner. The application of nanotechnology in the field of the food industry varies from the manufacture of essential food to the smart packaging that permits the buyers to alter food materials, depending upon the taste and nutritional needs. To explain the benefits of nanotechnology in the food sector and to overcome the negative thoughts and misconceptions, a new description of nanotechnology for food application is needed [1]. At present, the major use of nanotechnology within the food subdivision is mainly connected with food nanosensing and nanostructured food components. The nanostructured food components include the formulation of food and its packaging. Food nanodetection is used for the evaluation of the quality and protection of food. New nanomaterial-based approaches are the potential applications of nanotechnology for the distribution of essential nutrients and bioactive compounds in active foods.

To meet the current challenges of the food and bioprocessing sector, nanoceramics show excellent potential. The word "nanofood" defines food that has been cultivated, produced, processed, or packed by utilizing nanotechnology procedures, or instruments, or by the addition of nanomaterials [2]. Nanoceramics have potential applications in every process of the food chain beginning with storage, checking quality, processing, and packaging. For the past few years, nanoceramics have existed as a platform for the transformation of conventional material design to profitable product applications by adjusting their

Industrial Applications of Nanoceramics
DOI: https://doi.org/10.1016/B978-0-323-88654-3.00004-4

features through structural control at the nanoscale. The advancement of nanopackages, nanocapsules, nanosensors, nanotubes, and nanocarriers are leading to favorable features for nanoceramics in food manufacturing. Most of the nanostructured applications in consumable material consist of (1) improved quality of food, (2) bioactive fortification, (3) usage of nanocarrier encapsulation for controlled release of bioactive compounds, (4) food textures and structures alterations, and (5) the identification and neutralization of biochemical, microbiological, and chemical alterations using smart packaging systems. The objective of this book chapter is to provide information regarding agricultural food nanotechnology and a current account of feasible futuristic product outcomes of nanoceramics in the industries of food and bioprocessing.

3.2 Nanoceramics advancements for bioprocessing in the food industry

Nanobiocatalyst (NBC) science emerged as the result of the combination of nanotechnology and biotechnology, and it has rapidly developed in recent years as a modern field within nanoscience. The enhancement of enzyme activity, stability, and engineering performances have more scope within the field of NBC [3]. Several techniques are utilized for the manufacture of NBCs with enzymes for application in the bioprocessing technology in the food industry. β-Galactosidase, used in galactooligosaccharides (GOS) production or lactose hydrolysis, has been covalently bonded with glutaraldehyde in both polyaniline-magnetic and polysiloxane/polyvinyl alcohol-magnetic nanoceramics [4,5]. The β-galactosidase immobilization approach using concanavalin A-magnetic nanoceramics become a stepping stone in bioaffinity, as defined by Asari and Hussain [6]. An NBC for starch hydrolysis has been constructed with the usage of α-amylase connected to silver nanoceramics through thiol linkages [7]. α-Amylase binding via allosteric modulation to calcium hydrogen phosphate ($CaHPO_4$) nanocrystals used as nanocarriers efficiently improved enzymatic activity and positively changed the catalytical constant Km, showing greater affinity by the substrate than that showed by its nonbonded counterpart [8]. But the improvement and enhanced utilization of nanoceramic-dependent NBC's for bioprocesses was done within the small laboratory bioreactor or biotool structures. Nanobioceramics technology provides excellent perspectives along with interesting challenges and future outcomes via the interdisciplinary cooperation of researchers from the field of biochemical, biological, engineering, and materials science.

3.3 Nanoceramics in the modern food industry

In recent years there has been a broad application of nanoceramics in the production of food and beverages, which includes the targeted delivery of nutrients and/or bioactive molecules via nanoencapsulation, the use of biosensors to find out and quantify harmful microorganisms and changes in food composition, as well as fruit and vegetable preservation using edible biofilms [9]. In real time the important relationship between nanotechnology and the food industry is improving food protection, increasing shelf life, enhancing flavor and nutrient transport, detecting harmful microorganisms, and providing active foodstuffs [10]. Various nanostructures are used for new functionalities in food ingredients, including nanoemulsions, nanoliposomes, nanofibers, and nanoparticles [11]. The introduction of nanotechnology in the food sector enhances improvements in various aspects of food materials, which include shape, taste, sensory attributes, aroma, color tone, reduced use of fats, progressed absorption of vitamins, finer packaging traceability, protection, and stableness/shelf life of products [12].

Prospective nanoceramic application areas consist of natural and inorganic nanoadditives, food with nanoparticles (NPs), nanosensors for quality control and elegant packaging, nanocoating and nanofilms for utensils and foodstuffs, antipathogen, hygiene coatings, detection of pathogens in food and drinks, self-sanitizing surfaces, polymeric films for foodstuff packaging with high antibacterial properties, nanoscale freshness indicators, nanoemulsions for fat reduction, and so on. Nanoceramics hide the unwanted flavor of supplements in food and protect dietary supplements from degradation. Natural and artificial substances are the two main categories of nanomaterials used in foodstuff. Artificial nanomaterials aren't typically found in a food substance and can be mixed from engineered materials or naturally occurring resources [13]. For example, natural nanoscale components include casein micelles in the milk of animals and pectin nanostructures in fruits [14,15]. An example of engineered nanomaterial sources is nanocapsules that carry vitamins through the human stomach via the bloodstream [16].

3.3.1 Nanoencapsulation

Nanotechnology has advanced the food industry with the aid of innovative designs of transfer systems for the manufacturing of products of nanoformulated agricultural chemicals, enhancing nourishing values, and producing new output via bioactive encapsulation [17]. Nanoparticles show functionality for product encapsulation in addition to a higher release efficiency than normal encapsulation methods. Nanoceramics are frequently primarily based on nanoencapsulation of the determined compound in the form of micelles, liposomes, or protein-based carriers. Nanodelivery systems for nutraceuticals, enzymes, food additives, and antipathogens within the form of nanoliposomes, nanocochleate, and archaeosomes exist. In the model of primarily lipid-based nanoencapsulations, for example, liposomes exist as transport for nonhydrophilic particles, and nanoemulsions strengthen oral bioavailability [18].

Nanocochleates are used to deliver nutrients to cells without affecting the taste and coloration of foodstuffs [19]. Self-assembled nanoscale hydrostructures are involved in transportation and are labeled as dilated micelles; their size is less than 30 nm. They have been used for the delivery of nutraceuticals, such as lycopene, omega-3 fatty acids, β-carotene, and isoflavones in food and beverages. Consequently, nanoencapsulation lets in a precise (bio)availability of the product with a specific rate for the target time [20]. Nanoencapsulation stimulates the activity of antioxidants by enhancing their solubility, bioavailability, and stability while preventing undesired interaction with other food components [21]. Nanoencapsulation designs are used mostly for enhanced release of flavor along with maintaining its texture and design [22]. Nanoencapsulation is used for extremely reactive plant pigments that are volatile, like anthocyanins. Photostability and thermal balance can be enhanced by way of encapsulating precise particles as cyanidin-3-O-glucoside in the middle hollow space of recombinant soybean kernel. H-2 subunit ferritin (rH-2) enhances photostability and thermal stability. Rutin is taken into consideration to be a dietary flavonoid with extensive pharmacological properties, but owing to its diminished solubility, its use in the food enterprise is less common. The encapsulation of ferritin nanocages improved the thermal, UV radiation, and firmness function of ferritin-enclosed rutin as related to detached rutin [23].

Nanoencapsulation maintains the taste and the aroma control connections of potential elements through the food matrix, the release of active compounds, controls target time, and protects from different sources of contamination such as high temperature, moisture, biological and chemical degeneration [24], during storage, processing, and utilization. These nanoparticles also show compatibility compared with other compounds in the system [25]. Nanoparticles have increased capabilities such as release efficiency and encapsulation feature, compared with conventional encapsulation patterns. Also, these transmittal classifications keep the capability to have interaction in tissues owing to their narrowness, and thus permit the well-organized distribution of the active complexes to mark sites in the human body [26]. Diverse unnatural and common polymer-based encapsulation delivery classifications have been described for better quality bioavailability as well as the protection of functional components of food [27].

3.3.2 Packaging of food

Active packaging is one of the new terms used for packaging food components in which the condition of the packed nutrients changes to enhance the sensory quality as well as safety of food products, enhancing shelf existence by maintaining the standard of food products [28].

Mihindukulasuriya and Lim [29] noted the increasing involvement of nanotechnology in the representation of elegant or smart packaging to enhance the communication factor of the package. This smart packaging of food items may increase the performance of information transfer during distribution. The response generated due to changes associated with interior or exterior environmental elements might be noted by a particular sensor [29].

Packaging plays an important role in the contemporary food enterprise. But, the most important technique within the preservation of food merchandise is the development of high quality from manufacturing to eating [30]. Nanotechnology decreases contamination by eco-friendly packaging. A silicate coating nanostructure is used in the packaging of food products. Enhancements in the sensor technology present in the smart packaging of food material provide information on the quality, safety, and half-life of materials [31]. Nanocomposites are also involved in the packaging of food products. They have more suitable properties to resist thermal stress in food material processing, as well as the transportation and storage of food products. Currently, nanocomposites are utilized in beer bottles, enhancing their shelf life for up to 6 months. Carbon-based graphene nanoplates are resistant to high temperatures and have potential applications in the packaging of foodstuff in the food manufacturing sector [32].

Sonochemical coating is a versatile and simple method applied for coating materials using ultrasonication [33]. The coating has been established to be beneficial toward numerous bacterial lines (Gram-positive *Staphylococcus aureus* or Gram-negative *Escherichia coli*). The technique has been validated to enhance the manufacturing of materials that could help food to be conserved for a longer period, e.g., silver nanoparticles (Ag NPs). Numerous scientists have said that Ag NPs have importance in research regarding edible material packaging and food preservation, even though only a few techniques of Ag NPs are certified by the EFSA (European Food Safety Authority) to be used in food processing and packaging [34]. For everyday usage, zinc oxide (ZnO) is considered a safe material, licensed with the aid of the Food and Drug Administration, and is considered as an eatable additive [35]. Nanotechnology produces various antimicrobial factors with new characteristics together with ZnO, MgO, NiO, and Ag NPs. those NPs have proven potential applications and antimicrobial characters on the nanoscale. These NPs are included in matrices of polymers to provide desirable characteristics consisting of antimicrobial activity and improve the process of packaging [36]. Various NPs are manufactured and used in industry because of their encapsulation potential in functional components and the enhancement of bioavailability, stability and functionality [37]. Nanofillers also are utilized in food packaging because of their preferred functions. In biosensors, nanofillers not only protect food from environmental factors but also combine properties in packaging materials and permit various potential possibilities inside the food packaging industry [38].

The main degradation reaction that takes place in foodstuffs is oxidation, reducing its preservation time. Antioxidants are included within the packaging material in the active packaging approach to maintain the preservation of food products [39]. Nanoceramic composites and nanolaminates are efficiently used for the antimicrobial packaging of food

[40]. Bioactive packaging affords a high-quality impact on the health of consumers with the manufacturing of packed foods that are suitable for health [41]. An extensive range of nanoparticles have antimicrobial effects and have attracted tremendous attention in a range of applications in food packaging. Metal nanoparticles were incorporated in polymer films of metal nanoceramic composites, to allow antimicrobial active packaging [42]. The involvement of nanomaterials in dietary products enhances the packaging characteristics along with gas barrier properties, which is the most important application of nanotechnology in the food sector [43]. Active packaging also includes nanosensors that are meant to detect microbial and biochemical changes and provide signals [44]. The packaging of numerous food objects is progressing via nanotechnology. New nanopackages for food show special properties, that is, they are able to destroy microbes that might be present in food materials [45]. Nanosubstances used for the wrapping of food products enhance the shelf life of food products without causing any unwanted modification within the features of food products [46]. This application of nanotechnology gives solutions meant for the packaging of food via modification in the penetration activities of foils, increasing the mechanical, chemical and microbial impediment effects as well as resistance to high temperature [47].

3.3.3 Nanosensors

Early in the 21st century nanoceramics-based sensors and nanobiochips were applied in the food industry for quality assurance. In high-quality food packaging, an embedded digital tongue carries a set of nanosensors that can be especially sensitive to the gases emitted by waste food; a clean and visible sign is given indicating whether the food is fresh or spoiled via a sensor strip that modifies color [48]. For the rapid detection of food-borne harmful microbes, magnetic oxide nanoparticles are used in the food industry [49]. Nanosensors detect microorganisms, poisonous materials, and contaminants found in foodstuffs with their exceptional detection ability [50]. Nanoclay, Ag, ZnO, TiO_2, and TiN NPs are current nanoceramic substances utilized in food packaging [51,52]; they are giving new characteristic features to food substances, and thus paving the way to special new purposeful applications in the form of useful components [53]. Ruengruglikit et. al. developed a highly sensitive nanoceramic sensor that detects the gases released by foodstuffs as food spoilage occurs [54].

Immunosensing of *Staphylococcus* sp. enterotoxin B has been developed using poly (dimethylsiloxane) (PDMS) chips with strengthened, supported, fluid bilayer membranes (r-SBMs) and unique antibodies to the toxin [55]. Nanoceramic-based sensors are being developed with the aid of researchers at the Canadian Wheat Board Centre at the University of Manitoba, Canada for grain storage research. A G-liposomal nanoceramic vesicles-based immunomagnetic bead sandwich assay was developed to detect *E. coli* O157:H7, *Salmonella* sp., and *Listeria monocytogenes* in food. Various pathogens have been found with an unique type of immunosorbent assay with universal protein G-liposomal nanovesicles [56]. In agriculture, nanoceramic-based sensors are used for early warning of modifying conditions. Several biosensors primarily based on nanoceramics were invented for the identification of most common food pathogens, for example, *L. monocytogenes*, *E. coli*, and *Salmonella* species, as well as mycotoxins in consumables [57]. The application of nanoceramic particles for detection of foodborne pathogens and their toxins was reported [58]. Moreover, aflatoxins produced by *Aspergillus flavus* and *Aspergillus parasiticus* that contaminate food materials may be detected through the use of a magnetic nanogold immunosensor [59]. The benefit of this nanoceramic sensor gadget is that hundreds of nanoparticles can be positioned in one sensor to correctly locate the presence of microbes or fungi internally present in grain boxes. Due to miniaturization and the low electricity requirement, nanoceramic sensors that are small and lightweight can be fabricated. Nanoceramic material-based biosensors combine the fields of material science, molecular engineering, chemistry, and biotechnology and can increase the sensitivity and specificity of biomolecule detection. Surface-enhanced Raman spectroscopy incorporated with Ag nanosensors is one of the most efficient techniques for detecting pathogenic bacteria [60]. Other than Ag, nanosensors can be incorporated with other nanomaterials, including nanorods, nanocolloids, graphene oxide, carbon nanotubes, plasmonic gold, and magnetic beads, for the detection of foodborne pathogens [61–64]. In recent years, an array-based immunosorbent assay incorporated with liposomal nanovesicles has received a good reputation due to its efficacy and specificity in detecting *E. coli*, *L. monocytogenes*, and *Salmonella* spp. [65–67]. In addition, silicon-dependent nanosensors incorporated with proteins that vibrate at different frequencies depending on their biomass were used for pathogen detection in numerous blended liquid food structures [68].

In recent years, numerous new NP-enhanced detection systems, such as lateral-flow immune test strips with palladium NPs against *Klebsiella* and field-effect transistors with graphene-based NPs against *E. coli*, have been developed [69,70]. Different NPs show fluorescence at different colors upon encountering harmful microorganisms in foodstuffs. Oxygen-indicating nanoceramic sensors have been developed based on TiO_2 to photosensitize the reduction of methylene blue by using triethanolamine in polymer encapsulation using UVA light. Sensors show a color improvement within the presence of oxygen. The rate of color restoration is proportional to the amount of oxygen exposed to the sensor. Nanoceramic biosensors utilizing SnO_2 have also been developed as oxygen indicators in packaged food.

3.3.4 Nanoceramic additives

Nanoceramic-based additives are added to edible materials to utilize the dispersing ability of water-insoluble additives, such as colors, flavors, preservatives, and nutrients, without using any surfactant. These nanoceramic additives enhance the flavor, nutritional value, and intake of components into the body due to their large surface area. Nanoceramic-based additives that are currently available are vitamins, antioxidants, colors, flavors, and preservatives. Nanosized Ag, Fe, Ca, Mg, Se, and Si are finding application as nanoceramic composite additives in the food manufacturing sector. Nanoceramic composite salt has a better flavor than normal salt when used in a reduced amount. Nanopolylysine is added to the oil to protect it from oxidation by entering the gap between phytoglycogen octenyl succinate NPs. Some countries add nanoceramic composite ingredients to foodstuffs due to their demand within the health and sport food markets. Nanocomposite minerals with sizes less than 100 nm are added to food for better and quicker assimilation in the stomach than larger minerals.

3.3.5 Nanoemulsions

Nanoemulsions have been used to create lipid-soluble compounds that are bioactive. These have also found application in enhancing bioavailability and water dispersion [71].

3.3.6 Nanocarriers

A vast group of bioactive complexes, which include carbohydrates, lipids, vitamins, and proteins, are fragile in a low pH environment and due to enzyme action in the duodenum and stomach. Bioactive encapsulation not only permits them to resist that difficult scenario but also enables them to be adjusted for food items [72]. This phenomenon is quite hard to achieve by the use of noncapsulated bioactive compounds which have less solubility. To stimulate the delivery of vitamins, fragile micronutrients, and medicines, currently small edible capsules coated with NPs are used in everyday foodstuffs to provide useful health results [73].

Nowadays, nanoceramic carriers are used as delivery structures for different flavors of food in food yields without altering the texture. The size of the particle may affect the supply of bioactive complexes to various sites within the body, for example, it was determined that submicron molecules can only work on the nanoscale [74].

3.3.7 Nanofiltration

Nanoceramic particles were used as nanosieves to filter bacteria. Specially designed ceramic membranes for nanofiltration that can remove viruses from water have been developed [75]. Periodic mesoporous silica has been intentionally modified for use as an absorbent for pollutants in wastewater [76].

The uniquely arranged porous structure and the extremely high surface area provide mesoporous ceramics with highly selective and excellent adsorption capabilities, which enable them to act as both specific viral filters and adsorbents for potential virus removal applications.

3.3.8 Nanoceramic utensils

The American company, Fresh Corporation, launched innovative nanoceramics that allow the nanodrying of food, which reduces the use of oil in fast food and restaurants due to their large surface area. Nanoceramics added to deep-fryers reduce the oxidation of the frying oil by exposing a huge surface area to the oil and thus preventing the oil from thermal polymerization, resulting in healthy fried foods [77].

3.3.9 Nanoceramics in antimicrobial effects

Advanced food packaging and the nanobased structures minimize food losses due to various microbial contaminations. Different nanoparticles, such as titanium dioxide, magnesium oxide, zinc oxide, silver nanoparticles, carbon nanotubes, fullerene product, and zerovalent iron, have shown excellent impacts as antimicrobial agents [78]. Silver zeolite ceramics are used as an antimicrobial agent in many products, such as food preservation, disinfection of medical products, and decontamination of materials [79,80]. In comparison with zeolite-based material, silver-based nanocomposites enable constant antimicrobial activity and therefore they are better adapted for long-term foodstuff packaging [81].

Kumar et al. reported the preservation of macroporous, 3D cushioned scaffolds composed of alginate, gelatin, and polyvinyl alcohol where nanosilver hydroxyapatite was incorporated into the 3D spongy scaffolds. The composite scaffolds were 80% porous with interconnected pores with sizes between 75 and 90 μm. The scaffolds displayed antibacterial potential against *Bacillus* sp. and *E. coli* sp. and were not cytotoxic [82].

In addition, silver-containing HA coatings on Ti substrate by VPS exhibited significant antibacterial effects against *E. coli*, *Pseudomonas aeruginosa*, and *S. aureus*. Their antibacterial rate was greater than 95% [83]. Another study by Sahithi et al. combined nanohydroxyapatite, copper ions, and polyethylene glycol 400 (PEG 400) and it was found to be more potent against Gram-positive strains than Gram-negative strains [84].

3.4 Conclusion

Nanotechnology is functional in the development of emulsions, encapsulations, simple solutions, the association of colloids biopolymer matrices, and well-organized delivery organizations with the abovementioned abilities. Even though nanoceramics are proving to be outstandingly successful in the food and bioprocessing industry, there is an concern about the threat of toxicity to human health and the environment. The increasing regulatory problems have enforced many countries to set up regulatory systems capable of handling any dangers related to nanoceramic-encapsulated food. In reality, there is an urgent need for an international regulation system for the use of nanoceramics in the food and bioprocessing industry.

References

[1] F. Kampers, Food nanoscience in the Netherlands, in: IFT International Food Nanoscience Conference, 1 August 2007, Chicago, USA.

[2] T. Joseph, M. Morrison, Nanotechnology in Agriculture and Food, 2009.

[3] M. Misson, H. Zhang, B. Jin, Nanobiocatalyst advancements and bioprocessing applications, J. R. Soc. Interface 12 (2015) 08–91.

[4] D.F.M. Neri, V.M. Balcao, R.S. Costa, I.C.A.P. Rocha, E.M.F.C. Ferreira, et al., Galactooligosaccharides production during lactose hydrolysis by free Aspergillusoryzae [beta]-galactosidase and immobilized on magnetic polysiloxane-polyvinyl alcohol, Food Chem. 115 (2009) 92–99.

[5] D.F.M. Neri, V.M. Balcao, F.O.Q. Dourado, J.M.B. Oliveira, J.A. Teixeira, Immobilized [beta]-galactosidase onto magnetic particles coated with polyaniline: support characterization and galactooligosaccharides production, J. Mol. Catal. B Enzymatic 70 (2011) 74–80.

[6] S.A. Ansari, Q. Husain, Potential applications of enzymes immobilized on/in nanomaterials: a review, Biotechnol. Adv. 30 (2012) 512–523.

[7] P.C. Ashly, M.J. Joseph, P.V. Mohanan, Activity of diastase a-amylase immobilized on polyanilines (PANIs), Food Chem. 127 (2011) 1808–1813.

[8] L.B. Wang, Y.C. Wang, R. He, A. Zhuang, X. Wang, et al., A new nanobiocatalytic system based on allosteric effect with dramatically enhanced enzymatic performance, J. Am. Chem. Soc. 135 (2013) 1272–1275.

[9] N. Durán, P.D. Marcato, Nanobiotechnology perspectives. Role of nanotechnology in the food industry: a review, Int. J. Food Sci. Technol. 48 (2014) 1127–1134.

[10] S. Sun, C.B. Murray, D. Weller, L. Folks, A. Moser, Monodisperse FePt nanoparticles and ferro magnetnetic FePt nanocrystal super lattices, Science. 287 (5460) (2011) 1989–1992.

[11] B.S. Sekhon, Food nanotechnology—an overview, Nanotech. Sci. Appl. 3 (2010) 1.

[12] X. He, H.M. Hwang, Nanotechnology in food science: functionality, applicability, and safety assessment, J. Food Drug. Anal. 24 (2016) 671–681.

[13] S.M. Rodrigues, P. Demokritou, N. Dokoozlian, C.O. Hendren, B. Karn, M.S. Mauter, et al., Nanotechnology for sustainable food production: promising opportunities and scientific challenges, Environ. Sci. Nano 4 (2017) 767–781.

[14] J.M. Aguilera, Where is the nano in our foods? J. Agric. Food Chem. 62 (2014) 9953–9956.

[15] L. Zhang, F. Chen, H. An, H. Yang, X. Sun, et al., Physicochemical properties, firmness, and nanostructures of sodium carbonate-soluble pectin of 2 Chinese cherry cultivars at 2 ripening stages, J. Food Sci. 73 (2008) N17–N22.

[16] C. Thies, Nanotechnology as a delivery system in the food, beverage and nutraceutical industries. In: Huang Q (ed) Nanotechnology in the Food, Beverage and Nutraceutical Industries. Philadelphia, (2011), pp. 208.

[17] B.A. Magnuson, T.S. Jonaitis, J.W. Card, A brief review of the occurrence, use, and safety of food-related nanomaterials, J. Food Sci. 76 (2011) R126–R133.

[18] S.M. Yoo, K.C. Keum, S.Y. Yoo, J.Y. Choi, K.H. Chang, N.C. Yoo, et al., Development of DNA microarray for pathogen detection, Biotechnol. Bioprocess. Eng. 9 (2004) 93–99.

[19] C. Chellaram, G. Murugaboopathi, A. John, R. Sivakumar, S. Ganesan, S. Krithika, et al., Significance of nanotechnology in food industry, APCBEE Procedia 8 (2014) 109–113.

[20] P.J.P. Espitia, N.D.F.F. Soares, J.S. dos Reis Coimbra, N.J. de Andrade, R.S. Cruz, E.A.A. Medeiros, Zinc oxide nanoparticles: Synthesis, antimicrobial activity and food packaging applications, Food Bioprocess. Technol. 5 (2012) 1447–1464.

[21] R. Vidhyalakshmi, R. Bhakyaraj, R.S. Subhasree, Encapsulation, the future of probiotics: a review, Adv. Biol. Res. 3–4 (2009) 6–103.

[22] K. Nakagawa, Nano-and microencapsulation of flavor in food systems, Nano-and Microencapsulation for Foods, John Wiley & Sons, Inc, Hoboken, NJ, 2014, pp. 249–272.

[23] R. Yang, Z. Zhou, G. Sun, Y. Gao, J. Xu, P. Strappe, et al., Synthesis of homogeneous protein-stabilized rutin nanodispersions by reversible assembly of soybean (*Glycine max*) seed ferritin, RSC Adv. 5 (2015) 31533–31540.

[24] S. Drusch, S. Mannino, Patent-Based review on industrial approaches for the microencapsulation of oils rich in polyunsaturated fatty acids, Trends Food Sci. Technol. 20 (2009) 237–244.
[25] J. Weiss, P. Takhistov, D.J. Mc, Clements, functional materials in food nanotechnology, J. Food Sci. 71 (2006) R107–R116.
[26] A. Lamprecht, J.L. Saumet, J. Roux, J.P. Benoit, Lipid nanocarriers as drug delivery system for ibuprofen in pain treatment, Int. J. Pharm. 278 (2004) 407–414.
[27] R.K. Sastry, S. Anshul, N. Rao, Nanotechnology in food processing sector—an assessment of emerging trends, J. Food Sci. Technol. 50 (2013) 831–841.
[28] P. Ezhilarasi, P. Karthik, N. Chhanwal, et al., Techniques for food bioactive components: a review, Food Bioprocess. Technol. 6 (2013) 628–647.
[29] S.D.F. Mihindukulasuriya, L.T. Lim, Nanotechnology development in food packaging: a review, Trends Food Sci. Technol. 40 (2014) 149–167.
[30] H. Zhang, S. Chen, Nanoparticle-based methods for food safety evaluation, Evaluation Technologies for Food Quality, Elsevier, Amsterdam, The Netherlands, 2019, pp. 817–835.
[31] D. Schaefer, W.M. Cheung, Smart packaging: opportunities and challenges, Procedia CIRP 72 (2018) 1022–1027.
[32] H. Zhang, Gas Barrier Properties of Polymer Packaging: Influence on Food Shelf Life Following Microwave Assisted Thermal Sterilization. Ph.D. Thesis, Washington State University, Pullman, WA, USA, 2016.
[33] V. Morris, Emerging roles of engineered nanomaterials in the food industry, Trends Biotechnol. 29 (2011) 509–516.
[34] N.X. Dinh, N.V. Quy, T.Q. Huy, A.T. Le, Decoration of silver nanoparticles on multiwalled carbon nanotubes: antibacterial mechanism and ultrastructural analysis, J. Nanomater. 16 (2015) 63.
[35] E. Maestri, N. Marmiroli, J. Song, J.C. White, Ethical issues of engineered nanomaterials applications and regulatory solutions. In Exposure to Engineered Nanomaterials in the Environment; Elsevier: Amsterdam, The Netherlands, (2019); pp. 315–330.
[36] J.K. Momin, C. Jayakumar, J.B. Prajapati, Potential of nanotechnology in functional foods, Emir. J. Food Agric. 25 (2013).
[37] A. Abraham, Understanding the Effect of Phytochemical Coated Silver Nanoparticles on Mammalian Cells and the Protein Interactions with the Surface Corona of These Nanoparticles. Ph. D. Thesis, RMIT University, Melbourne, Australia, 2016.
[38] B.J. Arroyo, A.P. Santos, E.D.A. de Melo, A. Campos, L. Lins, L.C. Boyano-Orozco, Bioactive compounds and their potential use as ingredients for food and its application in food packaging, Bioact. Compd. (2019) 143–156.
[39] J.C. Clemente, E.C. Pehrsson, M.J. Blaser, K. Sandhu, Z. Gao, B. Wang, et al., The microbiome of uncontacted Amerindians, Sci. Adv. 1 (2015) e1500183.
[40] L.M. Couch, M. Wien, J.L. Brown, P. Davidson, Food nanotechnology: proposed uses, safety concerns and regulations, Agro Food Ind. Hi Tech. 27 (2016) 36–39.
[41] W.S. Abo-Elseoud, M.L. Hassan, M.W. Sabaa, M. Basha, E.A. Hassan, S.M. Fadel, Chitosan nanoparticles/cellulose nanocrystals nanocomposites as a carrier system for the controlled release of repaglinide, Int. J. Biol. Macromol. 111 (2018) 604–613.
[42] H. Jafarizadeh-Malmiri, Z. Sayyar, N. Anarjan, A. Berenjian, Nano-Sensors in food nanobiotechnology. In, Nanobiotechnology in Food: Concepts, Appl. Perspect, 2019, pp. 81–94.
[43] E. Fortunati, A. Mazzaglia, G.M. Balestra, Sustainable control strategies for plant protection and food packaging sectors by natural substances and novel nanotechnological approaches, J. Sci. Food Agric. 99 (2019) 986–1000.
[44] C. Ghosh, D. Bera, L. Roy, Role of nanomaterials in food preservation, Microb. Nanobionics (2019) 181–211.
[45] B. Kuswandi, M. Moradi, Improvement of food packaging based on functional nanomaterial. In Nanotechnology: Applied Energy, Drug and Food; (2019); pp. 309–344.
[46] M. Rai, A.P. Ingle, I. Gupta, R. Pandit, P. Paralikar, A. Gade, et al., Smart nanopackaging for the enhancement of food shelf life, Environ. Chem. Lett. 17 (2019) 277–290.

[47] J.W. Rhim, H.M. Park, C.S. Ha, Bio-Nanocomposites for food packaging applications, Prog. Polym. Sci. 38 (2013) 1629–1652.
[48] P.R. Salgado, L. Di Giorgio, Y.S. Musso, A.N. Mauri, Bioactive packaging: combining nanotechnologies with packaging for improved food functionality, Nanomaterials Food Appl. (2019) 233–270.
[49] S. Neethirajan, D.S. Jayas, Nanotechnology for the food and bioprocessing industries, Food Bioprocess. Technol. 4 (2011) 39–47.
[50] J.K. Patra, H.S. Shin, S. Paramithiotis, Application of nanotechnology in food science and food microbiology, Front. Microbiol. 9 (2018) 714.
[51] A.K. Mohanty, M. Misra, H.S. Nalwa, Packaging Nanotechnology, American Scientific Publishers, 2009.
[52] J. Tager, Nanomaterials in food packaging: FSANZ fails consumers again, Chain React. 122 (2014) 16–17.
[53] O. Rubilar, M.C. Diez, G.R. Tortella, G. Briceno, P.D. Marcato, et al., New strategies and challenges for nanobiotechnology in agriculture, J. Biobased Mater. Bioenergy 8 (2014) 1–12.
[54] C. Ruengruglikit, H. Kim, R.D. Miller, Q. Huang, Fabrication of nanoporous oligonucleotide microarrays for pathogen detection and identification, Polym. Prepr. 45 (2004) 526.
[55] Y. Dong, K.S. Phillips, Q. Cheng, Immuno sensing of *Staphylococcus exterotoxin* B (SEB) in milk with PDMS microfluidic systems using reinforced supported bilayer membranes (r-SBMs), Lab. a Chip 6 (675) (2006) 681.
[56] C.S. Chen, R.A. Durst, Simultaneous detection of *Escherichia coli* O157:H7, *Salmonella spp.* and *Listeria monocytogenes* with an array- based immuno sorbent assay using universal protein G-liposomal nanovesicles, Talanta. 69 (1) (2006) 232–238.
[57] N. Duran, P.D. Marcato, Nanobiotechnology perspectives. Role of nanotechnology in the food industry: a review, Int. J. Food Sci. Technol. 48 (2013) 1127.
[58] K.P. Burris, C. Neal Stewart, Fluorescent nanoparticles: sensing pathogens and toxins in foods and crops, Trends Food Sci. Technol. 28 (2012) 143–152.
[59] D. Tang, J.C. Sauceda, Z. Lin, S. Ott, E. Basova, I. Goryacheva, et al., Magnetic nanogold microspheres-based lateral-flow immunodipstick for rapid detection of aflatoxin B2 in food, Biosens. Bioelectron. 25 (514) (2009) 518.
[60] Z. Fang, Y. Zhao, R.D. Warner, S.K. Johnson, Active and intelligent packaging in meat industry, Trends Food Sci. Technol. 61 (2017) 60–71.
[61] A. Baranwal, K. Mahato, A. Srivastava, P.K. Maurya, P. Chandra, Phyto fabricated metallic nanoparticles and their clinical applications, RSC Adv. 6 (2016) 105996–106010.
[62] M. Holzinger, A. Le Goff, S. Cosnier, Nanomaterials for biosensing applications: a review, Front. Chem. 2 (2014) 63.
[63] M. Kahraman, M.M. Yazıcı, F. Sahin, M. Çulha, Convective assembly of bacteria for surface-enhanced Raman scattering, Langmuir 24 (2008) 894–901.
[64] P. Zuo, X. Li, D.C. Dominguez, B.C. Ye, A PDMS/paper/glass hybrid microfluidic biochip integrated with aptamer-functionalized graphene oxide nano-biosensors for one-step multiplexed pathogen detection, Lab. Chip 13 (2013) 3921–3928.
[65] S. Chen, L. Ma, R. Yuan, Y. Chai, Y. Xiang, C. Wang, Electrochemical sensor based on Prussian blue nanorods and gold nanochains for the determination of H_2O_2, Eur. Food Res. Technol. 232 (2011) 87–95.
[66] M. Ghasemi-Varnamkhasti, S. Mohtasebi, M. Rodriguez-Mendez, M. Siadat, H. Ahmadi, S. Razavi, Electronic and bioelectronic tongues, two promising analytical tools for the quality evaluation of non alcoholic beer, Trends Food Sci. Technol. 22 (2011) 245–248.
[67] I. Sondi, B. Salopek-Sondi, Silver nanoparticles as antimicrobial agent: a case study on E. coli as a model for Gram-negative bacteria, J. Colloid Interface Sci. 275 (2004) 177–182.
[68] L. Wang, J. Luo, S. Shan, E. Crew, J. Yin, C.J. Zhong, et al., Bacterial inactivation using silver-coated magnetic nanoparticles as functional antimicrobial agents, Anal. Chem. 83 (2011) 8688–8695.
[69] B. Thakur, G. Zhou, J. Chang, H. Pu, B. Jin, X. Sui, et al., Rapid detection of single *E. coli* bacteria using a graphene-based field-effect transistor device, Biosens. Bioelectron. 110 (2018) 16–22.

[70] T. Tominaga, Rapid detection of *Klebsiella pneumoniae, Klebsiella oxytoca, Raoultella ornithinolytica* and other related bacteria in food by lateral-flow test strip immunoassays, J. Microbiol. Methods 147 (2018) 43–49.
[71] L.Y. Kumar, Role and adverse effects of nanomaterials in food technology, J. Toxicol. Health. 2 (2015) 2.
[72] K. Oehlke, M. Adamiuk, D. Behsnilian, V. Gräf, E. Mayer-Miebach, E. Walz, et al., Potential bioavailability enhancement of bioactive compounds using food- grade engineered nanomaterials: A review of the existing evidence, Food Funct. 5 (2014) 1341–1359.
[73] A. Haider, I.K. Kang, Preparation of silver nanoparticles and their industrial and biomedical applications: A comprehensive review, Adv. Mater. Sci. Eng. (2015).
[74] A. Dube, K. Ng, J.A. Nicolazzo, I. Larson, Effective use of reducing agents and nanoparticle encapsulation in stabilizing catechins in alkaline solution, Food Chem. 122 (2010) 662–667.
[75] X.B. Ke, H.Y. Zhu, X.P. Gao, J.W. Liu, Z.F. Zheng, High-performance ceramic membranes with a separation layer of metal oxide nanofibres, Adv. Mat. 19 (6) (2007) 785–790.
[76] A.M. Liu, K. Hidajat, S. Kawi, D.Y. Zhao, A new class of hybrid mesoporous materials with functionalized organic monolayers for selective adsorption of heavy metal ions, Chem. Commun. 13 (2000) 1145–1146.
[77] B. Farhang, Nanotechnology and Lipids, 2007, Vol. 19, No. 6, Lipid Technology.
[78] T. Singh, S. Shukla, P. Kumar, V. Wahla, V.K. Bajpai, I.A. Rather, Application of nanotechnology in food science: perception and overview, Front. Microbiol. 8 (2017) 1501.
[79] K. Kawahara, K. Tsuruda, M. Morishita, M. Uchida, Antibacterial effect of silver zeolite on oral bacteria under anaerobic condition, Dent. Mater. 6 (2000) 452–455.
[80] Y. Matsumura, K. Yoshikata, S.I. Kunisaki, T. Tsuchido, Mode of bactericidal action of silver zeolite and its comparison with that of silver nitrate, Appl. Environ. Microbiol. 69 (7) (2003) 4278–4281.
[81] S. Egger, R.P. Lehmann, M.J. Height, M.J. Loessner, M. Schuppler, Antimicrobial properties of a novel silverss-silica nanocomposite material, Appl. Environ. Microbiol. 75 (2009) 2973–2976.
[82] R. Kumar Saini, L. Prasad Bagri, A.K. Bajpai, Nano-silver hydroxyapatite based antibacterial 3D scaffolds of gelatin/alginate/poly (vinyl alcohol) for bone tissue engineering applications, Colloids Surf. B Biointerfaces 177 (2019) 21118.
[83] Y.K. Chen, X.B. Zheng, Y.T. Xie, C.X. Ding, H.J. Ruan, C.Y. Fan, Anti-bacterial and cytotoxic properties of plasma sprayed silver-containing HA coatings, J. Mater. Sci-Mater. M. 19 (12) (2008) 3603–3609.
[84] K. Sahithi, M. Swetha, M. Prabaharan, A. Moorthi, N. Saranya, K. Ramasamy, et al., Synthesis and characterization of nanoscale hydroxyapatite-copper for antimicrobial activity towards bone tissue engineering applications, J. Biomed. Nanotechnol. 6 (4) (2010) 333–339.

CHAPTER 4

Nanoceramics: food and bioprocessing industry

Ramesh K. Guduru and Uttakantha Dixit
Department of Mechanical Engineering, School of Technology, Pandit Deendayal Energy University, Gandhinagar, Gujarat, India

4.1 Introduction

The food production processes usually start with natural/raw materials that undergo different food processing steps, typically sieving, separation, mixing, heating, and possibly biological and/or chemical processes [1]. Finally, they are packaged using shelf life-dependent packaging/storage materials before being getting to the consumer. During all the processing/packaging/storage/transportation stages of the food the areas that come in contact with food must be selected as well as designed to avoid health hazards [2]. The other important requirement for most of the materials and equipment used in the food industry should be wear resistance and high precision to minimize the product losses [1]. Thus, the food and bioprocessing enterprises deal with several challenges to maintain quality along with improved shelf life. Most of the materials and technologies employed in the food industry should be not only efficient from the performance point of view but also be economical, environment-friendly, and sustainable [3]. At the same time, maintaining clean surfaces is also extremely important from both the hygiene and economic point.

Thus, the drive for reduced costs in the food industry either for processing, packaging or storage and transportation should be well balanced with food safety, product integrity, and efficient logistics [4]. On top of these, the packaging sector demands distinctiveness, aesthetics, and ergonomic attributes as well. Hence, there is a constant demand for cost-effective food processing/packaging/storage/logistics that meet the customer's requirements and provide satisfaction, with health and safety being of the utmost importance. Along with these, the technologies implemented should have minimal impact on the environment [5]. In food/fast-moving consumer goods (FMCG) packaging the gas and moisture barrier characteristic is one of the key elements to provide a longer shelf life. The requirements for materials and the technologies implemented in the food and bioprocessing industry are absolutely connected to health hazards; however, they do not have any specific standards except for getting approvals from the Food and Drug Administration (FDA) before implementation at regular intervals [6].

Industrial Applications of Nanoceramics
DOI: https://doi.org/10.1016/B978-0-323-88654-3.00005-6

There are several materials from metals/alloys to ceramics and plastics used in the food industry for various purposes, ranging from separation to mixing, heating, and packaging. Among all these materials, ceramics are quite frequently employed due to their nonreactive nature toward food even at high temperatures, with their excellent biocompatibility [7]. They also possess superior wear-resistant characteristics compared to metals and plastics. Some of the ceramics exhibit self-cleaning or a hydrophobic nature too. They are extensively used in filtration (membranes), valve seats and guides, valve needles, dosing units and dosing sliders, packaging, and labeling. Depending on various application requirements, the following properties are highly in demand in ceramics [8]:

- Wear resistance
- High precision in geometries
- High mechanical strength and compressive strength
- Edge stability
- High-temperature and superheated steam sterilizability
- Thermal shock resistance
- Nonmagnetizability
- Easy/self-cleaning

High-purity oxide- and nonoxide-based ceramics can easily meet the above property requirements for low- and high-temperature applications. For example, the ceramics made of kaolin, quartz, or clay mixed with heavy metals are used in kitchenware, jars, containers, bottles, and trays [9] for food preparation/handling/storage purposes. Food-grade ceramic coatings are usually applied on the food processing components/appliances and are known for their durability, scratch resistance, easy cleaning, and nonstick nature. These food-grade ceramic coatings also possess high temperature resistance. The ceramics are used as coatings (e.g., silicon oxide) extensively for different applications, for example, injectable packaging, that is, as a protective material used to protect glass from acidic or basic products to avoid the product packaging interactions [10]. Ceramic-coated films can meet the demands in other sectors also and thus, making them very popular in several industrial applications. Ceramic-coated films are AlO_x- or SiO_x-coated films with excellent barrier properties, potentially reaching almost the same levels of O_2 and H_2O barriers [11]. Besides that, transparency is another important aspect of these films. Although the SiO_x layer is somewhat yellowish, both the AlO_x and SiO_x layers are transparent, providing the marketing advantage of allowing the consumer to see the product inside the package and the quality control advantage of more reliable visual inspection [2]. These films also have the advantages of metal detector capability and microwave ability. Unlike, aluminum foil or metalized films, the films are metal-free films that allow metal detectors to work correctly and do not cause arcing in the microwave oven [12]. The recent improvement also imparts these films with the exciting advantage of being retortable, with the durability to withstand the severe temperature and pressure conditions in the retorting process, e.g., sterilization (steam, pressure, dry heat) in injectable vials/syringes. In the past, only metal cans, glass containers, and aluminum–plastic pouches were retortable. Replacing the heavier and more bulky metal cans and glass containers with packaging made of ceramic-coated films can result in significant cost/energy savings in terms of transportation and waste reduction [13]. Thus, securing a great importance in the food industry.

In recent times, the food and bioprocessing industries have started to focus on adopting innovative materials and solutions in the food industry for enhanced efficiencies and reduced costs while upgrading the hygiene and food safety standards [11]. This is where nanomaterials and nanotechnologies have garnered great attention from various segments of the food industry. Nanotechnology has potential applications in all aspects of the food chain including storage, quality monitoring, food processing, and food packaging. Nanotechnology applications in the food industry range from intelligent packaging to the creation of on-demand interactive food that allows consumers to modify food, depending on the nutritional needs and tastes. For example, the applications of nanotechnologies include organic- and inorganic-based nanoscale food additives, nanosensors for food quality control and smart packaging, nanocoatings and films for kitchenware and foodstuffs, nanocoatings for antimicrobial applications with enhanced hygiene, detection of pathogens in food and beverages, self-sanitizing surfaces, nanoscale freshness indicators, nanoemulsions for fat reduction [14]. Naturally produced and intentionally man-made materials are the main categories of nanomaterials used in foods. Intentionally added nanomaterials are not generally present in a food substance and may come from engineered material or naturally occurring sources. Casein micelles in animal milk and pectin nanostructures in fruits are examples of natural nanoscale components, while nanocapsules that carry vitamins through the human stomach via the bloodstream are an example for engineered materials [11]. Functional additives like nanoclay, nanozinc oxide (nano-ZnO), nanotitanium dioxide (nano-TiO_2), and titanium nitride nanoparticles (nano-TiN) are engineered nanomaterials used in food packaging to give distinctive properties to the host, leading to different functional applications. On the other hand, quality assurance in the food and bioprocessing industry with regard to nanomaterials is an important nanoscience field of action because consumers demand safe and wholesome food. Therefore, stringent regulations to ensure food safety are nowadays in great demand. Indeed, tailor-made nanosensors for food analysis, flavors or colors, drinking water, and clinical diagnostics have already been developed. Since, early 2000s nanoscience applications through nanosensors and biochips for food quality assurance have captured the researcher's attention. A very sensitive nanosensor to detect gases released by foods was developed by Ruengruglikit et al. [15] Smart packaging to enhance the communication aspect of the package was developed by Mihindukulasuriya and Lim [16] in the year 2014. Nanosensors detecting bacterial pathogens in food have also been developed and used by various researchers and scientists. Another interesting application of biosensors other than pathogens detection, is monitoring the antimicrobial activity of different nanomaterials, which are able to act against foodborne pathogens and food spoilage bacteria. The antimicrobial nanoparticles mainly used in food packaging are MgO, Cu_2O, ZnO, and chitosan, as well as carbon nanotubes [16]. Titanium dioxide (nano-TiO_2) is generally used as a food additive and supplement in the following applications:

- White pigment, icing material, flavor enhancer
- Candies, confectionaries
- Chewing gum
- Food icing
- Dried vegetables, nuts, seeds,
- Soups, mustard, beer, wine

Nanotechnology and nanomaterials are commonly described as the design, production, and application of systems, devices, and structures at a scale of 10^{-9} m. Nanoscience/nanotechnology is an interdisciplinary area that stretches throughout the entire spectrum of science encompassing physics, chemistry, and biology in addition to engineering, along with microfabrication techniques. The physical, chemical, and biological properties of systems and structures at the nanoscale are appreciably different to their macroscale counterparts because of the interactions of individual atoms and molecules, thereby providing specific and novel useful applications [17]. As the size of the particles is reduced to nanoscale range the surface area to volume ratio increases resulting in increased chemical reactivity. Thus, there are changes to the mechanical, electrical, and optical properties of the particles. The nanostructure materials exhibit better catalytic performance than conventional materials with greater surface reaction activity, stronger adsorption ability, and thermal stability. On the other hand, the use of nanostructure carriers increases the life cycle of biocatalysts, consequently permitting their reuse and decreasing the bioprocessing cost. These capabilities are highly favored for commercial purposes. The nanomaterials with large surface area provide a higher matrix for enzyme immobilization and enhance the enzyme loading per nanoparticles mass unit and its catalytic properties. Nanotechnology is a newly rising technology and it is anticipated to continue growing since more than 400 corporations internationally have expanded various special programs of nanotechnology in food and food packaging sector [18].

Fig. 4.1 shows the distribution of food packaging in different segments of food industry, and shows the prospects for nanoceramic coatings in the packaging industry. In the last two decades the numbers of publications and patents have increased by 40% and 90%, respectively. The worldwide sales of nanoceramic products in the food and

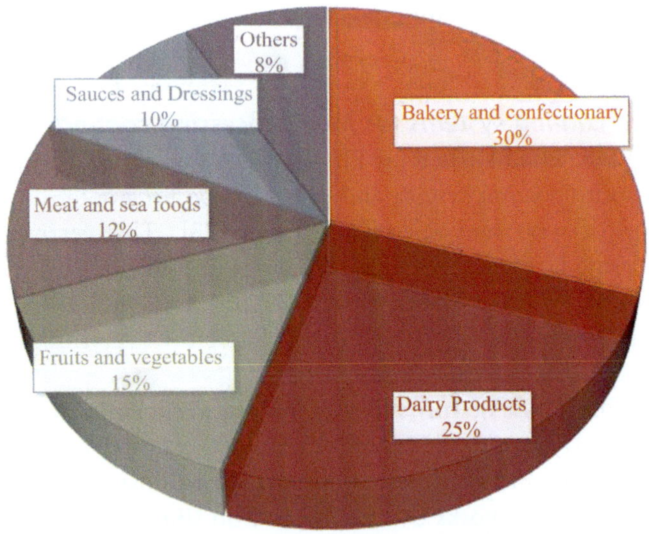

Figure 4.1 Distribution of segments of the food packaging industry.

beverage packaging sector are expected to have increased from \$16.2 billion in 2018 to \$24.5 billion by 2023 at a compound annual growth rate (CAGR) of 8.6% [19]. These numbers clearly reflect the potential interest of researchers and companies in using nanoceramics in food storage and processing industries. Moreover, more than a thousand companies have focused their research and development on nanoceramics-based products and the number is expected to grow in the next few years [20].

4.2 Use of ceramics in food and bioprocessing industries

Nanoceramics can improve the properties of materials in conact of food. Such innovations are gaining interest in food packaging sectors. The purpose of their use is to improve food storage and to guarantee both freshness and quality. A further goal is to improve the technical properties of materials in order to make them stronger, more resistant to abrasion, and easier to process [21].

4.2.1 Nanoceramic commercial products

The use of ceramics in the creation of the first food containers dates back to 7000 BCE. Since then, the importance of packaging food has never stopped increasing and today packaging is an essential part of a products identity. Packaging not only protects the food for consumption, it is also an essential marketing tool that creates an unforgettable customer experience. In this sense, the revival of ceramics for packaging food results from the need to find more sustainable and unique solutions [22].

This section briefly describes some examples of industrial food packaging solutions using nanoceramic coatings.

4.2.1.1 A. Endura coating systems for food processing coatings

- Endura 103 (FDA/USDA): It is an FDA/USDA compliant fluoropolymer-infused ceramic composite coating system and great choice for nonferrous alloys requiring a combination of wear and release properties [23]. This ceramic coating acts as backbone to provide advantages in aggressive washdown environments where aluminum may corrode and stainless steel is too cost-prohibitive an alternative. Endura 103 is a versatile food-safe coating used on everything from hamburger knockout cups to die-cast hard candy molds.
- Endura 203 (FDA/USDA): **Infused electroless nickel plating coating system,** used predominately to provide nonstick for applications where surface hardness, precision film builds, and sharpness is essential. This coating is used extensively in cutting applications and works extremely well on slitter blades and rotary roll cutters used to cut nougat, caramel, granola, and dry cereal.
- Endura 348 (FDA/USDA): It is a multicoat fluoropolymer coating system that affords unrivaled release characteristics across a vast array of food products. This coating is the perfect choice for commercial use food applications that requires release from multiple food products. It has been used successfully on coating grill rack platens, pot sticker molds, and a variety of commercial cookware components.
- Endura 350−3 (FDA/USDA): Highly durable and corrosion resistant, nonstick coating system often used on weigh buckets and scale hoppers.
- Endura 948 (FDA/USDA): Coating that combines the release properties of Endura 348 with a sinter metallic flame-spray base coat or backbone. The dual architecture of the coating provides better wear characteristics for more aggressive processing environments without sacrificing nonstick properties. This coating has worked famously on panini press sandwich platens to large taffy drums and taffy rollers.
- Endura 100R-V: It is a wear-resistant polymer-infused ceramic oxide coating system designed to provide improved surface hardness (up to 70Rc) and an extremely low coefficient of friction. This coating has been used with great success on transfer plates used in paper manufacturing for the printing industry.
- Endura 334GR: It ia a smooth, highly uniform, gray fluoropolymer coating used on a wide array of material conveying equipment to promote enhanced surface lubricity and antigalling properties. This coating is commonly applied to transfer plates, peel plates, and guide rail components.
- Endura 970: a textured, thermal spray coating system, consisting of a sintered metal layer sealed with a proprietary fluoropolymer resin designed for high release. This coating works extremely well in preventing adhesive buildup, particularly from glues and pressure-sensitive adhesives.

4.2.1.2 B. Cermer motto: ceramic food containers

The company is renowned for making ceramic containers that keeps the flavor and essence of food locked [24]. Following are some examples from a wide range of products they produce:

- R4020 ceramic container: This is an eco-friendly ceramic food container perfect for yogurts, desserts, ice creams, cheese, ready meals, and dipping [24]. The product has capacity of 235 cm^3. The Cermer containers are a nonpolluting inert material with a very low carbon footprint that complies with the strictest European and US regulations [25].
- R1027 ceramic container: An eco-friendly nanoceramic constituent food container used for yogurts, desserts, ice creams, cheese, ready meals and dipping. The product is available in various sizes, and in different colors.
  - The company sells similar products with the following model numbers:
  - R1027
  - R2036
  - R4007
  - R1043
  - R1026
  - R1006
  - R2049
  - R2050
  - B2037
  - R1069
  - B1022 and many more.

Compared to PTFE, sol–gel coatings are harder and can function at higher temperatures (up to 450°C). These coatings are also baked at high temperatures so any volatile compounds in the coating will be removed in the coater's ovens.

4.2.1.3 C. LiquiGlide Pvt. Limited

4.2.1.3.1 Food packaging

LiquiGlide's food packaging coatings specifically are made from common, food-safe ingredients that won't impact the quality of the product while unlocking new possibilities for ease of use and the value of getting every last drop [26]. Products like cream cheese that are sold in bricks and tubs can be used in user-friendly, no waste bottles. Their food containers have a coating of nanoceramics that ultimately won't contaminate the food, as it won't allow the food particle to rest on the surfaces of the container [26].

4.2.1.4 D. RBC Life Sciences Inc: nanotechnology in nutrition science

4.2.1.4.1 Slim shake vanilla

Silicon dioxide particles (4–6 nm) coated with flavor molecules [27]. RBC's nanoclusters are tiny particles, 100,000th the size of a single grain of sand, and they are designed to enhance conditions for the absorption of nutrients. During the process of creating nanoclusters, pure vanilla is added to the "cluster" formation to enhance the taste and the benefits of this food.

4.2.1.4.2 Chocolate slim shake

The first network marketing company to use nanotechnology into its product range is RBC Life Sciences. Their products are created with cutting-edge NanoCluster delivery technology and nanotechnology. These incredibly tiny particles, which can assist in delivering more nutrition to your cells, are around 100,000 times smaller than a grain of sand. One such product is Slim Shake, which has high-quality protein, vitamins, minerals, and it is one of the most effective high-value protein food.

4.2.1.5 E. Dorsch home appliances

4.2.1.5.1 Dorsch casserole set

The Dorsch ceramic premium casserole collection sets a high standard for nonstick performance [28]. Coated with a multilayer of nonstick ceramic materials that are absolutely free of PTFE, PFOA, and APEO, it is healthy and eco-friendly.

- Advanced formula with healthy nanoceramic coating from top manufacturer "Proglider."
- Proglider advanced formula ceramic-coating with even better nonstick performance.
- Silicon Rim glass lid which ensures that no dirt will get stuck under it.
- Exterior is color painted that makes cookware resistant to high temperature and easy to clean.

4.2.1.6 F. More Furniture Ltd

4.2.1.6.1 Silver nanoceramic marble fry pan

Silver nanocoating works efficiently at low temperature because of its high heat conductivity. One hundred percent nonstick and nontoxic [29].

4.2.1.7 G. Kitchen Faucets

4.2.1.7.1 Nanoceramic plating fully coating kitchen sink

The MENSARJOR 30 nanosink will not rust as it uses advanced bionic nanocoatings, which effectively isolate the sink from rust and dust [30].

- Prevents scratches during daily use.
- Prevents fingerprints on sink.
- Isolates dust and prevents rust.
- Solves the problem of slow drainage.

4.2.1.8 H. Joom online store

4.2.1.8.1 Nanoceramic sponge with handle cleaning brush for kitchen decontamination black

- Product material: silicon carbide PP + nano [31].
- Effectively removes stains and rust.
- PP material handle, comfortable to hold.
- Multipurpose, can clean kitchen utensils, but also as a pencil sharpener.

4.2.1.9 I. Nanogist Co. Ltd

4.2.1.9.1 Nanover toothpaste

- Silver (Ag) nanoparticle/nanopowder of diameter: 1−10 nm [32]
- Antibacterial
- Nontoxic
- Thermal stability
- Durability

4.2.1.10 J. Babydream

4.2.1.10.1 Silver nanonoble GS nursing bottle

Silver (Ag) nanoparticle/nanopowder is used to coat the surface of the nursing bottle.

- Antibacterial activity
- Antimicrobial activity
- Immunity Improvement
- Deodorization

4.2.1.10.2 Titanium dioxide (nano-TiO_2) used as a food additive and supplement

- White pigment, icing material, flavor enhancer
- Candies, confectionaries
- Chewing gum
- Food icing
- Dried vegetables, nuts, seeds
- Soups, mustard, beer, wine

4.3 Future prospects

Nanotechnologies are evolving quite rapidly in the consumer segments while serving different fields, ranging from biomedicine to pharmaceuticals and other technological applications. However, a major challenge still lies in their compatibility with human lives and environment in terms of both short- and long-term impacts. The above review certainly has indicated the spread of nanotechnologies with different salient features into the food and bioprocessing industries. Researchers are trying to develop better and more efficient nanocarriers with an increased level of bioavailability without compromising the appearance and taste of the food products in which these carriers are incorporated. The idea of "Smart Packaging" is slowly being realized and research is being carried out to develop antigen-specific biomarkers to be used in food packaging and also the incorporation of nanoceramics particles to make polymer nanocomposite films. The antigen-specific biomarkers will help to detect the presence of organisms responsible for the spoilage of the food material. The major challenge of food sector is the development of better food manufacturing, preservation, and storage techniques to provide healthier and nutritious food for human welfare. Hence, nanobiotechnology is an alternate opportunity to overcome global problems and help provide authenticate, nutritious, safe, secure, shelf-stable, high-quality, fortified, and therapeutic food products to future generations. Several previous studies have revealed the toxicity effects of nanoceramics used in food ingredients and packaging. For the evaluation of risk assessment of nanoceramics, it is essential to quantify various factors like biological molecules, commensal microbes, osmotic concentration, physical factors, chemical forces, osmotic concentration, absorption, distribution, metabolism, and excretion [33]. This is one of the reasons why the product development and commercialization sector in the field of nanotechnology is still lagging behind. However, it is slowly picking up with the trends of the world and devising newer methods and making its own contribution to this field. There is a lot of potential in India, and it can only be harnessed properly once there is proper collaboration with the scientists/universities and industries. Better marketing strategies and infrastructure can help heighten the present condition. In addition, it is essential to set strict regulatory guidelines of nanoparticles at global level for the safe use of nanoceramics in food products. The use of instrumentation and computational science can help researchers and scientists to gain unprecedented information about toxicological effect of nanomaterials on human cell lines. Hence, technological, scientific, and social considerations are required to enhance the applications of nanotechnology in different sectors in the coming years. As per the above market analysis done in this chapter, the nanoceramics market in food/bio industry is expected to grow rapidly in the next decade. Furthermore, some focus is required to engineer nanomaterials with modern and advanced technologies to make this technology more efficient in the food industry in coming years [20].

References

[1] S.J. Risch, Food packaging history and innovations, J. Agric. Food Chem. 57 (18) (2009) 8089–8092. Available from: https://doi.org/10.1021/jf900040r.

[2] F. Aguilar et al., Safety of aluminium from dietary intake 1 Scientific. Opinion of the Panel on Food Additives, Flavourings, Processing Aids and Food Contact Materials(AFC) (Question Nos EFSA-Q-2006-168 and EFSA-Q-2008-254), 2008.

[3] J. Rydz, M. Musiol, B. Zawidlak Wegrzyńska, W. Sikorska, Present and future of biodegradable polymers for food packaging applications, Biopolymers for Food Design, Elsevier Inc, 2018, pp. 431–467.

[4] V. Siracusa, M.D. Rosa, Sustainable packaging, Sustainable Food Systems from Agriculture to Industry, Elsevier, 2018, pp. 275–307.
[5] G.L. Robertson, Food Packaging Technology, in: Coles R, McDowellD, Kirwan MJ (eds). Blackwell: Oxford, UK. ISBN 1–84127-221-3. Publishedin the USA and Canada (only) by CRC Press LLC, Boca Raton, FL, USA. ISBN 0–8493-9788-X. 346 pages., Packag. Technol. Sci., 17, 6, pp. 333–335, Nov. 2004, doi: 10.1002/pts.655.
[6] D. Schaefer, W.M. Cheung, Smart packaging: opportunities and challengesinProcedia CIRP 72 (2018) 1022–1027. Available from: https://doi.org/10.1016/j.procir.2018.03.240.
[7] I. Franco, E. Falqué, Glass Packaging, in *Reference Module in Food Science*, Elsevier, 2016.
[8] K. Marsh, B. Bugusu, Food packaging – roles, materials, and environmental issues: scientific status summary, J. Food Sci. 72 (2007) 3. Available from: https://doi.org/10.1111/j.1750-3841.2007.00301.x.
[9] C. Struller, P. Kelly, N. Copeland, Conversion of aluminium oxide coated films for food packaging applications—from a single layer material to a complete pouch, Food Packag. Shelf Life 20 (2019). Available from: https://doi.org/10.1016/j.fpsl.2019.100309.
[10] M.J. Kirwan, D. McDowell, R. Coles, Food Packaging Technology, Blackwell, 2003.
[11] B. Geueke, K. Groh, J. Muncke, Food packaging in the circular economy: overview of chemical safety aspects for commonly used materials, J. Clean. Prod. 193 (2018) 491–505. Available from: https://doi.org/10.1016/j.jclepro.2018.05.005. Elsevier Ltd.
[12] P.J.P. Espitia, N. de, F.F. Soares, J.S. dos, R. Coimbra, N.J. de Andrade, et al., Zinc oxide nanoparticles: synthesis, antimicrobial activity and food packaging applications, Food Bioprocess. Technol. 5 (5) (2012) 1447–1464. Available from: https://doi.org/10.1007/s11947-012-0797-6.
[13] How will ceramic coating film help in the future growth of packaging? 2021. https://packagingguruji.com/how-will-ceramic-coating-film-help-in-the-future-growth-of-packaging/ (accessed Aug. 09, 2021).
[14] M. Dimitrijevic, et al., Safety aspects of nanotechnology applications in food packaging, Procedia Food Sci. 5 (2015) 57–60. Available from: https://doi.org/10.1016/j.profoo.2015.09.015.
[15] S. Neethirajan, D.S. Jayas, Nanotechnology for the food and bioprocessing industries, Food Bioprocess. Technol. 4 (1) (2011) 39–47. Available from: https://doi.org/10.1007/s11947-010-0328-2.
[16] S.D.F. Mihindukulasuriya, L.T. Lim, Nanotechnology development in food packaging: a review, Trends Food Sci. Technol. 40 (2) (2014) 149–167. Available from: https://doi.org/10.1016/j.tifs.2014.09.009.
[17] E. González-Soto, V. González-Rodríguez, C. López-Suárez, J.M. Castro-Romero, J. Pérez-Iglesias, J.M. Fernández-Solís, Migration of lead and cadmium from ceramic materials used in food preparation, Bull. Environ. Contam. Toxicol. 65 (5) (2000) 598–603. Available from: https://doi.org/10.1007/s001280000165.
[18] Nanoparticles released by quasi-ceramic pans | Food Packaging Forum. 2021. https://www.foodpackagingforum.org/news/nanoparticles-released-by-quasi-ceramic-pans (accessed Aug. 09, 2021).
[19] Advanced Ceramics Market | Nanoceramic Powders Research Report. 2021. https://www.bccresearch.com/market-research/nanotechnology/advanced-ceramics-nanoceramic-powders-report.html (accessed Aug. 10, 2021).
[20] Nanotechnology in the food industry: 'plenty of room' to innovate - New Food Magazine. 2021. https://www.newfoodmagazine.com/article/42338/nanotechnology-food-industry-plenty-room-innovate/ (accessed Aug. 10, 2021).
[21] Q. Huang, H. Yu, Q. Ru, Bioavailability and delivery of nutraceuticals using nanotechnology, J. Food Sci. 75 (1) (2010). Available from: https://doi.org/10.1111/J.1750-3841.2009.01457.X.
[22] Ceramics: The smart food packaging | PRODUCT | Cermer. 2021. https://cermer.com/magazine/ceramics-the-smart-food-packaging (accessed Aug. 07, 2021).
[23] Coatings for Packaging & Material Handling | Endura® Coatings. 2021. https://www.enduracoatings.com/packaging-material-handling.html (accessed Aug. 07, 2021).
[24] R4020 Ceramic food container – Classic Terracota Matt. 2021. https://cermer.com/ceramic-food-containers/product/1 (accessed Aug. 07, 2021).

[25] "Welcome to ceramics, welcome to Cermer."
[26] LiquiGlide is Revolutionizing the Way Liquids Move. 2021. https://liquiglide.com/ (accessed Aug. 10, 2021).
[27] B.S. Sekhon, Food nanotechnology – an overview, Nanotechnol. Sci. Appl. 3 (1) (2010) 1. Accessed: Aug. 10, 2021. [Online]. Available: /pmc/articles/PMC3781769/.
[28] Dorsch Premium Casserole Set – 4 Casseroles – Dorsch. 2021. https://dorschhome.se/en/product/dorsch-premium-casserole-set-4-casseroles?v = c86ee0d9d7 ed (accessed Aug. 09, 2021).
[29] P. Grayhurst, P.J. Girling, 6 Packaging of Food in Glass Containers.
[30] Kitchen Sink, MENSARJOR 30-inch Undermount Nanoceramic Plating fully Coating Kitchen Sink, 16 Gauge Single Bowl... | KitchenFaucets.com. 2021. https://www.kitchenfaucets.com/product/kitchen-sink-mensarjor-30-inch-undermount-nano-ceramic-plating-fully-coating-kitchen-sink-16-gauge-single-bowl-stainless-steel-handmade-kitchen-sink (accessed Aug. 09, 2021).
[31] Online Joom Buying at Low Price in India at desertcart.in. 2021. https://www.desertcart.in/brands/joom (accessed Aug. 10, 2021).
[32] Nanogist Co., Ltd. | Nanotechnology Company | NPD. 2021. https://product.statnano.com/company/nanogist-co., -ltd. (accessed Aug. 09, 2021).
[33] Y. Lugani, B.S. Sooch, P. Singh, S. Kumar, Nanobiotechnology applications in food sector and future innovations, Microb. Biotechnol. Food Heal. (2021) 197. Available from: https://doi.org/10.1016/B978-0-12-819813-1.00008-6.

CHAPTER 5

Applications of inorganic metal oxide and metal phosphate-based nanoceramics in dentistry

Aruna Jyothi Kora[1,2]
[1]National Centre for Compositional Characterisation of Materials (NCCCM), Bhabha Atomic Research Centre (BARC), ECIL PO, Hyderabad, Telangana, India
[2]Homi Bhabha National Institute (HBNI), Mumbai, Maharashtra, India

5.1 Introduction

In recent years, nanotechnology has played a pivotal role in developing novel and cost-effective dental materials and technologies and has augmented our knowledge of the structural, biomechanical, physical, chemical and biological properties of innovative dental materials [1]. Nanoceramics overcome the disadvantages of conventional ceramics, such as hardness, brittleness, and difficulty in machinability. The nanoceramics are bestowed with distinctive mechanical and surface characteristics, including superplasticity, machinability, strength, toughness, etc., arising from adequate grain size, ample grain boundaries, and governable crystallinity. Thus they gain the advantage in terms of ease of production, amenability to complex shapes, sintering at a lower temperature, low energy requirements, high manufacturing throughput, etc. Nanoceramics have revolutionized many industries, including dentistry in terms of material design, nanostructure control, and property tailoring [2–4]. The important nanoceramics include inorganic metal oxides, such as aluminum oxide, silicon dioxide, titanium dioxide, zinc oxide, zirconium oxide, cerium oxide, antimony oxide; mixed oxide, hydroxide, phosphate (calcium phosphate, hydroxyapatite), carbide, nitride, boride, phosphate, and carbonate [1,2,5,6]. The desirable characteristics of nanoceramics for dentistry are biocompatibility and enhanced mechanical properties such as hardness, fracture toughness, and wear resistance [3].

The current review focuses on various inorganic metal oxide and metal phosphate-based nanoceramics, their properties, and multifaceted applications in different branches/specializations of dentistry. The inorganic metal oxide-based nanoceramics include aluminum oxide, silicon dioxide, titanium dioxide, zinc dioxide,

Industrial Applications of Nanoceramics
DOI: https://doi.org/10.1016/B978-0-323-88654-3.00006-8

and zirconium oxide nanoparticles, and the metal phosphate-based nanoceramics include amorphous calcium phosphate and hydroxyapatite nanoparticles. The anti-bacterial, mechanical, and remineralization properties of nanoceramics are touched upon. The role of computer-aided design/computer-aided manufacturing in the production of the dental restorations is briefed. Nanoceramics are finding their utilization in various fields of dentistry such as preventive dentistry, prosthodontics, endodontics, periodontics, conservative and esthetic dentistry, implantology, regenerative dentistry, etc. [1,5,7].

5.2 Metal oxide nanoparticles

Aluminum oxide, magnesium oxide, silicon dioxide, titanium dioxide, zinc oxide, zirconium oxide, and other inorganic metal oxide nanoparticles are used in dentistry.

5.3 Aluminum oxide nanoparticles

The aluminum ceramics are known for their high gloss, hardness, wear resistance, chemical stability, biocompatibility, nonallergenicity, pleasing esthetics, etc. Nevertheless, they are limited by porcelain crack and crispiness [6]. The poly(methyl methacrylate (PMMA)matrix reinforced with aluminum oxide whiskers of 2–4 × 200–400 nm and aluminum oxide nanoparticles (Al_2O_3 NP) of <50 nm at 3% improved the mechanical characteristics of the composite in terms of glass transition temperature, storage modulus, modulus of elasticity, and hardness in comparison with the pure PMMA [8]. The silanized Al_2O_3 NP at 1% enhanced the flexural strength of PMMA by 24% [9]. The ceramics made with hybrid nanoparticles of Al_2O_3 (270 nm) and zirconium oxide (216 nm) produced by laser covaporization method showed increased mechanical characteristics in terms of flexural strength, fracture toughness in addition to superior low temperature degradation. Thus, these nanoceramics find applications as bioinert dental implants [10].

5.4 Magnesium oxide nanoparticles

The zinc polycarboxylate cements are widely used as water-based luting cement due to their biological and dentin adhesive properties. However, they lack the required mechanical characteristics of compressive strength, tensile strength etc. The amendment of zinc polycarboxylate dental cements with magnesium oxide nanoparticles (MgO NP) of 30–60 nm increased the compressive and tensile strength when compared to conventional zinc polycarboxylate cement [11].

5.5 Silicon dioxide nanoparticles

Silica or silicon dioxide nanoparticles (SiO_2 NP) are finding many applications in dentistry due to their virtues such as size, surface area, lower density, absorption capability, biocompatibility, and lower toxicity [4]. They are used as dental fillers and abrasive tooth polishing materials to prevent dental caries caused by cariogenic bacteria [12]. The SiO_2 NP surface modified with methacryloxypropyltrimethoxysilane enhanced the tensile strength and tensile modulus of PMMA [6]. The hybrid nanoparticles with 40-nm-sized SiO_2 NP as core and PMMA as shell in a ratio of 78/22 used as fillers in polymeric restorative dental resins (bis glycidyl dimethacrylate/triethylene glycol dimethacrylate, 40/60%) increased the flexural modulus of the composites [13]. The silanized silica microspheres of 400-nm-size used as fillers for the light curable resin mixtures (ethoxylated bisphenol A dimethacrylate/triethylene glycol dimethacrylate, 70/30%) exhibited superior light transmission, compression strength, depth of cure; and lower volume shrinkage in comparison with ordinary resins [14].

The SiO_2 NP of 20 nm sized adsorbed with 0.5% of silver nanoparticles showed outstanding antibacterial action against *Escherichia coli* and *Streptococcus faecalis*. Thus the silver-doped SiO_2 NP carrier can be incorporated and used to prepare antibacterial dental resins [15]. The AH plus (commercial epoxy resin-based root canal sealer) supplemented with quaternary ammonium epoxy silicate nanoparticles (2%–8%) of 120 nm showed antibacterial activity against *Enterococcus faecalis*, demonstrated through direct contact and biofilm viability tests. The study indicated its applications toward control of endodontic infection during canal filling and prevention of subsequent reinfection [16].

The mesoporous SiO_2 NP containing calcium carbonate nanoparticles were used as desensitizing agents for treating dentine hypersensitivity caused by enamel loss exposing the dentinal tubules. The dentine hypersensitivity is characterized by a sharp pain in response to various tactile, evaporative, thermal, osmotic, and chemical stimuli. The modified SiO_2 NP enables the permeation of Ca^{2+}, PO_4^{3-}, and HPO_4^{2-} ions into dentinal tubules, forms di-, tricalcium phosphate and hydroxyapatite crystals, and occludes the exposed dentinal tubules. Thus the ability to form a biomimetic crystalline dentin barrier by mesoporous SiO_2 NP biomaterial enables its application for the repair and restoration of hard dental tissues [17]. The mesoporous SiO_2 NP are bestowed with well-defined structure, high surface area, tunable pore size, high hydrothermal stability, and hydrocarbon sorption capability [4]. The infiltration of 12-nm-sized SiO_2 NP caused remineralization of demineralized dentin collagen when exposed to artificial saliva solution, leading to a new guided tissue regeneration strategy [18]. The composite scaffolds made with alginate/nanobioactive glass ceramic particles were used as a suitable bioactive matrix for periodontal regeneration. The scaffolds showed controlled porosity, swelling ability; limited degradation, enhanced biomineralization, protein adsorption; sufficient cell proliferation; and cytocompatibility toward human periodontal ligament fibroblast and osteosarcoma cells [19].

5.6 Titanium dioxide nanoparticles

The antimicrobial, hydrophilic, and self-cleaning properties of titanium dioxide nanoparticles (TiO_2 NP) are widely used in many fields. TiO_2 NP are employed as fillers for refining the physical properties of dental composite resins due to their biocompatible, antibacterial, and tooth-colored esthetic characteristics [4,7,20]. The TiO_2 NP surface modified with organosilane (allyltriethoxysilane) was used as filler for the light-cured dental resins. The data showed better dispersion of fillers within the matrix and enhanced microhardness and flexural strength of resin at 1% loading of TiO_2 NP [21]. The green synthesized, size-controlled (40 nm), and surface modified TiO_2 NP used as fillers at 5% demonstrated enhanced compressive stress and flexural stress; and reduced polymerization shrinkage for light-cured resin mixture (bisphenol A glycidyl methacrylate/triethylene glycol dimethacrylate/urethane dimethacrylate, 60%/20%/20%) [20]. The addition of TiO_2 NP of 21-nm-sized fillers at 1% to light-cured orthodontic composite increased the antibacterial action toward *Streptococcus mutans* without disturbing the shear bond strength [22]. The addition of TiO_2 NP into PMMA at 4% showed considerable dispersion of NP for the preparation of stereolithographic complete denture prototyping and antifungal action against *Candida scotti* [23]. The occlusion of TiO_2 NP into PMMA appreciably diminished the porosity and cell adhesion of *Candida albicans*, an oral commensal pathogenic fungus and found to be biocompatible toward embryonic fibroblast cells [24].

5.7 Zinc oxide nanoparticles

In dental cement, zinc is used as the main filler ingredient due to its osteogenic activity and the antimicrobial action of zinc allows its use in toothpastes in the form of citrate, sulfate and acetate for controlling the dental plaque formation [25]. The antibacterial activity of zinc is effective in the form of nanoparticles than bulk which are dependent on particle size, dose, surface coating and functionalization. The antibacterial activity of zinc oxide nanoparticles (ZnO NP) is attributed to Zn binding to proteins and lipids, membrane permeabilization and oxidative stress [5,26].

The dental caries is one of the regular infectious dental disease in children caused by cariogenic bacteria such as *S. mutans* due to acid production leading to demineralization of tooth tissues. The cavity filling used to be done by dental amalgam containing elementary mercury and metal alloys of silver, copper, tin, zinc, etc. The composite materials have replaced the use of amalgam due to poor esthetics and neurotoxic and renal toxic effects of mercury. The composites are usually made of polymerizable resins, photoinitiators and silane-coated fillers. The properties of dental composite resins are mainly reliant on filler size, shape, and concentration [20]. The carious teeth restored with composite resins are prone to polymerization shrinkage, recurrent caries, restoration failure, and microbial plaque. Thus dental composite resin incorporated with 50-nm-sized ZnO NP at 1% was checked for its antibacterial activity against *S. mutans*, the etiological agent of dental carries. The ZnO NP incorporated composite resins wholly inhibited the growth of *S. mutans* [27]. In another report, ZnO NP of 125-nm-size exhibited minimum inhibitory concentration (MIC) and minimum bactericidal concentration (MBC) value of 500 μg/mL toward *S. mutans* [25]. A flowable dental resin composite enriched with 20-nm-sized ZnO NP at 1% was inhibitory towards *S. mutans* without compromising its mechanical properties [28]. Also, the zinc polycarboxylate dental cement supplemented with 25–50-nm-sized ZnO NP showed exceptional mechanical strength compared to commercial zinc polycarboxylate cement [11].

In restorative dentistry, the artificial acrylic teeth surface coated with graphene/zinc oxide nanocomposite inhibited the biofilm film formation in cariogenic *S. mutans* and showed nontoxicity toward the human embryonic kidney cell line [29]. The ZnO NP of 10–120 nm were employed as a coating material for glass dental implants. The nanocoated implants were antibacterial against *Staphylococcus aureus* and biocompatible towards osteoblast cells [30]. The TiO_2 nanotubes decorated with 0.015 M ZnO NP of 20–50-nm-size suppressed the *S. mutans* and *Porphyromonas gingivalis* growth by 45% and 85% after 7 days, respectively, compared to unmodified Ti samples. They were compatible with mesenchymal stem cells, implying its application in dentistry [31]. A study carried out on antibacterial activity of 10–50 nm sized ZnO NP against bacteria associated with peri-implantitis including *P. gingivalis, Aggregatibacter actinomycetemcomitans, Fusobacterium nucleatum*, and *Prevotella intermedia* showed MBC values of 250, 250, 500, and 1000 μg/mL, respectively. The peri-implantitis is a site specific infectious disease characterized by inflammation of connective tissue and progressive bone loss around the dental implants [32].

5.8 Zirconium oxide nanoparticles

The zirconia or zirconium oxide is an insoluble, osseoconductive oxide with an array of uses in dentistry. The zirconium oxide nanoparticles (ZrO_2 NP) are one of the suitable nanofillers for dental resins due to their superior dispersion, biocompatibility, hardness, refractive index, scratch/fracture/abrasion, and corrosion resistance. In addition, the white coloration meets the required esthetic quality of dental ceramic restoratives [4,6,7,33,34]. The heat-cured acrylic resin, PMMA is the most extensively utilized matrix for the production of dentures/dental restorations due to its characteristics including biocompatibility, oral environment stability, refined esthetics, precise fit, amenable toward laboratory and clinical manipulation/processability, reparability, and reasonable cost. A photograph showing the image of PMMA resin block is given in Fig. 5.1. However, it lacks the required mechanical properties (abrasion resistance) for wider dental applications [4,6,34]. The modified ZrO_2 NP at a dosage of 3% were added as nanofillers to PMMA and found that the nanofillers improved abrasive wear resistance, tensile and fatigue strength; and decreased water sorption, solubility, and porosity of denture base resin [33]. The addition of 5–15-nm-sized ZrO_2 NP nanofiller to PMMA at 7% considerably increased the tribological characteristics, including flexural strength, fracture toughness, and hardness of PMMA. The enhancement in mechanical properties of dental restorative resin could be attributed to supramolecular bonding between nanofillers and resin matrix [34]. The ZrO_2 NP of <100-nm-size at 2% reinforcement were

Figure 5.1 A photographic image of poly(methyl methacrylate) (PMMA) resin block. *PMMA*, Poly (methyl methacrylate).

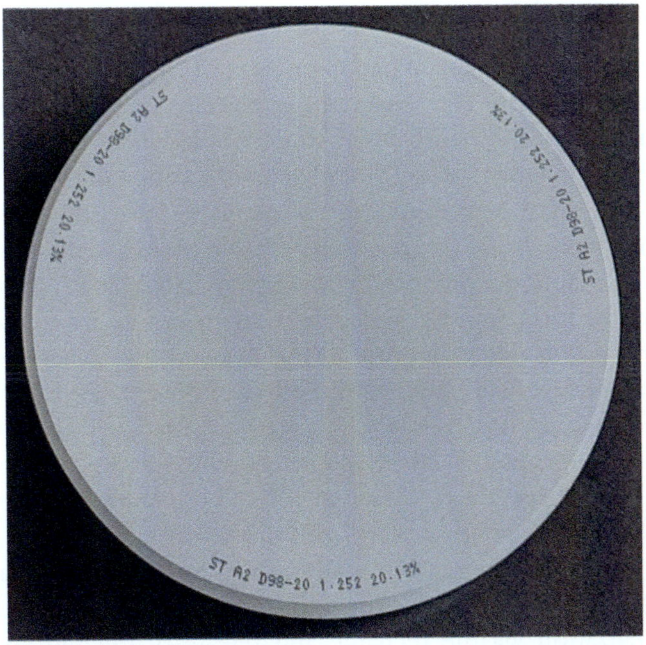

Figure 5.2 A photographic image of yttria-stabilized nanozirconia block.

known to augment the transverse strength of heat polymerized acrylic resin, PMMA, during prosthodontic/denture base repair [35].

The mechanical and microstructural characteristics of alumina ceramics toughened with nanozirconia at 20% were found to increase compared to control, implying its use as dental ceramic restoratives [36]. The 30-nm grain-sized nanozirconia powders doped with 3 at% of yttrium showed improved mechanical properties and low temperature resistance. Thus the yttria-stabilized nanozirconia has found application in fabrication of dental prosthetic devices intended for long-term durability (Fig. 5.2) [37]. The glass ceramics containing nanograin-sized zirconia showed superior corrosion resistance, high fracture toughness, and high translucency in comparison with commercial lithium disilicate-based glass ceramics, which are extensively utilized as dental veneers and crowns [38].

5.9 Metal phosphate nanoparticles

Metal phosphate nanoparticles, such as calcium phosphate and hydroxyapatite, are the subject of the current review. While the detailed dental applications of hydroxyapatite nanoparticles in dentistry are covered elsewhere in the book.

5.10 Amorphous calcium phosphate nanoparticles

The amorphous calcium phosphate nanoparticles (nACP) are widely used in conservative and esthetic dentistry [7]. The resin composite prepared with bisphenol-A-glycerolate dimethacrylate (BisGMA) and triethylene glycol dimethacrylate (TEGDMA) was supplemented with antibacterial nanoparticles of silver (0.02%–0.1%, 2.7 nm), quaternary ammonium dimethacrylate (7%–10%) and remineralizing (nACP, 112 nm, 30%–40%) agents enhanced the enamel/dentin remineralization and inhibited the viable count, metabolic activity, lactic acid production, biofilm formation and caries formation by *S. mutans* in the tooth cavity and along the tooth restoration margins, without compromising the mechanical characteristics [39,40]. The endodontic sealer made up of dimethylaminohexadecyl methacrylate and 2-methacryloyloxyethyl phosphorylcholine supplemented with nACP inhibited the endodontic biofilm forming bacteria, enhanced the remineralization process and increased the bond strength to dentine [7]. The nanocomposite made up of dimethylaminohexadecyl methacrylate nanoparticles and nACP could inhibit the metabolic activity of adherent biofilms and neutralize the bacterial acids responsible for secondary dental caries. In addition, the smart, dual agents offered notable remineralization via the recharge and release of calcium and phosphorous ions without affecting the bonding strength to dentine. Thus the results indicate the nACP incorporation into dental materials, including bonding agents and luting cements [2,3,7,41]. Further, the use of nACP (20%) as a reactive filler and bone grafting agent for the light curable, methacrylate resin matrix caused strong adherence of the grafting material to wet bone and recrystallization of nACP into hydroxyapatite [42].

A study carried out with fluoride (1100 μg/mL) toothpaste supplemented with nanoparticles of sodium trimetaphosphate (3%) significantly prevented enamel demineralization, promoted remineralization, reduced the dentin hypersensitity and exopolysaccharide concentration in situ formed biofilm, under a cariogenic challenge [43]. The antifungal activity of chlorhexidine coatings incorporated with sodium hexametaphosphate nanoparticles was investigated in dental silicones employed as denture soft liners and obturators. The coatings showed antifungal action against *C. albicans*, a causative agent of denture stomatitis. In addition, the coatings exhibited localized, slow and sustained delivery of chlorhexidine into the artificial saliva, without effecting the hydrophobicity [44]. In an in vitro study carried out with calcium carbonate nanoparticle containing dentifrice, the treatment caused remineralization of incipient enamel lesions via release of calcium ions into saliva [45,46]. The dentifrice is an agent in the form of powder, paste, gel or liquid used for cleaning and polishing teeth along with a toothbrush [47].

5.11 Hydroxyapatite nanoparticles

The hydroxyapatite is a natural mineral composed of calcium apatite and denoted as $Ca_{10}(PO_4)_6(OH)_2$. The hydroxyapatite nanoparticles (HA NP) are extensively in different fields of dentistry, including preventive, dental implantology, bone grafting/replacement, oral and maxillofacial surgery due to their similar composition with bone and teeth, biocompatibility, biomimetic functionality, higher surface area, higher biological activity etc. Especially, the tooth enamel is made up of 97% of HA nanocrystals as building blocks [4,47]. In endodontic treatments, the materials used for root end filling should exhibit biocompatibility, superior sealing capability, periapical tissue regeneration capacity, antimicrobial action, adequate radiopacity and superb workability. Studies on biocompatibility of nano-HA incorporated, bioceramic nanoparticulate cement showed nontoxicity and cytocompatibility toward human mesenchymal and osteoblast cells on par with mineral trioxide aggregate, a commercial, ceramic root end filler [48,49]. While the mineral linked gene expression in osteoblast cells was more than mineral trioxide aggregate [49]. The osseo integrative properties of nanoceramics implants depend upon chemical composition, surface topology, mechanical characteristics, grain size distribution etc. [50]. The use of HA coatings on metallic implants improves the biocompatibility of the implant while reinforcing elements improve the mechanical characteristics of the HA coatings. The titanium plates coated with 60-nm-sized nanocrystallite hydroxyapatite via a biomimetic process by simulated body fluid showed substantial mechanical properties. Also, excellent adhesion and proliferation of human fibrosarcoma cells indicate its utilization as an artificial bone substitute [51].

Compared with traditional glass ionomer cement, the incorporation of hydroxyapatite nanoparticles into glass ionomer cements substantially enhanced the compressive, tensile and biaxial flexural strength [7]. The applications of HA NP in preventive dentistry are increasing. For example, the inclusion of HA nanoparticles into sports drink effectively prevented dental erosion, evidenced from an in vitro study carried out with bovine enamel [52]. The glucolytic biofilm mediated dental caries, a chronic dental disease that could be controlled using nano-HA incorporated toothpowder, toothpaste, and mouthwashes. The HA nanoparticles are known for reducing dentine hypersensitivity and oral biofilm formation and enamel demineralization loss because of abrasion, erosion, and attrition [5,53].

5.12 Computer-aided design/computer-aided manufacturing

The nanoceramics are primarily fabricated as a disk made up of evenly distributed ceramic nanoparticles bound in the resin matrix, with definite/precise specifications.

Figure 5.3 A photographic image of CAD/CAM milling machine. *CAD/CAM*, Computer-aided design/computer-aided manufacturing.

From these disks, the dental restorations, such as inlays, onlays, veneers, crowns, and bridges, are milled and manufactured through computer-aided design/computer-aided manufacturing (CAD/CAM) [54,55]. The photograph of CAD/CAM milling machine is shown in Fig. 5.3. The CAD/CAM restorative dental materials offer many advantages, including esthetics, durability, high color stability, superior fracture resistance, high modulus elasticity, high hardness, high flexural strength, high translucency, in comparison with traditional dental ceramics Fig. 5.4 [6,7,56,57].

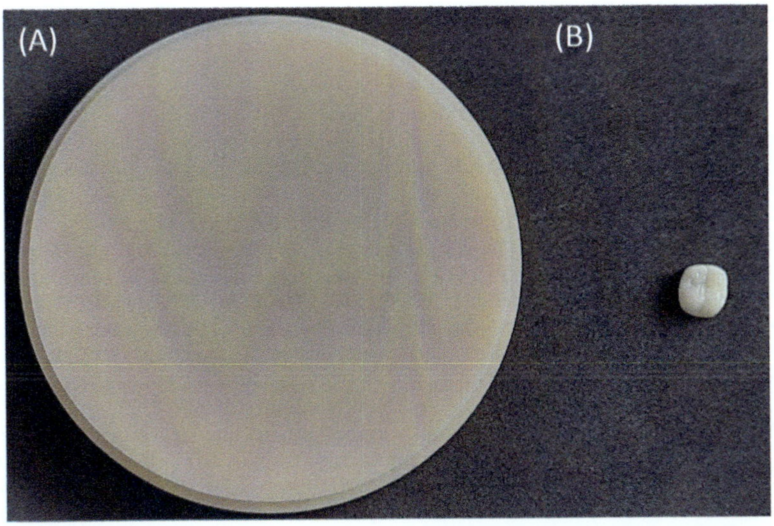

Figure 5.4 Photographic images of (A) CAD/CAM suitable restorative dental material, graphene-reinforced PMMA resin block and (B) CAD/CAM milled molar dental crown of graphene-reinforced PMMA resin. *CAD/CAM*, Computer-aided design/computer-aided manufacturing.

5.13 Conclusion

The nanoceramics, specifically the inorganic metal oxide and metal phosphate nanoparticles are ingredients of various dental materials, including composite dental resins, luting cement, dental fillers, tooth polishers, dentine hypersensitivity desensitizers, dental scaffolds, dentifrices, endodontic sealers, and dental implants, in various specializations of dentistry. The cost-effective nanoceramics function as antimicrobial agents toward various oral pathogenic bacteria and are employed to treat dental caries, gingivitis, peri-implantitis, etc. However, more research on the behavior of nanoceramics in the oral cavity in terms of pH, saliva buffer action, mucosal contact and dental tissue spread; mechanism of action, in vitro toxicity, long-term clinical studies on restorative materials and applications of nanoceramics toward the treatment of various oral diseases is required [4,5].

Acknowledgments

The author would like to thank Dr. M. V. Balarama Krishna, Head, Environmental Analytical and General Services Section and Dr. A.C. Sahayam, Head, NCCCM/BARC, for their constant support and encouragement throughout the work. The author would like to acknowledge Dr. Seelam Akash Chakravarthy, Saritha Dental Clinic LLP, Hyderabad for allowing the photography of their nanoceramic dental composites and facilities.

References

[1] L.K. Foong, M.M. Foroughi, A.F. Mirhosseini, M. Safaei, S. Jahani, M. Mostafavi, et al., Applications of nano-materials in diverse dentistry regimes, RSC Adv. 10 (26) (2020) 15430–15460. Available from: https://doi.org/10.1039/D0RA00762E.

[2] A. Kiani, M. Rahmani, S. Manickam, B. Tan, Nanoceramics: synthesis, characterization, and applications, J. Nanomater. 2014 (2014) 1–2. Available from: https://doi.org/10.1155/2014/528348.

[3] R.A. McIntyre, Common nano-materials and their use in real world applications, Sci. Prog. 95 (Pt 1) (2012) 1–22. Available from: https://doi.org/10.3184/003685012x13294715456431.

[4] S. Priyadarsini, S. Mukherjee, M. Mishra, Nanoparticles used in dentistry: a review, J. Oral. Biol. Craniofacial Res. 8 (1) (2018) 58–67. Available from: https://doi.org/10.1016/j.jobcr.2017.12.004.

[5] G.C. Padovani, V.P. Feitosa, S. Sauro, F.R. Tay, G. Duran, A.J. Paula, et al., Advances in dental materials through nanotechnology: facts, perspectives and toxicological aspects, Trends Biotechnol. 33 (11) (2015) 621–636. Available from: https://doi.org/10.1016/j.tibtech.2015.09.005.

[6] W. Wang, S. Liao, Y. Zhu, M. Liu, Q. Zhao, Y. Fu, Recent applications of nanomaterials in prosthodontics, J. Nanomater. 2015 (2015) 1–11. Available from: https://doi.org/10.1155/2015/408643.

[7] R.N. AlKahtani, The implications and applications of nanotechnology in dentistry: a review, Saudi Dental J. 30 (2) (2018) 107–116. Available from: https://doi.org/10.1016/j.sdentj.2018.01.002.

[8] S.A.B. Hasan, M.M. Dimitrijević, A. Kojović, D.B. Stojanović, K. Obradović-Đuričić, R.M.J. Heinemann, et al., The effect of the size and shape of alumina nanofillers on the mechanical behavior of PMMA matrix composites, J. Serbian Chem. Soc. 79 (10) (2014) 1295–1307.

[9] B. Jasim, I. Ismail, The effect of silanized alumina nano - fillers addition on some physical and mechanical properties of heat cured polymethyl methacrylate denture base material, J. Baghdad Coll. Dent. 26 (2014) 18–23. Available from: https://doi.org/10.12816/0015190.

[10] J.F. Bartolomé, A. Smirnov, H.-D. Kurland, J. Grabow, F.A. Müller, New ZrO_2/Al_2O_3 Nanocomposite fabricated from hybrid nanoparticles prepared by CO_2 laser co-vaporization, Sci. Rep. 6 (1) (2016) 20589. Available from: https://doi.org/10.1038/srep20589.

[11] M.A. Karimi, S. Roozbahani, A. Reza, A. Hatefi, M.H. Mashhadizadeh, R. Behjatmanesh-Ardakani, et al., Synthesis and characterization of nanoparticles and nanocomposite of ZnO and MgO by sonochemical method and their application for zinc polycarboxylate dental cement preparation, Int. Nano Lett. 1 (2011).

[12] R.M. Gaikwad, I. Sokolov, Silica nanoparticles to polish tooth surfaces for caries prevention, J. Dental Res. 87 (10) (2008) 980–983. Available from: https://doi.org/10.1177/154405910808701007.

[13] G. Canché-Escamilla, S. Duarte-Aranda, M. Toledano, Synthesis and characterization of hybrid silica/PMMA nanoparticles and their use as filler in dental composites, Mater. Sci. Eng.: C. 42 (2014) 161–167. Available from: https://doi.org/10.1016/j.msec.2014.05.016.

[14] X. Miao, Y. Li, Q. Zhang, M. Zhu, H. Wang, Low shrinkage light curable dental nanocomposites using SiO_2 microspheres as fillers, Mater. Sci. Eng.: C. 32 (7) (2012) 2115–2121. Available from: https://doi.org/10.1016/j.msec.2012.05.053.

[15] H. Jia, W. Hou, L. Wei, B. Xu, X. Liu, The structures and antibacterial properties of nano-SiO_2 supported silver/zinc–silver materials, Dental Mater. 24 (2) (2008) 244–249. Available from: https://doi.org/10.1016/j.dental.2007.04.015.

[16] S.Q. Gong, Z.B. Huang, W. Shi, B. Ma, F.R. Tay, B. Zhou, In vitro evaluation of antibacterial effect of AH Plus incorporated with quaternary ammonium epoxy silicate against *Enterococcus faecalis*, J. Endod. 40 (10) (2014) 1611–1615. Available from: https://doi.org/10.1016/j.joen.2014.03.010.

[17] Y.-C. Chiang, H.-P. Lin, H.-H. Chang, Y.-W. Cheng, H.-Y. Tang, W.-C. Yen, et al., A Mesoporous silica biomaterial for dental biomimetic crystallization, ACS Nano 8 (12) (2014) 12502–12513. Available from: https://doi.org/10.1021/nn5053487.

[18] A. Besinis, R. van Noort, N. Martin, Remineralization potential of fully demineralized dentin infiltrated with silica and hydroxyapatite nanoparticles, Dent. Mater. 30 (3) (2014) 249–262. Available from: https://doi.org/10.1016/j.dental.2013.11.014.

[19] S. Srinivasan, R. Jayasree, K.P. Chennazhi, S.V. Nair, R. Jayakumar, Biocompatible alginate/nano bioactive glass ceramic composite scaffolds for periodontal tissue regeneration, Carbohydr. Polym. 87 (1) (2012) 274–283. Available from: https://doi.org/10.1016/j.carbpol.2011.07.058.

[20] D.V. Raorane, R.S. Chaughule, S.R. Pednekar, A. Lokur, Experimental synthesis of size-controlled TiO_2 nanofillers and their possible use as composites in restorative dentistry, Saudi dental J. 31 (2) (2019) 194–203. Available from: https://doi.org/10.1016/j.sdentj.2019.01.008.

[21] Y. Xia, F. Zhang, H. Xie, N. Gu, Nanoparticle-reinforced resin-based dental composites, J. Dent. 36 (6) (2008) 450–455. Available from: https://doi.org/10.1016/j.jdent.2008.03.001.

[22] M. Poosti, B. Ramazanzadeh, M. Zebarjad, P. Javadzadeh, M. Naderinasab, M.T. Shakeri, Shear bond strength and antibacterial effects of orthodontic composite containing TiO_2 nanoparticles, Eur. J. Orthod. 35 (5) (2013) 676–679. Available from: https://doi.org/10.1093/ejo/cjs073.

[23] E.E. Totu, A.C. Nechifor, G. Nechifor, H.Y. Aboul-Enein, C.M. Cristache, Poly(methyl methacrylate) with TiO_2 nanoparticles inclusion for stereolitographic complete denture manufacturing – the fututre in dental care for elderly edentulous patients? J. Dent. 59 (2017) 68–77. Available from: https://doi.org/10.1016/j.jdent.2017.02.012.

[24] L.S. Acosta-Torres, L.M. López-Marín, R.E. Núñez-Anita, G. Hernández-Padrón, V.M. Castaño, Biocompatible metal-oxide nanoparticles: nanotechnology improvement of conventional prosthetic acrylic resins, J. Nanomater. 2011 (2011) 941561. Available from: https://doi.org/10.1155/2011/941561.

[25] J.F. Hernández-Sierra, F. Ruiz, D.C. Cruz Pena, F. Martínez-Gutiérrez, A.E. Martínez, A. de Jesús Pozos Guillén, et al., The antimicrobial sensitivity of *Streptococcus mutans* to nanoparticles of silver, zinc oxide, and gold, Nanomed.: Nanotechnol. Biol. Med. 4 (3) (2008) 237–240. Available from: https://doi.org/10.1016/j.nano.2008.04.005.

[26] Z. Huang, X. Zheng, D. Yan, G. Yin, X. Liao, Y. Kang, et al., Toxicological effect of ZnO nanoparticles based on bacteria, Langmuir 24 (8) (2008) 4140–4144. Available from: https://doi.org/10.1021/la7035949.

[27] S. Kasraei, L. Sami, S. Hendi, M.-Y. Alikhani, L. Rezaei-Soufi, Z. Khamverdi, Antibacterial properties of composite resins incorporating silver and zinc oxide nanoparticles on *Streptococcus mutans* and *Lactobacillus*, Restor. Dent. Endod. 39 (2) (2014) 109–114. Available from: https://doi.org/10.5395/rde.2014.39.2.109.

[28] S. Tavassoli Hojati, H. Alaghemand, H. Alaghemand, F. Hamze, F. Ahmadian Babaki, R. Rajab-Nia, et al., Antibacterial, physical and mechanical properties of flowable resin composites containing zinc oxide nanoparticles, Dent. Mater. 29 (5) (2013) 495–505. Available from: https://doi.org/10.1016/j.dental.2013.03.011.

[29] S. Kulshrestha, S. Khan, R. Meena, B.R. Singh, A.U. Khan, A graphene/zinc oxide nanocomposite film protects dental implant surfaces against cariogenic *Streptococcus mutans*, Biofouling 30 (10) (2014) 1281–1294. Available from: https://doi.org/10.1080/08927014.2014.983093.

[30] K. Memarzadeh, A.S. Sharili, J. Huang, S.C.F. Rawlinson, R.P. Allaker, Nanoparticulate zinc oxide as a coating material for orthopedic and dental implants, J. Biomed. Mater. Res. Part. A 103 (3) (2015) 981–989. Available from: https://doi.org/10.1002/jbm.a.35241.

[31] W. Liu, P. Su, S. Chen, N. Wang, Y. Ma, Y. Liu, et al., Synthesis of TiO_2 nanotubes with ZnO nanoparticles to achieve antibacterial properties and stem cell compatibility, Nanoscale 6 (15) (2014) 9050–9062. Available from: https://doi.org/10.1039/C4NR01531B.

[32] M.A. Vargas-Reus, K. Memarzadeh, J. Huang, G.G. Ren, R.P. Allaker, Antimicrobial activity of nanoparticulate metal oxides against peri-implantitis pathogens, Int. J. Antimicrobial Agents 40 (2) (2012) 135–139. Available from: https://doi.org/10.1016/j.ijantimicag.2012.04.012.

[33] D. Mohammed, M. Mudhaffar, Effect of modified zirconium oxide nano-fillers addition on some properties of heat cure acrylic denture base material, J. Baghdad Coll. Dent. 24 (2012) 4.

[34] M. Ashour, M. Ebrahim, Effect of Zirconium oxide nano-fillers addition on the flexural strength, fracture toughness, and hardness of heat-polymerized acrylic resin, World J. Nano Sci. Eng. 04 (2014) 50–57. Available from: https://doi.org/10.4236/wjnse.2014.42008.

[35] M. Gad, A.S. ArRejaie, M.S. Abdel-Halim, A. Rahoma, The reinforcement effect of nano-zirconia on the transverse strength of repaired acrylic denture base, Int. J. Dent. 2016 (2016) 7094056. Available from: https://doi.org/10.1155/2016/7094056.

[36] G.K. Wang, H. Kang, G.J. Bao, J.J. Lv, F. Gao, Influence on mechanical properties and microstructure of nano-zirconia toughened alumina ceramics with nano-zirconia content, West China J. Stomatol. 24 (5) (2006) 404–406.

[37] I.G. Tredici, M. Sebastiani, F. Massimi, E. Bemporad, A. Resmini, G. Merlati, et al., Low temperature degradation resistant nanostructured yttria-stabilized zirconia for dental applications, Ceram. Int. 42 (7) (2016) 8190−8197. Available from: https://doi.org/10.1016/j.ceramint.2016.02.026.
[38] C. Persson, E. Unosson, I. Ajaxon, J. Engstrand, H. Engqvist, W. Xia, Nano grain sized zirconia−silica glass ceramics for dental applications, J. Eur. Ceram. Soc. 32 (16) (2012) 4105−4110. Available from: https://doi.org/10.1016/j.jeurceramsoc.2012.06.028.
[39] L. Cheng, M.D. Weir, H.H. Xu, J.M. Antonucci, A.M. Kraigsley, N.J. Lin, et al., Antibacterial amorphous calcium phosphate nanocomposites with a quaternary ammonium dimethacrylate and silver nanoparticles, Dental Mater. 28 (5) (2012) 561−572. Available from: https://doi.org/10.1016/j.dental.2012.01.005.
[40] M.A. Melo, L. Cheng, M.D. Weir, R.C. Hsia, L.K. Rodrigues, H.H. Xu, Novel dental adhesive containing antibacterial agents and calcium phosphate nanoparticles, J. Biomed. Mater. Res. Part. B, Appl. Biomater. 101 (4) (2013) 620−629. Available from: https://doi.org/10.1002/jbm.b.32864.
[41] J. Wu, H. Zhou, M.D. Weir, M.A.S. Melo, E.D. Levine, H.H.K. Xu, Effect of dimethylaminohexadecyl methacrylate mass fraction on fracture toughness and antibacterial properties of CaP nanocomposite, J. Dent. 43 (12) (2015) 1539−1546. Available from: https://doi.org/10.1016/j.jdent.2015.09.004.
[42] Y. Pradeepkumar, K. Panishankar, P. Saraswathi, A. Saravanan, Current research in nano periodontics, SRM J. Res. Dental Sci. 3 (1) (2012) 46−50.
[43] N.G. Emerenciano, A.C. Botazzo Delbem, J.P. Pessan, G.P. Nunes, F.N. Souza Neto, E.R. de Camargo, et al., In situ effect of fluoride toothpaste supplemented with nano-sized sodium trimetaphosphate on enamel demineralization prevention and biofilm composition, Arch. Oral. Biol. 96 (2018) 223−229. Available from: https://doi.org/10.1016/j.archoralbio.2018.09.019.
[44] S.J. Garner, A.H. Nobbs, L.M. McNally, M.E. Barbour, An antifungal coating for dental silicones composed of chlorhexidine nanoparticles, J. Dent. 43 (3) (2015) 362−372. Available from: https://doi.org/10.1016/j.jdent.2014.12.005.
[45] M.F. Morks, N.F. Fahim, A. Kobayashi, Structure, mechanical performance and electrochemical characterization of plasma sprayed SiO_2/Ti-reinforced hydroxyapatite biomedical coatings, Appl. Surf. Sci. 255 (5, Part 2) (2008) 3426−3433. Available from: https://doi.org/10.1016/j.apsusc.2008.09.088.
[46] S. Nakashima, M. Yoshie, H. Sano, A. Bahar, Effect of a test dentifrice containing nano-sized calcium carbonate on remineralization of enamel lesions in vitro, J. Oral. Sci. 51 (1) (2009) 69−77. Available from: https://doi.org/10.2334/josnusd.51.69.
[47] H.A. Adeola, S. Sabiu, T.A. Adekiya, R.T. Aruleba, C.E. Aruwa, B.E. Oyinloye, Prospects of nanodentistry for the diagnosis and treatment of maxillofacial pathologies and cancers, Heliyon 6 (9) (2020) e04890. Available from: https://doi.org/10.1016/j.heliyon.2020.e04890.
[48] G. De-Deus, A. Canabarro, G. Alves, A. Linhares, M.I. Senne, J.M. Granjeiro, Optimal cytocompatibility of a bioceramic nanoparticulate cement in primary human mesenchymal cells, J. Endod. 35 (10) (2009) 1387−1390. Available from: https://doi.org/10.1016/j.joen.2009.06.022.
[49] Z. Yuan, B. Peng, H. Jiang, Z. Bian, P. Yan, Effect of bioaggregate on mineral-associated gene expression in osteoblast cells, J. Endod. 36 (7) (2010) 1145−1148. Available from: https://doi.org/10.1016/j.joen.2010.03.025.
[50] T. Webster, R. Siegel, R. Bizios, Nanoceramic surface roughness enhances osteoblast and osteoclast functions for improved orthopaedic/dental implant efficacy, Scr. Materialia 44 (2001) 1639−1642.
[51] J. Ma, H. Wong, L.B. Kong, K.W. Peng, Biomimetic processing of nanocrystallite bioactive apatite coating on titanium, Nanotechnology 14 (6) (2003) 619−623. Available from: https://doi.org/10.1088/0957-4484/14/6/310.
[52] J.H. Min, H.K. Kwon, B.I. Kim, The addition of nano-sized hydroxyapatite to a sports drink to inhibit dental erosion—in vitro study using bovine enamel, J. Dent. 39 (9) (2011) 629−635. Available from: https://doi.org/10.1016/j.jdent.2011.07.001.
[53] C. Hannig, M. Hannig, Natural enamel wear—a physiological source of hydroxylapatite nanoparticles for biofilm management and tooth repair? Med. Hypotheses 74 (4) (2010) 670−672. Available from: https://doi.org/10.1016/j.mehy.2009.11.007.

[54] R. Arif, B. Yilmaz, W.M. Johnston, In vitro color stainability and relative translucency of CAD-CAM restorative materials used for laminate veneers and complete crowns, J. Prosthet. Dent. 122 (2) (2019) 160–166. Available from: https://doi.org/10.1016/j.prosdent.2018.09.011.
[55] Shuman, 2020. Nanoceramic CAD/CAM restorations. Ben Zev. https://www.dentistrytoday.com/restorative-134/10088-nanoceramic-cad-cam-restorations. Accessed 15th December 2020 2020.
[56] A.B. Venturini, C. Prochnow, G.K.R. Pereira, R.D. Segala, C.J. Kleverlaan, L.F. Valandro, Fatigue performance of adhesively cemented glass-, hybrid- and resin-ceramic materials for CAD/CAM monolithic restorations, Dental Mater. 35 (4) (2019) 534–542. Available from: https://doi.org/10.1016/j.dental.2019.01.013.
[57] N. Ceren, V. Turp, F. Emir, G. Akgungor, S. Ayyildiz, D. Sen, Nanoceramics and hydrid materials used in CAD/CAM systems, Aydın. Dental J. 2 (3) (2016) 55–62.

CHAPTER 6

Nanoceramics: novel and benign materials in prosthodontics

Shikha Gulati, Sanjay Kumar, Arikta Baul and Harish Neela Lingam B.
Department of Chemistry, Sri Venkateswara College, University of Delhi, Delhi, India

6.1 Introduction

Prosthodontics is the branch of dentistry which mainly deals with dental defects, occlusal trauma, edentulism, bruxism, osteonecrosis, treatments after tooth loss (such as lays, crowns, and dentures), and the use of artificial prostheses for periodontal disease, temporomandibular joint disease, and maxillofacial tissue defects [1–4]. One of the main eight branches of dentistry, prosthodontics, contrary to popular perception, is an ancient skill that was practiced by Egyptian and Etruscan societies. After years of disuse, it was revived by Frenchman Pierre Fauchard, the "father of modern dentistry." Since then, the field has undergone massive developments as a result of widespread research and practice of dentistry. And perhaps no other field of dentistry has enjoyed the fruits of this revolutionary development as much as dental materials. What makes prosthodontics essential is the fact that dental prostheses restore the patient's oral function and facial features and maintains their health. The different subbranches are fixed prosthodontics (which deals with permanent prostheses), removable prosthodontics (which deals with removable prostheses), and maxillofacial prosthodontics, which treats much more complicated ailments like cleft lips, cleft palates, speech correction etc. Prosthodontic treatment is not one single treatment during the lifetime of the patient, in fact it is a continuation of several diverse kinds of restorative and prosthodontics dental treatments.

Traditionally, the prostheses used in prosthodontics were made of some conventional materials like hydroxyapatite, metal, resins, ceramics, glass ionomers, alumina, and zirconia. These materials stay in the oral environment for a very long period of time and are in direct contact with oral mucosa, and hence must exhibit some favorable characteristics, like high fatigue strength, hardness, good castability, chemical stability, biocompatibility, and comparatively less shrinkage deformation [2–4]. Due to these properties, prostheses like crowns are popularly made of zirconium oxide and alumina ceramic, and veneers are made of composite resins and glass-ceramics. However, despite these preferable qualities, these materials suffer from several drawbacks like pigment adhesion, problems due to

Industrial Applications of Nanoceramics
DOI: https://doi.org/10.1016/B978-0-323-88654-3.00007-X

their porous and brittle nature, color change, aging fracture, etc., which have restricted the use of these compounds, inhibiting their development and turning these materials obsolescent. Hence, a need has arisen to develop novel materials which can substitute for these conventional materials and overcome their drawbacks.

In this context, with the application of nanotechnology to prosthodontics, it has been found that the efficacy of many conventional materials has been enhanced after reducing their scale to nanometers. Nanomaterials have a small size, large surface area, high surface energy, a large proportion of surface atoms, and four unique effects: small size effect, quantum size effect, quantum tunneling effect, and surface effect [5]. Therefore, the limitations of conventional materials can be significantly overcome when their scale is reduced to the nanosize. Nanomaterials have some unique structures and properties which have vast applicability in the medical field [6–9], especially in the field of dentistry, and they have attained massive recognition for their ability to improve clinical, physical, and mechanical properties of dental materials [4]. "Nanoceramic refers to the ceramic material with nanoscale dimensions in the microstructures phase." These nanocomposites show great enhancement in desirable properties like surface hardness, modulus elasticity, polymerization shrinkage, filler loading [10,11], superplasticity, machinability, strength, bioactivity, fracture toughness, corrosion resistance, and translucency [12,13]. Hence, we can say that, in prosthodontics, nanoceramics have almost unlimited potential waiting to be tapped into. Nanoceramics can be divided into four categories: nanopowder, nanofiber, nanomembrane, and nanoblock, in which the development of nanopowder has been in progress longest, and as such its technology is the most mature [14]. In this chapter, the general synthesis and applications of nanoceramics in prosthodontics are discussed briefly. In addition, significant properties of nanoceramics are compared with those of the conventional ceramics that are currently in use. Also, some important nanoceramic materials such as nano-TiO_2, nano-ZrO_2, and carbon nanotubes are discussed.

6.2 Prosthodontics

One of the dental specialties, prosthodontics is also known as dental prosthetics or prosthetic dentistry. The American Dental Association defines prosthodontics as "Prosthodontics is the dental specialty pertaining to the diagnosis, treatment planning, rehabilitation and maintenance of the oral function, comfort, appearance and health of patients with clinical conditions associated with missing or deficient teeth and/or oral and maxillofacial tissues using biocompatible substitutes" [15]. Prosthodontics employs a number of techniques, prostheses, and materials to treat dental, maxillofacial, and temporomandibular defects and can be fixed prosthodontics, removable prosthodontics, implant prosthodontics, and maxillofacial prosthodontics.

6.3 Conventional ceramic materials used in prosthodontics

Over the last three centuries, dentists have tried their hand at a plethora of materials ranging from silver and gold to nanosized materials. Some of the materials have stood the test of time and have become the go-to for various dental defects and have come to be known as "conventional materials" or "traditional materials" in dentistry. Out of the many various conventional materials like amalgams, composites, adhesives, cements, etc., ceramics as a conventional material is discussed here, as it has likely had the greatest impact on the profession and oral health of the people [16].

Based on the microstructure, ceramics are classified as follows (Fig. 6.1) [17]:

6.3.1 Glass-based ceramics (mainly silica)

The first known instance of synthesis of glass-based ceramics is of feldspathic ceramics by Stanley Stookley in 1953 [18] and then W.T. McCulloch proposed its use as a dental restorative material in 1968. Since then, glass-ceramics have attracted a lot of interest in dentistry, which is evident from the number of papers published (Fig. 6.1) [19]. Dental glass-ceramics are favored by dentists and patients owing to their favorable physical and chemical properties, such as outstanding aesthetics, translucency, low thermal conductivity, adequate strength, biocompatibility, wear resistance, and chemical durability [20,21]. These materials are generally composed of silicon dioxide (silica or quartz) with varying amounts of aluminum oxide [22]. Strictly speaking, glass-ceramics are a special group of materials consisting of crystalline phase and glassy matrix; the crystalline phase(s) is embedded in the glass matrix. Generally, glass-ceramics are used to make dental crowns mainly, along with veneers, multilayered dental prostheses, and other monolithic prostheses. Powder condensation, lost wax/heat pressed technique, slip casting, and CAD/CAM

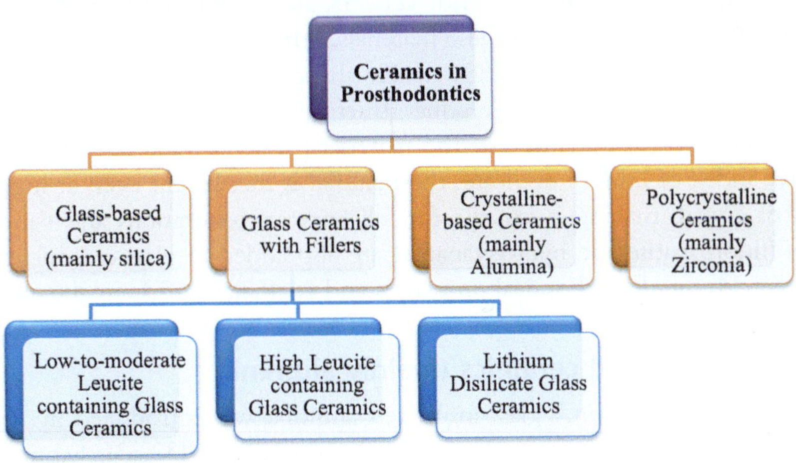

Figure 6.1 Conventional ceramic materials used in prosthodontics.

techniques are some of the methods used to manufacture glass-ceramics [23]. Despite favorable aesthetics, chemical resistance, fracture toughness, and microhardness, glass-ceramics suffer from a major setback: they are brittle in nature and susceptible to the development of cracks under stress [24]. Also, the increasing hardness of glass-ceramics results in the abrasion of adjacent tissues, including enamel.

6.3.2 Glass-based ceramics with fillers (mainly leucite and apatites)

This category of ceramics was created by modifications to general glass-ceramics, to overcome some of the shortcomings of the latter. Generally, these are prepared by adding or growing varying amounts of crystal on traditional glass-ceramics, popularly leucite, lithium disilicate, and apatites. This category covers a large range of glass—crystal ratios and crystal types, so much so that, it can be divided into three subcategories [25], which are:

- *Low-to-moderate leucite-containing feldspathic glass:* popularly known as "feldspathic porcelains," these are glass-ceramics to which varying amounts of leucite have been added, primarily to increase the coefficient of thermal expansion of the glass-ceramic. This results in the low fracture resistance and abrasive properties, which enables it to be applied on metals and zirconia. Manufactured using heat pressing technique [26] and dispersion strengthening [27,28], these materials are used as crowns and veneers.
- *High leucite containing (50%) glass:* these materials are glass-ceramics, in which leucite crystals have been grown to a considerable concentration in the glass matrix using a technique known as "controlled crystallization of glass." The main differences between this subcategory and the preceding one are the concentration of leucite and leucite is added in the latter whereas they are grown in the former. Available in powder/liquid, machinable, and pressable forms, these materials are used to make posterior inlays, onlays, anterior veneers, and crown restorations.
- *Lithium disilicate glass-ceramics:* in these materials, lithium disilicate crystals are grown in the glass matrix which are peculiarly needle-shaped and comprise more than 70% of the material by volume. This unique structure results in almost double flexural strength and fracture toughness of the material [29—31]. Due to the relatively low refractive index of lithium disilicate, the material is translucent even at high concentration of crystal, making it suitable for full-contour restorations and veneers. Also, apatites (fluoroapatite and hydroxyapatite) are also added to glass-ceramics to prepare materials with the desired optical properties and coefficient of thermal expansion.

6.3.3 Crystalline-based ceramics (mainly alumina)

Under the name "In-Ceram," glass-infiltrated alumina was introduced in 1988, as an alternative to metal-ceramics and has achieved great success [32,33]. Fabricated using slip casting, it generally consists of alumina, alumina-magnesia and alumina-zirconia mixtures. It has extremely high flexural strength due to the structure of the material which consists of sintered crystalline matrix of high modulus (~85%). Due to the large difference in the refractive index of alumina and glass filler, the material is opaque in nature, which makes it unsuitable for dental anterior restorations. This property, coupled with its high strength, makes In-Ceram suitable for posterior restorations involving molars, which has been a success due to the longevity of the material. However, the alumina—magnesia mixture could be used for anterior restorations, due to its translucent nature, although it has lower strength comparatively.

6.3.4 Polycrystalline ceramics (mainly zirconia)

These are monophase materials without any matrix which are dense, air-free, glass-free, polycrystalline monostructures. Used to make crowns and fixed partial dentures, these materials exhibit high strength and toughness. In particular zirconia shows high flexural strength and fracture toughness, which makes it efficient for making posterior partial dentures. The availability of a highly controlled starting powder and the application of computers for ceramic processing catapulted the development and usage of these materials. Zirconium oxide exists in three different states: tetragonal, monoclinic, and cubic. Initially, due to this, this material was unstable. However, after adding Yt or Ce, the material settled into a metastable state in the tetragonal form. When a crack occurs, the stress triggers the conversion into a monoclinic state, following which the 4.4% volume increase squeezes the crack close, resulting in the end of crack propagation.

6.4 Drawbacks of conventional ceramic materials

Though ceramics present many favorable properties for use in dental prostheses, they have certain disadvantages which hinder their further development and restrict their current use. Various researches have been conducted over the past years to overcome these restrictions with zero or partial success. Some of the main disadvantages presented by conventional ceramics are the brittleness, fabrication technique sensitivity, wear of opposing teeth, high cost, time involved, and low repair potential. The survival rates of dental materials have been documented to be 88%—100% after 2—5 years and 84%—97% after 5—14 years [34]. Also, the reported modes of failure are fracture, delamination of veneers, root fracture, caries, and endodontic treatment [35].

In general, the glass phase results in translucency and brittleness whereas the crystalline phase is the reason for opacity and strength. Since strength and translucency are preferred in dental materials, it is crucial to strike the right balance between the two, and more often than not either one of the properties is compromised. Mechanically speaking, our jaws can exert force ranging from 700—4800 N [36] and this stress has to be endured by our teeth, in particular the enamel [37] and the dental material used in the prostheses also must be able to withstand it. However, while using "stronger" material, the aesthetics is compromised and the adjacent teeth are abraded comparatively higher. Aesthetically pleasing materials do not offer the required fracture toughness.

Considering glass-ceramics, the major disadvantages associated are (Fig. 6.2):

- Inability to color internally is a major limitation. Although they can be stained and glazed, they are monochromatic in nature [38,39] which is also not desired.

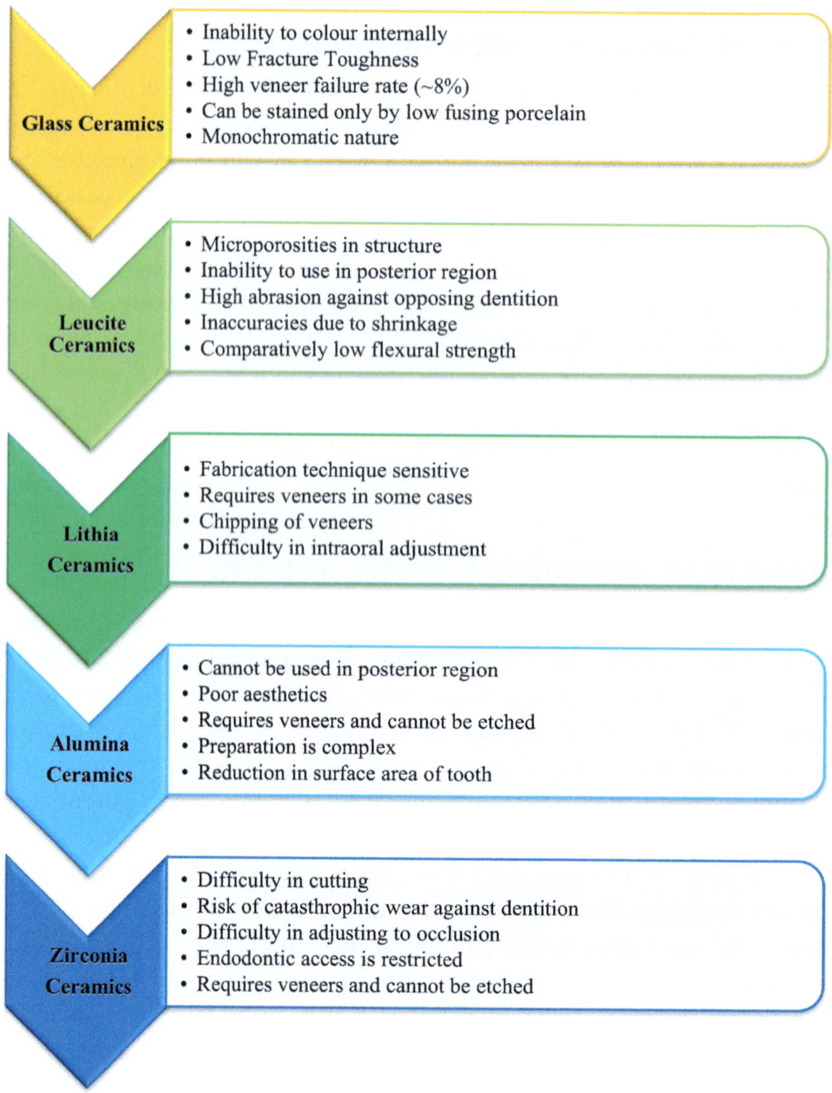

Figure 6.2 Drawbacks of conventional ceramic materials used in prosthodontics.

- The limited usability in high stress regions. Though they can be strengthened by a number of techniques and by changing the composition of the crystalline phase [38,39], the material is not strong enough, and hence other alternatives must be considered for molar restorations.
- Other disadvantages associated are high veneer failure rate (~8%), expensive equipment for fabrication, and it can be stained only using a low fusing feldspathic porcelain.

In leucite reinforced-glass-ceramics, some disadvantages posed are:

- Sintering can result in microporosities which can lead to crack development during use. This makes it prone to fracturing in the posterior region [40]. Moreover, marginal inaccuracies might occur due to the shrinkage caused by sintering and this must be accounted for accurately.
- Comparatively less flexural strength, a high abrasive effect on the adjacent teeth and inability to cover the color of a darkened tooth or post and core (due to high translucency) are other drawbacks.

Problems associated with lithium disilicate (Lithia) glass-ceramics are:

- The properties of the materials are fabrication technique sensitive and their properties vary according to the technique used.
- Though some variants are highly translucent, some exhibit lesser translucency and in those cases, a veneering layer, preferably by an apatite ceramic, is required. Often times, glaze coating and the veneer can increase wear.
- Difficulty in intraoral adjustment and chipping of veneer are other associated disadvantages [41].

Disadvantage of using apatite ceramics is:

- Relatively low fracture toughness which results in fracturing and chipping of veneers.

Alumina ceramics pose certain difficulties such as:

- Cannot be used in posterior region because the strength of the material is not enough to withstand the occlusal trauma.
- The opaque alumina core shows poor aesthetics and must be veneered.
- It is incapable of being etched and the technique results in reduction in the tooth surface.
- The preparation technique of slip casting is a complex procedure and requires specialized equipment.
- Magnesia-reinforced ceramics exhibit excessive translucency which presents a glassy, low-value appearance.

Zirconia ceramics pose:

- Risk of catastrophic wear of opposing dentition along with difficulty in cutting and shaping.
- A difficulty in adjusting occlusion when significant premature contacts are present.
- Due to generation of heat, endodontic access is restricted.

- Also, in some cases, there is a requirement of veneering due to increased opacity.
- Like other reinforced ceramic cores, there is an inability to etch the surface and difficulty of bonding with resins, along with chipping of veneers.

In conclusion, almost all ceramics used as dental materials fail due to the propagation of cracks at a critical strain of 0.1% [42]. Moreover, chewing results in the creation of millions of subcritical loads and materials ultimately fail due to this cyclic fatiguing. All these disadvantages have led to the search for the synthesis of an ideal material with no disadvantages, which is still continuing with the end out of our reach.

6.5 Nanoceramics: novel and benign materials in prosthodontics

6.5.1 What are nanoceramics?

Nanoceramics refer to the ceramic materials with nanoscale dimensions in the microstructures phase. Nanoceramics were first discovered in the early 1980s. As the name suggests, they are fabricated from particles that are ultrafine with diameters less than 100 nm. Nanoceramics were first formed by the sol–gel process. Later on, they were also prepared in a sintering process using heat and pressure. Use of nanoceramics has been growing worldwide in all fields of biomedical research and prosthodontics is no exception.

6.5.2 Merits of nanoceramics in prosthodontics

Nanoceramics are needed to take over from conventional ceramic materials for their innately unique properties like superplasticity, machinability, better toughness, strength, etc. and the drawbacks of conventional ceramic materials can be significantly overcome by the use of these materials. The advantages of nanoceramics for application in prosthodontics are as follows:

- Firstly, they have superior mechanical properties such as strength and hardness. Hardness and strength of nanoceramics is believed to be 4–5 times higher than those of traditional ceramic materials.
- While ceramic is brittle and has low ductility, nanoceramics show good ductility.
- The toughness of nanoceramics is higher than that of traditional ceramic materials.
- Nanoceramics may offer translucency of dental restoration making them even more favorable.
- They show resistance to corrosion and form deformation.

Researchers in the past have conducted studies to see the effect of the addition of silver and platinum nanoparticles in the dental porcelain and results have only favored this modification. Uno et al. [43] in his research found out that the fracture toughness and Vickers hardness of dental porcelain increased upon the addition of silver nanoparticles. On the other hand, Fujieda et al. [44] concluded that the incorporation of silver and platinum nanoparticles increased Young's modulus and fracture toughness of ceramics (Fig. 6.3).

Talking in specific terms, glass-ceramics are undoubtedly better than all-ceramics for they deliver translucency in dental restoration, making it look aesthetically pleasing. But their fracture toughness and flexural strength is generally less than half of the all-ceramics [24]. Thus a translucent glass-ceramic with improved mechanical properties may be needed for dental restoration applications. The later part of this chapter talks about how nanograin-sized zirconia—silica glass-ceramics prepared by Persson et al. [13] showed better mechanical properties than glass-ceramics based on lithium disilicate. Zirconia ceramic materials are preferred for medical restoration purposes, the reason being their excellent biocompatibility, abrasion resistance, corrosion resistance, etc. However, their clinical application is limited due to their inherent brittleness which is significantly resolved by the use of nanoceramics [45].

6.5.3 General synthetic approaches for nanoceramics

In the search for more suitable materials for dental materials, nanoceramics are a milestone, as they exhibit the favorable properties of conventional ceramics in far greater proportions without most of the latter's disadvantages. Currently, the research in

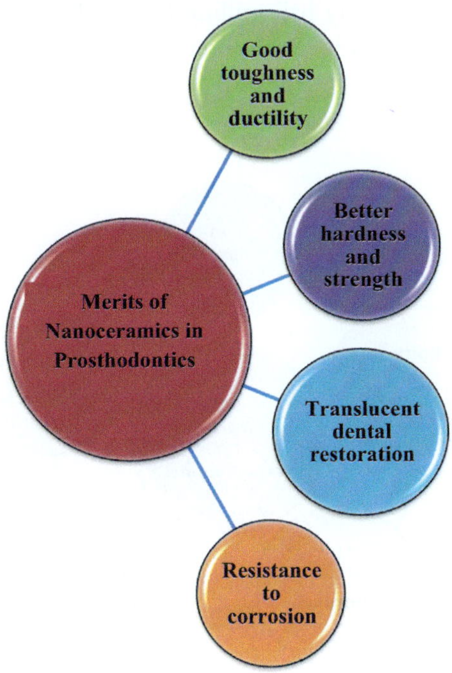

Figure 6.3 Merits of nanoceramics in prosthodontics.

nanomaterials is concentrated on two different directions, namely (1) preparation of new inorganic nanoparticles; and (2) surface modification of conventional materials with novel nanoparticles. Considering nanoceramics, research is primarily concerned with the latter approach where nanofillers are used to enhance the properties of the entire material [46]. Taking the entirety of nanomaterials in consideration, there are some different approaches to synthesis, some of them are detailed below (Fig. 6.4).

6.5.3.1 Top-down approach

In this approach, a large piece of material is reduced to particles with nanoscale dimensions. In this method, there is a high consistency and controllability on the shape, size, and structure of the particle, though it suffers from drawbacks like high cost of equipment, low fabrication capacity, and limited access to facilities [47]. This is a physical method for the synthesis of nanomaterials [48]. Despite this, most of the nanomaterials in prosthodontics, including nanoceramics [49] are synthesized by this approach. Some popular methods of this approach are mechanical grinding, wet chemical synthesis, lithography, etching, ball milling, and micromachining. In particular, mechanical grinding is preferred for the synthesis of nanocrystalline particles [48] for use as fillers in nanoceramics and nano-ZrO_2 ceramics. This has been developed by Li et al. using precision grinding [45]. It is noteworthy that nano-ZrO_2 toughened alumina ceramics was prepared by Wang et al. using ball milling [12] and the sol–gel method (a part of wet chemical synthesis) was used by Persson et al. to synthesize glass-ceramics with zirconia–silica nanograins [13].

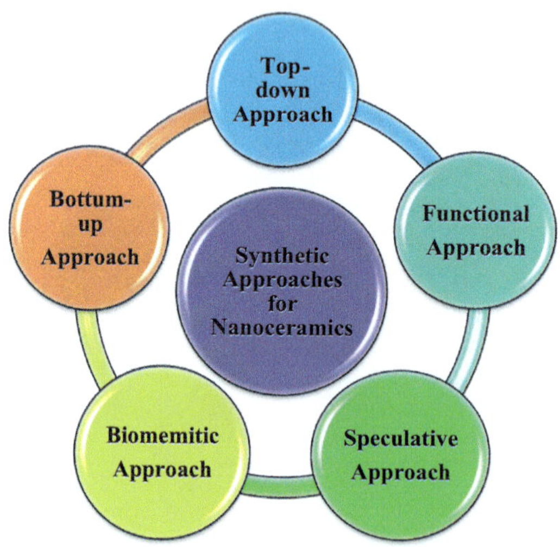

Figure 6.4 General synthetic approaches for nanoceramics.

6.5.3.2 Bottom-up approach

The bottom-up approach is a chemical approach in which smaller units assemble to form larger structures by use of physical and chemical forces functioning at nanoscale [50]. Being a low-cost method and providing a yield with less defects, bottom-up approaches are way more advantageous than the top-down methods of synthesis. A few examples of the bottom-up approach are chemical synthesis, molecular fabrication, and self-assembly. Just like while building a house, lots of building blocks are taken and put together to construct the final structure. Similarly, in this approach, several custom-made molecules self-assemble to form higher-order complex structures. Polymerization, precipitation, and micellar arrangements are the most commonly used bottom-up approaches in nanodentistry (Fig. 6.5).

6.5.3.3 Biomimetic approach

This approach is based on the use of concepts and techniques found in biological systems to create nanoparticles. This approach, also known as biomimicry, bionics, or biognosis, is still under research and has not been significantly applied [51].

6.5.3.4 Functional approach

In this approach, desired functionality components are generated without giving importance to their assembly.

6.5.3.5 Speculative approach

This approach is concerned more with the effects of fabrication rather than the process of fabrication. It takes societal and environmental impact of nanoparticles into consideration. This seeks to anticipate what inventions nanotechnology might yield and attempts to propose an agenda along which inquiry might progress [47].

Apart from the techniques mentioned above, various other techniques like flame pyrolysis, flame spray pyrolysis, etc. have also been used to synthesize nanoceramics for dental applications. Raj et al. devised a facile one-pot synthesis of alumina ceramics with nano-TiO_2 fillers [52] and An et al. synthesized alumina ceramics reinforced with carbon nanotubes using hot pressing [53]. This shows that more varied and diverse techniques are being used to synthesize nanoceramics, particularly for application in prosthodontics.

6.6 Applications of nanoceramics in prosthodontics

The nanoceramics thus synthesized have a plethora of applications in dentistry and this is particularly so in the case of prosthodontics. The favorable properties exhibited by nanoceramics like superplasticity, greater flexibility, and greater strength, along with "special" properties like biocompatibility, make them highly suitable for use as dental

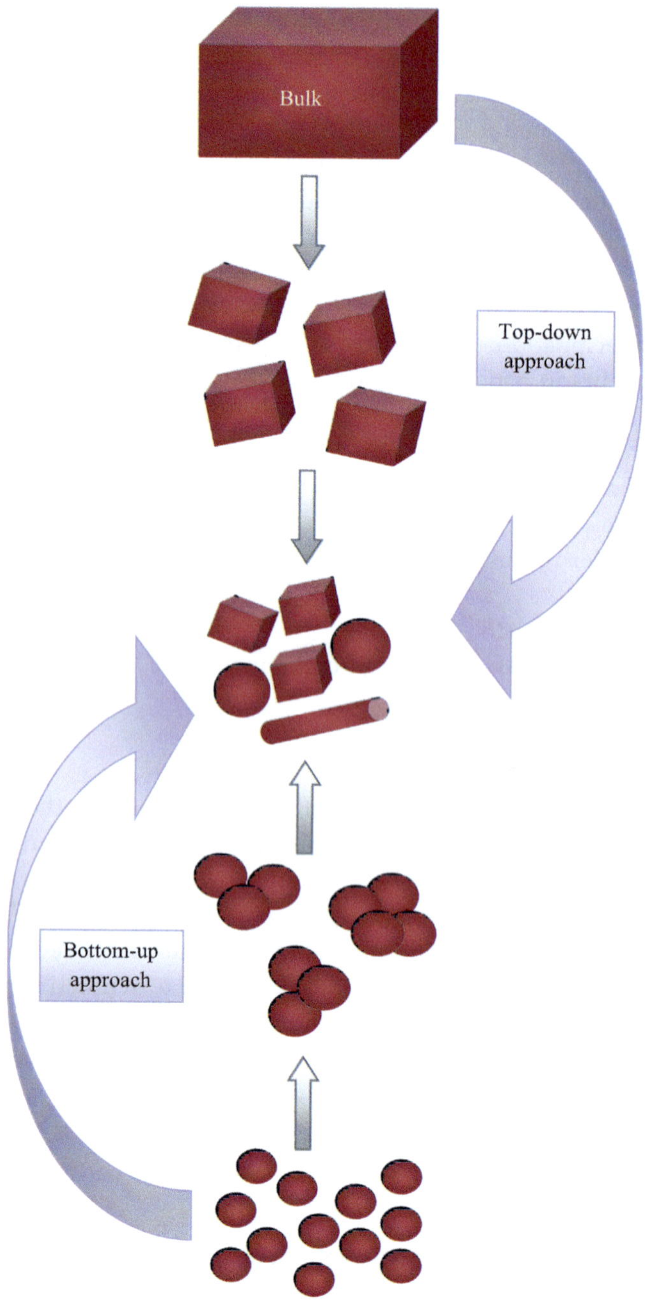

Figure 6.5 Top-down and bottom-up approach of nanoceramic synthesis.

material [54]. Like traditional ceramics, nanoceramics also can be used for making crowns, bridges, and other restorations. In particular, ceramics with lithium disilicate nanofillers can be used to prepare crowns and bridges.

It is noteworthy that nanoceramics can also be used for antibacterial nanotherapy. Nanoparticles of silver [55,56], zinc oxide [57,58], and polyethylenimine [59] can be incorporated in nanoceramics to impart antibacterial properties to the substance. These nanoceramics can be used as coatings to fight against microbes and control their growth in the oral environment. These nanoparticles exhibit antibacterial properties by different mechanisms like active transport, metabolism, disruption of cell envelope [60] etc. This can be used to make veneers and other coatings with antibacterial properties (Fig. 6.6).

Considering prosthodontic implants, the implant would be more desirable if it had biocompatibility, that is, osseointegration with the underlying bone. However, metallic (mainly Ti) implants are not up to the mark in this regard. But on coating with nanoceramics, the osseointegration is found to increase significantly. Coating of implants with nanoporous alumina ceramic improves the implant properties by a process known as "anodization of aluminum." In this method, alumina is rendered by loading the porous nanoceramics with bioactive agents, hence improving cells response and facilitating osseointegration [61,62] (Fig. 6.7).

High-strength nanoceramics like that of hydroxyapatite and calcium phosphate apatite can be made into paste which can be used for bone and enamel regeneration and repair. This is due to the tendency of the ceramics to interlock with bone. As natural bone is about 70% calcium phosphate apatite (by weight), the biocompatibility of these ceramics is very high and due to their dense three-dimensional (3D) crystalline structure, they can be used in both anterior and posterior regions as well as for weight-bearing and nonweight-bearing bones [63].

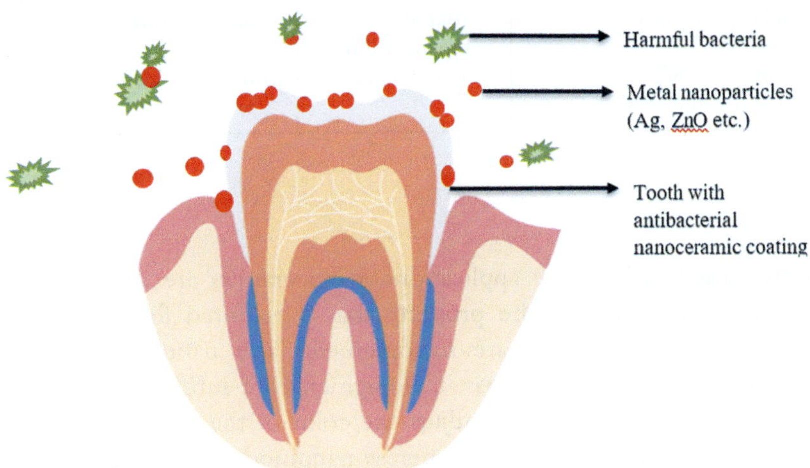

Figure 6.6 Ceramic coating doped with nanometallic particles showing antibacterial properties.

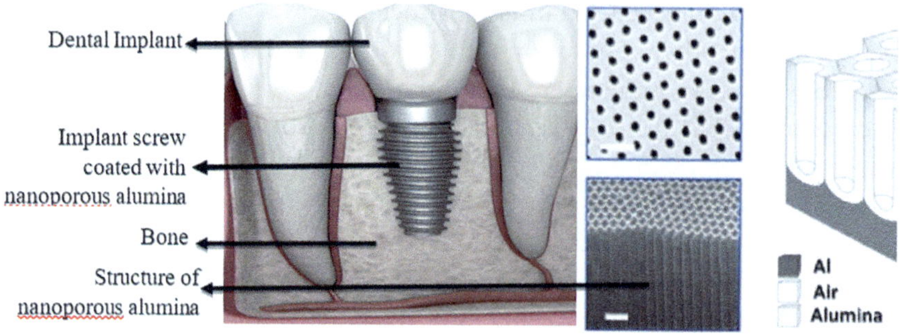

Figure 6.7 Enhanced osseointegration of implants through nanoporous alumina silica coating.

Table 6.1 Applications of various nanoceramics in prosthodontics.

S.no	Nanoceramic composition	Application	Merits	References
1.	Nano-TiO_2	Additive to conventional ceramics	Enhances favorable properties like microhardness	[43]
2.	Nano-ZrO_2	Dental material	Higher fracture toughness	[46]
3.	Nanoporous alumina	Coating on dental implants	Anodization of Al results in better osseointegration	[61,62]
4.	Nanoceramics with nanoparticles of Ag, ZnO, and polyethylenimine	Antibacterial nanotherapy	–	[55–59]
5.	Hydroxyapatite, calcium phosphate apatite	Bone and enamel regeneration	High strength, Very high biocompatibility	[63]
6.	Alumina–CNT nanocomposites	Dental material	Enhanced mechanical properties	[64,65]

For more varied and diverse applications, nanoceramics are doped with different substances which can produce the property that is required for any particular use. Generally, the substances added are nanopigments and nanomodifiers. Nanofillers enhance polishability and reduce wear, nanopigments are substances added to achieve a particularly desired shade and they adjust the color of the restoration to that of the surrounding teeth (chameleon effect), whereas nanomodifiers increase the stability of the material and reduce its tendency to stick to instruments (Table 6.1).

6.7 Comparison of some nanoceramic materials with conventional ceramics

6.7.1 Nanotitanium dioxide (nano-TiO_2) over conventional TiO_2

Studies indicate that at 100°C, the microhardness of nanotitanium dioxide ceramics is 13,000 kN/mm^2, while that of ordinary titanium oxide ceramic is lower than 2000 kN/mm^2. Nano-TiO_2 shows high toughness at room temperature. It doesn't break and remains intact even when compressed to 1/4 of the original length [43].

6.7.2 Nanozirconia (nano-ZrO_2) over conventional ZrO_2

The different physical properties of nano-ZrO_2 ceramic materials from the conventional ones were reported by Li et al. [46]. Hardness of conventional ZrO_2 is around 1500, while hardness of nanozirconia reaches more than 1750, thus increasing its hardness by 20%. While processing, conventional ZrO_2 breaks easily as its fracture toughness is also quite low, which too is greatly increased in nano-ZrO_2. Wang et al. [12] found that Al_2O_3 showed better toughness with 20% nano-ZrO_2 as he reported the influence of nano-ZrO_2 content on the mechanical properties.

6.7.3 Nanograin-sized zirconia–silica glass-ceramics over conventional lithium disilicate-based glass-ceramics

Lithium disilicate-based glass-ceramics lacking in mechanical properties are used very commonly in dental veneers and crowns. However, failures have been reported often in clinical cases, the reasons being their insufficient mechanical properties. Thus alternative materials with better mechanical properties may be preferred. With the intent to improve mechanical properties of lithium silicate-based glass-ceramics, Persson et al. produced glass-ceramics in zirconia–silica system with nanosized grains using a sol gel method. They not only produced glass-ceramics with 30%–40% ZrO_2 but also evaluated their corrosion resistance, crystallinity, elastic modulus, hardness, and fracture toughness. The produced glass-ceramics were found to be translucent, with a transmittance of over 70%. It possessed excellent corrosion resistance, higher extent of hardness, and a somewhat lower modulus of elasticity compared to the conventional lithium disilicate. They achieved a higher fracture toughness for their zirconia–silica glass-ceramic with the proper heat treatment. Corrosion resistance and translucency were preferable for requirements of dental materials [13].

6.7.4 Alumina–CNT composites over conventional alumina

For improvement in mechanical and electronic properties, carbon nanotubes are known to be used as nanofiller in nanomaterials [64]. Carbon nanotubes, owing to their unique mechanical properties, have been considered as reinforcing elements in ceramic matrix [64,65]. An et al. [53] via hot pressing, produced alumina–carbon nanotube (CNT) composites to examine the effect of carbon nanotube addition on wear resistance. On investigating the mechanical and tribological properties of alumina–CNT composites, they found that, in the range of 0%–4% CNT content, wear and mechanical properties were enhanced as a result of CNT acting as a reinforcing filler in the alumina composite and the reinforcement effect increased about 30% upon the addition of CNTs up to 4% (Fig. 6.8).

6.8 Conclusions and future outlook

The persistent progress in material sciences has been recognized as one of the major pillars catalyzing development in the sector of prosthodontics technology. Nanoscience has been significantly important for a number of basic scientific innovations and in bringing changes into prosthodontics making it a more clinical

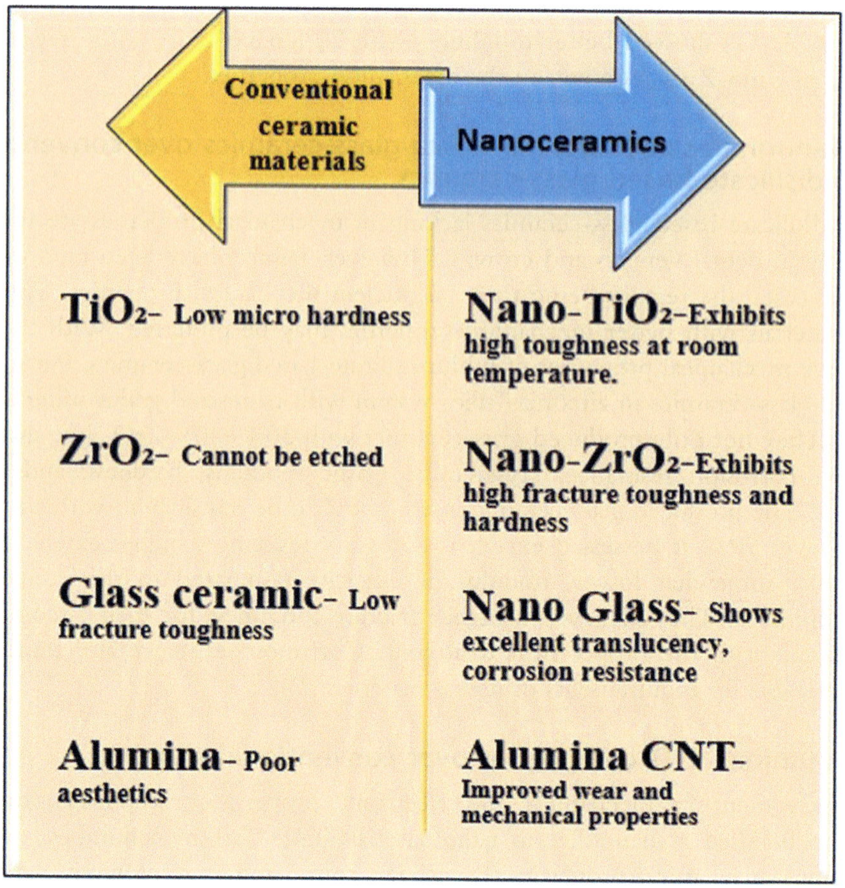

Figure 6.8 Comparison of nanoceramics with conventional ceramic materials.

technology. The immense potential of nanotechnology in the field of dentistry is quite evident in the latest research progress on applications of nanoceramics in prosthodontics, an important branch of dentistry dealing with dental prosthesis. In this chapter, we have discussed the advantages that nanoceramics can potentially provide over conventional ceramic materials and how properties of dental materials can be significantly improved by reducing their size to the nanoscale. The unique characteristics of nanoceramics make them a requisite to supersede the conventional ceramic materials. Thus nanoceramics have been recognized as novel and benign materials for prosthodontics and provide massive scope for future development in this field.

References

[1] C.S. Petrie, M.P. Walker, K. Williams, A survey of U.S. prosthodontists and dental schools on the current materials and methods for final impressions for complete denture prosthodontics, J. Prosthodont. 14 (4) (2005) 253–262. Available from: https://doi.org/10.1111/j.1532-849x.2005.00051.x.

[2] M. Mehra, F. Vahidi, R.W. Berg, A complete denture impression technique survey of postdoctoral prosthodontic programs in the United States, J. Prosthodont. 23 (4) (2013) 320–327. Available from: https://doi.org/10.1111/jopr.12099.

[3] G. Saavedra, L.F. Valandro, F.P.P. Leite, R. Amaral, M. Özcan, M.A. Bottino, et al., Bond strength of acrylic teeth to denture base resin after various surface conditioning methods before and after thermocycling, Int. J. Prosthodont. 20 (2) (2007).

[4] J.L. Cuy, A.B. Mann, K.J. Livi, M.F. Teaford, T.P. Weihs, Nanoindentation mapping of the mechanical properties of human molar tooth enamel, Arch. Oral Biol. 47 (4) (2002) 281–291. Available from: https://doi.org/10.1016/s0003-9969(02)00006-7.

[5] X. Li, R. Cui, W. Liu, L. Sun, B. Yu, Y. Fan, et al., The use of nanoscaled fibers or tubes to improve biocompatibility and bioactivity of biomedical materials, J. Nanomater. 2013 (2013) 1–16. Available from: https://doi.org/10.1155/2013/728130.

[6] S. Gulati, P. Singh, A. Diwan, A. Mongia, S. Kumar, Functionalized gold nanoparticles: promising and efficient diagnostic and therapeutic tools for HIV/AIDS, RSC Med. Chem. 11 (11) (2020) 1252–1266.

[7] S. Kumar, A. Mongia, S. Gulati, P. Singh, A. Diwan, S. Shukla, Emerging theranostic gold nanostructures to combat cancer: novel probes for combinatorial immunotherapy and photothermal therapy, Cancer Treat. Res. Commun. 25 (2020) 100258. Available from: https://doi.org/10.1016/j.ctarc.2020.100258.

[8] S. Kumar, A. Diwan, P. Singh, S. Gulati, D. Choudhary, A. Mongia, et al., Functionalized gold nanostructures: promising gene delivery vehicles in cancer treatment, RSC Adv. 9 (41) (2019) 23894–23907.

[9] S. Gulati, S. Kumar, P. Singh, A. Diwan, A. Mongia, Biocompatible chitosan-coated gold nanoparticles: novel, efficient, and promising nanosystems for cancer treatment, Handbook of Polymer and Ceramic Nanotechnology (2020) 1–29. Available from: https://doi.org/10.1007/978-3-030-10614-0_56-1.

[10] R.B. Huang, S. Mocherla, M.J. Heslinga, P. Charoenphol, O. Eniola-Adefeso, Dynamic and cellular interactions of nanoparticles in vascular-targeted drug delivery (review, Mol. Membr. Biol. 27 (4–6) (2010) 190–205. Available from: https://doi.org/10.3109/09687688.2010.499548.

[11] S.K. Kim, S.J. Heo, J.Y. Koak, J.H. Lee, Y.M. Lee, D.J. Chung, et al., A biocompatibility study of a reinforced acrylic-based hybrid denture composite resin with polyhedraloligosilsesquioxane, J. Oral Rehabil. 34 (5) (2007) 389–395. Available from: https://doi.org/10.1111/j.1365-2842.2006.01671.x.

[12] G.K. Wang, H. Kang, G.J. Bao, J.J. Lv, F. Gao, West China J. Stomatol. 24 (5) (2006) 404–406.

[13] C. Persson, E. Unosson, I. Ajaxon, J. Engstrand, H. Engqvist, W. Xia, Nano grain sized zirconia–silica glass ceramics for dental applications, J. Eur. Ceram. Soc. 32 (16) (2012) 4105–4110. Available from: https://doi.org/10.1016/j.jeurceramsoc.2012.06.028.

[14] I. Roy, M.K. Stachowiak, E.J. Bergey, Nonviral gene transfection nanoparticles: function and applications in the brain, Nanomedicine 4 (2) (2008) 89–97. Available from: https://doi.org/10.1016/j.nano.2008.01.002.

[15] The Glossary of Prosthodontic Terms, J. Prosthet. Dent. 117 (5) (2017) e1–e105. Available from: https://doi.org/10.1016/j.prosdent.2016.12.001.

[16] S.C. Bayne, J.L. Ferracane, G.W. Marshall, S.J. Marshall, R. van Noort, The evolution of dental materials over the past century: silver and gold to tooth color and beyond, J. Dent. Res. 98 (3) (2019) 257–265. Available from: https://doi.org/10.1177/0022034518822808.

[17] A. Shenoy, N. Shenoy, Dental ceramics: an update, J. Conserv. Dent.: JCD 13 (4) (2010) 195.

[18] E.D. Zanotto, Bright future for glass-ceramics, Am. Ceram. Soc. Bull. 89 (8) (2010) 19–27.

[19] L. Fu, H. Engqvist, W. Xia, Glass–ceramics in dentistry: a review, Materials 13 (5) (2020) 1049. Available from: https://doi.org/10.3390/ma13051049.

[20] C. Ritzberger, E. Apel, W. Höland, A. Peschke, V. Rheinberger, Properties and clinical application of three types of dental glass-ceramics and ceramics for CAD-CAM technologies, Materials 3 (6) (2010) 3700–3713. Available from: https://doi.org/10.3390/ma3063700.

[21] M. Montazerian, E.D. Zanotto, Bioactive and inert dental glass-ceramics, J. Biomed. Mater. Res. A 105 (2) (2016) 619–639. Available from: https://doi.org/10.1002/jbm.a.35923.

[22] R. Giordano, E.A. McLaren, Ceramics overview: classification by microstructure and processing methods, in: Compendium of Continuing Education in Dentistry (Jamesburg, N.J.:1995). 31 (9) (2010) 682–684, 686, 688 passim; quiz 698, 700.

[23] M.J. Santos, M.D. Costa, J.H. Rubo, L.F. Pegoraro, G.C. Santos Jr, Current all-ceramic systems in dentistry: a review, Compend. Contin. Educ. Dent. 36 (1) (2015) 31–37.

[24] H. Yilmaz, C. Aydin, B.E. Gul, Flexural strength and fracture toughness of dental core ceramics, J. Prosthet. Dent. 98 (2) (2007) 120–128.

[25] E.A. McLaren, P.T. Cao, Ceramics in dentistry—part I: classes of materials, Inside Dent. 5 (9) (2009) 94–103.

[26] J.A. Sorensen, C. Choi, M.I. Fanuscu, W.T. Mito, IPS Empress crown system: three-year clinical trial results, J. Calif. Dent. Assoc. 26 (2) (1998) 130–136.

[27] K.J. Anusavice, C. Shen, H.R. Rawls (Eds.), Phillips' Science of Dental Materials, Elsevier Health Sciences, 2012.

[28] J.R. Kelly, U.S. Patent No. 4,978,640. Washington, DC: U.S. Patent and Trademark Office, 1990.

[29] M. Albakry, M. Guazzato, M.V. Swain, Fracture toughness and hardness evaluation of three pressable all-ceramic dental materials, J. Dent. 31 (3) (2003) 181–188.

[30] M. Albakry, M. Guazzato, M.V. Swain, Biaxial flexural strength, elastic moduli, and x-ray diffraction characterization of three pressable all-ceramic materials, J. Prosthet. Dent. 89 (4) (2003) 374–380.

[31] M. Guazzato, M. Albakry, S.P. Ringer, M.V. Swain, Strength, fracture toughness and microstructure of a selection of all-ceramic materials. Part I. Pressable and alumina glass-infiltrated ceramics, Dent. Mater. 20 (5) (2004) 441–448.

[32] L. Pröbster, Survival rate of In-Ceram restorations, Int. J. Prosthodont. 6 (3) (1993).

[33] R. Scotti, S.A.D.E. Catapano, A clinical evaluation of In-Ceram crowns, Int. J. Prosthodont. 8 (4) (1995).

[34] H.J. Conrad, W.J. Seong, I.J. Pesun, Current ceramic materials and systems with clinical recommendations: a systematic review, J. Prosthet. Dent. 98 (5) (2007) 389–404.

[35] P. Marquardt, J.R. Strub, Survival rates of IPS empress 2 all-ceramic crowns and fixed partial dentures: results of a 5-year prospective clinical study, Quintessence Int. (Berl) 37 (4) (2006).

[36] The Power of the Human Jaw, Sci. Am. 105 (23) (1911) 493. Available from: https://doi.org/10.1038/scientificamerican12021911-493.

[37] R.W. Thresher, G.E. Saito, The stress analysis of human teeth, J. Biomech. 6 (5) (1973) 443–449. Available from: https://doi.org/10.1016/0021-9290(73)90003-1.

[38] S. Pollington, R. van Noort, An update of ceramics in dentistry, Int. J. Clin. Dent. 2 (4) (2009) 283–307.
[39] D. Grossman, H. Walters, The chemical durability of dental ceramics, J. Dent. Res. 63 (574) (1989) 234.
[40] W. Höland, M. Schweiger, M. Frank, V. Rheinberger, A comparison of the microstructure and properties of the IPS Empress® 2 and the IPS Empress® glass-ceramics, J. Biomed. Mater. Res. 53 (4) (2000) 297–303.
[41] F. Zarone, M.I. Di Mauro, P. Ausiello, G. Ruggiero, R. Sorrentino, Current status on lithium disilicate and zirconia: a narrative review, BMC Oral Health 19 (1) (2019) 1–14.
[42] W. Bömicke, P. Rammelsberg, T. Stober, M. Schmitter, Short-term prospective clinical evaluation of monolithic and partially veneered zirconia single crowns, J. Esthet. Restor. Dent. 29 (1) (2017) 22–30.
[43] M. Uno, M. Kurachi, N. Wakamatsu, Y. Doi, Effects of adding silver nanoparticles on the toughening of dental porcelain, J. Prosthet. Dent. 109 (4) (2013) 241–247.
[44] T. Fujieda, M. Uno, H. Ishigami, M. Kurachi, N. Wakamatsu, Y. Doi, Addition of platinum and silver nanoparticles to toughen dental porcelain, Dent. Mater. J. 31 (5) (2012) 711–716.
[45] C.H. Li, Y.L. Hou, Z.R. Liu, Y.C. Ding, Investigation into temperature field of nano-zirconia ceramics precision grinding, Int. J. Abrasive Technol. 4 (1) (2011) 77–89.
[46] A. Mohan, A.J. Hines, New era in prosthodontics—nanotechnology, J. Prosthet. Dent. 3 (1) (2019).
[47] X. Fu, J. Cai, X. Zhang, W.D. Li, H. Ge, Y. Hu, Top-down fabrication of shape-controlled, monodisperse nanoparticles for biomedical applications, Adv. Drug Deliv. Rev. 132 (2018) 169–187.
[48] H. Aeran, J. Seth, Nanomaterials: the changing phase of prosthodontics, Dental Applications of Nanotechnology, Springer, Cham, 2018, pp. 121–132.
[49] C. Praveena, P. Manne, L. Kalluri, R. Anne, Nanobiomaterials and their application in prosthodontics, Dent. Appl. Nanotechnol. (2018) 177–201.
[50] S. Das, A.J. Gates, H.A. Abdu, G.S. Rose, C.A. Picconatto, J.C. Ellenbogen, Designs for ultra-tiny, special-purpose nanoelectronic circuits, IEEE Transactions on Circuits and Systems I: Regular Papers 54 (11) (2007) 2528–2540.
[51] Z. Ghalanbor, S.A. Marashi, B. Ranjbar, Nanotechnology helps medicine: nanoscale swimmers and their future applications, Med. Hypotheses 65 (1) (2005) 198–199.
[52] V. Raj, M.S. Mumjitha, Formation and surface characterization of nanostructured Al_2O_3-TiO_2 coatings, Bull. Mater. Sci. 37 (6) (2014) 1411–1418.
[53] J.W. An, D.H. You, D.S. Lim, Tribological properties of hot-pressed alumina–CNT composites, Wear 255 (1-6) (2003) 677–681.
[54] W. Wang, S. Liao, Y. Zhu, M. Liu, Q. Zhao, Y. Fu, Recent applications of nanomaterials in prosthodontics, J. Nanomater. 2015 (2015).
[55] K. Blecher, A. Nasir, A. Friedman, The growing role of nanotechnology in combating infectious disease, Virulence 2 (5) (2011) 395–401.
[56] Q. Li, S. Mahendra, D.Y. Lyon, L. Brunet, M.V. Liga, D. Li, et al., Antimicrobial nanomaterials for water disinfection and microbial control: potential applications and implications, Water Res. 42 (18) (2008) 4591–4602.
[57] H. Gu, D. Fan, J. Gao, W. Zou, Z. Peng, Z. Zhao, et al., Effect of $ZnCl_2$ on plaque growth and biofilm vitality, Arch. Oral Biol. 57 (4) (2012) 369–375.
[58] Y. Xie, Y. He, P.L. Irwin, T. Jin, X. Shi, Antibacterial activity and mechanism of action of zinc oxide nanoparticles against Campylobacter jejuni, Appl. Environ. Microbiol. 77 (7) (2011) 2325–2331.
[59] D.K. Shvero, M.P. Davidi, E.I. Weiss, N. Srerer, N. Beyth, Antibacterial effect of polyethyleneimine nanoparticles incorporated in provisional cements against Streptococcus mutans, J. Biomed. Mater. Res. Part B Appl. Biomater. 94 (2) (2010) 367–371.
[60] N. Kawabata, M. Nishiguchi, Antibacterial activity of soluble pyridinium-type polymers, Appl. Environ. Microbiol. 54 (10) (1988) 2532–2535.
[61] E.P. Briggs, A.R. Walpole, P.R. Wilshaw, M. Karlsson, E. Pålsgård, Formation of highly adherent nano-porous alumina on Ti-based substrates: a novel bone implant coating, J. Mater. Sci.: Mater. Med. 15 (9) (2004) 1021–1029.

[62] M. Wei, A.J. Ruys, M.V. Swain, S.H. Kim, B.K. Milthorpe, C.C. Sorrell, Interfacial bond strength of electrophoretically deposited hydroxyapatite coatings on metals, J. Mater. Sci.: Mater. Med. 10 (7) (1999) 401–409.
[63] K. Subramani, W. Ahmed, Emerging Nanotechnologies in Dentistry: Processes, Materials and Applications (Micro and Nano Technologies), 2nd ed., Elsevier, 2017.
[64] E.T. Thostenson, Z. Ren, T.W. Chou, Advances in the science and technology of carbon nanotubes and their composites: a review, Compos. Sci. Technol. 61 (13) (2001) 1899–1912.
[65] A. Peigney, C. Laurent, E. Flahaut, A. Rousset, Carbon nanotubes in novel ceramic matrix nanocomposites, Ceram. Int. 26 (6) (2000) 677–683.

CHAPTER 7

Bioresorbable nanoceramics: novel and efficient drug delivery vehicles

Shikha Gulati[1], Sanjay Kumar[1], Nandini Sharma[2], Kartika Goyal[1] and Tanvi Agarwal[1]
[1]Department of Chemistry, Sri Venkateswara College, University of Delhi, Delhi, India
[2]Department of Biological Science, Sri Venkateswara College, University of Delhi, Delhi, India

7.1 Introduction

"Nanostructured materials" refers to delicate structures having sizes within a range of 1−100 nm. As a result of this size and their versatile surface properties, extensive developments in nanotechnology have taken place over the last decade in the field of materials science, engineering, and targeted drug delivery. Particularly, research has been emphasized toward a special material known as nanoceramics, which are mostly inorganic systems with porosity as a characteristic feature. The history of ceramics goes back thousands of years to when it was used in the form of ceramic pottery using clay and fire [1]. With the ever-changing demand of industries and consumers, another revolution occurred by reducing the size to the nanoscale, which enabled them to be applied into various biosystems. For example, layered double hydroxides (LDHs) have been studied recently for the effective delivery of drugs or genes into biological systems. This is attributed to its controlled-release properties under different pH conditions. Nanoscale ceramics such as alumina (Al_2O_3), hydroxyapatite (HA), and calcium carbonate ($CaCO_3$) were made from new synthetic techniques to improve their properties with reduced cytotoxicity in biological systems. These nanoparticles have properties in-between those of metals and nonmetals. More significantly, nanobioceramics are also effective as body interactive materials, helping the body to heal or supporting the regeneration of tissues [2].

Fig. 7.1 classifies the bioceramic materials according to their bioactivity into the following categories:

Bioinert—These have enough fracture toughness, for example, partially stabilized zirconia and alumina.

Bioresorbable—These are biodegradable nanoceramic materials that solubilize and finally get incorporated in tissue, for example, calcium sulfate and coral calcium carbonate.

Bioactive—They usually have low fracture toughness and support fixation of implants in the skeletal system, for example, bioglass and glass ceramic.

Industrial Applications of Nanoceramics
DOI: https://doi.org/10.1016/B978-0-323-88654-3.00008-1

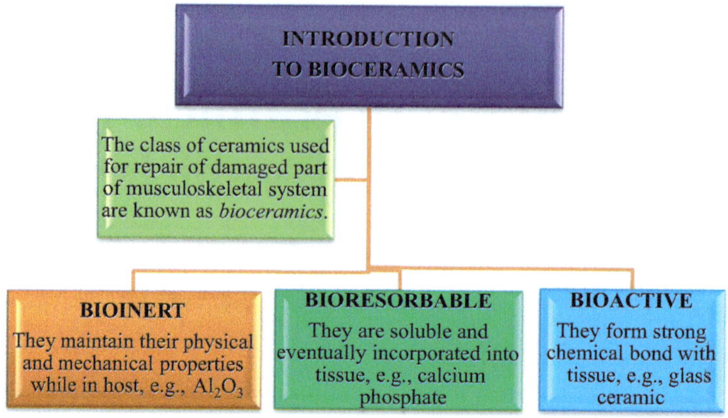

Figure 7.1 Significant properties of bioceramics.

When these materials are incorporated in the human body, tissues react in a variety of ways [3]. The utilization of these ceramic nanoparticles with entrapped biomolecules has applications in various frontiers of modern material science, especially as drug delivery vehicles. Also, these advancements in science and technology have opened the doors to a possibility of employing implants in the human body. Over other bioceramic materials, resorbable bioceramics have an edge because of the fact that they degrade in a set time frame followed by replacement of natural host tissue. Advancements in the development of bioresorbable implants for fields like traumatology and sports injuries are continuously evolving and moving forward. For example, polymers are used as bioresorbable materials for clinical applications like bone implants, and as drug delivery vehicles [4,5]. The term ceramics includes electrically conducting, insulating, and semiconducting materials, such as chromium oxide, aluminum oxide and silicon carbide. This chapter will focus on these bioresorbable materials and their application in drug delivery in comparison with conventional strategies.

7.2 Drug delivery systems

Drug delivery vehicles are usually inert systems used as solvents in which a medically active agent is administered. In principle, it shows reduced toxicity for multiple medical agents. Also, they accurately help in targeted delivery to tissue or cell of choice. Traditional strategies have the possible disadvantage of not being sufficiently effective due to the high concentration of drugs. For example, oral administration commonly requires very high effective dosages to reach the required drug concentrations in the poorly irrigated bone tissues. Some additional examples since late 1950s are:

1. Transdermal and transmucosal controlled-release delivery systems
2. ml6 nasal and buccal aerosol sprays

3. Drug-impregnated lozenges
4. Encapsulated cells
5. Oral soft gels
6. iontophoretic devices to administer drugs through skin, and other programmable, implanted drug delivery devices.

These conventional drug administration methods, while widely utilized, have many problems that may be potentially overcome by new emerging technologies. Equally important, these advances may appear attractive as compared to the costs of new drug development. Humankind has depended throughout the ages on nature for improving human health but the ever-increasing population and their demands are not fulfilled due to the limited availability of resources. This implies the persistent need for the involvement of synthetic compounds so as to cure diseases plaguing life on earth. The discovery of nanoparticles is likely to have a significant impact on the drug delivery sector due to the fascinating physiochemical properties attributed to their small size [6]. The primary goals accomplished by nanotechnology in drug delivery include:

Targeted drug delivery

Biocompatibility

Reduced toxicity

Development of safer medicines at fast pace

These goals have envisioned the use of nanotechnology as the plausible future of novel drug delivery technology due to their potential to become useful therapeutic and diagnostic tools in upcoming times. To date, numerous types of delivery systems have been produced and successfully employed in the treatment of diseases worldwide. Fig. 7.2 provides insight into a few of the drug delivery vehicles primarily based on nanotechnology [5].

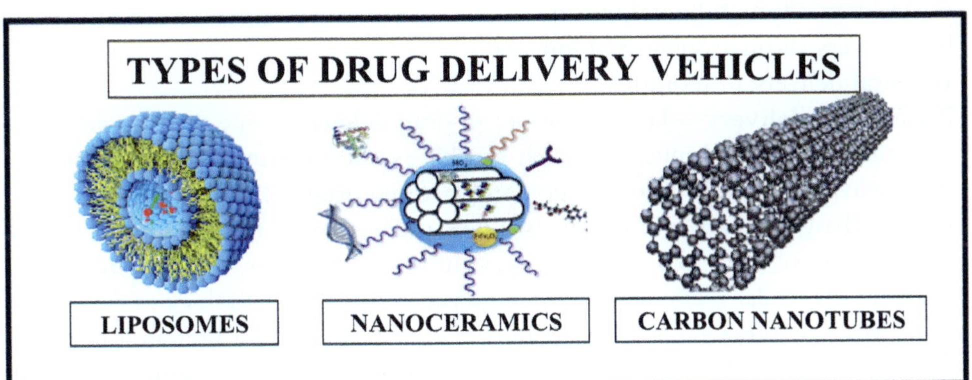

Figure 7.2 Common drug delivery vehicles.

There are not only these, but various more carriers like quantum dots, microspheres, lipoproteins, and polymeric micelles. Examples of the importance for involving nanotechnology in drug delivery include:

Various liposome formulations loaded with anticancer agents have been reported to exhibit lower cytotoxicity as compared to free drugs [7].

Bone repair was observed at a faster rate for a critical-sized calvarial vault defect in rabbits after the implantation of apatitic cement loaded with bioceramic material, bone morphogenetic protein-2 [5].

MCF-7, breast cancer cells evaluated for easy and increased absorption of doxorubicin incorporated certain polymeric nanoparticles [8].

7.3 Various pathways of drug delivery

There are several pathways through which drugs can be delivered to the body system. This ensures the efficacy of the drugs being administered. The development of diverse drug delivery pathways ensures the possibility of better control of pharmacology and biochemistry of the body system, as follows:

Pharmacokinetics—the movement of therapeutic agents into, through and out of the body.

Pharmacodynamics—the study of effects of drugs and their mechanisms.

Toxicity—the harmful effects of drugs in a pervasive way.

Immunogenicity—the effect of drug as therapeutic antigen on provoking an immune response in the body.

Some pathways of drug delivery are briefly discussed below:

1. **Buccal drug delivery**—Buccal controlled drug delivery system (DDS) has been developed because the surroundings of the oral cavity provide potential sites for drug delivery. Here, release of drug can be affected by continuous secretion of saliva [9]. Buccal control drug delivery can be achieved in three ways-
   a. Delivery through buccal mucosa.
   b. Delivery through sublingual mucosa.
   c. Local delivery to mouth.
2. **Nasal drug delivery**—The nasal route of drug delivery is one that provides access to highly vascularized mucosa, which can be exploited as interesting sites for local drug delivery, systemic drug delivery, and targeted drug delivery [10].
3. **Ocular drug delivery**—In the ocular DDS, there is rapid and extensive elimination of conventional eye drops from the eye. Currently, scientists are facing many challenges in the ocular DDS due to the unique anatomy and physiology of eye [11].
4. **Oral drug delivery**—Oral drug administration is one of the preferred routes and widely used formulation for new existing and new drugs. It might be due to its ease of delivery [12].

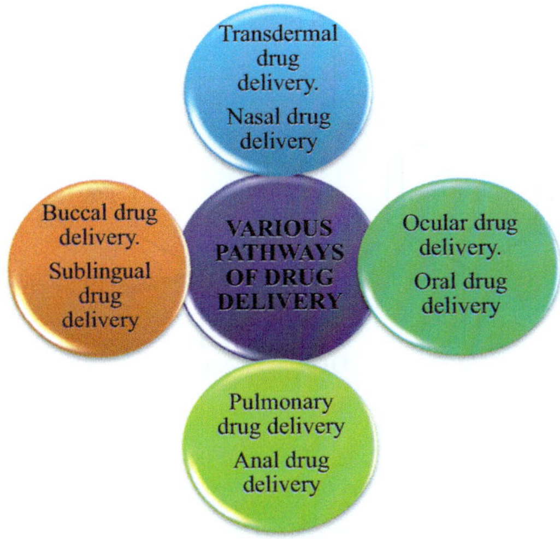

Figure 7.3 Schematic for various pathways of drug delivery.

Apart from these pathways there are others like pulmonary drug delivery, sublingual drug delivery, transdermal drug delivery, and anal drug delivery. A schematic for various drug delivery pathways is shown in Fig. 7.3 [13].

7.4 Targeted drug delivery systems

Nanotechnology involves the utilization of nanometer-length particles for better healthcare and improved pharmaceutical applications. Over other methods, nanotechnology offers an edge in that it helps targeted drug delivery to the site of disease. This can be achieved either through passive targeting or by active targeting of drugs to the disease site:

1. **Passive targeting**—Passive targeting, also known as physical targeting, is based on the fact that it inhibits the excretion of drug from human body in such a way that it remains circulating in the bloodstream permitting its accurate action on the desired receptor. Diseased tissues often show enhanced permeability and retention effect, this further make accumulation of nanoparticles more in diseased tissue as compared to normal tissues. For example, direct drug injection involves the principle of passive targeting. Doxil, a sterically stabilized liposome encapsulating doxorubicin, acts as a good drug retention carrier in a liposomal formulation with enhanced circulation time compared to free doxorubicin [14]. Fig. 7.4 shows the chemical structure of a widely used chemotherapeutic drug known as doxorubicin [15].

Another strategy can be designed by direct intratumor delivery of anticancer agents using nanoparticles. It is reported in recent literature that transferrin-conjugated paclitaxel-loaded biodegradable nanoparticles have shown enhanced antiproliferative activity.

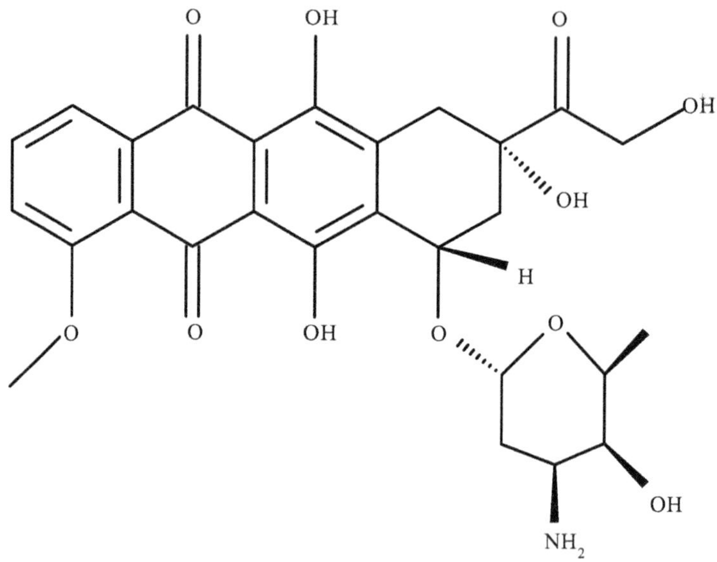

Figure 7.4 Chemical structure of doxorubicin drug that can be encapsulated into nanoparticles.

Thus, these nanoparticles serve not only as effective drug delivery vehicles but also help to protect them from undesirable environmental conditions and controlled release at their target sites. These biodegradable nanoparticles have potential and are promising tools for intracellular delivery [16].

2. **Active targeting**—An active targeted DDS is made by coupling moieties like antibodies and peptides with DDSs acting as a carrier to receptor structures expressed at target sites. Most active targeting DDSs take antibodies as a targeting moiety and polymer or lipid as a carrier or active chemotherapeutic drug. For successful and efficient drug delivery targeting the moiety must have high affinity to cell surface receptors [14,15].

   Certain examples of active targeting are stated below-

   PEGylated gold nanoparticles are incorporated with different concentration of human transferrin for increased active targeting [17].

   In another study, cancer cells were actively targeted actively using polylactic-*co*-glycolic acid (PLGA) nanoparticles [18].

   Aptamer-coated paclitaxel–polylactide nanoconjugates also showed enhanced active targeting.

Table 7.1 summarizes ligands showing potential for active and passive targeting. For example, polysaccharides like hyaluronic acid show active targeting. In the following sections, the conventional DDSs will be discussed, and newer methods and their merits will be focused upon and described in detail.

Table 7.1 Ligands for active and passive targeting of nanoparticle drug delivery system.

Type	Ligands	Active/passive targeting	References
Proteins	Transferrin	Active	[19]
Polysaccharides	Hyaluronic acid	Active	[17]
Biodegradable nanoparticle	Conjugated transferrin	Passive	[16]
Lipid	Liposome	Passive	[14]

7.5 Conventional strategies of drug delivery and their limitations

Drugs are highly specialized chemical substances that possess the ability to correct or mitigate the consequences of any disease or discomfort to the body while at the same time they can prevent and cure such diseases as well. According to their purpose and mode of action they can be divided into a variety of categories like analgesics, antibiotics, muscle relaxants, chemotherapeutics, etc. [20]. There are numerous requirements and purposes of drugs that are kept in mind during their administration into the body, like time of persistence and the concentrations that have to be maintained during their course of action, mode of delivery, targeted or slow delivery based on the type of affliction suffered by the patient, etc. The process of developing new drug molecules for various diseases and testing them pharmacologically is a very time-consuming and tedious process, requiring a lot of capital investment. To improve the already existing drugs by changing the safety efficacy ratio has also been attempted by many techniques like personalizing or individualizing drug administration, therapeutic monitoring of drugs, and even dose titrations [21]. The most common method of drug delivery and the most commonly used route of delivering drugs is the "oral route," because of its convenience, safety, ease of administration, cost-effectiveness, and patient acceptance. The drugs used here follow the conventional DDS of "immediate release and rapid absorption" type that have usually one peak in their plasma concentration graph. This type of a conventional method is used in many existing technologies of DDSs like swallowing, inhalation, absorption through skin, or by intravenous injections [22]. These conventional DDSs have many wide-ranging advantages, but it should be understood that nowadays, their limitations cannot be ignored. Their disadvantages are as follows:

1. The drugs utilized in such DDS have a very short half-life as they operate on immediate release activity, which decreases their efficiency.
2. Such conventional DDS require frequent and multiple administrations of drugs, thus, increasing the chances of toxicity and accumulation of harmful substances inside the body.

3. In multiple applications or administrations of drug in such systems, there is an increased risk of missing turns and creating matters worse.
4. This also results in poor patient compliance and not a very patient-friendly mechanism of drug delivery.
5. Since the drugs used here have a typical peak valley in their plasma concentration, it gets difficult to attain the drug's steady-state condition.
6. It has also been observed that in the conventional DDS, unavoidable fluctuations in the concentration of drugs inside the patient are present which are highly harmful.
7. It has also been seen that the drugs used have a low/narrow therapeutic index, which is a form of quantitative measurement used to determine the relative safety of the therapeutic agent.
8. The pattern of either overmedication or undermedication has also been perceived in this type of DDS.
9. The concentration of steady state of the therapeutic agents used here is known to fall or increase beyond the therapeutic range, making them ineffective and inefficient.
10. There is also the lack of a mechanism which maintains the dose administered or controls the release rate of drugs to make them more efficient and potent.
11. Since there are no regulations for site targeting of drugs, there are several cases of adverse effects of the drugs in other parts of the human body
12. These limitations are also represented in Fig. 7.5 [23].

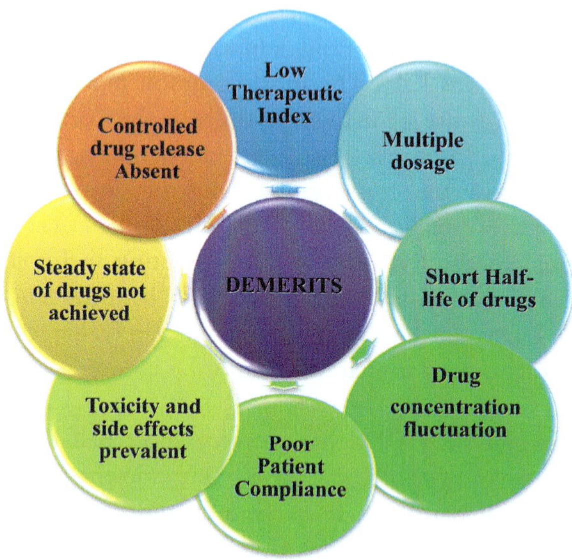

Figure 7.5 Limitations of conventional drug delivery systems.

To overcome the vast disadvantages of conventional methods of drug delivery, many newer and efficacious techniques are being explored, like microneedles, targeted drug delivery, or controlled-release technology involving organ-specific and even intracellular-targeted delivery of drugs in the form of sprays, soft gels, implantables, etc. The research on DDSs comprises four main arenas: delivery vehicles, delivery routes, targeting strategies, and cargoes. It has been observed that to revolutionize the existing DDS, the most effective pathway is to alter the delivery vehicle being used and the latest research has proven that the application of nanotechnology shows promising results. In the following sections, the advantages and type of nanoscale materials that show most potential as drug delivery vehicles are discussed.

7.6 Need of nanoformulations in drug delivery systems

Nanotechnology is a fast-emerging field of science with numerous and wide-ranging applications in almost all walks of life. Its utilization in medicine is ground-breaking and so is the ongoing research of this field. One of the most promising aspects of nanotechnology can be seen in DDSs where it can be employed in vehicular drug delivery for a myriad of diseases. Nanoformulations are being probed to be used as potent vehicles for drug delivery because of their small size ranging between 10 and 1000 nm. The therapeutic agents (drugs) can:

1. be adsorbed on their outer surface, or
2. be chemically attached with the help of temporary bonds, or
3. be bound to them in the form of a solid solution or dispersion.

The nanoformulations, only over a very short period of time have proven to be highly successful in drug delivery, including in vivo delivery. For example, in the brain, the first drug to be administered using nanoparticles was hexapeptide dalargin, which has opioid activity [24]. To fully understand the working of nanotechnology, it is vital to understand what exactly its basic unit, that is, the nanoparticle, is and how they are so efficient as vehicles for drug administration.

7.6.1 Nanoparticles: what are they?

Discovered more or less 50 years ago by Richard Feynman ("The Father of Nanotechnology"), nanoparticles are described to be submicroscopic, ultrafine units which are measured in dimensions of nanometers (nm; $1\ nm = 10^{-9}\ m$). The definition of a "nanoparticle" was initially given in 2008 by the International Organization for Standardization (ISO) as an individual nanoobject where all three Cartesian dimensions are less than 100 nm. After this, in 2011 the Commission of the European Union approved a more technical as well as all-encompassing definition: "a natural, incidental, or manufactured material containing particles, in an unbound state or as an aggregate or as an agglomerate and where for 50% or more of the particles in the number size distribution, one or more external dimensions are in the size range 1 nm−100 nm" [25]. Under this distinction a nanoscale object requires only one of its dimensions to be in the range 1−100 nm so as to classify as a nanoparticle, even if its other dimensions are outside the specified range.

Nanoparticles can be categorized into multiple types on a variety of bases, like size, shape, material properties, etc. Some classifications differentiate between inorganic and organic nanoparticles; the former includes quantum dots, metallic nanoparticles (gold, silver, silicon, etc.), and metallic oxide nanoparticles, and the latter accounts for nanoparticles made of polymers like ceramics, lipids, chitosan, dendrimers, and PLGA. Nanoparticles can show a variety of effects and unique properties in different compositions due to their three main attributes:

1. Possess high mobility in free state.
2. Have exceptionally enormous specific areas.
3. Known to exhibit unique quantum effects in different compositions and conditions.

The thermal, mechanical, and catalytic properties can be altered by increasing or decreasing surface area/volume ratio and the new properties generated allow applications of nanomaterials in many biological fields, like drug delivery and novel drug development, diagnostic tests and biosensors, etc. The nanoparticles of chemically manufactured structures like polysaccharides, polymers, metals, incorporated with active drugs can effectively fight against human pathogens like bacteria and even viruses and can be used to treat various pathological conditions, making them highly efficacious as delivery vehicles for drug administration.

There are some requirements which need to be kept in mind while using nanoparticles as delivery vehicles for therapeutic agents. These mainly include the understanding of drug release and incorporation, targeting of specific place of action, biocompatibility, stability of formulation and its shelf life, biodistribution, and possible adverse reactions. Nanoformulations have become a prime choice as drug delivery vehicles because their miniature dimensions help them to penetrate through cellular apparatus like membranes and also stabilize the protein structure of drugs while at the same time helping in lysosomal escape when endocytosis occurs.

7.7 Role of nanoformulations in drug delivery

With the recent advancements of nanotechnology in drug delivery, a myriad of nanocarriers with different properties, mode of formulations, sources, etc. have been identified that not only help in delivering the therapeutic agents but also in their stabilization, persistence in the target organ, and enhancement of their actions. As discussed before, some major types of nanocarriers are ceramic, liposomal, polymeric, dendrimers, solid lipid, nanocrystals, nanotubes, etc. Their detailed description can be found in Table 7.2.

Table 7.2 Major nanocarriers used as vehicles in drug delivery.

S. no.	Type of nanocarrier	Characteristics	Material used	Applications	References
1.	Ceramic nanoparticles	Mostly consist of oxides, carbides or nitrides; chemically inert; heat resistant; high mechanical strength, load capacity, stability	Silica, alumina, calcium phosphates, zirconium and titanium dioxide	Drug delivery, tissue engineering, drug carrier for photodynamic therapy	[26]
2.	Liposomal nanoparticles	Concentrically bilayered vesicles with monolayer of lipids surrounded by membrane of phospholipids; good biocompatibility; high shelf life	Lecithin	Metastatic breast cancer, protozoal and fungal infections, malignant lymphomatous meningitis	[27]
3.	Polymeric nanoparticles	Macromolecules with numerous repeating units in chain-like fashion; biodegradable; biocompatible	Chitosan, gelatins, PLGA, PGA, etc.	Drug delivery, gene therapy in breast cancer and multiple sclerosis	[28]
4.	Dendrimers	Nanosized 3D polymeric structures with tree like branching and spherical geometry in solution	Branched polymers like PAMAM etc.	Delivery of genetic transfection agents, diagnostics like CT scan, delivery of chemotherapeutic drugs	[29]
5.	Solid lipid nanoparticles	Spherical structures with solid lipid core stabilized by surfactant; diameter between 10 and 1000 nm	Fatty acids, waxes, surfactants, acylglycerols, sphingomyelins, bile salts, sterols	Oral and parenteral drug delivery, clinical medicine and research	[30]
6.	Nanocrystals	Aggregates of nanoparticles combined into a crystalline form of drug with help of surfactant	Dispersion media—aqueous (water, ethanol, etc.) and nondispersion media (PEG or oil)	Material research, chemical engineering, bio-imaging, immunosuppressant	[31,32]
7.	Nanotubes	Organic/inorganic composition, self-assembling sheets of atoms of mostly carbon in tubes	Soluble fullerene derivatives like C_{60}	Biopersistent fibers	[33]

PAMAM, Polyamidoamine; *PLGA*, polylactic-*co*-glycolic acid; *PGA*, Polyglycolic acid; *PEG*, Polyethylene glycol; *CT*, Computed tomography.

These nanoparticles that act as drug carriers are the preferred choice and more fruitful than the conventional methods because of the following advantageous properties:

1. Highly penetrable and can easily penetrate inside cells because of their small size.
2. Large payload capacity because of a high surface to volume ratio resulting in exceptionally high surface area for adsorption of large amounts of drugs.
3. Affordable mass production and high purity of obtained product.
4. Proper control over particle shape, size, porosity, morphology, etc. making them extremely targetable and precise in action.
5. Enhance cellular uptake by supporting fenestrations in the cells.

All these attributes of nanoscale particles, which make them a superior choice, are summarized in Fig. 7.6.

Because of these advantages and the abovementioned properties nanoparticles are a promising novel future alternative for drug delivery vehicles in enhanced and controlled DDS. Recent trends have shown that ceramic nanoparticles have tremendous scope and applications in this particular field and the following sections are focused on the importance and scope of nanoceramics as well as bioresorbable nanoceramics in drug delivery.

7.8 Nanophase ceramics: emerging choice for drug delivery

Numerous nanocarriers have shown their mettle and proven to be efficient and preferable solutions to overcome the limitations of conventional DDSs. These nanoscale particles include the likes of nanotubes, dendrimers, polymeric nanoparticles, etc. [34].

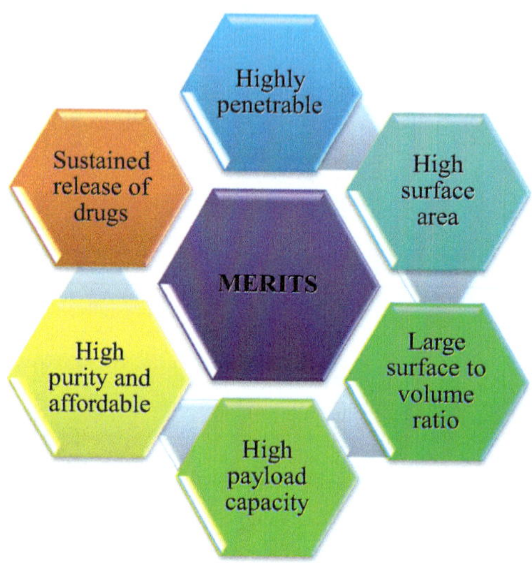

Figure 7.6 Merits of nanoparticles as drug delivery vehicles.

Out of these examples, a newer and better class of nanoparticles has evolved, known as ceramic nanoparticles. Ceramics are nanoparticles which are extremely resistant to environmental changes and were first discovered during the 1980s. They can be described as particles in the nanoscale dimensions with a definitive solid core, achieved by application of both heat and pressure to make them heat and pressure resistant. Ceramics comprise both metals and nonmetals, in such a way that the combinations are of metal and nonmetallic elemental solid or a nonmetal, and so on [35]. The solid core of these nanoceramics is predominantly made of either alumina or silica, but they are also known to exist with cores of metals and their oxides and sulfides as well. The different cores decide a number of properties and produce various nanosystems with different shape, size, porosity, etc. Nowadays ceramics are mostly produced from silica, alumina, calcium phosphates, iron oxides, zirconium, titanium dioxide, and carbonates for a variety of medical applications like in dentistry as filling material for endodontic filling and crowns, in orthopedics for administration on joints and vertebral column, in plastic surgery to treat cranial defects and what not. The ceramics are usually made from inorganic origins as they are designed to dodge the reticuloendothelial system by making alterations in their size and surface composition. They are also porous in nature which protects them physically from degranulation, degradation or decomposition. Some characteristic features of nanoceramics are prominent mechanical strength, low or nonexistent biodegradable nature, suitable and good body response [36]. All the major characteristic features are summarized and illustrated in Fig. 7.7.

Out of the huge class of nanoceramics, there are many types of nanophase ceramics, which are classified on the basis of differences in their architectural design:

1. **Ceramic nanoparticles**—Usually made of inorganic substances like alumina, titania, and silica, these ceramic nanoparticles provide full protection to encapsulated molecules like enzymes, proteins, and most importantly drugs, preventing denaturing because of unsuitable pH and temperature changes. They are known to be quite efficient and sensitive to the drugs incorporated for targeted delivery.

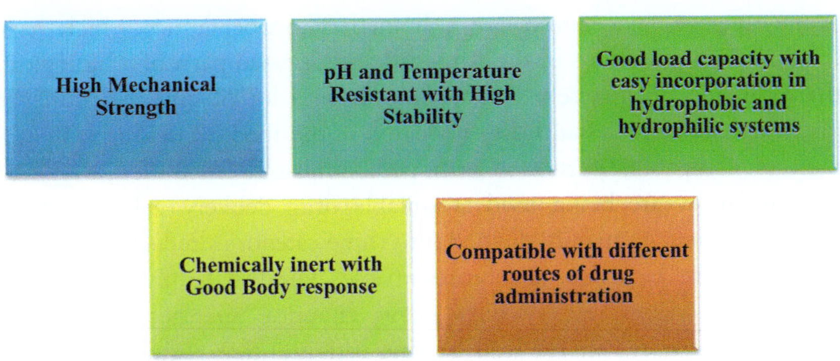

Figure 7.7 The characteristic features of nanoceramics which are useful in drug delivery.

2. **Ceramic nanoscaffolds**—A scaffold is a substance that provides mechanical support to growing cells and also enables interactions between cells and extracellular matrices; when present in nanoscale dimensions it is called a nanoscaffold. The ceramic nanoscaffolds have myriad properties like exceptionally high porosity, increased structural stability, large surface area, and a long degradation time span. These properties help them to become highly suitable for storage and controlled drug delivery, like in the case of targeted treatment of prostate cancer [37].
3. **Nanoclay**—Nanoclay architecture is the newest addition to nanoceramics and involves the presence of thin layers with each layer being a few nanometers thick. These clay nanoceramics have some unique properties like high internal surface area, large adsorption capacity, low/negligible toxicity, interlayer interactions, great cation exchange ability, and chemical inertness. Some examples of widely used ceramic nanoclays are zeolites, titanium dioxide, porous silica, etc. It has also been reported that controlled delivery of encapsulated drugs in nanostructures allows high concentrations of therapeutic agents to be released in the target organ over a specific time period [38].

The main thing to be kept in mind when selecting a suitable drug delivery vehicle is to focus on not only organ/cell targeting or adsorption of drugs or biocompatibility of vehicle and agent, but to also analyze the fate of the nanoscale substance being used when they are inside the cells. In the intracellular apparatus, the foreign substances are engulfed by lysosomes and then finally degraded. So, for the therapeutic agent to show its effect, it is vital that its release happens before degradation and this is achieved by surface modifications of nanoceramics that help in drug release, targeting, cellular uptake and binding, and even transport. Recent studies and research have shown that a certain type of nanoceramics provides all the suitable and required conditions for proper and efficient drug delivery, and these are called bioceramics.

7.9 Bioceramics: prime choice in drug delivery systems

As the name suggests, bioceramics are a branch of nanoceramics which have useful remedial and repair applications in the damaged and diseased parts of a biological system, mainly the musculoskeletal system. Bioceramics have a lot of superior properties compared to normal nanoceramics and they appeared in the medical field mainly as an effort to overcome biocompatibility issues in orthopedic implants. Since then, they have come a long way and now are being used as suitable drug delivery vehicles. As discussed previously, bioceramics are mainly divided into three categories reflecting their biological performance: bioinert, bioactive, and bioresorbable. The first family comprises zirconia, alumina, pyrolytic carbon, etc., which show chemical inertness in the biological environment and are excellent for load-bearing activities that require good mechanical attributes like strength, wear resistance, high fracture resilience, etc. [5]. The second family consists of mainly glass nanoceramics of various types like silicate, borate, etc. These have unique "ion-release" properties, which help in processes like osteogenesis and angiogenesis. Usually bioactive nanoceramics are widely known for bone repair and regeneration but their utility is also extended toward skin repair and wound healing [39]. The third and final family is called the bioresorbable nanoceramics and comprise calcium phosphate (CaP) substances, although they are not just limited to HA, tricalcium phosphate, etc. These numerous CaP compounds are highly biocompatible and bioresorbable (which means readily absorbed in the biosystem) because of their degree of similarity with the chemical composition of tissues in the body. Most of the time they are used in combination with polymers of either synthetic or natural origin [40].

The bioresorbable nanoceramics also have the unique ability to behave as "biomimetic systems" which are interfaces that are designed in such a way that they resemble the biochemical and biophysical properties as well as they mechanical aspects of a specific tissue. These bioresorbable biomimetic systems are intended to showcase temporary cell attachment, differentiation, proliferation, and matrix deposition in the target organ for therapeutic action. These systems are also incorporated with active therapeutic molecules or drugs to increase efficiency and attain maximum results. Such systems of drug delivery are known to overcome majority of the limitations faced by conventional DDS and provide additional benefits as they aim to promote biological repair, decrease postsurgical complications, help in tissue regeneration, and many more meritorious applications. Because of mimicking the multidimensional structure of tissues, these systems have enhanced performance by *in situ* delivery of therapeutic agents, which decreases side effects and provides long-term relief to the patient because of the targeted application. They also enable the administration of poor water-soluble drugs without any additives [5]. These DDSs involving bioresorbable nanoceramics open pathways for the dual advantages of coupling drug release and ceramic resorption, which can pave the way for smart release profiles that are suitable and customized to a particular tissue [41]. The advantages of bioresorbable nanoceramics and their superior properties are discussed in detail in the following sections.

7.10 Significant properties of bioceramics over other nanoformulations

Nanoscaled materials display distinguishable physical, chemical, electrical, optical, and mechanical properties when compared to macroscopic and other formulations of drug delivery carriers. In this interdisciplinary science, a bit of size variation can make various differences in particle properties within the nanoscale range, for example, nanoparticles with diameter range between 10 and 70 nm differ from the particles ranging with diameter range of 70–200 nm. Such variations provide bioceramic material the significant properties with some added advantages over other nanoformulations.

Even, within the bioceramic materials, different categories are there differentiated on the basis of specific properties and synthesis methods. For example:

1. **Silica-based mesoporous bionanoceramics**—Silica-based mesoporous materials are ordered porous structures with narrow pore-size distributions. They have good stability because of the presence of thick walls. They have large surface area and pore volumes that enable the drug to be incorporated in them easily. The exceptional and outstanding properties of silica-based mesoporous bioceramics make them best suited as drug delivery carriers for different biomolecules. In brief, these properties are:
   - **a.** Stable mesoporous structure.
   - **b.** High surface area.
   - **c.** Large pore volume.
   - **d.** Regular and tunable pore size (2–50 nm).
   - **e.** Open texture.
   - **f.** Can undergo chemical functionalization of pore walls with organic groups.

   The ordered pore network permits a controlled molecular load and drug release kinetics. Their high pore volume allows hosting of a large amount of determined biologically active biomolecules. The high surface area provides a good surface for the molecules to get adsorbed on these bioceramic materials, this too allows a large number of molecules to be adsorbed in one go [42]. Fig. 7.8 shows the structural morphology of mobile composition of matter (MCM-41) that is part of silica-based mesoporous bioceramics. Another representation includes that of layered doubled hydroxides where the blue portion represents the hydroxide layer with an interlayer of anions (red) and water molecules (green) in between [43].
2. LDHs—These are another type of bioceramic, possessing bioresorbable properties that could be reduced to nanosize, and thus become nanoceramic materials serving as drug delivery carriers. LDHs are also known as hydrotalcite-like materials or anion

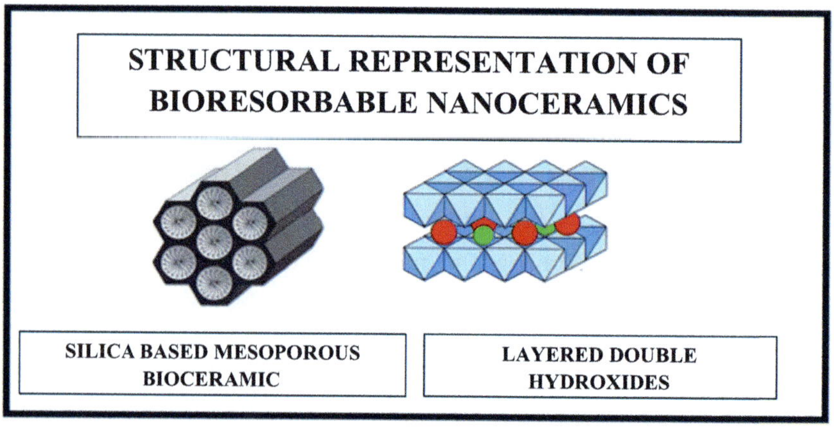

Figure 7.8 Structure of some important bioresorbable nanoceramics.

exchanging clays. They have the property of anion exchange. The interlayer gallery between the hydroxide layers can have direct loading of anionic drugs and biomolecules due to this anion-exchange property. The intercalation of anionic molecules reduces positive charges on these materials and this facilitates cellular uptake [44].

The hydroxide layers of LDH are hydrophilic in nature, which allows anionic drugs to be incorporated with them easily. The more comprehensive approach was taken for the properties of cell viability and proliferation in LDH using independent assays. The results showed that the $Mg_2Al\ (OH)_6NO_3$ causes 50% growth inhibition of HEK293T cells at 0.125 mg/mL concentrations, and there is no negative effect on cell viability. After being exposed for 3 days, it remains around 94%, whereas the viability efficiency of lipofectamine reduced to 70% [44]. Fig. 7.9 shows some important industrial application of LDHs and their derivatives that further highlights their importance as bioresorbable nanoceramic [45].

7.11 Synthesis of bioresorbable nanoceramics for drug delivery

Nanotechnology has introduced innovative approaches for drug delivery. The different materials are being tested to estimate the efficacy of drug delivery. Ceramics reduced to nanosize have been considered as one of the greatest findings in the field of drug delivery. Ceramic particles in nanoscale dimensions, like LDHs, have been developed to preserve and deliver drugs or genes to biological cells.

The synthesis of these novel materials should be done keeping in mind the goals of drug delivery, which are: maximizing the therapeutic effects, lowering the toxic activities, site-specific targeting, controlled drug release, and overcoming the limitations of biomacromolecular therapeutics [45]. The syntheses with distinctive methods of different bioceramic materials are as follows:

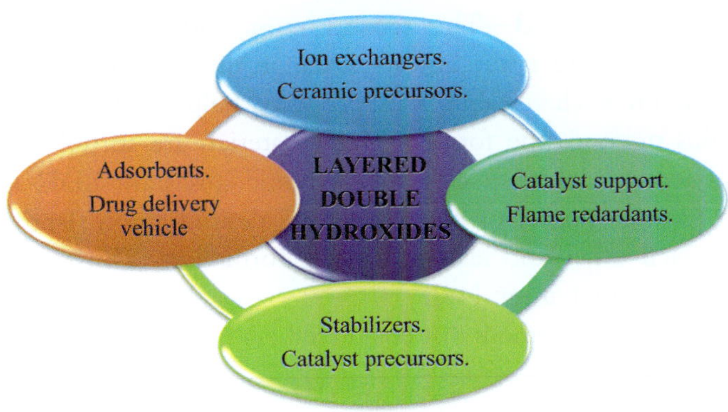

Figure 7.9 Some industrial applications of layered double hydroxides (LDHs).

7.11.1 Synthesis of bioresorbable layered double hydroxide nanoceramics

LDHs have emerged as nonviral delivery vectors. LDHs are two-dimensional nanoceramic materials of cationic layers incorporating anions. The positively charged layers are made up of magnesium and aluminum hydroxides, whereas the interlayer spaces are filled with negative species, such as nitrate, carbonate, etc.

The synthesis of bio-LDH nanohybrids is done through intercalation; the different kinds of inorganic or organic anions are introduced between the hydroxide layers by the process of coprecipitation or can be done using simple ion-exchange reactions. Bio-LDH nanohybrids can be formed by intercalation of functional biomolecules, such as nucleoside monophosphates, ATP, DNA, antisense oligonucleotides, etc. X-ray diffraction was done for pristine LDH and bio-LDH nanohybrid. The substitution of nitrate (NO_3^-) ions with anionic biomolecules results in expansion of double hydroxide layers due to the intercalation of bulky anionic biomolecules. LDH materials are available in nature, for example, the precipitation of saline water resources or basalt weathering cause the formation of LDH materials [45]. They can be synthesized in the laboratory and the two most popular and preferred methods are:

Coprecipitation method—This is widely used for the synthesis of LDH. This is a direct synthesis method that necessitates the nucleation. The mixture of an aqueous solution incorporating the salts of two metal ions with a base solution in the presence of the preferable anion is used for the growth of metal hydroxide layers. This method of coprecipitation varies in terms of titration at constant or different pH and buffered precipitation. The simplest buffer used here has surplus amount of divalent metal ions that are found during titration at varied Ph. This method is considered as one of the advantageous methods because precipitation will occur at a lower pH than in the absence of excess divalent captions. This reduces the risk of amalgamation of nonrequired hydroxide anions and low uptake of carbon dioxide by the reaction mixture. If there is good uptake of CO_2 molecules, it gives rise to carbonate that have has ability to be strongly held inside the LDH structure.

Anion-exchange method—The second most widely used method for LDH nanohybrids manufacture is the anion-exchange method. The intercalating spaces in LDH are anion exchangeable, this property helps in the formation of LDH hybrids. In this method, the formed LDH materials are stirred in a solution that contains replacement anion species. The anionic surfactant LDH is handled with the salt of a cationic surfactant containing the desired anion. Both the surfactants interact with each other and lead to micelle formation. Thus, it leaves the LDH free to arrest the desired anion [44].

There are many more methods that could synthesis LDH materials, they are as follows:

1. Sol–gel techniques
2. Homogeneous preparation

3. preparation via oxidation
4. Preparation through precipitation via aluminate
5. Preparation from metal oxides/hydroxides

After the manufacturing of LDH bioresorbable nanoceramic materials, some post-preparative treatments are given:

1. Hydrothermal treatment
2. Microwave treatment
3. Heating at gentle reflux

These treatments enhance their crystalline structure and maintain uniformity. The more precise molecular structure of interlayer DNA can be elucidated using circular dichroism (CD) spectra. This technique has the ability to probe the changes of secondary structure of DNA because of the high sensitivity of the CD spectrum toward variations in conformations of nucleic acids. The interlayer spacing in DNA–LDA hybrid was observed to be around 19 angstrom and are stable with the thickness of double helical conformational DNA molecule. In the DNA–LDH hybrid, the negative charge of DNA is compensated by positively charged LDH surface and are electrostatically interacted. This results in a more tight, precise, and narrowly wound secondary structure of DNA.

7.11.2 Synthesis of mesoporous silica particles

The synthesis of novel mesostructured bioceramic materials was first time reported by Japanese researchers in the early 1990s. They have the characteristic property of regular arrays of uniform channels, thus their dimensions can be tailored using surfactants, additives, and chemicals. These materials are synthesized using the liquid crystal templating mechanism. The liquid crystals are formed in mixture of polar solvents with a nonpolar tail group. These crystals are formed following some steps:

1. More and more amount of surfactant molecules is being dissolved in the aqueous solution.
2. When the concentration of surfactant becomes equal to critical micelle concentration, it results in micelle formation.
3. The surfactant molecules will get organized in the micellar liquid crystals; this serves as a template for the formation of mesoporous materials.
4. The final step includes the removal of surfactant by thermal degradation or solvent extraction after the silica source gets condensed around the micelles.

The surfactant removal forms a network of cavities within the silica framework. The more functionalized materials are being synthesized using two major methods:

1. Direct functionalization: During the synthesis process, the trialkoxysilane is being added with the selected functional group to the reaction mixture
2. Post synthesis functionalization: This involves grafting of the desired functional group on mesoporous material after the removal of surfactant.

7.11.3 Synthesis of titanium dioxide nanoceramics

Lopez and coworkers synthesized a TiO_2 nanostructure bioceramic device. This was done using a sol–gel process that controls the pore-size distribution and size of the particle.

Titania is considered as one of the best options as the manipulation of the structure and number of hydroxyl groups is possible in titania. The desired interaction between high polar drug and titania device was obtained by the generation of point defects in titania meshwork. The titania device comprised an anticonvulsant drug, valproic acid (VPA). This drug could be targeted directly to the temporal lobe of the brain at a constant rate.

A complete homogeneous solution was created by addition of VPA at the initial stage of bioceramic synthesis. The constant stirring to the solution was done until a gel was formed. The titanium dioxide underwent self-organization to form a new structure in such conditions; TiO_2 was restructuring around the VPA simultaneously. The low pressure and temperature were maintained which was considered well-controlled conditions. With the preserved molecular and electronic properties of ceramic material, a dry gel was obtained ultimately. Now, the anticonvulsant drug was closed within the pore structure in the formed dry gel.

7.11.4 Synthesis of silica-based bioceramics

Byrne and Deasy discovered porous aluminosilicate ceramics. These were made using extrusion–spherization or by cryopelletization. The structure has a highly interconnected porous microstructure where the pore-size distribution depends on the product used. The vacuum impregnation technique was used for loading of drugs in the pellets. The concentration of drug loading solution and the porosity of material ensures the loading. The rate-determining step of the incorporated drug release depends upon the pellet size, porosity, pore-size distribution, porous microstructure, and electrostatic interactions between drug and the pellet. The drug gets diffused from the interior of the porous pellet into the bulk dissociation medium.

7.12 Advantages of bioresorbable nanoceramics as drug delivery carriers

In the fields of pharmacology, medicine, and chemical research, drugs play an important role in curing diseases. Nanoscale DDSs with various properties are tailored to provide added advantages. The bioresorbable nanoceramics as a novel formulation has revolutionized the development of drug delivery [5].

The critical pore and grain size of the ceramic material usually vary from nanometers to microns. This property of ceramics results in controlled drug delivery and better-targeting efficiency as compared to other nano-based particles. Nano-based bioceramic particle size is so confined that they allow easy transportation of drug entities in volume-restricted administration pathways. These routes consist of blood vessels, cell membranes, digestive tract, etc. The efficacy of bioresorbable nanoceramic drug is increased along with the significant expense of such drugs. These drugs are highly precise with the specific targeted delivery. The large surface area-to-volume ratio of these drug carriers allows for a high drug payload and a prolonged drug release profile. For example, studies have shown that nanoparticles with sizes from 10 to 70 nm in diameter can penetrate capillaries, and those with sizes 70–200 nm have the most prolonged circulation time compared with other sizes. The nanoceramic particles have a low rate of degradation that allows prolonged drug release and thus they possess excellent biocompatibility. The titanium dioxide nanoparticles regulate drug release by UV irradiation time to further control the anticancer effect [46].

As compared with other nanoformulations, the intercalated antiinflammatory drugs in LDH nanoceramic results in different drug release profiles. This formulation enhanced the solubility of poorly water-soluble drugs and highly reduces the side effects of drugs. There are two possible reasons that determine the high efficiency of the delivery of molecular drugs, such as ibuprofen, fenbufen, etc. using LDH materials. The reasons are:

1. The methods of coprecipitation and anion exchange for the synthesis of LDH nanohybrids cause quantitative incorporation of the drug within LDH materials.
2. The drugs released from LDH conjugates after being aided in crossing the cell membrane have the desired effect because most of these drugs do not require delivery to the nucleus.

The most used LDH in drug delivery is Mg_2Al (OH)6-LDHs. During the uptake process, the elements present get dissolved, which leads to the formation of free Al^{3+}, Mg^{2+}, and NO^{3-} or Cl^- ions. The concentration used very low and thus they do not have any detrimental effects on cell viability and proliferation [5]. The silica-based mesoporous bioceramics have enhanced properties. The property of large surface area stems the possibility of functionalizing the silanol-containing surface with the desired organic group that allows better control over the molecule and controlled drug release.

7.13 Applications of nanoceramics in drug delivery

The potential of ceramic nanoparticles as drug carriers with great properties provides an opportunity to apply them in various fields of medical problems and diseases. Cancer is one of the major research fields and ceramic nanoparticles have been under the spotlight for the treatment of different cancers.

7.13.1 Calcium phosphate nanoparticles

The calcium phosphate nanoparticles have several applications in drug delivery. They are being explored as one of the drug carriers for the treatment of cancer and many other diseases. These nanoparticles contain folic acid and successfully target tumor cells expressing folate receptors. The release behavior of these drugs showed a biphasic pattern that is characterized by an initial burst release followed by slower or continuous release. This showed tumor inhibition effect on hepatocarcinoma with very less side effects. HA nanoscale crystals can be modulated to produce HA/ biomolecule conjugates that are suitable for many therapeutic applications [46].

7.13.2 Titanium dioxide nanoparticles

PEG and FA-modified HA and titanium dioxide nanoparticles loaded with paclitaxel have higher anticancer activity than pure paclitaxel, due to the targeting of the drug to the folate receptor in cancer cells. In PH-responsive titanium dioxide nanoparticles loaded with daunorubicin, drug was released more rapidly at pH 5.0 and 6.0 than at pH 7.4. The intracellular concentration of daunorubicin was increased by these particles that enhance its anticancer efficiency by inducing apoptosis in a caspase-dependent manner [46]. Doxorubicin−titanium dioxide nanocomposites markedly increased the efficiency of drug per dosage and decreased IC50 in human SMMC-7721 hepatocarcinoma cells. This results in anticancer efficacy enhancement and side effect attenuation.

7.13.3 Silica nanoparticles

In avidin-capped cisplatin-loaded mesoporous nanoparticles MMP9-mediated release of cisplatin was achieved. This induces apoptotic cell death in lung tumor regions. These particles also permit effective combinatorial drug delivery of cisplatin and proteasome inhibitor bortezomib that leads to a synergistic effect on the toxicity. Lactosaminated, docetaxel-loaded mesoporous silica nanoparticles have the potential to inhibit HepG2 and SMMC-7721 cells efficiently. VEGF121-conjugated sunitinib-loaded mesoporous silica nanoparticles efficiently target a higher amount of sunitinib and are delivered to U87MG tumor cells [46].

The antiepileptic drugs can be introduced at constant drug release rate directly to the central nervous system using titania nanoceramic. It has great applications in the pharmaceutical industry. A thermosensitive nanogel with the ability to target tumors was developed recently. The poly (*N*-isopropylacrylamide-co-propyl acrylic acid) nanogel, when conjugated with arginine-glycine-aspartic acid containing peptide and transferrin, shows a dual ability that it can target tumor cells and at the same time release embedded doxorubicin—a DNA-interacting anticancer drug [3].

7.13.4 Layered double hydroxides

In the last 10 years, the LDH materials reduced to nanosizes of 50−300 nm were innovated as drug and gene carriers for controlled and targeted DDSs. The synthesis of LDH materials as carriers for drug delivery is of great interest. They have the potential to serve as catalysts, confiners for anionic pollutants, ceramic precursors, ion exchanges, catalyst supports, and as additives for polymers. They have significant properties of biocompatibility and bioresorbable, and thus have various pharmaceutical applications.

LDHs can have beneficial effects on antiinflammatory therapy. The intercalation of antiinflammatory drugs such as indomethacin with LDH nanoceramics reduces the side effects of ulceration, bleeding, or holes in the stomach. According to the World Health Organization, the primary cause of death worldwide is cardiovascular disease. Heparin with its anticoagulant property has the potential to cure heart diseases. The anticoagulant low-molecular-weight heparin (LMWH), a highly sulfated glycosaminoglycan, can be intercalated into LDHs. Heparin has a few pharmaceutical limitations, such as:

Very short half-life (2–4 hours)

Low efficiency of cellular delivery

Lack of oral absorption

Twice-daily injections needed

Therefore the half-life of LMWH in blood plasma can be extended by their intercalation with LDH. The strong negative charge of LMWH allows for the anion exchange and intercalation into the LDH interlayer that protects the respective anticoagulant. This also ensures the sustained release of LMWH.

7.13.5 Silica-based ordered mesoporous materials

These bioceramic candidates are used as bone implants along with targeted drug delivery properties. The two reasons are the bioactive behavior of these materials when in contact with simulated body fluids and their potential to act as drug delivery vehicles for a large number of biomolecules. This includes drugs for the treatment of bone infections. They can also carry molecules involved in the promotion of bone tissue regeneration, such as growth factors. They can be used as starting materials and can initiate the formation of scaffolds for bone tissue engineering [3].

One of the best biomedical applications of mesoporous bioceramic particle is that they can carry albumin. Albumin is a plasma protein that acts as a multifunctional transport protein and can carry various polypeptides of desired interest. The nanoceramic carriers for albumin can target them to specific places and could help in bone regeneration and many other bone-related diseases.

7.14 Challenges faced in nano-based drug delivery

Drug delivery through nanoparticles has become a major topic of interest in the field of medical and chemical research. The various types of nanoparticles have different advantages and disadvantages. The unique material properties of nanoceramic drugs have the ability to deal with drug delivery problems. The challenges faced by the researchers in efficient delivery systems include targeted drug delivery, drug release kinetics, and controlled drug release [46].

The ceramic-based drugs have a nondegradable or slow biodegradable nature. From this stems the problem of the removal of these drugs out of the body. The metabolic pathways and elimination route of nanoparticles is not yet properly known. The size of nanoceramics is so small that it allows easy penetration through biological membranes and the blood–brain barrier. This could lead to the accumulation of such materials in unwanted cells. The study of the removal of nanoceramics is still ongoing and in progress. The next concern is the variations in physical and chemical properties of nanoceramic materials. Such variations could lead to inconsistent studies that could lead to possible risks due to the use of nanomaterials on human health [3].

There are many drugs that have an alkaline or acidic nature and cannot be administered orally, for example, indomethacin, owing to alkalinity. This provides a challenge for LDH hybrids to have a more promising approach for better drug delivery pathway systems. The untreated and basic LDH hybrids could have a burst release profile for the incorporated drug. The coatings of LDH with enteric polymers, such as Eudragit S 100, showed a controlled drug release.

In in vitro conditions, the LDH nanoparticles have been shown to be very effective. The extended studies of in vivo conditions are yet to be done. In vivo cytotoxicity and biodistribution are not well understood. The way these particles behave in the bloodstream, including their degradation and aggregation, is not fully understood. The new innovative approaches are to be performed for surface-functionalization of carriers with antibodies or other targeting groups. The maintenance of the original properties of LDH nanoparticles is very important even after their intercalation with different biomolecules. This provides these drug carriers an advantage compared to other nanoformulations.

Specific intracellular targeting is another approach that has to be completed. The treatment of neurodegenerative diseases is also a major concern; the research on the ability of LDH carriers to cross the blood–brain barriers is of crucial need. The intercalation of supercoiled structure of plasmids and synthesis of such LDH nanohybrids has to be seen. Plasmids get wrapped around the nanoparticles and do not get properly intercalated in-between the hydroxide layers, which results in an aggregation of different particles and restricts their ability to cross the biological membranes [5].

With these challenges, scientists and researchers continue to discover materials with more advanced properties that could meet the requirements without affecting any system.

7.15 Conclusion and future perspectives

Nanoceramics are emerging nanomaterials that have been prudentially well-chosen for wide-ranging applications in drug delivery by researchers around the globe. The above discussion in this chapter covers the basics of the potential applications of nanoceramics in drug delivery, which itself serves as evidence that these nanoparticles hold a great interest as drug carriers to deliver and transfer active pharmaceutical agents to the targeted site in a controlled manner, resulting in the achievement of a concentration of the therapeutic drug at the target site. Ceramic nanoparticles offer a plethora of technical advantages as vehicles for drug delivery. Compared with the large number of available conventional DDSs, this novel approach involving bioresorbable nanoceramics has an edge as it is environment friendly, biocompatible, has high stability and mechanical strength, and also mimics the biosystems for better penetration and cellular attachment, resulting in an efficacious system of drug administration. They often solubilize in the human body without causing any significant residue accumulation and in turn reduce the risk of side effects and toxicity. The future discoveries in nanoceramic materials need further exploration regarding their safety and toxicity to the tissue environment while using them as drug delivery agents. Plenty of research has already been done, yet factors like their stability and pharmacokinetic properties need more detailed investigation studies, so that biomimetic bioresorbable nanoceramics can serve as the prominent future approach for efficient drug delivery.

Important websites

1. https://www.feedspot.com/infiniterss.php?_src = feed_title&followfeedid = 4712720&q = site:https%3A%2F%2Fwww.azonano.com%2Fsyndication.axd%3Fformat%3Drss
2. https://www.feedspot.com/infiniterss.php?_src = feed_title&followfeedid = 11514&q = site:http%3A%2F%2Fwww.nanotech-now.com%2Frss%2Frecentnews
3. https://www.feedspot.com/infiniterss.php?_src = feed_title&followfeedid = 1051282&q = site:https%3A%2F%2Fwww.nanowerk.com%2Fnwfeedcomplete.xml
4. https://www.feedspot.com/infiniterss.php?_src = feed_title&followfeedid = 11514&q = site:http%3A%2F%2Fwww.nanotech-now.com%2Frss%2Frecentnews
5. https://www.feedspot.com/infiniterss.php?_src = feed_title&followfeedid = 4713146&q = site:https%3A%2F%2Fwww.nano.gov%2Frss.xml
6. https://www.feedspot.com/infiniterss.php?_src = feed_title&followfeedid = 151288&q = site:https%3A%2F%2Fwww.nature.com%2Fnnano.rss
7. https://www.feedspot.com/infiniterss.php?_src = feed_title&followfeedid = 4714780&q = site:https%3A%2F%2Fwww.journals.elsevier.com%2Fnano-today%2Frss
8. https://www.feedspot.com/infiniterss.php?_src = feed_title&followfeedid = 205471&q = site:http%3A%2F%2Ffeeds.feedburner.com%2Facs%2Fancac3
9. https://www.azonano.com/article.aspx?ArticleID = 5143
10. https://what-when-how.com/nanoscience-and-nanotechnology/nanoceramics-nanotechnology/

References

[1] L.L. Hench, J. Wilson, An Introduction to Bioceramics, An Introduction to Bioceramics, World Scientific, 1993. Available from: https://doi.org/10.1142/2028.

[2] A. Kiani, M. Rahmani, S. Manickam., B. Tan, Nanoceramics: synthesis, characterization, and applications, J. Nanomater. 2014 (2014) 2–4.

[3] A. Hasan, et al., Nanoparticles in tissue engineering: applications, challenges and prospects, Int. J. Nanomed. 13 (2018) 5637–5655.

[4] M. Prakasam, et al., Biodegradable materials and metallic implants—a review, J. Funct. Biomater. 8 (4) (2017) 1–15.

[5] M. Vallet-Regí, D. Arcos, Bioceramics for drug delivery, Acta Mater. 61 (3) (2013) 890–911.

[6] N. Martinho, C. Damgé, C. Pinto Reis, Recent advances in drug delivery systems, J. Biomater. Nanobiotechnol. 2 (2011) 510–526.

[7] K.Y. Lee, D.J. Mooney, Alginate: properties and biomedical applications, Prog. Polym. Sci. (Oxf.) 37 (1) (2012) 106–126.

[8] R.A. Jain, The manufacturing techniques of various drug loaded biodegradable poly(lactide-*co*-glycolide) (PLGA) devices, Biomaterials 21 (23) (2000) 2475–2490.

[9] P.S. Patel, et al., Buccal drug delivery system: a review, Int. J. Drug Dev. Res. 5 (3) (2013) 35–48.

[10] S. Suresh, S. Bhaskaran, Nasal drug delivery: an overview, Indian J. Pharm. Sci. 67 (1) (2005) 19–25.

[11] R. Gaudana, H.K. Ananthula, A. Parenky, A.K. Mitra, Ocular drug delivery, AAPS J. 12 (3) (2010) 348–360.

[12] S. Tiwari, N. Batra, Oral Drug Delivery System: A Review 2 (2014) 27–35.

[13] M. Moreno-Sastre, et al., Pulmonary drug delivery: a review on nanocarriers for antibacterial chemotherapy, J. Antimicrob. Chemother. 70 (11) (2015) 2945–2955.

[14] V. Kumar Khanna, Targeted delivery of nanomedicines, ISRN Pharmacol. 2012 (2012) 1–9.

[15] S. Parveen, R. Misra, S.K. Sahoo, Nanoparticles: a boon to drug delivery, therapeutics, diagnostics and imaging, Nanomedicine 8 (2) (2012) 147–166.

[16] S.K. Sahoo, W. Ma, V. Labhasetwar, Efficacy of transferrin-conjugated paclitaxel-loaded nanoparticles in a murine model of prostate cancer, Int. J. Cancer 112 (2) (2004) 335–340.

[17] C.H.J. Choi, A.A. Christopher, W. Paul, E.D. Mark, Mechanism of active targeting in solid tumors with transferrin-containing gold nanoparticles, Proc. Natl. Acad. Sci. U.S.A. 107 (3) (2010) 1235–1240.

[18] P. Kocbek, et al., Targeting cancer cells using PLGA nanoparticles surface modified with monoclonal antibody, J. Controlled Rel. 120 (1–2) (2007) 18–26.

[19] A.J. Clark, E.D. Mark, Increased brain uptake of targeted nanoparticles by adding an acid-cleavable linkage between transferrin and the nanoparticle core, Proc. Natl. Acad. Sci. U.S.A. 112 (40) (2015) 12486–12491.

[20] O. Kayser, A. Lemke, N. Hernandez-Trejo, The impact of nanobiotechnology on the development of new drug delivery systems, Curr. Pharm. Biotechnol. 6 (1) (2016) 3–5.

[21] G. Tiwari, et al., Drug delivery systems: an updated review, Int. J. Pharm. Invest. 2 (1) (2012). Available from: www.jpionline.org.

[22] K.M. Shakesheff, Drug delivery systems, Handbook of Biodegradable Polymers: Isolation, Synthesis, Characterization and Applications, Wiley-VCH, 2011, pp. 363–378.

[23] A. Rosina, A. Mathew, O. Mohammed, Conventional vs Controlled Drug Delivery Systems (2018).

[24] M.R. Deo, et al., Proliposome-based transdermal delivery of levonorgestrel, J. Biomater. Appl. 12 (1) (1997) 77–88.

[25] E.C. Wang, A.Z. Wang, Nanoparticles and their applications in cell and molecular biology, Integr. Biol. (U.K.) 6 (1) (2014) 9–26.

[26] D. Singh, P. Dubey, M. Pradhan, M.R. Singh, Ceramic nanocarriers: versatile nanosystem for protein and peptide delivery, Expert Opin. Drug Deliv. 10 (2) (2013) 241–259. Available from: https://pubmed.ncbi.nlm.nih.gov/23265137/.

[27] R.H. Müller, M. Radtke, S.A. Wissing, Solid lipid nanoparticles (SLN) and nanostructured lipid carriers (NLC) in cosmetic and dermatological preparations, Adv. Drug Deliv. Rev. 54 (Suppl. 1) (2002) S131–S155. https://pubmed.ncbi.nlm.nih.gov/12460720/.
[28] R.N. Alyaudtin, et al., Interaction of poly(butylcyanoacrylate) nanoparticles with the blood-brain barrier in vivo and in vitro, J. Drug Target 9 (3) (2001) 209–221. Available from: https://pubmed.ncbi.nlm.nih.gov/11697206/.
[29] S.H. Medina, M.E.H. El-Sayed, Dendrimers as carriers for delivery of chemotherapeutic agents, Chem. Rev. 109 (7) (2009) 3141–3157. Available from: https://pubs.acs.org/doi/abs/10.1021/cr900174j.
[30] R.H. Müller, K. Mäder, S. Gohla, Solid lipid nanoparticles (SLN) for controlled drug delivery – a review of the state of the art, Eur. J. Pharm. Biopharm. 50 (1) (2000) 161–177.
[31] Bt Dubertret, et al., In vivo imaging of quantum dots encapsulated in phospholipid micelles, Science 298 (5599) (2002) 1759–1762. Available from: https://pubmed.ncbi.nlm.nih.gov/12459582/.
[32] A. Shiohara, et al., On the cyto-toxicity caused by quantum dots, Microbiol. Immunol. 48 (9) (2004) 669–675. Available from: https://pubmed.ncbi.nlm.nih.gov/15383704/.
[33] A.E. Porter, et al., Direct imaging of single-walled carbon nanotubes in cells, Nat. Nanotechnol. 2 (11) (2007) 713–717. Available from: https://www.nature.com/articles/nnano.2007.347.
[34] J. Shi, et al., PEGylated fullerene/iron oxide nanocomposites for photodynamic therapy, targeted drug delivery and MR imaging, Biomaterials 34 (37) (2013) 9666–9677. Available from: https://pubmed.ncbi.nlm.nih.gov/24034498/.
[35] C. Wu, H. Zreiqat, Porous bioactive diopside (CaMgSi2O6) ceramic microspheres for drug delivery, Acta Biomater. 6 (3) (2010) 820–829. Available from: https://www.meta.org/papers/porous-bioactive-diopside-camgsi2o6-ceramic/19800428.
[36] J.E. Block, M.R. Thorn, Clinical indications of calcium-phosphate biomaterials and related composites for orthopedic procedures, Calcif. Tissue Int. 66 (3) (2000) 234–238.
[37] S.T. Lo, A. Kumar, J.T. Hsieh, X. Sun, Dendrimer nanoscaffolds for potential theranostics of prostate cancer with a focus on radiochemistry, Mol Pharm. 10 (3) (2013) 793–812. Available from: https://pubs.acs.org/doi/abs/10.1021/mp3005325.
[38] B.D. Kevadiya, H.C. Bajaj, The layered silicate, montmorillonite (MMT) as a drug delivery carrier, Key Eng. Mater. 571 (2013) 111–132. Available from: https://www.scientific.net/KEM.571.111.
[39] A. Hoppe, N.S. Güldal, A.R. Boccaccini, A review of the biological response to ionic dissolution products from bioactive glasses and glass-ceramics, Biomaterials 32 (11) (2011) 2757–2774.
[40] N. Kohli, E. García-Gareta, Bioactive nanoceramics, in: F. Rossi, A. Rainer (Eds.), Nanomaterials for Theranostics and Tissue Engineering, Elsevier, 2020, pp. 233–257. Available from: https://linkinghub.elsevier.com/retrieve/pii/B9780128178386000097.
[41] M. Parent, et al., Design of calcium phosphate ceramics for drug delivery applications in bone diseases: a review of the parameters affecting the loading and release of the therapeutic substance, J. Controlled Rel. 252 (2017) 1–17. Available from: https://pubmed.ncbi.nlm.nih.gov/28232225/.
[42] M. Colilla, M. Manzano, M. Vallet-Regí, Recent advances in ceramic implants as drug delivery systems for biomedical applications, Int. J. Nanomed. 3 (4) (2008) 403–414.
[43] P.N.E. Diagboya, E.D. Dikio, Silica-based mesoporous materials; emerging designer adsorbents for aqueous pollutants removal and water treatment, Microporous Mesoporous Mater. 266 (2018) 252–267.
[44] K. Ladewig, P.X. Zhi, Q.L. Gao, Layered double hydroxide nanoparticles in gene and drug delivery, Exp. Opin. Drug Deliv. 6 (9) (2009) 907–922. Available from: https://doi.org/10.1517/17425240903130585.
[45] W.M. Kriven, S.Y. Kwak, M.A. Wallig, J.H. Choy, Bio-resorbable nanoceramics for gene and drug delivery, MRS Bull. 29 (1) (2004) 33–37. Available from: https://link.springer.com/article/10.1557/mrs2004.14.
[46] S. Thomas, B. Science Publisher Harshita, P. Mishra, S. Talegaonkar, Ceramic nanoparticles: fabrication methods and applications in drug delivery, Curr. Pharm. Des. 21 (42) (2015) 6165–6188.

CHAPTER 8

Applications of nanoceramics in the biomedical industry

Shashi Chawla[1], Prateek Rai[1], Tanya Garain[1] and Chaudhery Mustansar Hussain[2]
[1]Department of Chemistry, Amity Institute of Applied Sciences, Amity University, Noida, Uttar Pradesh, India
[2]Department of Chemistry and Environmental Sciences, New Jersey Institute of Technology, Newark, NJ, United States

8.1 Introduction

Nanoceramic is defined as a particle having its dimensions in nanometers and its composition is that of ceramic. **Composites** or combinations of polymeric substances, metallic substrates, and ceramics are chosen for fabricating materials for biomedical purposes, as they are characterized broadly and are readily available [1]. Recently, researchers across the globe have been focusing on the development and production of novel nanoceramics because they have been found to be relevant for a wide range of biomedical applications [2]. These **biomedical applications** include medical devices that are modified and are implantable, devices that increase bioactivity, treatment of cancer, gene or drug delivery, tissue engineering targeting viral and bacterial diseases, diagnosis and imaging techniques, and development of minimal tools for surgery [3].

A human body allows only those substances to completely conform with itself that are autogenous, that is, are produced by the human body itself [4]. A foreign particle, when introduced into a living body, never becomes completely compatible and might get rejected. Hence, when implanting a synthetic substance in a body, such as nanoceramics, it should comply with three key features [5]. These three features that define a biomaterial are that it is bioinert (acceptable by the human body), bioresorbable (dissolving in the physiologic surrounding of the body), and bioactive (generating required and good chemical reactions or bonds) [6].

To improve the medical applications of nanoceramics, the development and research of novel biomaterials should be encouraged and promoted so that they have better mechanical properties and enhanced bioactivity [1].

8.2 Efficacy of nanoceramics in biological systems

In modern times nanotechnology has become one of the most important and efficient techniques to be used in the biomedical industry [7]. The development of

Industrial Applications of Nanoceramics
DOI: https://doi.org/10.1016/B978-0-323-88654-3.00009-3

nanoceramics for medicine is accelerating bringing about sophisticated and novel substances that are economical, lightweight, smart, and strong [8]. Various chemical, physical, mechanical, and magnetic properties of nanoceramics are different from other conventional materials. These properties are enhanced as well as unique. Properties like ferro- or dielectricity, piezo- or pyroelectricity, magnetoresistance, and conductivity all depend on the type, amount, and the size of the material [8,9].

Nanoceramics are known to be resistant to bending or compression and hence are strong. Even at higher temperatures, ceramics maintain their robustness (Wei et al., n.d.). Nanoceramics are inert, that is side reactions that are not required or might be toxic for the physiological conditions of the body do not occur spontaneously [10]. Bioactive ceramics have properties that are remarkably like that of a bone, and hence nanoceramics play a vital role in bone regeneration. Apart from that, recent research and studies have proved that nanoceramics have the potential to generate biomaterials that are not only effective but also safe [9].

8.3 Applications of nanoceramics in the biomedical industry

Various applications of nanoceramics in the biomedical industry are as follows.

8.4 Nanoscaffolds

A scaffold is a substance that helps maintain interaction between the cells and the extracellular region or the matrix and provides mechanical strength to the tissues and cells that are growing. Scaffolds are porous and are known to have two diverse types of porosity [6]: macroporous, which is scaffolds having a pore size greater than 50 nm; and microporous, having a pore size less than 10 nm. Using nanoscaffolds made of ceramics has several advantages due to their high porosity, higher degree of structural stability, large surface area, and slow degradation [11]. Nanoscaffolds have nanofibers that cause the stimulation of the extracellular matrix (Fig. 8.1). This is achieved because of the high fiber-to-diameter ratio [12]. These characteristics of nanoscaffolds

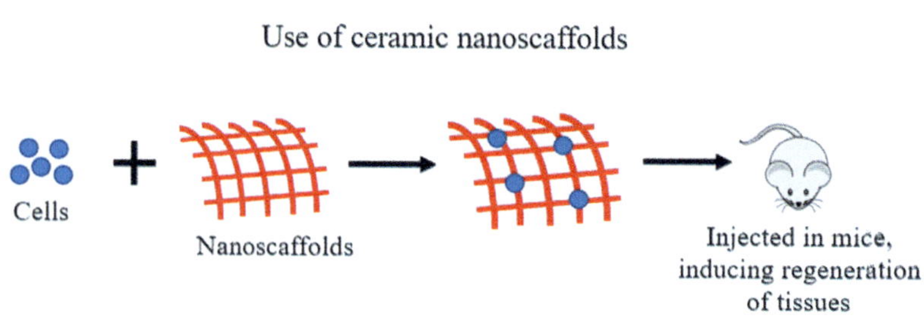

Figure 8.1 Use of nanoceramic scaffolds.

make them suitable for storage purposes and controlled drug release, and they are known to make a large contribution to cellular proliferation and adhesion.

Scaffolds are developed using proteins that are osteoinductive, ceramics, and composites of polymer–ceramic for tissue engineering [13]. Nanoscaffolds have been reported to be developed as potential systems for the treatment of prostate cancer. Ceramic nanoscaffolds also provide the scope of surface modifications by various processes such as plasma treatment. This enhances their properties and broadens the variety of their uses [14].

8.5 Bone tissue engineering

Bones are a type of connective tissue that is very dynamic. Bones consist of osteogenic proteins and cells. This is the bioorganic phase of a bone that is present above calcium phosphate ($CaPO_4$), a biological appetite, and is regarded as the inorganic phase [15]. There are several pathological circumstances or agents that can cause injury to the bone. These agents can be metabolic, inflammatory, or mechanical. These injuries can cause several conditions like osteolysis, cysts, osteomyelitis, or fractures. These conditions disrupt the bone structure and hence disturb the locomotion and other functions of the human body [9].

Bone tissue engineering can be defined as the procedure of analyzing the structure of the bone, mechanics of the bone, and formation of the tissue. It also aims at inducing new functional bone tissues [16]. Tissue engineering takes advantage of the combination of live cells and ceramic nanoscaffolds to transport important cells to a damaged location in a body [5]. The fusion of implants made up of nanoceramics and the differentiating property of osteogenic cells have produced feasible strategies for tissue engineering. These novel bone tissues are regenerated synergistically by combining biomaterials, some cells, and a therapy factor [17]. Scaffolds (bioactive ones), anchor cells, and signals that are osteogenic are applied for the restoration of impaired or contaminated tissues. **Bionanoceramics** that are *bioactive* (hydroxyapatite), *bioresorbable* (tricalcium phosphates), or *bioinert* (alumina and zirconia), are used for bone tissue engineering [18]. **Polymers** that can undergo biodegradation and **ceramics** that are bioactive are chosen to make a **composite**, and these composites are an ideal candidate for creating scaffolds [19]. The properties of such scaffolds are remarkable to that of a bone. These scaffolds are bioceramic in nature and have a pore size like that of a macroporous scaffold [20]. They are fabricated using the technique of foam replica. In this process, polymeric foam is impregnated with a glass slurry. After that, they undergo heat treatment slowly. This step permits the foam to dry, and the polymer present is burned out [5]. The glass particles of glass are also sintered out. Fig. 8.2 These **bioactive ceramic glass scaffolds** have generated a lot of focus and it is regarded as the most favorable biomaterial for bone tissue engineering. To regenerate

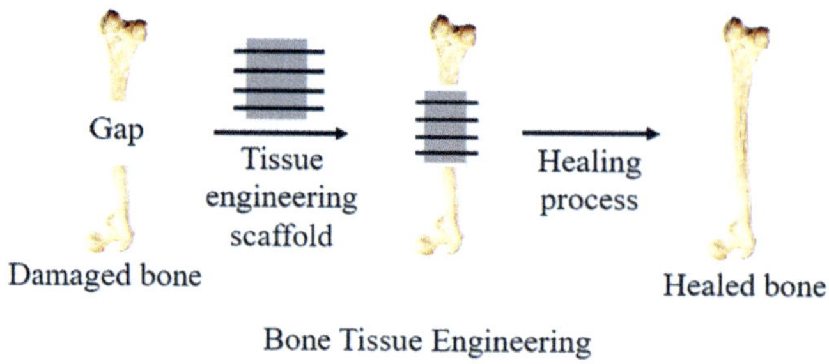

Figure 8.2 Bone tissue engineering.

bone tissue, these ceramic materials are widely used as filler substrates. Bioactive ceramic glass engenders responses, once implanted. Apart from that, it has a high potential of forming strong interactions with hard and soft tissues. This is done by the dissolution of phosphate and calcium ions [10].

8.6 Incorporation of stem cells

A cell that can reproduce for long durations and is extracted from an adult, a fetus, or a fertilized embryo is called a **stem cell**. Stem cells obtained from an adult body are not specialized or differentiated [3]. However, these cells are present in tissues that are specialized or differentiated. Due to this reason, such stem cells can renew themselves and become specialized. Specialized cells formed from these stem cells later form the vital tissues and organs of a body [21]. Cultured adult stem cells are known to be mesenchymal precursor cells as they can also yield lineages like chondrocytes, myocytes, osteoblasts, and adipocytes [22]. Stem cells were known to form bone tissues that were thoroughly vascularized when implanted onto mice. These mice were immunodeficient. It was seen that stem cells combined with mineralized scaffolds to form the bone tissues [23].

Nanoceramics along with cultured stem cells can treat bone defects due to the incorporation of ceramic scaffolds and stem cells with the bone and can lead to excellent recovery.

8.7 Bone replacement applications

The composition of a bone material consists of nanoplatelets, known as hydroxyapatite [24]. These have remarkable similarities with dahllite, a mineral. The apatite present in the bone has the molecular formula of $(Ca, Mg, Na)_{10}(PO_4CO_3)_6(OH)_2$ and is commonly known as carbonate hydroxyapatite [6]. Bones have pores that have their size from 1 to 100 μm (cortical bone) and 200 to 400 μm [8]. There are interconnections between these pores. Nanotechnology can be used for producing synthetic hydroxyapatite (Hap) coatings and nanopowders, which are bone-like in structure and function. Hydroxyapatites produced commercially have a composition like that of bone

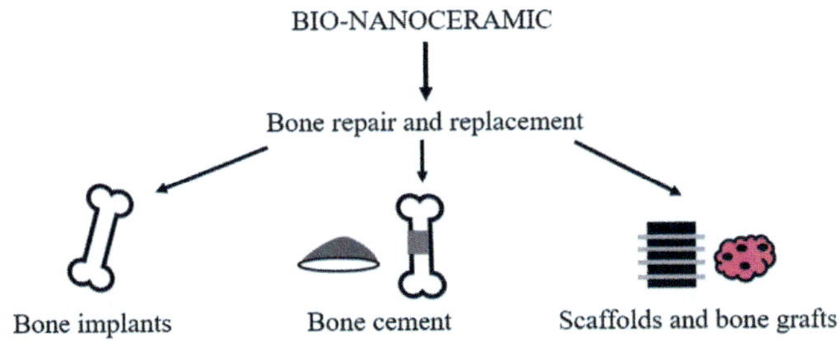

Figure 8.3 Uses of bionanoceramics.

apatite [4]. Microporosity is developed in synthetically produced apatite by adding volatile substances like H_2O_2 or naphthalene.

Nanoparticles of hydroxyapatites can be used to create biocompatible implant coatings and orthopedic nanocomposites which are highly tensile. In their experiments, and with the help of their experience in the clinical field, Oonishi et al. stated that bionanoceramics are playing and will continue to play an especially key role in the durability of complete replacement of joints [2]. Currently many manufacturers are producing nanocomposites with a mixture of nano- and macroceramics. These ceramic nanocomposites are known to provide mechanical strength which otherwise wouldn't be acquired by using only nanoceramics [25]. There are various orthopedic implants made up of nanoceramics that can be used solely or in permutations/combinations of each other. These include screws, nails, plates, metallic implants, etc. (see Fig. 8.3). Such implants have the required resistance against wear and tear and are easily attached to a bone. Once implanted, they are strong, ductile, and elastic [19].

At the interface of a bone and an implant load, the interface will carry a greater load if the implant material is stiff, as compared to the tissues present in the surrounding. This causes an imbalance in the load [26]. This disparity has been termed stress-shielding. Stress-shielding can cause the resorption of bone tissue. If the rigidness of the implant material is too high, the likelihood of a **fracture** increases, because of bone thinning or osteoporosis. Osteoporosis is observed because the stress-shielding effect generates protection beyond required. This limitation might be answered by using implants having nanocoating of calcium phosphate and having the required design, bonding, and bioactivity [27]. Also, surface modifications and appropriate production strategies can form hydroxyapatites having excellent bonding ability and increased adhesiveness between the HAp and the bone material [28].

8.8 Simulated body fluid

Physiological functions of a body can be promoted using nanoceramic surface modifications. Simulated body fluid or SBF has a variety of organic and inorganic substances present, on which layers of hydroxyapatite can be generated. Simulated body fluid has various ions present in different concentrations: Na^+ (142 mM), K^+ (5 mM), Mg^{2+} (1.5 mM), Ca^{2+} (2.5 mM), Cl^- (147.8 mM), HCO_3^- (4.2 mM), HPO_4^{2-} (1.0 mM), and SO_4^{2-} (0.5 mM) [29]. Experiments conducted by Kokubo and Takadama demonstrated that biomaterials, when immersed in simulated body fluid, displayed crystalline structures like carbonate ion apatite [30]. Proliferation and differentiation of bone cells were observed when these cells were placed on this apatite. They also showed the formation of novel bone cells and increased biological activity. Apatite of calcium phosphate has been induced in various ceramic, metal, or polymer substrates that were soaked in simulated body fluid. SBF is known to resemble the HBBS (Hank's balanced salt solution). Their main purpose is to simulate ionic concentrations in the plasma of the human body [31].

Under normal circumstances, SBF is supersaturated with Ca^{2+} and PO_4^{2-} ions, in comparison to an apatite. This is the reason a substance having appropriate functional groups for initiating the nucleation for apatite on the surface of the fluid can easily do so. Apatite formation rate can be increased by adding excess Ca^{2+}, Ti-OH, Ta-OH, Zr-OH, Nb-OH, Si-OH, PO_4H_2, etc. [32]. **Nucleation of an apatite layer on SBF** is also used widely to estimate bioactivity levels by studying the ability of apatite formation on the SBF surface. Various research has shown that SBF solutions can be used to coat apatite on a two-dimensional and a three-dimensional scaffold. For example, a **brediglite scaffold** having the molecular formula $Ca_7MgSi_4O_{16}$ was prepared for bone–tissue engineering [33]. This scaffold was new, biologically active, biocompatible, and biodegradable. The polymer sponge technique was employed for the preparation of this scaffold. This scaffold was then deposited with a BTAp by immersing them in simulated body fluid. Investigations carried out on the brediglite scaffolds revealed that they were highly porous, having a pore size of 300–500 μm. They had biomimetic properties that led to the development of an apatite layer like a bone [34].

8.9 Nanocoatings and surface modifications

8.9.1 Calcium phosphate coatings

Synthetic calcium phosphates are an ideal candidate for the replacement of a diseased or fractured bone. However, the mechanical properties of synthetic calcium phosphate are quite different from natural bone, and hence they have limited use in terms of bearing load. Due to this drawback, metallic implants are seen as a superior choice for load-bearing applications. Implants made up of metals are biologically inert and hence do not form any chemical bond with a bone. For fixing such implants with the bones, they should be produced in a manner that they have appropriate surface bristliness, achieved by macro- or microtexturing. As the roughness of a surface increases, the surface area is enhanced, and this results in increased fixation [20].

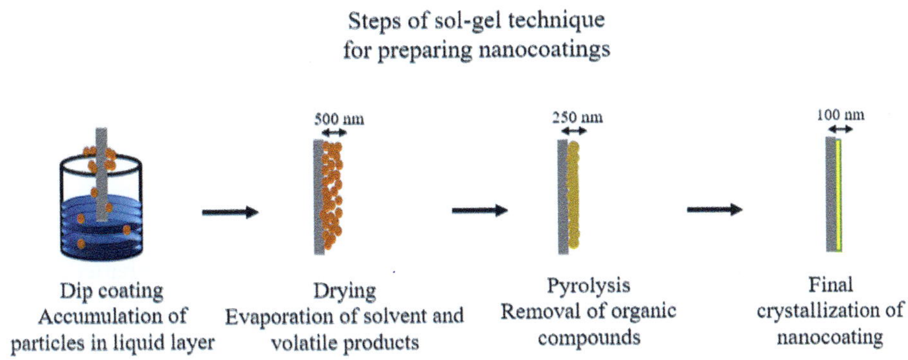

Figure 8.4 Steps of sol–gel technique for preparing nanocoating.

Nanoceramics, when used as a coating on an implant, provides quick bone attachment or biological fixation. **Biological fixation** is the phenomenon in which synthetic components strongly bind to the human bone, without using mechanical or adhesive techniques [35]. **The performance and compatibility of biomaterials** are enhanced by using nanoceramic coatings. Implants coated with nanoceramic coatings demonstrated effective bone collocation [36]. *The strength of the bone-implant interfacial surface* improved and this was attributed to the released phosphate and calcium ions that established biological interactions. Nanocoated implants were seen to promote healing rapidly and were attached properly with the bone. Properties like porosity, crystalline structure, phase, thickness, surface roughness, and complete design led to the **implant's durability** [37].

The manufacturing process of such biologically active coatings has adapted to the following generalized methods [38]:

1. Ducheyne and his team used thick coatings (100 μm to 2 mm) of calcium phosphate for promoting bone growth. This method is known as **spray coating** [21,39].
2. Hench and the team are credited with initiating the method of thick bioglass coating, used specifically for surface activity.
3. **Biomimetics**, that is, biomaterials mimicking natural processes or compounds that have undergone self-assembly in the solution of SBF.
4. The sol–gel technique of preparing nanocoating (see Fig. 8.4).

8.10 Nanocoated coralline apatite

Coralline hydroxyapatite is the main source of producing bone grafts currently. The major limitation associated with this process is that commercially produced HAp retains coral or calcium carbonate and due to this nanopores are present in the structure of this apatite [24]. This increases the rate of dissolution of hydroxyapatite, compromising its strength and durability. As a result, in places where high structural strength is needed, this apatite can't be used there. Ben-Nissan and the team developed a novel technique to overcome these drawbacks [40]. This procedure has an application route that is double staged. The first stage involves completely converting the coral to a pure form of hydroxyapatite. The second stage involves nanocoating of HAp so that it covers the nano- and mesopores. For the maintenance of bone development, the larger pores are left uncovered. This two-step treatment is unique and improves fracture resistance, Young's modulus, and strength of the hydroxyapatite. Bioactivity is enhanced and this coating technique can be used for load-bearing purposes [33].

8.11 Surface modifications

Various physical, chemical, or biological processes can be employed to modify and functionalize the surface of nanomaterials with several reagents [15]. **Surface modification** enhances stability and solubility along with characteristics and properties of nanomaterials in an aqueous medium [41]. Some **physical methods** that are used for this purpose involve molecular coating, physical treatment with plasma, surface entrapment, UV light, or ozone. These physical methods are known to introduce a range of active chemical groups, charges, functional atoms, or molecules, etc. on the surface of nanomaterials [42]. **Chemical methods** have the advantage of not only activating the surface but also contributing to developing stronger chemical bonds between the linked molecules and the surface of the material [43]. Modifying the surface of a nanomaterial using **biological techniques** is more of a necessity for functionalizing the nanoparticles. Physical or chemical methods introduce biospecific entities into the nanomaterials, which is further responsible for the immobilization of biospecific ligands to those entities. Biological methods of surface modifications are responsible for this immobilization [26]. There is a vast range of biomedical devices available today, hence their surface modification varies with their required characteristics or uses. Pergolio and their team are credited for producing polycaprolactone-coated alumina scaffolds, which are a **polymer—nanoceramic composite** and show improved resistance to fracture [42]. This surface modification changed the mechanical properties of the scaffold, and its apparent fracture energy was reported to increase by 13-fold. Surface modifications of nanoparticles should also be done to overcome technical limitations and toxicity [44].

8.12 Nanoceramics for removal of virus

Various nanoceramics, such as nanostructured zirconia, titania, and alumina, have recently been known for their use in the removal of viruses from a living body. Removing viruses from blood and hence establishing thermal or chemical stability can be achieved by designing a system using **nanoporous ceramic membranes, fibers,**

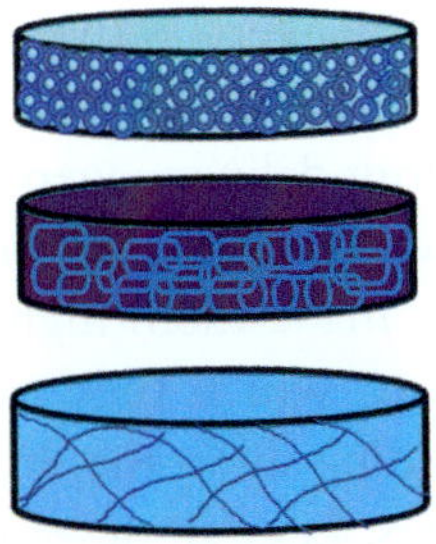

Figure 8.5 Layers of a ceramic membrane.

or nanoparticles [45]. The conventional methods of removal of viruses involve **nanofiltration or microfiltration** using membranes made up of organic polymers. Their limitation is that they are prone to flux reduction due to an increase in pressure associated with this technique, making the membranes compact. If certain acids, bases, or oxidizing agents are present in the medium, they also affect these organic membranes adversely [46].

Recent research has concluded that **membranes made up of nanoceramics** might help overcome these drawbacks, owing to their stability. They are biocompatible and durable. They function normally in biological or chemical systems and can undergo easy sterilization by steam treatment. **The structure of a ceramic membrane** is layered and is unsymmetrical [47]. They have a macroporous layer for support, having a pore size greater than 400 nm. A middle layer is present which is also porous, and the topmost layer of this membrane is nanoporous (Fig. 8.5). Ceramic membranes are selective, and this depends on the size of the pores present on the top layer. The top layer's thickness affects the flux that passes through the filter. The different layers of a ceramic membrane can be prepared using the sol–gel technique [48]. These nanoceramic membranes can easily filter the pathogenic virus, purifying the bloodstream. The controlled pore size and the design of these membranes absorb the unwanted microorganisms for their removal from the body. This paves the way for a novel technique to tackle and monitor the spread of viruses in living bodies. These membranes have no *neurological toxicity* and are highly stable in organic solvents or other mediums, as compared to the organic membranes used previously. To change the electrostatic or chemical/physical properties of a pore and membrane, diverse types of materials can be used to develop nanoceramic membranes such as **titania, zirconia, and alumina**.

8.13 Nanoceramics in drug delivery

Nanocarriers such as nanoparticles, fullerenes, micelles, nanotubes, dendrimers, etc. have shown excellent results when used in drug delivery. Currently, nanoceramics have gained much attention as carriers for drug delivery as they have shown high resistance to changes in the environment [39].

The advantages of using nanoceramics as drug delivery agents are as follows [37]:

- Easy penetration in cells and can undergo drug release for prolonged periods.
- Economic to produce. Advancements have made them highly pure, and they have a large surface area to volume ratios, due to which they can be easily fabricated, and their particle size can be altered, depending upon the drug and the target.
- They are reported to have high efficacy rates. Li et al. developed a nanofibrous system with polyvinyl alcohol as the polymer. It was designed to deliver caffeine and riboflavin. Caffeine release was 100% while riboflavin release was more than 40%.
- Deep penetration in capillaries was seen when nanoceramics were used as nanocarriers. This increased the cellular uptake of drugs.
- Many topical problems have been resolved by using nanoceramics as topical drug delivery agents. Attributed to their ultrafine structure, large area to volume ratio, increased porosity, and small pores, they are highly suitable for this purpose.
- The area to volume ratio of nanoceramics is high which leads to high adsorption rates and better drug release (Fig. 8.6).

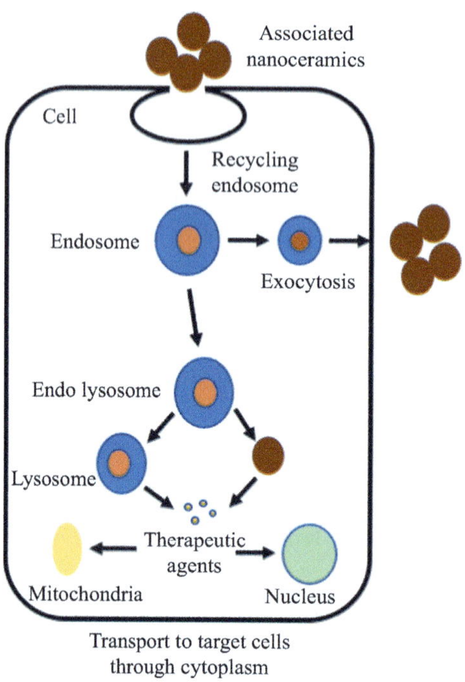

Figure 8.6 Drug delivery mechanism using nanoceramics.

8.14 Nanoceramics in radiotherapy

Many anticancer drugs have exhibited excellent potential, but their efficacy is reduced due to the absence of targeted delivery of drugs. However, nanoceramics have proved to be a milestone in radiotherapy. Removing the defective parts from the body is one of the most ideal procedures for treating cancer. Unfortunately, complete recovery of biological functions is far from possible, and it happens in exceedingly rare cases. In 1980, a novel technique, which was noninvasive in nature, was introduced that destroyed only the targeted cancerous cells. To treat liver cancer, **in situ radiotherapy** was done and $17Y_2O_3-19Al_2O_3-64SiO_2$ (mol%) glass was found to be effective for it. The diameter of this ceramic microsphere was equal to or less than 20–30 μm. The element yttrium (mass number 80) present in this glass is not radioactive, but after bombardment with neutrons it changed into the radioactive form of yttrium, which emits betta rays, and has a half-life of about 64.1 hours [49].

A cancer-affected liver is injected with these microspheres through an artery known as the hepatic artery. These are then trapped in tiny blood vessels, blocking the supply of blood to the cancerous cells, and hence radiating them with beta rays [45]. Since the penetration of a beta ray in living tissue is only about 2.5 mm, and these glass microspheres are chemically durable, adjacent normal cells or tissues are not harmed by the beta-irradiation. Countries like Canada, the United States, and Australia produce these ceramic microspheres commercially and use them in in situ radiotherapy [46].

8.15 Limitations of nanoceramics in the biomedical industry

There are many advantages of using nanoceramics that outweigh their limitations. However, their drawbacks cannot be ignored. Their toxicity is one of the major disadvantages that limit their use globally. Due to their small size, reactivity, and higher ratio of area to volume, they do provide several medical benefits but at the same time may cause hazardous implications on cells and tissue [50].

The size of nanoceramics is comparable with the size of a cell organelle. Hence, they often cause disturbance to cellular processes, becoming potentially toxic. Some researchers have pointed out that nanoparticles may cause inflammation and oxidative stress through the reticuloendothelial system [47]. The toxicity of a nanoceramic varies from a cell to a tissue. They can affect the immunity system of a human in a way that it harms the functioning of vital organs such as the lungs, liver, brain, and heart. Prethrombosis in the functioning of the heart, toxicity of genes, carcinogenicity, etc. are some of the dangerous side effects. They may also cross the blood–brain barrier and thus affect the brain negatively [43]. When compared to natural bone, nanoceramic scaffolds lack elasticity and their magnitude of strength is less. Some **crucial factors that determine the toxicity of a nanoceramic** are its size, surface chemistry, chemical composition, and dosage [26].

8.16 Conclusion and future aspects

Ceramic nanoparticles have gained superiority over other synthetic or organic counterparts and have proved that they are better candidates for *medical applications*, as compared to conventional systems. External conditions like temperature and pH have no or negligible effect on nanoceramics. They can be fabricated in the desired size or structure. Their unique and enhanced properties mean they are ideal options for drug delivery, bone–tissue engineering, radiotherapy, etc., but the ongoing research on nanoceramics needs to accelerate to make them more viable clinically.

Toxicity is the biggest hurdle in their widespread use. Despite their advantages, their side effects continue to be of concern, and hence there is a lot of scope for further studies and improvement. However, with the right direction of research and modification of their properties, nanoceramics can be considered a boon for the biomedical industry.

References

[1] W. Wang, S. Liao, Y. Zhu, M. Liu, Q. Zhao, Y. Fu, Recent applications of nanomaterials in prosthodontics, J. Nanomateri (2015)(2015). Available from: https://doi.org/10.1155/2015/408643, http://www.hindawi.com/journals/jnm/.

[2] S. Nayar, A. Guha, Waste utilization for the controlled synthesis of nanosized hydroxyapatite, Mater. Sci. Eng. C 29 (4) (2009) 1326–1329. Available from: https://doi.org/10.1016/j.msec.2008.10.002.

[3] Y. Zhao, S. Sugiyama, T.J. Miller, X. Miao, Nanoceramics for blood-borne virus removal, Expert. Rev. Med. Dev. 5 (3) (2008) 395–405. Available from: https://doi.org/10.1586/17434440.5.3.395Australia, http://www.expert-reviews.com/doi/pdf/10.1586/17434440.5.3.395.

[4] R.H. Doremus, Bioceramics, J. Mater. Sci. (1992) 285–297. Available from: https://doi.org/10.1007/bf00543915.

[5] T. Traykova, C. Aparicio, M.P. Ginebra, J.A. Planell, Bioceramics as nanomaterials, Nanomedicine. 1 (1) (2006) 91–106. Available from: https://doi.org/10.2217/17435889.1.1.91.

[6] N. Ceren, V. Turp, F. Emir, S. Ayyildiz, D. Sen, Nanoceramics and hybrid materials used in CAD/CAM SYSTEMS, Aydin. Dent. J. (3)(2016).

[7] A. Sionkowska, Current research on the blends of natural and synthetic polymers as new biomaterials: review, Prog. Polym. Sci. (Oxf.) 36 (9) (2011) 1254–1276. Available from: https://doi.org/10.1016/j.progpolymsci.2011.05.003.

[8] M. Ni, B.D. Ratner, Nacre surface transformation to hydroxyapatite in a phosphate buffer solution, Biomaterials. 24 (23) (2003) 4323–4331. Available from: https://doi.org/10.1016/S0142-9612(03)00236-9, http://www.journals.elsevier.com/biomaterials/.

[9] K. Abdelrazek, Advanced sintering of nano-ceramic materials, Ceram. Mater. - Prog. Mod. Ceram. (2012). Available from: https://doi.org/10.5772/38287.

[10] J. Fan, N.A. Kotov, Chiral nanoceramics, Adv. Mater 32 (41) (2020). Available from: https://doi.org/10.1002/adma.201906738, http://onlinelibrary.wiley.com/journal/10.1002/(ISSN)1521-4095.

[11] A.M.M. Amin, A.A.M. El-Amir, G. Karunakaran, D. Kuznetsov, E.M.M. Ewais, In-vitro evaluation of wollastonite nanopowder produced by a facile process using cheap precursors for biomedical applications, Ceram. Int. 47 (13) (2021) 18684–18692. Available from: https://doi.org/10.1016/j.ceramint.2021.03.201, https://www.journals.elsevier.com/ceramics-international.

[12] A.P. Tiwari, S.S. Rohiwal, Synthesis and Bioconjugation of Hybrid Nanostructures for Biomedical Applications Hybrid Nanostructures for Cancer Theranostics, Elsevier, India, 2018, pp. 1–10.

Available from: https://doi.org/10.1016/B978-0-12-813906-6.00002-0, https://www.sciencedirect.com/book/9780128139066.

[13] A. Carvalho, A.R. Fernandes, P.V. Baptista, Elsevier BV, 2019 257–295. Available from: https://doi.org/10.1016/b978-0-12-814029-1.00010-7.

[14] K. Taylor-Smith, What are nanoceramics and their applications? AZONano. (2019) 1–4.

[15] S.V. Dorozhkin, Biocomposites and hybrid biomaterials based on calcium orthophosphates, Biomatter. 1 (1) (2011) 3–56. Available from: https://doi.org/10.4161/biom.1.1.16782.

[16] N Anoceramics in Biomedical Applications. 2019 (2004) 28–32.

[17] M. Winterer, H. Hahn, Nanoceramics by chemical vapour synthesis, Z. fuer Metallkunde/Materials Res. Adv. Tech. 94 (10) (2003) 1084–1090. Available from: https://doi.org/10.3139/146.031084.

[18] Z. Singh, Nanoceramics in bone tissue engineering: the future lies ahead, Trends J. Sci. Res. 3 (3) (2018) 120–123. Available from: https://doi.org/10.31586/nanomaterials.0303.03.

[19] A. Kiani, M. Rahmani, M. Sivakumar, B. Tan, Nanoceramics: Synthesis, characterization, and applications, J. Nanomater 2014 (2014). Available from: https://doi.org/10.1155/2014/528348, http://www.hindawi.com/journals/jnm/.

[20] D. Singh, S. Singh, J. Sahu, S. Srivastava, M.R. Singh, Ceramic nanoparticles: recompense, cellular uptake and toxicity concerns, Artif. Cells Nanomed. Biotechnol. 44 (1) (2016) 401–409. Available from: https://doi.org/10.3109/21691401.2014.955106, http://www.tandfonline.com/loi/ianb20#.VmugQbfovcs.

[21] C.S. Diercks, M.J. Kalmutzki, O.M. Yaghi, Covalent organic frameworks – organic chemistry beyond the molecule, Molecules. 22 (9) (2017). Available from: https://doi.org/10.3390/molecules22091575, http://www.mdpi.com/1420-3049/22/9/1575/pdf.

[22] H. ien Mohamed Darweesh, Nanoceramics: materials, properties, methods and applications—Part II, Nanoscience (2018) 40. Available from: https://doi.org/10.31058/j.nano.2018.11004.

[23] S. Chin, Carbon Nanotube Reinforced Composites Nanostructured Materials in Electrochemistry Block Copolymers in Nanoscience Bulk Nanostructured Materials Nanomaterials for the Life Sciences Nanomaterials Nanotechnologies for the Life Sciences Nanoparticles and Cat. Bioceramics-The Changing Role Ceram. Nanoceramics Med. Appl., 2007.

[24] W. Paul, C.P. Sharma, Nanoceramic matrices: biomedical applications, Am. J. Biochem. Biotechnol. (2006) 41–48. Available from: https://doi.org/10.3844/ajbbsp.2006.41.48.

[25] S. Saranya, M.P. Rani, In vitro studies of Ce-doped hydroxyapatite synthesized by sol–gel method for biomedical applications, J. Pharm. Innov. 16 (3) (2021) 493–503. Available from: https://doi.org/10.1007/s12247-020-09472-y, https://link.springer.com/journal/12247.

[26] S. Sai Nievethitha, N. Subhapradha, D. Saravanan, N. Selvamurugan, W.B. Tsai, N. Srinivasan, et al., Nanoceramics on osteoblast proliferation and differentiation in bone tissue engineering, Int. J. Biol. Macromol 98 (2017) 67–74. Available from: https://doi.org/10.1016/j.ijbiomac.2017.01.089, http://www.elsevier.com/locate/ijbiomac.

[27] Z.Z. Zyman, I.G. Ivanov, V.I. Glushko, Possibilities for strengthening hydroxyapatite ceramics, J. Biomed. Mater. Res. 46 (1) (1999) 73–79. Available from: https://doi.org/10.1002/(SICI)1097-4636(199907)46:1<73::AID-JBM8>3.0.CO;2–4.

[28] Z. Yao, S. Tan, M. Xia, Y. Ye, J. Li, Synthesis, characterization and sintering behaviour of indialite ceramic from fly ash, Waste Manage. Res. 29 (10) (2011) 1090–1097. Available from: https://doi.org/10.1177/0734242X10378329.

[29] R.D. Goodridge, K.W. Dalgarno, D.J. Wood, Indirect selective laser sintering of an apatite-mullite glass-ceramic for potential use in bone replacement applications, in: Proceedings of the Institution of Mechanical Engineers, Part H: Journal of Engineering in Medicine. 220 (1) (2006) 57–68 Available from: http://journals.pepublishing.com/(nagj2045rnxkkcbvwxbzxr2m)/app/home/contribution.asp?referrer = parent&backto = issue,6,7;journal,2,56;linkingpublicationresults,1:119779,1.10.1243/095441105X69051United Kingdom.

[30] S. Dinesh Kumar, K. Mohamed Abudhahir, N. Selvamurugan, S. Vimalraj, R. Murugesan, N. Srinivasan, et al., Formulation and biological actions of nano-bioglass ceramic particles doped with Calcarea phosphorica for bone tissue engineering, Mater. Sci. Eng. C. 83 (2018) 202–209. Available from: https://doi.org/10.1016/j.msec.2017.08.077.

[31] D. Kumar, V. Arya, R. Kaur, Z.A. Bhat, V.K. Gupta, V. Kumar, A review of immunomodulators in the Indian traditional health care system, J. Microbiol. Immunol. Infect. 45 (3) (2012) 165–184. Available from: https://doi.org/10.1016/j.jmii.2011.09.030.

[32] P.N. Lisboa-Filho, P.H.S. Gomes-Ferreira, F.R.S. Batista, G.A.C. Momesso, L.P. Faverani, R. Okamoto, Bone repair with raloxifene and bioglass nanoceramic composite in animal experiment, Connect. Tissue Res. 59 (2018) 97–101. Available from: https://doi.org/10.1080/03008207.2018.1430143, http://www.tandfonline.com/loi/icts20.

[33] A. Bhowmick, N. Pramanik, P. Jana, T. Mitra, A. Gnanamani, M. Das, et al., Development of bone-like zirconium oxide nanoceramic modified chitosan based porous nanocomposites for biomedical application, Int. J. Biol. Macromol. 95 (2017) 348–356. Available from: https://doi.org/10.1016/j.ijbiomac.2016.11.052, http://www.elsevier.com/locate/ijbiomac.

[34] H. Shokrollahi, F. Salimi, A. Doostmohammadi, The fabrication and characterization of barium titanate/akermanite nano-bio-ceramic with a suitable piezoelectric coefficient for bone defect recovery, J. Mech. Behav. Biomed. Mater. 74 (2017) 365–370. Available from: https://doi.org/10.1016/j.jmbbm.2017.06.024, http://www.elsevier.com/wps/find/journaldescription.cws_home/711005/description#description.

[35] B.D. Chithrani, A.A. Ghazani, W.C.W. Chan, Determining the size and shape dependence of gold nanoparticle uptake into mammalian cells, Nano Lett. 6 (4) (2006) 662–668. Available from: https://doi.org/10.1021/nl052396o.

[36] J. Ai, E. Biazar, M. Jafarpour, M. Montazeri, A. Majdi, S. Aminifard, et al., Nanotoxicology and nanoparticle safety in biomedical designs, Int. J. Nanomed. 6 (2011) 1117–1127.

[37] D.A. LaVan, T. McGuire, R. Langer, Small-scale systems for in vivo drug delivery, Nat. Biotechnol. 21 (10) (2003) 1184–1191. Available from: https://doi.org/10.1038/nbt876.

[38] J.M. Slocik, M.O. Stone, R.R. Naik, Synthesis of gold nanoparticles using multifunctional peptides, Small. 1 (11) (2005) 1048–1052. Available from: https://doi.org/10.1002/smll.200500172.

[39] O.C. Farokhzad, R. Langer, Impact of nanotechnology on drug delivery, ACS Nano 3 (1) (2009) 16–20. Available from: https://doi.org/10.1021/nn900002m.

[40] R. Sharma, P.K. Sharma, R. Malviya, Polysaccharide-based scaffolds for bone marrow regeneration: recent work and commercial utility (Patent), Curr. Smart Mater. (2019) 29–35. Available from: https://doi.org/10.2174/2405465804666190326151131.

[41] S.V. Dorozhkin, Bioceramics of calcium orthophosphates, Biomaterials. 31 (7) (2010) 1465–1485. Available from: https://doi.org/10.1016/j.biomaterials.2009.11.050.

[42] R. Niranjan, C. Koushik, S. Saravanan, A. Moorthi, M. Vairamani, N. Selvamurugan, A novel injectable temperature-sensitive zinc doped chitosan/β-glycerophosphate hydrogel for bone tissue engineering, Int. J. Biol. Macromol. 54 (1) (2013) 24–29. Available from: https://doi.org/10.1016/j.ijbiomac.2012.11.026.

[43] S. Canello, G. Gasparini, P. Luisetto, A. Di Cerbo, F. Pomerri, Bone computed tomography mineral content evaluation in chickens: Effects of substances in homeopathic concentration, Homeopathy. 105 (1) (2016) 92–95. Available from: https://doi.org/10.1016/j.homp.2015.07.003.

[44] A. Hoppe, N.S. Güldal, A.R. Boccaccini, A review of the biological response to ionic dissolution products from bioactive glasses and glass-ceramics, Biomaterials. 32 (11) (2011) 2757–2774. Available from: https://doi.org/10.1016/j.biomaterials.2011.01.004.

[45] K. Ando, K. Takahashi, T. Osada, Structural ceramics with self-healing properties, Handbook of Smart Coatings for Materials Protection, Elsevier Inc, Japan, 2014, pp. 586–605. Available from: https://doi.org/10.1533/9780857096883.3.586, http://www.sciencedirect.com/science/book/9780857096807.

[46] W. Nakao, K. Takahashi, K. Ando, Threshold stress during crack-healing treatment of structural ceramics having the crack-healing ability, Mater. Lett. 61 (13) (2007) 2711–2713. Available from: https://doi.org/10.1016/j.matlet.2006.04.122.

[47] K. Takeo, Y. Aoki, T. Osada, W. Nakao, S. Ozaki, Finite element analysis of the size effect on ceramic strength, Materials. 12 (18) (2019) 2885. Available from: https://doi.org/10.3390/ma12182885.

[48] J. Sedláček, D. Galusek, P. Švančárek, A. Brown, R. Brydson, Alumina-carbon composites with high hardness, Key Eng. Mater. 264–268 (2004) 841–844. https://www.scientific.net/KEM.

[49] V. Miikkulainen, M. Leskelä, M. Ritala, R.L. Puurunen, Crystallinity of inorganic films grown by atomic layer deposition: overview and general trends, J. Appl. Phys. 113 (2) (2013). Available from: https://doi.org/10.1063/1.4757907.
[50] A. Nakahira, K. Niihara, Sintering behaviors and consolidation process for Al2O3/SiC nanocomposites, Nippon. Seramikkusu Kyokai Gakujutsu Ronbunshi/Journal Ceram. Soc. Jpn. 100 (1160) (1992) 448–453. Available from: https://doi.org/10.2109/jcersj.100.448.

CHAPTER 9

Industrial applications of nanoceramics: from lab to real-time utilization in the biomedical industry

Darul Raiyaan G.I.[1], Mohammed Junaid Hussain Dowlath[2], Sathish Kumar Karuppannan[3] and Kantha D. Arunachalam[4]
[1]Department of Biotechnology, College of Science and Humanities, SRM Institute of Science and Technology, Kattankulathur, Tamil Nadu, India
[2]Department of Anatomy, SRM Medical College Hospital and Research Centre, SRM Institute of Science and Technology, Kattankulathur, Tamil Nadu, India
[3]School of Materials Science and Engineering, Gwangju Institute of Science and Technology, Gwangju, South Korea
[4]Faculty of Sciences, Marwadi University, Rajkot, Gujarat, India

9.1 Introduction

The increase in life expectancy of people has led to an increase in new medicinal devices. The application of new devices in the field of healthcare is the most prominent need in the field of science [1]. Bioceramics have experienced significant development over the last 50 years [2]. Ceramic devices such as bioactive glasses (BAG) have been involved in the healthcare sector for bone repair or tissue engineering, muscular implant applications, artificial organs, drug delivery, eyeglasses, fiber-optics for endoscopy, radiotherapy, medical devices, diagnostic instruments, etc. [2–7].

9.2 Nanoceramics in bone fixation

Human bone transplantation using various nanoceramics among genetically identical twins or individuals and genetically nonidentical twins consists of cells from bone marrow, bone cells, or osteocytes, and bone matrix, which helps in healing. However, the autograft technique is complicated due to the lack of availability, high cost, and problems related to pain and infection. Together with this, issues such as immunological rejection and increased vulnerability to the transmission of diseases make allograft techniques sensitive [8,9].

The artificial or engineered material used for implantation should attain certain prerequisite conditions such that:

- it should be biocompatible;
- it should not induce any inflammatory reaction; and
- it should endure external forces without causing any rapid fracture or any dimensional changes from tiredness, or stress corrosion over a longer period [10,11].

Industrial Applications of Nanoceramics
DOI: https://doi.org/10.1016/B978-0-323-88654-3.00010-X

Nanoceramics help to induce osteoconductivity, osseointegration, osteogenesis, and osteoinduction [8,9]. At present bioceramics are specially manufactured for bone repair and human bone regeneration. Various bioceramics are supplied for commercial purposes by orthopedic and maxillofacial surgeons [12,13]. Osseous fixation or articulation devices are manufactured using ceramics and ceramic-coated products [12].

In the past few years, various types of bioceramics have been used in bone repair, and they are classified medically into three types based on the reactions they cause in the neighboring tissue: (1) bioinert, (2) bioactive, and (3) bioresorbable ceramics

9.2.1 Bioinert nanoceramics

The materials that have minimal contact with their neighboring tissues and are inactive are known as bioinert. It leads to a decrease in the immune response against a foreign particle after attaining a suitable combination of physical properties similar to the replaced tissue [14]. Metal-based bioceramics such as titanium, alumina, and zirconia-based materials are the best examples of bioinert nanoceramics. They are characterized by their bioinertness, resistance to cracks, and mechanical rigidity in the biological organisms [15].

9.2.1.1 Titanium-based ceramics

Titanium and its alloys are corrosion-resistant materials, and hence they play an important part in the regeneration of bone tissue [16]. Nanotubes titanium disks (NTD) are used in the human osteoblast culture.

The genes that are the indicator of osteodifferentiation of osteoblasts and responsible for bone formation are ENG, SP7, SPP1, and FOSL1. These genes were upregulated by nanotubes titanium disks (NTD) cultured human osteoblast, thus promoting osseointegration and persistent expression [17]. Nanoporous titanium (30 nm size) cultured cells enhance $\beta 1$ integrin mRNA expression which in turn triggers the adhesion of osteoblasts [18].

9.2.1.2 Aluminum oxide-based nanoceramics

The first hip prostheses with Al_2O_3 ceramic bearings were implanted by Dr. Pierre Boutin in the year 1970 [19] since then various research has been carried out on bone surgery using alumina ceramics. Total hip arthroplasty (THA) is mainly done by using ceramic materials like Al_2O_3 which is also used in other joints of the human body including elbow, spinal joint, knees, wrists, ankle, and phalanges, as well as in bone reconstruction [20–24].

9.2.1.3 Zirconia-based nanoceramics

For articulation implants, alumina matrix composites and alumina-toughened zirconia ceramics are used in orthopedic applications as they have good mechanical properties [1].

9.2.2 Bioactive nanoceramics

The materials that interact with neighboring tissues are known as bioactive. As soon as they are placed inside the human body, they act as an anchor or scaffold for cells, which can attach and differentiate. The best example for bioactive nanoceramics is bioglass which is made up of Na, Si, Mg, K, O_2, P, and Ca and is readily absorbed by the cells [25].

Bioactive nanoceramics give a large bioactive surface and enhance osseointegration at the tissue–implant interface. Also, they elicit a controlled action and reaction in the physiological environment [14].

9.2.2.1 Silicon-based nanoceramics

In orthopedics, silicon nitride (Si_3N_4) is used as an implant material as it is a strengthened, hardened, biocompatible, and toughened industrial ceramic material [26,27]. In spinal surgery, it also promotes bone fusion, prosthetic hip and knee joints, and also it is used in osteofixation where plates and screws are used as it is partially radiolucent in maxillofacial surgery where the interaction of the implant can be seen in radiography [27,28].

Soluble silica and other inorganic ions released from glass also play a central role in achieving direct bonding with bone. Xynos et al. showed that ionic species, such as soluble silica and calcium ions, increased human osteoblast proliferation by gene activation [29]. Reffitt et al. studied the effects of orthosilicic acid on collagen type I synthesis and osteoblastic differentiation using four different types of cells, such as primary cells, and found that the acid enhances the COL I synthesis in all treated cells [30].

As a trace element, silicon plays a vital role in the calcification and development of bone matrix [31]. The silicate bioceramics, such as wollastonite ($CaSiO_3$) as binary oxides of silicon and akermanite ($Ca_2MgSi_2O_7$) as ternary oxides and quaternary oxides, also have been developed for bone regeneration. Because of their complex chemical composition, they show bioactivity, mechanical strength, and degradation properties [32].

9.2.2.2 Silicone–hydroxyapatite ceramics

Si-HA enhances osteoblast functions, which contributes to its higher bone apposition level in vivo compared to hydroxyapatite (HA) [33]. Human osteoblasts exhibited significantly higher proliferation and ALP production on Si-HA disks prepared using the granules than their HA equivalents. These effects of silicon on osteoblasts were also observed for Si-HA thin coatings prepared by magnetron cosputtering. Thin coatings containing more silicon than the aforementioned granules were developed as bioactive coatings for metallic implants, such as titanium [34]. Human osteoblast-like cells were cultured on titanium plates without or with the Si-HA coating containing

0.8, 2.2, and 4.9 wt.% of Si. These cells showed improved adhesion and mineralization on Si-HA coatings with the increasing silicon content.

Aizawa et al. [35] developed Si-HA fiber scaffolds for bone tissue engineering. These scaffolds showed an enhanced mouse osteoblast proliferation (MC3T3-E1 cells) compared to their HA equivalents [35].

9.2.2.3 Siloxane-containing vaterite

The calcium carbonate polymorph vaterite exhibits higher dissolution ability than calcite and aragonite. They released calcium and soluble silica species by dissolving in an aqueous solution [36]. Composites comprising SiV particles and poly(L-lactic acid) (PLLA) or PLGA, which are biodegradable polymers, have been developed by the same research group [37,38] Nonwoven [37] and cotton-wool-like fibrous composites [38], benefit from easy handling for bone-filling surgery and the ability to release calcium and soluble silica species, inducing cell migration in vitro and enhancing mineral tissue formation in vivo.

The composite fibers induced biological apatite precipitation in simulated body fluid (SBF). Culture tests using MC3T3-E1 cells showed a faster cell proliferation and differentiation on SiV/PLLA composites than on their vaterite/PLLA analogs. Therefore, nonwoven and cotton-wool-like SiV/PLLA and SiV/PLGA composites may be useful for enhancing bone regeneration [36].

9.2.2.4 Bioactive glass

Hench, from the University of Florida, invented the first bioactive glass, known as 45S5 or Bioglass, in 1969. This material, which comprised 46.1 mol% SiO_2, 24.4 mol% Na_2O, 26.9 mol% CaO, and 2.6 mol% P_2O_5, formed a bond with bone [39]. BAG bonds tightly with the native bone in the absence of fibrous tissue at the implant site, which results in its biocompatibility [40]. The 45S5 BAG shows higher bioactivity and takes a shorter duration for bone attachment in vivo than that of HA [41].

The glass-ceramic Bioverit-I is used widely in bone replacements such as middle ear implants, stomatology, and orthopedics [42]. Hoppe et al. reviewed the biological response elicited by inorganic ions derived from bioactive glasses and glass-ceramics. These ions enhance osteogenesis, angiogenesis, and antibacterial activity [43]. BAG converts to bioactive carbonated apatite, which attaches to the bone firmly [44]. BAG, such as 45S5 or 1393, involves in vitro differentiation and proliferation of bone marrow stromal cells and osteoblast precursor cell lines [45,46]. Yuan et al. reported osteoinductivity of BAG 45S5, which stimulates native stem cells to differentiate into bone cells [47]. The osteogenic ability of BAG 45S5 could result from its dissolution products, which induce osteoprogenitor cells at the genetic level [46]. BAG can stimulate skin repair by accelerating wound healing with the released ionic products [43].

9.2.3 Bioresorbable nanoceramics

A material that dissolves, resorbed, and is replaced slowly by the advanced tissues (e.g., bone) of the human body is known as bioresorbable. Examples are calcium-phosphate (CaP) materials such as HA, tri-CaP, di-CaP dihydrate (brushite), octa-CaP, or biphasic-CaP [48]. They are bioactive, osteoconductive, biocompatible, and osteoinductive because of their chemical similarity with the tissues of human bones. These materials form an attachment with bone tissue by forming a layer of bioactive apatite on their surfaces, which stimulates their osteointegration [48,49]. CaP materials regulate the osteogenic cell function by releasing HPO224, PO324, and Ca21 ions on exposure to biological fluids due to hydration [15].

Apatite−Al_2O_3 ceramics have also been used in bone remodeling. b-Tri-CaP $Ca_3(PO_4)_2$, and calcium HA $Ca_{10}(PO_4)_6(OH)_2$ have been used in corrective therapies for maxillofacial reconstruction, bone repair, as burrhole buttons, spinal fusion devices ossicles, replacement and repair of teeth, orbital implants, alveolar ridge augmentation, and bone scaffolds [3,5,50,51].

9.2.3.1 Hydroxyapatite coatings

Femoral stems treated with thermal sprayed HA coatings were reported in 1980 [52]. Plasma sprayed HA coatings on hard tissue implants such as knee prostheses, dental and hip implants have given the best results in short-term trials. At the initial stage, HA coating has proved its capacity of fast attachment with native bone, although with the existence of a gap [53].

9.2.4 Porous scaffold bioceramics

The current bioceramics are used for making scaffolds to support the cells involved in the regeneration process. The scaffolds should have the following criteria:

- Mechanical support while being biocompatible.
- Not causing any adverse effects.
- Load-bearing capacity may be temporary but the degradation rate of scaffolds should match the regeneration rate of tissue.
- Interconnected porosity with an average pore size distribution.
- Promote cell and tissue colonization.
- Metabolite transit while offering a high surface area for cell anchoring.
- Intertwined porosity and suitable pore size [54].

Nanoporous ceramics induce osteogenesis and vascularization, which leads to osteochondral formation [55−57]. In the tissue engineering of bones, various pore sizes such as macropores, mesopores, and micropores play a vital role in new tissue formation. Macropores induce cell infiltration, mesopores act as a storehouse for nutrients and growth factors, and micropores act as nucleation sites for mineralization [58]. Macroporous bioceramics act as scaffolds for tissue engineering. The improved bioactive behavior along with their controlled delivery capability makes silica mesoporous bioceramics a suitable candidate to produce 3D-scaffolds for tissue engineering of bones [54]. The 2–50nm-sized mesoporous porous glass-ceramics present special properties, such as size distribution, specific surface area, narrow pore, high pore volume, adjustable size of pores, and tunable composition [58,59]. Similarly, scaffolds with nanosized topography on the surface reduce the chances of rejection and enhance the osteoblast fixing with the potential to initiate new tissue in bone tissue regeneration [60]. Hence mesoporous bioglass is being studied extensively to manufacture scaffolds for bone repair because of its capability to induce tissue formation in bones and to work simultaneously in preventing inflammation and infection, as well as drug delivery around the implant [59].

Hexagonally ordered biocompatible mesoporous materials like SiO_2, CaO and P_2O_5-based mesoporous BAG have also been studied for the delivery of drugs [61,62].

9.2.4.1 Nanoceramics combined collagen/gelatin scaffolds for bone tissue engineering

HA interacts with tissue collagen to carry considerable load and stiffness [63]. Biomaterial tissue engineers and scientists are working on imitating the structure and composition of host tissues. Hence the combination of collagen/gelatin with ceramics is the best option for application in bone tissue imitation.

Similar growth of carbonated HA on the nanofibrillar collagen gel was confirmed by immersing it in SBF for 3 days, and a 13 × development in the scaffold compressive modulus was observed after 7 days. The MC3T3-E1 pre-osteoblasts cells culture remained active after 4 weeks in vitro and accelerated osteogenic differentiation was seen in the absence of osteogenic supplements. There was no cell-induced contraction of the gels. Hence the collagen/nBG gels were potentially suitable as osteoinductive cell delivery scaffolds for bone regeneration [64]. The chitosan/nBG/gelatin scaffolds were fabricated with nBG content using a sol–gel method followed by lyophilizing and chemical cross-linking with glutaraldehyde to increase its mechanical strength [65].

Gelatin/nBG scaffolds stimulate bone formation via intramembranous formation in a rabbit ulnar critical-sized defect model [66]. The amalgam of materials—nHA, collagen, and PLA—was used by Liu et al. along with recombinant human bone morphogenetic protein 2 (rhBMP-2)-mediated dental pulp stem cells for the reconstruction of alveolar bone defects [67]. The scaffold was made of gelatin, chitosan, and nHA, incorporated with nanocopper–zinc alloy through lyophilization. This new scaffold is biocompatible with the potential for bone tissue engineering [68].

9.2.4.2 Nanoceramics combined chitin and chitosan scaffolds for bone tissue engineering

The natural polysaccharides–chitin and chitosan polymers have various applications in tissue engineering (e.g., skin, bone, cartilage), drug delivery, and wound healing.

An in vivo study carried out by Lee et al. on tibial defects of rabbits showed that bone surface, total volume, trabecular thickness, bone volume, trabecular separation, and trabecular number, increased in chitosan–nanohydroxyapatite (nHA) mixed scaffolds in comparison with chitosan–microhydroxyapatite (mHA) scaffolds. The marine fish bone-derived nHA and mHA were used in the preparation of scaffolds by the lyophilization method [69]. Ma et al. studied the cross-linked glutaraldehyde with chitosan–nHA scaffold in a standard critical-sized calvarial bone defect (6.5 mm) in Sprague-Dawley rats. Seven days later, the histology showed several cells attached to the pores of the chitosan–nHA scaffolds. Fourteen days of observation showed new bone formation, both at the edge and in the center of implants. Thirty-five days later, a high mineralized volume of bone tissue was formed in the damaged site with implanted scaffolds, when compared to control (with no damages) [70]. Chitosan–nHA fucoidan combined scaffolds showed higher biocompatibility and excellent mineralization, resulting as a potential candidate for bone tissue engineering [71].

β-Chitin hydrogel and nanobioglass combined scaffolds show a lower pore size and low degradation rate when immersed in lysozyme PBS for 4 weeks, compared to the control (β-chitin scaffold), hence they have been applied in periodontal bone regeneration using the freeze-drying method [72]. The chitosan and 1% nBG scaffolds are nontoxic, biocompatible, and suitable for alveolar bone tissue engineering. They showed a lower degradation when prepared by combining the ceramic nanoparticles with a chitosan solution, chemical cross-linking with 0.25% glutaraldehyde, followed by freeze-drying. Also it showed a 5% loss in weight after a 1-week immersion in lysozyme PBS, when compared with chitosan scaffolds (weight loss of 25%). It also showed bioactivity when immersed in SBF for a week, and was cell compatible when seeded with MG-63 cells in vitro [73].

When the natural polymer fucoidan is mixed with HA, there is an increase in differentiation of adipocyte-derived stem cells into osteoblasts with matrix mineralization [74]. The chitosan, nHA, and Ag NPs with a mean pore size ranging from 50 to 100 μm have the capability of supporting osteogenesis and vascularization, and are essential for optimal osseointegration of the implant along with antibacterial activity to prevent any postoperative infections [75].

9.2.4.3 Nanoceramics combined alginate-based scaffolds for bone tissue engineering

Alginate-based hydrogels mimic the structure of native ECM of tissues. They have the characteristics of biodegradability, low toxicity, gentle gelling kinetics, and biocompatibility [76]. They are mixed with bioresorbable and/or bioactive nanoceramics as reinforcing agents, which gives osteoconductive and osteogenic properties to these composites, to make them the best solution for treating bone damage [77]. The osteoconductive component nHA played a vital role in promoting cell proliferation and osteogenic differentiation of microencapsulated osteoblast-like cells using factorial experimental design, due to its bioactivity and contribution toward the improvement of the alginate-based hydrogel's mechanical strength [78].

Mesenchymal stem cells (MSCs) derived from the bone marrow of rat were encapsulated into alginate dialdehyde/gelatin hydrogel with nanobioglass and showed high cell survival in vitro for up to 4 weeks, showing it is a cyto-friendly encapsulation process. After subcutaneous implantation into rats, high cell survival was observed after 7 days of implantation; although a decrease in cell survival was noticed after 28 days. The studied immune reaction was very mild, which reveals the biocompatibility of the scaffold. The nanobioglass encapsulated alginate–gelatin hydrogel showed an increase in the number of viable mesenchymal stem cells and lectin-positive endothelial cells, thus showing higher angiogenesis [79].

9.3 Nanoceramics in dentistry

Nanoceramics have also found application in dentistry due to the similar mechanical properties of bones and biocompatible surface properties. Zirconia-based ceramics are used for next-generation implants due to their notifiable properties compared to Al_2O_3 [80]. Zirconia-based ceramics possess properties such as outstanding toughness, high fracture resistance [81,82], and inertness, which make them suitable for possible applications in implant-retained and tooth-retained fixed dental prosthesis, calvarium prosthesis, etc. [83–87]. Toothpaste and mouthwashes containing nHA formulations are used for mineralizing the demineralized enamel [88,89].

9.4 Nanoceramics in skin repair and regeneration/wound healing

Although natural and synthetic polymers are gaining attention for wound healing treatments, silicate-based nanoceramics and BAG received considerable attention in the last decade for their ability to stimulate soft tissue regeneration by easy dissolution and release of bioactive compounds to provide antibacterial, antiinflammatory, or angiogenic properties [63,90]. Bioactive glasses and glass-ceramics can also stimulate neovascularization, nerve reconstruction, and cardiac tissue regeneration [91]. This recent trend can heal chronic wounds and helps to meet the challenges in wound healing. 45S5 Bioglass ionic extracts showed efficiency in promoting fibroblast proliferation and migration. This also speeds up wound healing by enhancing the secretion of collagen type I. In a rat model, borate-based bioactive glasses sol–gel was developed and evaluated for antibacterial and wound-healing properties. This could significantly enhance the wound-healing process and the closure of wounds [92,93]. Chitosan/polyvinyl alcohol (PVA)/nBG nanofibers accelerated the wound-healing process by promoting regeneration with excelling biocompatibility and antibacterial activity. In vivo wound-healing experiments conducted in rats and mice models evidently showed the upregulation of growth factors like TGF-b and downregulation of TNF-a and IL-b [94]. In another study, the combination of nBG and polymer scaffolds synergistically enhanced the effectiveness and reepithelialization in the mice model [95]. These studies show the activity of mixing nanobioglass and polymeric matrices for wound-healing applications.

Similar many other bioactive metallic ions can be incorporated into the silicate network to provide antiinflammatory, antibacterial, and angiogenic properties. These properties help for wound healing activities [91]. The nanostructured silicate coatings over Ti-6Al-4V substrates promoted cell proliferation and differentiation with no cytotoxicity [96]. Mesoporous bioactive glass (MBG) microspheres enhanced the blood coagulation rate and reduced the time taken for blood clotting [97]. Nanoscale bioactive glasses doped with silver ions had antibacterial activity and could speed up clotting time [98].

9.5 Nanoceramics in peripheral nerve regeneration

Instead of autologous nerve grafting, nerve regeneration can be achieved medically by a concept called nerve guide conduits (NGCs). Functional artificial NGCs are developed by combining natural and synthetic polymers. An nBG/gelatin NGC was used to regenerate a 10 mm gap in the right sciatic nerve of male Wistar rats. The histological and functional evaluation indicated nerve regeneration suggesting that the nBG/gelatin NGC is an apt agent for repairing peripheral nerves [99]. In another study, PCL/collagen/nBG electrospun nanofibers filled with human endometrial stem cells (hEnSC) improved nerve regeneration in a rat model [100]. Lin et al. tested a combination of polypyrrole (PPY), collagen, and nanostrontium substituted bioactive glass (nSrBG) for nerve regeneration in the rodent model and found a substantial recovery of sciatic nerve wounds after 24 weeks of treatment [101].

9.6 Nanoceramics in treatment of cancer

Nanoceramics are used in treating cancers of bone. Surgeries are done for treating bone tumors, but still some cells remain around the bones. Nanoscaffolds are used nowadays for tumor therapy and regeneration of bones after treatment. In a study to address the disadvantages, collagen and nHA nanoscaffolds mixed with poly lactic-co-glycolic acid (PLGA) nanoparticles and adriamycin were produced. These nanoscaffolds were osteoconductive and inhibited osteosarcoma, as a chemotherapy drug Adriamycin was present and released over the long term. Similar nHA/collagen scaffolds loaded with Adriamycin showed excellent biocompatibility in rat muscles as well as enhanced antineoplastic with fewer side effects compared to intraperitoneal administration of Adriamycin [102].

MSNs can be employed in treating cancers for localized drug deliveries into the cancerous tissues and to avoid side effects to healthy tissues. The passive targeting along with abnormalities like wide interendothelial junctions and fenestrations facilitates the accumulation of MSNs from bloodstream and tumor vessels into tumor interstitium [103]. Whereas in active targeting, the internalization and retention of NPs occur when the NPs are factionalized with some ligands as the tumor cells are known to express receptors [54]. Further, the MSNs can be used as prodrugs, in which the MSNs will transport the drugs into the tumor cells which get activated by the enzymes coated on the surface of MSNs [104].

9.7 Antiviral ceramic membranes

Various nanoceramics such as alumina, zirconia, and titania are being used for antiviral applications. At present, virus separation by efficient filtration and adsorption system using nanoporous/mesoporous ceramic membranes, ceramic nanofibers, and ceramic nanoparticles is applied for the removal of viruses from blood [105]. Ceramic nanoparticles synthesized by the flame-spray method can bind with nucleic acids, thereby carrying therapeutic agents into the virus-infected cells to combat them. Amorphous $Ca_3(PO_2)_2$ is one such harmless material that can be used for virus disinfection. This can be combined with filtration technology to remove pathogens from contaminated water in a cost-effective manner [106].

9.8 Drug delivery from silica mesoporous nanoparticles

Nanoceramics such as mesoporous silica nanoparticles (MSNs) are novel carriers for nanodrugs. They have properties like high surface area, high adsorption to a wide range of molecules, and high pore capacity with variable pore size. These properties open routes for its application in drug delivery, diagnosis, and gene transfection [107]. MSNs can withstand high temperatures, contain numerous functionalizing agents and can be endocytosed into mammalian cells for intracellular release [108,109]. Moreover, these MSNs are biocompatible and less toxic at concentrations $>100\ \mu g/mL$ in vitro experiments [110] and >200 mg/kg in vivo animal models [111] and results are promising as they are administered through intravenous injection [112].

nHA is also employed for controlled drug deliveries. For treating bone infections, zinc-doped nHA has shown enhanced efficiency in releasing antibiotic ciprofloxacin [113]. Similarly, promising results were observed in nHA nanofiber composites loaded with several antibiotics such as gentamicin, amoxicillin, and doxorubicin [114–116].

9.9 In diagnosis using MRI

Some nonmagnetic ceramics such as Si_3N_4 ceramic enable magnetic resonance imaging (MRI) scanning of tissues proximal to Si_3N_4 implants compared to metal implants [27].

9.10 Conclusions and future scope of nanoceramics

This chapter summarizes the application of bioactive nanoceramics for various medical applications. Nanoceramics in the fields such as tissue regeneration, drug delivery, therapeutic, and diagnosis applications are promising as their structure and chemistry can be tailor-made with an appropriate combination of growth factors and physicochemical conditions as per the requirements of the patient to provide enhanced activity. Further improvements in the techniques for the synthesis and drug holding capacity can enhance their applications in terms of antibacterial activities, gene delivery, and vascularization, which cannot be rendered by any other materials. Research on the interaction of nanoceramics with cells can explain their performance upon internalization and reduce the associated risks. However, further understanding of the controlled release of therapeutic agents is required and investigation of the dissolution kinetics has to be carried out for ceramics doped with metallic bioactive ions.

References

[1] E. Vunain, S.B. Mishra, A.K. Mishra, B.B. Mamba, Nanoceramics: fundamentals and advanced perspectives, in: A. Mishra (Ed.), Sol-gel Based Nanoceramic Materials: Preparation, Properties and Applications, Springer, Cham, 2017. Available from: https://doi.org/10.1007/978-3-319-49512-5_1.

[2] I. Izquierdo-Barba, M. Colilla, M. Vallet-Regí, Zwitterionic ceramics for biomedical applications, Acta Biomater. 40 (2016) 201–211.

[3] M. Vallet-Regi, Comptes Rendus Chim. 13 (2010) 174–185.

[4] M. Vallet-Regi, D. Arcos, Acta Mater. 61 (2013) 890–911.

[5] S. Dorozhkin, Biomaterials 31 (2010) 1465–1485.

[6] S.V. Dorozhkin, Ceram. Int. 42 (2016) 6529–6554.

[7] Q.Z. Chen, I.D. Thompson, A.R. Boccaccini, Biomaterials 27 (2006) 2414–2425.

[8] G. Chiara, F. Letizia, F. Lorenzo, S. Edoardo, S. Diego, S. Stefano, et al., Nanostructured biomaterials for tissue-engineered bone tissue reconstruction, Int. J. Mol. Sci. 13 (2012) 737–757.

[9] Z.G. Zhang, Z.H. Li, X.Z. Mao, W.C. Wang, Advances in bone repair with nano biomaterials: mini-review, Cytotechnol 63 (5) (2011) 437–443. Available from: https://doi.org/10.1007/s10616-011-9367-4.

[10] G.C. Wang, H. Zreiqat, Functional coatings or films for hard-tissue applications, Materials 3 (2010) 3994–4050.

[11] R.B. Heimann, Thermal spraying of biomaterials, Surf. Coat. Technol. 201 (5) (2006) 2012–2019.

[12] B.J. Mcentire, B.S. Bal, M.N. Rahaman, J. Chevalier, G. Pezzotti, J. Eur. Ceram. Soc. 35 (2015) 4327–4369.

[13] M. Vallet-Regí, Bioceramics with Clinical Applications, John Wiley & Sons, Chichester, 2014.

[14] L.L. Hench, J.M. Polak, Third-generation biomedical materials, Science 295 (5557) (2002). 1014 + 6–7.

[15] S.N. Sethu, S. Namashivayam, S. Devendran, S. Nagarajan, W.-B. Tsai, S. Narashiman, et al., Nanoceramics on osteoblast proliferation and differentiation in bone tissue engineering, Int. J. Biol. Macromol. 98 (2017) 6774.
[16] M.S. Laranjeira, A. Carvalho, A. Pelaez-Vargas, D. Hansford, M.P. Ferraz, S. Coimbra, et al., Modulation of human dermal microvascular endothelial cell and human gingival fibroblast behavior by micropatterned silica coating surfaces for zirconia dental implant applications, Sci. Tech. Adv. Mater. 15 (2014) 025001.
[17] A. Pozio, A. Palmieri, A. Girardi, F. Cura, F. Carinci, Titanium nanotubes activate genes related to bone formation in vitro, Dent. Res. J. 9 (2012) S164−S168.
[18] Y. Hao, S. Li, X. Han, Y. Hao, H. Ai, Effects of the surface characteristics of nanoporous titanium oxide films on Ti-24Nb-4Zr-8Sn alloy on the initial adhesion of osteoblast-like MG-63 cells, Exp. Ther. Med. 6 (2013) 241−247.
[19] P. Boutin, Orthop. Traumatol. Surg. Res. 100 (2014) 15−21.
[20] S. Kosugi, A. Taniguchi, K. Tomiwa, H. Kurokawa, Y. Tanaka, Clin. Res. Foot Ankle 02 (2014) 141.
[21] J.J. Yoo, P.W. Yoon, Y.K. Lee, K.H. Koo, K.S. Yoon, H.J. Kim, J. Arthroplasty 28 (2013) 132−138.
[22] B. Denkena, J. Köhler, A. Turger, P. Helmecke, T. Correa, C. Hurschler, Procedia CIRP 5 (2013) 179−184.
[23] C. Zietz, D. Kluess, P. Bergschmidt, M. Haenle, W. Mittelmeier, R. Bader, Semin. Arthroplasty 22 (2011) 258−263.
[24] L. Ta, X. Yu, P. Wan, K. Yang, J. Mater. Sci. Technol. 29 (2013) 503−513161.
[25] D. Durgalakshmi, S.P. Subhathirai, S. Balakumar, Nano-bioglass: a versatile antidote for bone tissue engineering problems, in: Procedia Engineering, MRS Singapore − ICMAT Symposia Proceedings, 7th international Conference on Materials for Advanced Technologies 92, 2014, pp. 2−8.
[26] K. Bodisova, M. Kasiarova, M. Domanicka, M. Hnatko, Z. Lences, Z.V. Novakova, et al., Ceram. Int. 39 (2013) 8355−8362.
[27] B.S. Bal, M.N. Rahaman, Acta Biomater. 8 (2012) 2889−2898.
[28] R.M. Bock, B.J. McEntire, B.S. Bal, M.N. Rahaman, M. Boffelli, G. Pezzotti, Acta Biomater. 26 (2015) 318−330.
[29] I.D. Xynos, A.J. Edgar, L.D.K. Buttery, L.L. Hench, J.M. Polaks, Gene-expression profiling of human osteoblasts following treatment with the ionic products of Bioglass® 45S5 dissolution, J. Biomed. Mater. Res. 55 (2) (2001) 151−157.
[30] D.M. Reffitt, N. Ogston, R. Jugdaohsingh, H.F.J. Cheung, B.A.J. Evans, R.P.H. Thompson, et al., Orthosilicic acid stimulates collagen type 1 synthesis and osteoblastic differentiation in human osteoblast-like cells in vitro, Bone 32 (2) (2003) 127−135.
[31] A. Hoppe, A.R. Boccaccini, Biological impact of bioactive glasses and their dissolution products, Front. Oral. Biol. 17 (2015) 22−32.
[32] C. Wu, J. Chang, A review of bioactive silicate ceramics, Biomed. Mater. 8 (2013) 032001.
[33] C.M. Botelho, R.A. Brooks, S.M. Best, M.A. Lopes, J.D. Santos, N. Rushton, et al., Human osteoblast response to silicon-substituted hydroxyapatite, J. Biomed. Mater. Res. Part. A 79A (3) (2006) 723−730.
[34] E.S. Thian, J. Huang, S.M. Best, Z.H. Barber, W. Bonfield, Silicon-substituted hydroxyapatite: the next generation of bioactive coatings, Mater. Sci. Eng. C. 27 (2) (2007) 251−256.
[35] M. Aizawa, N. Patel, A.E. Porter, S. Best, W. Bonfield, Syntheses of silicon-containing apatite fibres by a homogeneous precipitation method and their characterization, Key Eng. Mater. 309−311 (2006) 1129−1132.
[36] A. Obata, T. Kasuga, Chapter 2, Silica-based Nanoceramics, in: Nanobiocceramics for Healthcare Applications, 2016. Downloaded from: <http://www.worldscientific.com>.
[37] K. Fujikura, S. Lin, J. Nakamura, A. Obata, T. Kasuga, Preparation of electrospun fiber mats using siloxane-containing vaterite and biode-gradable polymer hybrids for bone regeneration, J. Biomed. Mater. Res., B 101 (8) (2013) 1350−1358.
[38] A. Obata, H. Ozasa, T. Kasuga, J. Jones, Cotton wool-like poly(lactic acid)/vaterite composite scaffolds releasing soluble silica for bone tissue engineering, J. Mater. Sci-Mater Med. 24 (7) (2013) 1649−1658.

[39] J.R. Jones, Review of bioactive glass: From Hench to hybrids, Acta Biomater. 9 (1) (2013) 4457–4486.
[40] L.L. Hench, J. Wilson, Surface-active biomaterials, Science 226 (1984) 630–636.
[41] P. Ducheyne, Bioceramics: material characteristics versus in vivo behavior, J. Biomed. Mater. Res. B 21 (A2) (1987) 219–236.
[42] E. Verne, M. Ferraris, C. Jana, L. Paracchini, Bioverit I base glass/Ti particulate biocomposite: "in situ" vacuum plasma spray deposition, J. Eur. Ceram. Soc. 20 (4) (2000) 473–479.
[43] A. Hoppe, N.S. Gu Ldal, A.R. Boccaccini, A review of the biological response to ionic dissolution products from bioactive glasses and glass-ceramics, Biomaterials 32 (2011) 275774.
[44] G. Kaur, O.P. Pandey, K. Singh, D. Homa, B. Scott, G. Pickrell, A review of bioactive glasses: their structure, properties, fabrication and apatite formation, J. Biomed. Mater. Res., A 102 (2014) 25474.
[45] Q. Fu, M.N. Rahaman, B. Sonny Bal, R.F. Brown, D.E. Day, Mechanical and in vitro performance of 1393 bioactive glass scaffolds prepared by a polymer foam replication technique, Acta Biomater. 4 (2008) 185464.
[46] S. Radin, G. Reilly, G. Bhargave, P.S. Leboy, P. Ducheyne, Osteogenic effects of bioactive glass on bone marrow stromal cells, J. Biomed. Mater. Res. Part. A 73A (2005) 219.
[47] H. Yuan, J.D. de Bruijn, X. Zhang, C.A. Van Blitterswijk, K. de Groot, Bone induction by porous glass ceramic made from Bioglass (45S5), J. Biomed. Mater. Res. 58 (2001) 2706.
[48] R.Z. Le Geros, Calcium phosphate-based osteoinductive materials, Chem. Rev. 108 (2008) 474253.
[49] E. Garcia-gareta, M.J. Coathup, G.W. Blunn, Osteoinduction of bone grafting materials for bone repair and regeneration, Bone 81 (2015).
[50] V.V. Anusha Thampi, M. Prabhu, K. Kavitha, P. Manivasakan, P. Prabu, V. Rajendran, et al., Ceram. Int. 40 (2014) 14355–14365.
[51] S. Ghosh, N. Dandapat, V.K. Balla, Mater. Today Proc. 2 (2015) 1326–1331.
[52] R.J. Furlong, J.F. Osborn, Fixation of hip prostheses by hydroxyapatite ceramic coatings, J. Bone Jt. Surg. 73 (5) (1991) 741–745.
[53] K. Soballe, E.S. Hansen, H. Brockstedtrasmussen, C. Bunger, Hydroxyapatite coating converts fibrous tissue to bone around loaded implants, J. Bone Jt. Surg. 75 (2) (1993) 270–278.
[54] M. Vallet-Regi, Bioceramics: from bone substitutes to nanoparticles for drug delivery, Pure Appl. Chem. (2019). aop.
[55] R.A. Martin, S. Yue, J.V. Hanna, P.D. Lee, R.J. Newport, M.E. Smith, et al., Characterizing the hierarchical structures of bioactive sol-gel silicate glass and hybrid scaffolds for bone regeneration, Phil. Trans. R. Soc. A370 (2012) 1422–1443.
[56] Z. Hong, R.L. Reis, J.F. Mano, Preparation and in vitro characterization of novel bioactive glass ceramic nanoparticles, J. Biomed. Mater. Res. 88 (2009) 304–313.
[57] E. Zeimaran, S. Pourshahrestani, I. Djordjevic, B. Pingguan-Murphy, N.A. Kadri, M.R. Towler, Bioactive glass reinforced elastomer composites for skeletal regeneration: a review, Mater. Sci. Eng. C. 53 (2015) 175–188.
[58] R. Das, Application of nanobioceramics in bone tissue engineering, Nanobiomaterials in Hard Tissue Engineering, Elsevier, 2016, pp. 353–379.
[59] C. Wu, J. Chang, Mesoporous bioactive glasses: structure characteristics, drug/growth factor delivery and bone regeneration application, Interface Focus. 2 (2012) 292–306.
[60] O. Salata, Applications of nanoparticles in biology and medicine, J. Nanobiotechnol. (2004).
[61] Y.F. Zhao, S.C.J. Loo, Y.Z. Chen., F.Y.C. Boey, J. Ma, In situ SAXRD study of sol–gel induced well-ordered mesoporous bioglasses for drug delivery, J. Biomed. Mater. Res. 85 (4) (2008) 1032–1042. Available from: https://doi.org/10.1002/jbm.a.31545.
[62] W. Xia, J. Chang, Well-ordered mesoporous bioactive glasses (MBG): a promising bioactive drug delivery system, J. Control. Release. 110 (2006) 522–530.
[63] E. Garcia-gareta, Collagen, from Tissue Culture to Biomaterials. Tissue Engineering and Beyond, Ist edition, Cambridge Scholars Publishing, 2019.
[64] B. Marelli, C.E. Ghezzi, D. Mohn, W.J. Stark, J.E. Barralet, A.R. Boccaccini, et al., Accelerated mineralization of dense collagen-nano bioactive glass hybrid gels increases scaffold stiffness and regulates osteoblastic function, Biomaterials 32 (2011) 891526.

[65] K. Maji, S. Dasgupta, K. Pramanik, A. Bissoyi, Preparation and evaluation of gelatin-chitosan-nanobioglass 3D porous scaffold for bone tissue engineering, Int. J. Biomater. 2016 (2016) 9825659. Available from: https://doi.org/10.1155/2016/9825659.

[66] F. Hafezi, F. Hosseinnejad, A.A.I. Fooladi, M.S. Mafi, A. Amiri, M.R. Nourani, Transplantation of nano-bioglass/gelatin scaffold in a nonautogenous setting for bone regeneration in a rabbit ulna, J. Mater. Sci. Mater Med. 23 (2012) 278392.

[67] H.-C. Liu, L.-L. E, D.-S. Wang, F. Su, X. Wu, Z.-P. Shi, et al., Reconstruction of alveolar bone defects using bone morphogenetic protein 2 mediated rabbit dental pulp stem cells seeded on nano-hydroxyapatite/collagen/Poly (L-lactide), Tissue Eng., A 17 (2011) 241733.

[68] J.C. Forero, E. Roa, J.G. Reyes, C. Acevedo, N. Osses, Development of useful biomaterial for bone tissue engineering by incorporating nano-copper-zinc alloy (nCuZn) in chitosan/gelatin/nano-hydroxyapatite (Ch/G/nHAp) scaffold, Materials (Basel), 10, 2017.

[69] J.S. Lee, S.D. Baek, J. Venkatesan, I. Bhatnagar, H.K. Chang, H.T. Kim, et al., In vivo study of chitosan-natural nano-hydroxyapatite scaffolds for bone tissue regeneration, Int. J. Biol. Macromol. 67 (2014) 3606.

[70] X. Ma, Y. Wang, H. Guo, J. Wang, Nano-hydroxyapatite/chitosan sponge-like biocomposite for repairing of rat calvarial critical-sized bone defect, J. Bioact. Compat. Polym. 26 (2011) 33546.

[71] B. Lowe, J. Venkatesan, S. Anil, M.S. Shim, S.K. Kim, Preparation and characterization of chitosan-natural nano hydroxyapatite-fucoidan nanocomposites for bone tissue engineering, Int. J. Biol. Macromol. 93 (2016) 147987.

[72] S. Sowmya, P.T. Sudheesh Kumar, K.P. Chennazhi, S.V. Nair, H. Tamura, R. Jayakumar, Biocompatible β-chitin hydrogel/nanobioactive glass ceramic nanocomposite scaffolds for periodontal bone regeneration, Trends Biomater. Artif. Organs 25 (2011) 1−11.

[73] M. Peter, N.S. Binulal, S.V. Nair, N. Selvamurugan, H. Tamura, R. Jayakumar, Novel biodegradable chitosan−gelatin/nano-bioactive glass ceramic composite scaffolds for alveolar bone tissue engineering, Chem. Eng. J. 158 (2010) 353−361.

[74] A.T. Young, J.H. Kang, D.J. Kang, J. Venkatesan, H.K. Chang, I. Bhatnagar, et al., Interaction of stem cells with nano hydroxyapatite-fucoidan bionanocomposites for bone tissue regeneration, Int. J. Biol. Macromol. 93 (2016) 1488−1491.

[75] S. Saravanan, S. Nethala, S. Pattnaik, A. Tripathi, A. Moorthi, N. Selvamurugan, Preparation, characterization and antimicrobial activity of a bio-composite scaffold containing chitosan/nano-hydroxyapatite/nano-silver for bone tissue engineering, Int. J. Biol. Macromol. 49 (2) (2011) 188−193.

[76] Y. Zhang, D. Zhou, J. Chen, X. Zhang, X. Li, W. Zhao, et al., Biomaterials based on marine resources for 3D bioprinting applications, Mar. Drugs 17 (2019).

[77] R. Kumar Saini, L. Prasad Bagri, A.K. Bajpai, Nano-silver hydroxyapatite based antibacterial 3D scaffolds of gelatin/alginate/poly(vinyl alcohol) for bone tissue engineering applications, Colloids Surf. B Biointerfaces 177 (2019) 21118.

[78] M. Nabavinia, A.B. Khoshfetrat, H. Naderi-Meshkin, Nano-hydroxyapatite-alginate-gelatin microcapsule as a potential osteogenic building block for modular bone tissue engineering, Mater. Sci. Eng. C 97 (2019) 6777.

[79] U. Rottensteiner, B. Sarker, D. Heusinger, D. Dafinova, S.N. Rath, J.P. Beier, et al., In vitro and in vivo biocompatibility of alginate dialdehyde/gelatin hydrogels with and without nanoscaled bioactive glass for bone tissue engineering applications, Materials 7 (2014).

[80] T. Masaki, J. Am. Ceram. Soc. 69 (1986) 638−640.

[81] I. Sailer, B.E. Pjetursson, M. Zwahlen, C.H.F. Hämmerle, Clin. Oral. Implant. Res. 18 (2007) 86−96.

[82] B.E. Pjetursson, U. Brägger, N.P. Lang, M. Zwahlen, Clin. Oral. Implant. Res. 18 (2007) 97−113.

[83] Y. Damestani, C.L. Reynold, J. Szu, M.S. Hsu, Y. Kodera, D.K. Binder, et al., Nanomed. Nanotechnol. Biol. Med. 9 (2013) 1135−1138.

[84] C. Ângela, M. Volpato M, L. Gustavo, D.A. Garbelotto, M.C. Fredel, F. Bondioli, Application of zirconia in dentistry: biological, mechanical and optical considerations, Adv. Ceram. (2011) 397−420.

[85] I.K. Konstantinidis, S. Jacoby, M. Rädel, K. Böning, J. Dent. 43 (2015) 87−93.
[86] D. Fabris, J. Mesquita-Guimarães, P. Pinto, J.C.M. Souza, M.C. Fredel, F.S. Silva, et al., Mechanical properties of zirconia periodic open cellular structures, Ceram. Int. 45 (13) (2019) 15799−15806.
[87] Z. Pang, A. Chughtai, I. Sailer, Y. Zhang, Dent. Mater. 31 (2015) 1198−1206.
[88] K.L. Lv, J.X. Zhang, X.C. Meng, X.Y. Li, Remineralization effect of the nano-HA toothpaste on artificial caries, Key Engineering Materials, 330, Trans Tech Publications, 2007, pp. 267−270.
[89] R. Haghgoo, M.B. Rezvani, M.S. Zeinabadi, Comparison of nano-hydroxyapatite and sodium fluoride mouth rinse for remineralization of incipient carious lesions, J. Dent. (Tehran, Iran.) 11 (4) (2014) 406.
[90] S. Naseri, S.N. Nazhat, Bioactive and soluble glasses for wound-healing applications, Bioactive Glasses, second ed., Elsevier, 2018, pp. 381−405.
[91] V. Miguez-Pacheco, L.L. Hench, A.R. Boccaccini, Bioactive glasses beyond bone and teeth: emerging applications in contact with soft tissues, Acta Biomater. 13 (2015) 1−15.
[92] H. Li, Z. Wu, Y. Zhou, J. Chang, 8 − Bioglass for skin regeneration, in: E. Garcıa-Gareta (Ed.), Biomaterials for Skin Repair and Regeneration, Woodhead Publishing, 2019, p. 22550.
[93] Shen, W. Gu, X. Yang, Y. Li, Z. Gou, Z. Jin. Systematic investigation of a new nanoscale bioactive glass on wound healing in vivo in comparison with the clinically applied 45S5 bioglass, Int. J. Regen. Med. (2018) 2613−5914. http://dx.doi.org/10.31487/j.RGM.2018.10.001.
[94] Q. Chen, J. Wu, Y. Liu, Y. Li, C. Zhang, W. Qi, et al., Electrospun chitosan/PVA/bioglass Nanofibrous membrane with spatially designed structure for accelerating chronic wound healing, Mater. Sci. Eng. C. 105 (2019) 110083.
[95] H. Xu, F. Lv, Y. Zhang, Z. Yi, Q. Ke, C. Wu, et al., Hierarchically micro-patterned nanofibrous scaffolds with a nanosized bio-glass surface for accelerating wound healing, Nanoscale 7 (2015) 1844652.
[96] X. Wang, Y. Zhou, L. Xia, C. Zhao, L. Chen, D. Yi, et al., Fabrication of nano-structured calcium silicate coatings with enhanced stability, bioactivity and osteogenic and angiogenic activity, Colloids Surf. B Biointerfaces 126 (2015) 358−366.
[97] T.A. Ostomel, Q. Shi, C.K. Tsung, H. Liang, G.D. Stucky, Spherical bioactive glass with enhanced rates of hydroxylapatite deposition and hemostatic activity, Small 2 (2006) 1261−1265.
[98] G. Hu, L. Xiao, D. Bi, H. Wang, H. Ma, G. Zhu, Antibacterial hemostatic dressings with nanoporous bioglass containing silver, Int. J. Nanomed. 7 (2012) 2613.
[99] M.F. Koudehi, A.A.I. Fooladi, K. Mansoori, Z. Jamalpoor, A. Amiri, M.R. Nourani, Preparation and evaluation of novel nano-bioglass/gelatin conduit for peripheral nerve regeneration, J. Mater. Sci. Mater Med. 25 (2014) 36373.
[100] F. Mohamadi, S. Ebrahimi-Borough, M.R. Nourani, K. Mansoori, M. Salehi, A.A. Alizadeh, et al., Enhanced sciatic nerve regeneration by human endometrial stem cells in an electrospun poly (ε-caprolactone)/collagen/NBG nerve conduit in rat, Artif. Cell Nanomed. Biotechnol. 46 (2018) 173143.
[101] B. Lin, G. Dun, D. Jin, Y. Du, Development of polypyrrole/collagen/nano-strontium substituted bioactive glass composite for boost sciatic nerve rejuvenation in vivo, Artif. Cell Nanomed. Biotechnol. 47 (2019) 342330.
[102] Z.-J. Rong, L.-J. Yang, B.-T. Cai, L.-X. Zhu, Y.-L. Cao, G.-F. Wu, et al., Porous nano-hydroxyapatite/collagen scaffold containing drug-loaded ADMPLGA microspheres for bone cancer treatment, J. Mater. Sci. Mater Med. 27 (2016) 89.
[103] H. Nakamura, J. Fang, H. Maeda, Expert. Opin. Drug. Deliv. 12 (2015) 53.
[104] A. Baeza, E. Guisasola, A. Torres-Pardo, J.M. González-Calbet, G.J. Melen, M. Ramírez, et al., Adv. Funct. Mater. 24 (2014) 4625.
[105] Y. Zhao, S. Sugiyama, T. Miller, X. Miao, Nanoceramics for blood-borne virus removal, Expert. Rev. Med. Dev. 5 (3) (2008) 395−405.
[106] N. Link, T.J. Brunner, I.A.J. Dreesen, W.J. Stark, M. Fussenegger, Inorganic nanoparticles for transfection of mammalian cells and removal of viruses from aqueous solutions, Biotech. Bioeng. 98 (5) (2007) 1083−1093.

[107] B. González, E. Ruiz, M.J. Feito, C. Lopez, D. Arcos, C. Ramírez, et al., J. Mater. Chem. 21 (2011) 4598.
[108] J.L. Paris, P. de la Torre, M. Manzano, M.V. Cabañas, A.I. Flores, M. Vallet-Regí, Acta Biomater. 33 (2016) 275.
[109] Y. Chen, H. Chen, J. Shi, Adv. Mater. 25 (2013) 3144.
[110] S.P. Hudson, R.F. Padera, R. Langer, D.S. Kohane, Biomaterials 29 (2008) 4045.
[111] J. Lu, M. Liong, Z. Li, J.I. Zink, F. Tamanoi, Small 6 (2010) 1794.
[112] M. Joglekar, R.A. Roggers, Y. Zhao, B.G. Trewyn, RSC Adv. 3 (2013) 2454.
[113] G.D. Venkatasubbu, S. Ramasamy, V. Ramakrishnan, J. Kumar, Nanocrystalline hydroxyapatite and zinc-doped hydroxyapatite as carrier material for controlled delivery of ciprofloxacin, 3 Biotech. 1 (3) (2011) 173–186.
[114] P. Shi, Y. Zu, X. Li, Q. Zou, H. Liu, L. Zhang, et al., Gentamicin-impregnated chitosan/nanohydroxyapatite/ethylcellulose microspheres granules for chronic osteomyelitis therapy, J. Biomed. Mater. Res., A 93 (3) (2010) 1020–1031.
[115] F. Zheng, S. Wang, S. Wen, M. Shen, M. Zhu, X. Shi, Characterization and antibacterial activity of amoxicillin-loaded electrospun nano-hydroxyapatite/poly(lactic-co-glycolic acid) composite nanofibers, Biomaterials 34 (4) (2013) 1402–1412.
[116] F. Zheng, S. Wang, M. Shen, M. Zhu, X. Shi, Antitumor efficacy of doxorubicin-loaded electrospun nano-hydroxyapatite–poly (lactic-co-glycolic acid) composite nanofibers, Polym. Chem. 4 (4) (2013) 933–941.

CHAPTER 10

Role and importance of hydroxyapatite in the healthcare sector

Rusham Das and Amit K. Goyal
Department of Pharmacy, School of Chemical Sciences and Pharmacy, Central University of Rajasthan, Ajmer, Rajasthan, India

10.1 Introduction

Given their biological applications, calcium phosphates (CaP) are among the best bioceramics. CaP gives rise to a collection of compounds known as "apatites," which refers to compounds having an identical hexagonal configuration with the space group P63/m. Werner first coined the term "apatite" in 1786. The word "apatite" originated from the Greek term "apataw" [1]. Since 1970, three bone replacement materials have been invented, namely, bioglass, sintered hydroxyapatite (HA) (molecular formula: $Ca_{10}(PO_4)_6(OH)_2$), and crystallized glass A-W. The latter two have been extensively investigated. Dr. H. Aoki of Tokyo Medical and Dental University invented sintered HA and its commercial manufacturing process in 1972. Understanding the potential of HA having a high affinity to real bones could not be done until 1975. Ceramics with a calcium phosphate origin have a chemical makeup that is very close to that of bones, making them an excellent bone transplant material [2,3]. Prof Nakamura of Kyoto University specifically suggests that there are three generations of artificial bones. The first generation comprises sintered HA and A-W glass ceramics, the second one has porous material and CaP cement, while the third generation consists of a unique complex material coupled with bone growth factors and bone cells. It is a challenge to increase the strength and reduce the Young's modulus of artificial bone to the comparable standards of natural bone. Coating HA with titanium alloy or the use of collagen−apatite composites are solutions to enhance strength. The development of next-generation apatites is being addressed by using composites and surface modification. [4]. Among the different varieties of calcium phosphate ceramics adopted in dentistry and orthopedics, HA holds good promise. It possesses good stability and has broad applications, such as in bone defect repair, maxillofacial reconstruction, as a pulp capping material, for replacement of immediate tooth root, and alveolar ridge augmentation. [1] HA is a nanomaterial of natural and biological origin which is found mostly in the hard tissues of mammals. These are either rodlike structures with diameters of about 20−40 nm as found in enamel, or long platelet structures 20 nm long and 5 nm thick as in bone and dentin. Though having many similar properties to

Industrial Applications of Nanoceramics
DOI: https://doi.org/10.1016/B978-0-323-88654-3.00011-1

other biomaterials, HA has some very unique properties which makes it a potential candidate for its applications in the medicinal fields [5].

On chemical evaluation, HA qualifies as the most stable candidate in the CaP fraternity having the highest degree of similarity in terms of composition and structure with the CaP apatitic phase seen in the teeth and bone [6]. Application of HA is observed in various forms in terms of composition and structure, which includes porous and dense, along with ionically substituted and calcium deficient, microcrystalline and nanocrystalline, granular, monolithic and colloidal and their biological responses also varied.

Regardless of the fact that HA is utilized to repair and regenerate bone deformities as well as in tissue engineering, mechanical problems like insufficient fracture toughness and failure caused by brittleness and fatigue are prevalent. On the other hand the resorption rate of HA is found to be very long and may extend up to decades [7], which is obviously a great drawback of HA. It has osteoinduction deficiency, poor loading capacity, and lacks targeting ability [8]. The current trend in the growing research on HA indicates that it is a material which holds immense potential that is yet to be utilized in the medicinal fields.

10.2 Major properties of hydroxyapatite

For identifying, manufacturing, and working with a compound and utilizing it to its fullest potential, it is necessary to know its physicochemical properties. A lot of investigations have been conducted in order to learn more about the characteristics of HA. There is also a calcium-deficient type of HA. Some of the major properties of HA and calcium-deficient HA are given below Table 10.1 [9,10]:

10.3 Synthesis of hydroxyapatite

The properties of HA powder is very important when preparing implants from them. Controlling the factors like stoichiometry, grain size, particle shape, orientation, purity

Table 10.1 Major properties of HA and calcium deficient HA .

Name of compound	Formula	Ca/P Molar ratio	pH stable range of aqueous medium at 25°C	Solubility at 25°C, g/L	Solubility at 25°C, $-\log(K_s)$
Hydroxyapatite (HA, Hap or OHAp)	$Ca_{10}(PO_4)_6(OH)_2$	1.67	9.5–12	~0.0003	116.8
Calcium-deficient hydroxyapatite (CDHA or Ca-def HA)[a]	$Ca_{10-x}(HPO_4)_x(PO_4)_{6-x}(OH)_{2-x}$ $(0<x<1)$	1.5–1.67	6.5–9.5	~0.0094	~85

[a]It's also known as "precipitated HA (PHA)" in some cases.

of phase, crystallinity, the nature of agglomeration of the powder, and its homogeneity are very important [11]. For the determination of the mechanical, physical, and biological properties, the abovementioned factors play an important role, as the powdered HA will be subjected to further addition to polymers for the production of biocomposites which are either applied as implant coatings or sintered to form the green bodies [11]. Various solid-state reactions and wet chemistry procedures were used to establish HA manufacturing techniques. Precipitation [12,13], hydrothermal interactions [14–16], sol–gel production [17–19], mechanochemical modification [11,20,21], emulsion and microemulsion formation [22] are all common methods used in today's research. A few of the more common and effective strategies are briefly addressed.

10.3.1 Precipitation

Precipitation is by far the most popular method for making HA powder. The extremely slow synthesis of HA occurs when dilute calcium hydroxide with a pH of 9.0 reacts with orthophosphoric acid in the range of 25°C–90°C. With steady agitation, the orthophosphoric acid is poured drop by drop to the diluted calcium hydroxide solution. It is also observed that the influence of higher temperature in the reaction results in the higher formation of the crystalline product. The following is the reaction that occurs during the procedure [23,24]:

$$3Ca_3(PO_4)_2 + Ca(OH)_2 \rightarrow Ca_{10}(PO_4)_6(OH)_2$$

Using the process of precipitation, ammonium hydroxide, di-ammonium hydrogen phosphate, and calcium nitrate can be considered to make HA. Here the use of ammonium hydroxide is done to maintain a constant pH which aids the faster rate of precipitation. Once the precipitate is obtained it should be washed properly for removal of ammonium hydroxides and nitrates [11,24].

While carrying out the precipitation process, continuous stirring is undertaken to enable slow calcium integration into the apatite framework to achieve the necessary stoichiometric Ca/P ratio. Crystals undergo morphological changes at this point and the needle-shaped crystals tend to become a block-like conformation. During calcium deficient HA preparation, the method could be done at a lower pH than 9.0 [11,23,24].

10.3.2 Hydrothermal method

A simple hydrothermal reaction of phosphate along with solutions of calcium are done in an environment where excessive temperature and pressure are provided in order to obtain HA [14,16,25–28]. There are a huge number of phosphate and calcium salts used as starting materials for HA that have been reported to date. Calcium carbonate, calcium hydroxide, calcium chloride, calcium nitrate, diammonium hydrogen phosphate, calcium hydrogen phosphate, and dipotassium hydrogen phosphate are all examples of calcium compounds. The reaction is carried out for 24 hours at a temperature ranging from 60°C−250°C and the end product is obtained in agglomerated form. The development of nanoparticles, nanowhiskers, and nanorods is reported using the hydrothermal technique. The reaction involved in this process is as follows [28,29]:

$$4Ca(OH) + 6CaHPO_4 \bullet 2H_2O \rightarrow Ca_{10}(PO_4)_6(OH)_2 + 18H_2O$$

10.3.3 Sol−gel synthesis

This type of synthesis is done with the aid of sol−gel materials, which are generally obtained by any one of the following techniques: hypercritical drying, gel formation of colloid-based powders, regulating the condensation and hydrolysis in the precursor followed by incorporation of a drying process at ambient temperature [30−32]. This method provides some advantages like a low temperature requirement, more control on the formation of the particular kind of phase, and better phase purity. There are some drawbacks in this technique like the costly starting materials [33] and it shows difficulty to hydrolyze phosphates.

10.4 Solid-state method

This method is not a very common method to be adopted for the preparation of HA but the main advantages that make it effective are its simplicity and cost effectiveness. In this method, $Ca(OH)_2$ and β-TCP are mixed in the ratio of 3:3.4, followed by wet milling after mixing properly in water. Then the casting of this mixture is done to prepare the bodies. Then follows the drying, and finally the sintering. The application of high heat at a temperature at around 1000°C or above is mandatory and sintering time should be a minimum of 8 hours in order to obtain HA with high crystallinity and phase purity. From the studies it was also interpreted that the most critical factor of this process is the sintering temperature in order to get pure HA. The pH was observed to be associated with particle agglomeration [34]. The following is the reaction that occurs during the procedure:

$$3Ca_3(PO_4)_2(\beta\text{-TCP}) + Ca(OH)_2 \rightarrow Ca_{10}(PO_4)_6(OH)_2$$

10.4.1 Mechanochemical method

This technique was used to generate nano-HA particles. The dry powders of diammonium hydrogen phosphate and calcium hydroxide were used in this experiment. The mixture was then dry milled at different speeds [20,21]. Following 2 hours of processing of the powders, HA was generated, and then after 6 hours of milling of the powder combination, nano-HA particles with particle size between 10 and 50 nm were obtained.

10.4.2 Gel casting method

A method was devised using the gel casting technique to develop dense HA ceramics. Uniform green HA ceramic bodies were prepared using a simple process of gel casting. The sintering of these compacts was found to be more consistent than that of the dry pressing procedure [35]. Sintering took place over 2 hours at temperatures ranging from 1100°C to 1250°C. The compacts' sintering properties were evaluated using density measurements, microhardness changes, and scanning electron microscopy. The maximum density for gel cast specimens was attained after being sintered for over 2 hours at 1200°C. The microstructure of the content was observed to be very consistent.

The experiment was designed in two stages where the preparation of the HA powder was done at first, followed by the gel casting to develop a uniform and high-density ceramic.

The filter cake drying method was adopted for this study and proper evaluation was done, where it was revealed that this new method was good for the production of green bodies. The homogeneous stacking of the particles, combined with the delayed evaporation of the solvent from the gel interstitials, is a benefit of this approach. These compacts are better sintered with respect to the ones produced by the commonly used solid-state route. The microhardness and microstructural uniformity values were found to be better for the samples produced by this route. It has only a small drawback which is its prolonged drying time on which work is ongoing [36].

Prashant N. Kumta et al. developed a unique synthesis and characterization of nanostructured CaP for biological applications in their study. A strategy was devised to use CaP as genome engineering vectors. There are a variety of high and low temperatures available for the synthesis of HA or Brushite [37,38]. However, TCP is mostly produced at high temperatures [39]. For the production of nanostructured HA, TCP, and Brushite phases, a novel low-temperature approach was established. This new approach under physiological conditions formed nanosized and stoichiometric HA. Keeping in mind the study of CaP as gene delivery entities, the method of synthesis was developed in such a way that it is biocompatible with the biological systems like DNA, cells and proteins. Though HA type CaP phases are well known for their use in gene delivery [40,41], due to the unavailability of a suitable biocompatible method for synthesis, little knowledge is present about the other forms of CaP as gene delivery entities. Ionic substitution was performed in addition to the physiologically acceptable manufacturing approach for CaP structures to give adequate resources for studying DNA-to-particle relationships so that assessment can be done for the effect of the substitutions of the ions on the extent of cellular DNA utilization following its release to the cellular nucleus.

Crystalline brushite platelets (pH 7.5 stable) are formed by substituting calcium with 14% magnesium. Unique spherical nanostructured morphological structures of brushite are formed due to further substitution. Nanocrystals (80 nm) of Magnesium substituted Tricalcium Phosphate (TCMP) in rosette form with a large specific surface area ($200m^2/g$) can be created. The uniqueness of this method can be considered in the high stability of the generated phases of HA, β-TCMP and brushite in various physiological conditions, enabling them to be potential entities to be used more commonly in biological systems and also as vectors that can be used for the delivery of the nonviral genetic materials. The obtained nanocrystalline phosphate was tested for composition, structure, morphology, and thermal stability. The in vitro transfection results were also discussed [1,40,41].

10.5 Applications of hydroxyapatite

10.5.1 Hard tissue repair/engineering

The structure of the bones has a role in a variety of metabolic and mechanical operations. Creating a scaffold with the ability to possess and exhibit the mechanical features of osseous matter along with the quality to serve the additional features associated with bone is of prime importance. Therefore, crafting scaffolds mimicking the pleiotropic property of bone is a target of research in bone tissue engineering.

HA shows similarity with the inorganic component of teeth and bone for which it is extensively used in the orthopedic and dental industry. HA bioceramics are used as granules and powders in filling up tooth defects. It is also used in bone cements and toothpastes, small unloaded implants, like the one used in middle ear and also in scaffolds with a porous nature which function as temporary substrates for cell and new bone formation in nonload-bearing sites. HA finds its use as a biologically active and compatible coating used on metal implants in the dental regions and also in prosthetics used in hip joints where the need for load properties is a must [42,43].

Research has revealed that the integrity of synthetic HA grafts with the adjoining host bone is found to be promising as they also encourage the development of new osseous materials, which leads to the establishment of bonds with the bones that are freshly formed, paving the way for regenerating osseous tissues that are affected. Bone grafts and allografts with low HA content were found to be inferior to the ones with high HA content, which showed superior bonding, integration, and induction with host bones [44]. HA nanoparticles are extensively used as a matrix in bone tissue grafting and also as a bioactive component for the fabrication of biocomposite grafts of bone [45]. It was also revealed in a recent study that new bone formation along with mineralization is promoted by nano-/microstructured HA bioceramics when compared to HA with smooth surfaces. The findings of 8 weeks of implantation of macroporous HA bioceramics scaffolds in a major calvarial defect in rats indicated that the percentage of new bone formation in a micro–nano hybrid surface of the pore wall was $14.79 \pm 2.22\%$ while that of the smooth surface was $5.08 \pm 1.09\%$. This showed that the micro–nano hybrid nature can be a very influential parameter to be considered for the design and development of bone grafts [46,47]. Factors like micropore size and porosity, ion substitution, shape, crystallinity, and size control the biological responses like proliferation, differentiation, cell adhesion, spreading, and bone regeneration [48,49].

Nitta Gelatin Inc. collaboratively designed an artificial bone with composition of protein-I type collagen and HA with a flexural strength of 50 MPa, which is similar to human bone (40–200 MPa). The designed composite when transplanted to the cranial bone of a rat started getting replaced with real bone within 2 weeks [2].

Enamel is composed of around 97 wt.% needle-like nanocrystals of apatites which are bundled in the conformation of parallel prisms which give its biological protection and mechanical strength. Enamel has its ability to regenerate in living organisms but this process may be disturbed due to substantial mineral loss. Hence a study has reported that HA with a size of around 20 nm is better than any other materials used to date for repairing the eroded enamel perfectly. The remineralization of the eroded enamel takes place by the strong absorption of the HA to its surface.

Studies have also stated that HA nanocrystals shows promise to enhance cartilage regeneration by acting as a bioactive component [48]. HA nanoparticles were compared with PLGA scaffolds and it was observed that the HA showed apparent initiation of adhesion and proliferation of chondrocytes, and without any gene transfer or growth factors had better cartilage regenerative capability [50,51]. Apart from giving good bioactivity, the addition of HA components results in the improvement of mechanical properties and also an antifriction property in cartilage regeneration [50]. The most effective form is a 2D HA crystal which is used for composite material stiffening. A 1D structure is less effective than a 2D structure but is more effective than the spherical-shaped ones [52].

Transitions to polymers have markedly resulted in a new era of biocomposites where HA is used for adding bioactivity to the scaffolds followed by adding compressive strength. The unique properties of silk polymers like elasticity, biocompatibility, biodegradability, and flexibility makes them good candidates to be used in medical fields when combined with HA [53,54]. This approach will take care of their individual weaknesses. The low toughness and pliability of HA would be enhanced by silk polymers, while HA will compensate for their low osteoconductivity [55,56]. These composite materials exhibit a higher cell adhesive nature along with the capacity of estrogenic differentiation compared to their individual components [57,58]. Nanofibers made from ε-polycaprolactone strengthened with HA were found to be a convenient approach as they help to improve fibroblast proliferation and adhesion capability, as well as tensile strength, elastic modulus, and fracture toughness, when compared to typical HA-free scaffolds. [59].

For enhancement of angiogenesis and access of cells to nutrients, the choice of right synthesis method to provide the desired porosity of the scaffold is very important. Addition of HA in scaffolds made of sphere-shaped entities of chitosan in a system where sodium tripolyphosphate is present and where it is functioning like a cross linker, resulted in the formation of a material with greater strength for ultimate compression, particle size around 2−10 μm, and wider particle distribution. It was seen to facilitate angiogenesis along with access of cells to nutrients [56].

Biocompatibility is of utmost importance in scaffold engineering usually achieved by the incorporation of HA into a comparatively poor biocompatible polymer. A study revealed the design of a composite which enhanced the porosity, biocompatibility and mechanical strength of graphene hydrogels by the addition HA and chitosan to graphene oxide [60].

Some research has looked at the addition of metallic or oxides particles to the HA scaffold. The adhesion and proliferation of osteoblasts [61] are improved by the presence of magnesium oxide in polylactic acid nanocomposites [62].

Control of bond strength and the interface between the polymer and the bioceramic phase is one of the most challenging issues. Impact modifiers added to the HA-PLA nanocomposite lead to an improvement in the impact properties. Similarly, GO use shows enhancement in interfacial bond strength amongst the HA of scaffolds and polyetheretherketone (PEEK) [63].

Impregnation of drugs like growth factors and bone morphogenetic proteins to HA scaffolds for enhancing cell differentiation and other effects are now being studied extensively. β-Cyclodextrin was used in one study to coat HA in order to enhance the drug binding of simvastatin and adamantane. The drug−HA association was observed to be arbitrated by β-cyclodextrin. The synthesized composite was shown to improve osteogenic differentiation in adipose-derived stromal cells. [64,65].

HA particles impregnated with ions are also coming into the field of current research. HA doped with cobalt was used for triggering angiogenesis of osseous tissues, and hence has been described as good for in vivo uses [66]. However, bone therapies involving strontium are connected to vascular fatalities and hence it has lost its approval for therapeutic uses in the EU [67].

CaP cements [68,69] are considered a good form of HA-based entity which generally set themselves and then harden into HA after being placed into the void of the bone. The cements that are based on a combination of di-CaP [70] and tetra-CaP set in the form of HA, while that of solely tri-CaP ones [71−73] sets like di-CaP cements. These are ideal to fill small-sized defects which are less prone to wear and tear, mostly in the craniofacial region. These are also being used, in addition to screw fixation and titanium implant, in reinforcing burst fractures and for augmentation of osteoporotic vertebral bodies [74]. The main drawback of these cements is not being able to control their internal porosity, which is a very important parameter controlling the transfer of nutrient, ingrowth of cells, and vascularization. To address this specific issue, a method was developed where silica and gelatin was added to the HA-forming cement [75]. This method was beneficial as the gelatin, after dissolution in the cytological environment, formed internal pores. The gelatin imparts its thermosetting properties to the cement before leaving the hardened HA graft which shows setting at 36.5°C is quick, whereas setting at room temperature takes longer.

10.5.2 Soft tissue repairs

HA has also generated importance in the regeneration of soft tissue such as muscle, skin, and gums due to their tremendous biocompatibility. Studies have revealed that HA can activate fibroblasts and has the capacity to accumulate vessel endothelial cells, which aids skin wound healing [76]. HA bioceramics have strong adherence and tight contact with skin tissues, which does not allow tunnel bacterial infections, and hence it has been concluded that HA can be used in percutaneous devices [77]. A successful development of composite products from an HA component was used in soft tissue augmentation [78]. Another study indicated that HA nanoparticles can initiate axonal outgrowth, which shows its potential as a novel therapy for nerve injury [79].

10.5.3 Implants

Implants are devices usually intended for use as lasting substitutes for different types of body tissues, specifically for bones and teeth. Implants made of metals like stainless steel, cobalt or titanium or their alloys, though providing good mechanical strength, lower elastic modulus, and good corrosion resistance have poor biocompatibility and low osteointegration [80]. Though HA has many advantages, the poor creep resistance, low Weibull modulus, low fatigue strength, brittleness, and poor fracture toughness give limitations to its application in terms of load bearing [81]. A study was conducted to evaluate the use of implants made of HA ceramic for cranioplasty in children [82]. As HA implants have poor mechanical strength, they are prone to get damaged even with a minor injury, and many investigators have used only HA to prepare implants. However, the composite forms were far better than these traditional HA implants [83].

K. Yamashita et al. investigated electrically polarized HA and found that electrical polarization is possible for apatite and the stored energy can be measured. On implantation of this charged apatite on a dog to monitor its ability to form new bones, they discovered in less than a week, fresh bone development was taking place on the negative surface. It was confirmed that for 6 months, the polarization can be retained in vivo and hence it can find immense use in dental and orthopedic sectors.

Dr. T. Matsumoto et al. from Osaka Dental School carried out research on the surface modification of HA with a carboxyl group with the vision that bone growth factors BMP and b-FGF possess affinity toward negatively charged surfaces. They successfully produced and implanted it in rats and came to a very important conclusion that the amount of the carboxyl group located on the surface affects cell adhesion and the load of the attached protein. Prof K. Shinomiya from Tokyo Medical and Dental University implanted a block prepared from HA and hemp attached collagen composite in a beagle. Generation of bone was observed [2].

HA-infused metallic implants hold more promise. Titanium and related alloys are commonly utilized to prepare such implants for their morphological and mechanical qualities. On metallic implants, HA is also employed as a coating. Plasma spray [84], electrophoretic deposition [85], pulsated laser deposition [86], sputtering techniques [87], biomimetic precipitation from sol gel [88,89], and simulated body fluid (SBF) [90] are just a few of the technologies that may be used to cover the surface. Though there are other methods like collagen coating [91] and hydrofluoric acid treatment [92], coating a metallic implant with HA is by far the easiest and cheapest method.

TiO_2 nanotubes coated with HA using sol–gel method showed an improvement in cellular response, osteointegration, and biological characteristics [93]. It also showed extended cell survival which is hypothesized to be an effect of the addition of HA in the fabrication and also the improvement of roughness of surface.

Another study reported that a coating of HA with silver and zinc oxide nanoparticles was done for implants made from titanium alloy (Ti6Al4V). The laser cladding method was used for this process. It showed quick osseointegration ability, good osteogenesis, and resistance to bacteria [94]. A synergistic effect was observed due to the presence of silver and magnesium oxide showing antimicrobial activity and increased osteoblast activity resulting in enhanced osseous matter growth. Generation of fresh bones was observed surrounding implants incorporated into the femurs of rabbits for 5 weeks.

In vivo and in vitro studies were done for a composite coating with HA with polycaprolactone and graphene oxide which was deposited on Ti6Al4V with the aid of the electrodeposition method [95]. The addition of the polycaproloactone and graphene oxide showed a distinct effect on fracture toughness, biodegradability, biocompatibility, corrosion resistance, and other mechanical properties in the SBF medium. It was also observed that these coated implants showed more activity against *Staphylococcus aureus* and *Escherichia coli* rather than the uncoated ones. They also induced higher proliferation of osteoblasts and growth rates. The coating acted as a proper substrate for the growth of bones and integration for its porosity which resulted in bone growth in rats within 28 days of the implantation [95].

The mechanical properties and morphological features of nonstoichiometric HA coating materials sintered upon air were investigated. The study involved the development of two different types of nonstoichiometric apatite coatings on Ti6Al4V substrates using hydrothermal synthesis and electrodeposition [96]. One of the apatite coatings was made of needle-shaped highly pure HA which was calcium deficient. The other one was made of the calcium-deficient HA, which was needlelike in shape, and brushite, which was plate-like in appearance. The assessment of phase transformation, structure, and binding capabilities of coatings sintered from 450°C to 950°C in an air atmosphere was given in the report. A hole drifting technique was employed to measure the residual stress of the calcium-deficient HA variety of coating and also the variety which was sintered at around 800°C. The structural features of the coating (Ti6Al4V) samples were similar to dense TiO_2 film/Ti6Al4V and biphasic [HA + $-Ca_3(PO_4)_2$] coatings, with the residual stresses of the coatings increasing significantly when sintered at a greater temperature. Bonding strength was also shown to be significantly enhanced as a result of this. Hence it can have a potential for new bone growth and also has a role in the bonding strength of the coating to the bone [97].

Titanium alloy was coated with a coating of HA/polydopamine on the inner sides with a three-dimensional porous structure. Polydopamine was applied to the implant before a layer of HA was applied on top. This HA coating increased the mechanical stability and tolerance, cell proliferation, gene expression related to osteoblasts, surface roughness, and formation of new bones in rabbits was observed around 4 to 12 weeks after inserting the implants. This was observed using a laser scanning microscope, micron X-ray 3D imaging system, CCK-8 assay, and polymerase chain reaction [98,99].

Metallic implants are coated with HA coatings mostly using the electrospraying method. This method was adopted to coat the titanium implants using HA which is loaded with biphosphonate. In osteoporotic rats [100] it was observed that this type of implant exhibited growth in osseous matter and peri-implant bone volume from their surfaces compared to the controlled ones [101].

The electrophoretic deposition process was used to deposit the HA coating on the Mg−3Zn alloy [85]. This alloy is considered to be one of the best choices for slowly resorbable, temporary orthopedic implants, as zinc prevents the high corrosion rate of Mg and Mg itself is biodegradable in nature. HA coating was also done on titanium implants using a method of radiofrequency magnetron sputtering [102,103]. The tilt angle is very crucial for this method for proper deposition of HA [104].

Apart from composite coatings, research has also been done on coatings of HA impregnated with foreign ions. Studies showed that HA having 10% strontium, magnesium, and zinc ions showed better formation of bones after implantation in female rats for 12 weeks [105]. It also exhibited better osseointegration and mechanical strength. Over all the other coatings, the strontium HA coating had the best results. The vacuum plasma spray method was used to spray silver ion-impregnated HA on stainless steel and titanium implants [106]. Antibacterial assays revealed that the silver ions emitted out from the implants had a decreased bacterial cell adherence and inhibited the development of biofilms on the surface of the implant. Human primary bone cells showed no signs of cytotoxicity. Another study found that implantable devices made of titanium plated with Ag ion HA had high antimicrobial property against *S. aureus.* [107]. A biomimetic method was developed where a supersaturated solution with cerium and collagenous materials fused with HA was plated on implantable devices made of titanium [108]. Cerium ions reduced the size of HA particles while also increasing their crystallinity. This improved dissolution and biodegradability, as well as antimicrobial efficacy against *S. aureus* and *E. coli*. As a result, this coating has been proven to be acceptable for biomedical applications, particularly in orthopedic implants.

The fixation and stability of strontium-fused HA coated titanium implants by the effect of human parathyroid hormone was studied and it was deduced that the Sr^{2+} and PTH showed increased formation of bone and improved fixation of the implant in a rat model [109]. A study indicated that silicon/silver-fused HA coated on titanium nanotube implants was found to be better than uncoated and HA-only-coated implants [110]. The implants' antimicrobial property and biocompatibility were both boosted by the discharge of Si and Ag ions. The resistance to corrosion, higher strength in adhesion, the viability in cell, the inhibition of the formation of biofilm, and bacterial proliferation were observed in this coating. Though this coating showed good osseointegration and enhanced the osteoblasts growth, it showed lower mechanical properties.

Bone morphogenetic protein 2 (BMP2), a growth factor, was fused to silver-impregnated HA and was coated on titanium implants with the help of the electrochemical deposition method. It was observed that the implant showed sustained release of BMP2 for over 25 days and Ag for over 30 days. It was also observed that the bone formation was enhanced along with antibacterial effect due to the stimulatory action of BMP2 [111].

Summarizing, we can conclude that the application of HA in implants shows positive effects and results, specifically in in vivo models. However, to gain the approval of these type of implants for use in patients needs more target-oriented research. It is necessary also to enhance the mechanical qualities of HA. It is still challenging to use on metallic implants of other metals like chromium/cobalt alloys. Detachment and delamination of the coatings on the exterior of metallic implants is a significant disadvantage, resulting in metal oxide or biologically inert material being exposed to the tissues. [112].

A study was focused in the precipitation of apatite after incubation in various solutions to see the influence of seed species. Apart from HA, there is β-tricalcium phosphate (β-TCP) which possesses good reactivity for use for the same purpose. Biphasic calcium phosphate (BCP), a mixture of both HA (60%) and β-TCP (40%) with the qualities of both HA and β-TCP, has the ability to offer optimum bioactivity as well as a scaffold framework for implants. [113–116].

After implantation in osseous or nonosseous sites [115,117–119] or postincubation in different solutions [115,120,121], dissolution of ceramic results in the supersaturation of Ca^{2+} and PO_4^{3-} ions in the microenvironments [115], causing the precipitation of carbonated apatitic microcrystals in the microporosity of the ceramics [115]. This formation is caused by the secondary nucleation process which takes place due to the dissolution–precipitation process [115].

Precipitation is only not the outcome of the precipitation of inorganic salt from a supersaturated solution [115], but it also includes biomolecules such as enzymes and proteins that have a function in the alignment, form, size, and density of microcrystals [116,117,122–124]. Proteins aid mineralization and can either enhance crystal nucleation as nucleating agents or hinder crystal development by attaching to specific crystallographic surfaces [125,126].

Although the role of phosphoproteins in bone mineralization is well understood, subsequent research has showed that two glycoproteins, fibronectin and vitronectin, exhibit strong affinity for calcium ions and HA crystals [115,127–129]. Fibronectin and vitronectin both have an importance as a ligand in tissue remodeling, cell adhesion, and cell differentiation [35].

Incubation of blocks made of BCP was done in different biological fluid solutions to study the dissolution precipitation process. All these are ionic SBF solutions [115] containing or not including different proteins and macromolecules (fibronectin, vitronectin, albumin, and poly-L-glutamic acid) [115]. The research aimed to find out the structural and morphological characteristics of precipitated microcrystals, along with evaluation of the impact of various proteins and macromolecules on spatial distribution, shape, and size of the precipitates. Secondary nucleation resulted in the generation of apatite-precipitated microcrystals all around ceramic crystals, according to TEM examinations. In HA crystals, the microcrystals were parallel to the crystallographic planes, but in β-TCP, the orientation was reversed. In the case of the fibronectin, vitronectin, and poly-L-glutamic acid-containing SBF [115], smaller sized precipitates were obtained when compared to the SBF with no protein. Vitronectin and fibronectin in the empty spaces of the ceramic crystals started the crystal nucleation process. [115].

J. Barralet and colleagues used Wistar Rats to investigate the dissolving process of dense carbonate apatite subcutaneous implants. Their research was the first to show that subcutaneous dissolution of dense sintered HA, TCP, and carbonate apatite (CHA) in rats could be directly compared. Because bone mineral includes a significant quantity of carbonate [40,41], CHA has a chemical structure that is remarkably identical to that of bone mineral. Research agreed that HA and bones form a bond which led to the debate of whether reabsorption of HA takes place in vivo or not.

Comparative studies revealed that β-TCP gets rapidly removed from the site of implant due to an unknown combined effect of dissolution and/or resorption [130–133]. The neutralization process of $Ca(OH)_2$ by H_3PO_4 was used to make TCP and HA, whereas the precipitation procedure of $(NH_4)_3PO_4$ and $Ca(NO_3)_2$ was used to make TCA [134,135]. The ceramics were sintered at high temperature and cut into equal sized squares followed by polishing, cleaning, and soaking in 70% alcohol for a span of 24 hours before implantation. Three 10-week-old female Wistar Rats were selected as subjects and subcutaneous implantation of four pieces of each material was done into their dorsal region. The rats were killed after 4, 10, and 23 weeks, respectively, and the results were compared. HA lost 60 $mg/m^2/day$, β-TCP lost 520 $mg/m^2/day^1$, and 3.2 wt.% CHA lost 320 $mg/m^2/day$ by weight. No signs of giant cells was observed in histological sections while all the samples were fibrous tissue encapsulated.

The results clearly indicated that all the sintered ceramics tested dissolve to some extent after implantation in vivo in rats. The rate of dissolution for CHA is between HA and β-TCP. Initially the rates are not much different but over time the difference of the dissolution rates are clearly visible. TCP and CHA were also shown to be resilient to disintegration along the sample margins, but to a lesser extent. After X-ray investigation, no phase transition was discovered in the dissolution-resistant zone. The reason behind this was not discovered [3].

Norihisa Goto et al. inspected the adjuvant activities and the effects of local tissue irritation of aluminum hydroxide [136–138] and calcium phosphate [138] along with their other physical properties [138]. Their effects on local histopathological reactions, the adjuvanticity of total IgG antibody response to tetanus toxoid (TT) and ovalbumin (OVA) administered subcutaneously and the electron microscopical changes in macrophages, were studied in female guinea pigs (Hartley Strain). Active inflammatory responses including neutrophils (pseudoeosinophils) and foamy macrophages [138], which are connected to numerous multinuclear giant cells [138], were created by the association of CaP gel over a period of 4 weeks. Small lymphocytes, large cells, and macrophages with foamy cytoplasm were observed in granulomatous inflammatory responses at the injection site, which were induced by aluminum hydroxide gel (Al-Gel) during an 8-week period. Both Ca-Gel and Al-Gel were found to show the same severity of irritation of local tissues, only their durations varied. By the end of the 4th week, the local responses caused by calcium phosphate suspension (Ca-Sus) had entirely ended, but the local reactions generated by aluminum hydroxide suspension (Al-Sus) continued until the end of the 8th week. As evident by the electron microscopical evaluation, macrophages are damaged by both the Al-Gel and Al-Sus. Al-Gel shows distinctive strong adjuvant activity for TT or OVA when compared to any other adjuvant material. Up to a concentration of 3 mg/mL calcium phosphate, the adjuvant efficacy of calcium phosphate gel and calcium phosphate suspension was not seen. On Al-Sus-TT [138], a comparable dosage of 3 mg/mL of aluminum hydroxide expressed extremely poor levels of antibodies. The reports led to the conclusion that calcium phosphate adjuvant is not a good substitute for aluminum adjuvant [138].

Marie M. Griffiths et al. in their work on biomaterials and granulomas described that implants are extensively used for traumatized/diseased tissue repair but it is seen that most of the times they cause acute or chronic inflammation which results in rejection and tissue damage. It is usually found that inflammation takes place at the interface of biomaterial tissue. It represents plasma protein surface adsorption, neutrophil and macrophage invasion, complement activation, proteolytic enzyme production, inflammatory mediator release, and hyperplasia. Degradation of biomaterials can occur spontaneously or as a result of biological activity, resulting in component leakage into surrounding tissues and the likelihood of it entering the bloodstream. It has the potential to produce systemic poisoning as well as have an impact on distant places. Inflammation of the connective tissues activates the immune system, which leads to the breakdown of tolerance to tissue-specific antigens in genetically vulnerable people. Induced animal model having chronic arthritis, which is a localized destructive inflammation in the peripheral joints, was taken. Other models that were taken had autoimmune arthritis involving hyperimmunization with heterologous compounds like proteoglycan [139] or type II collagen [140] resulting in self-cartilage cross-reactivity. Some other nonimmune models like pristane arthritis [141] and oil-induced arthritis [142,143] were considered as they would show uncontrolled production of cytokines and also the formation of granuloma. For testing the biocompatibility of the probable immunogenic cationic substances, the antigen-induced arthritis models, which are developed by hyper immunization of a foreign antigen and injection to the joint cavity, are found to be ideal. The different types of biomaterial grafts used were HA, zirconium, aluminum, Teflon, and dental amalgam. The selection of these various models gave a proper idea for understanding the formation of granuloma and adverse interactions between implants and tissues in order to ensure the manufacture of safer biomaterials in the days to come [144].

10.6 Drug carriers

Having excellent biocompatibility, suitable morphology, surface composition, porous structure, size, low production cost and low toxicity, good storage stability, microbiological degradation inertia along with pH-dependent dissolution, protein/drug delivery and gene therapy might both benefit from the use of HA material. The rate of degradation of HA increases drastically with the lowering of pH from alkaline to acidic medium. The release of drug from the surfaces of HA is increased as a result of this. The pH dependent dissolution property is very intriguing as normal tissues show a pH of 7.4 while within the endosome −lysosome cell compartment, that is, the extracellular environments of solid tumors, the pH is found to be 5. This property enables favored drug release from the HA surface in a pH dependent manner in the tumor area. Release properties, therapeutic properties and the amount of drug loading are some important features which are dependent on the properties like shape, size, crystallinity, porous structure, aspect ratio, chemical composition, morphology, surface chemistry, etc. As indicated by some recent studies, 3D designed and structurally porous HA fabrication are the potential approaches to address the disadvantages like faster rate of release and capacity of drug loading of HA as drug carriers. HA has been widely explored for medication delivery across a wide range of treatments. Antibiotic loaded HA is found to be used in curing and preventing bone infections. *S. aureus*, *Pseudomonas aeruginosa*, methicillin-resistant *S. aureus* and *E. coli* microorganisms, the causative pathogen for osteomyelitis, are employed to check the efficiency of antibiotic-doped HA [145].

A chemical precipitation approach was used to create ciprofloxacin-loaded HA. Following a review of the in vitro release patterns, it was determined that the medication was distributed by diffusion. In this case, the drug ciprofloxacin was delivered in two novel carrier systems having high capacity for drug loading and properties of controlled releasing [146]. Solid HA nanoparticles and mesoporous hollow HA nanoparticles were two types of HA nanoparticles. The former nanocarrier showed promise in respect to its release which was pH dependent and followed the Fickian diffusion while its drug encapsulation efficiency was also found to be very high due to its porous and hollow structure [147]. Nanoparticles of HA loaded with ciprofloxacin, synthesize by in situ precipitation method showed prolonged and sustained release along with adjustable dose loading [148]. A different study was conducted for the development of a double sustained release system where ciprofloxacin was released along with gentamicin and norfloxacin from a composite made from anionic collagen and nonstoichiometric HA. These Composites were found to show sustained release of these drugs in levels much higher in concentration than that of the minimum inhibitory concentrations (MICs) resulting in the prevention of the growth of Gram-negative bacteria. Gentamicin was observed to be released within 48 hours from the matrix while norfloxacin and ciprofloxacin showed a release mechanism in 2 phases due to their complexation within the nonstoichiometric HA with the calcium ions. Hence it was concluded that dissociation and resorption of these complexes would deliver most of the ciprofloxacin and gentamicin drugs [149].

Good results like proper biocompatibility, pH sensitive and sustained drug release, enhanced flexibility, recyclability and improved capacity for drug loading [150] were observed in ultralong nanowires of HA doped with ciprofloxacin and silver nanoparticles. Bone grafts made from modified ciprofloxacin and strontium-fused HA showed controlled as well as sustained release of the drug for till 45 days. Assays showed that in osteoblasts MG-63 cells, the cell functionality and viability were enhanced with the addition of strontium which indicated the use of strontium-fused HA as a matrix for drug delivery [151].

Beads impregnated with Gentamicin was prepared using HA and evaluated for in vitro microbiological studies in opposition to bacteria which forms biofilms in infections of implants. Strains like *Staphylococcus epidermidis*, *E. coli*, *Streptococcus agalactiae*, *Streptococcus faecalis*, and *S. aureus* were evaluated and the developed beads showed the release of Gentamicin in therapeutic concentrations from the 1st hour itself proving itself to be effective specifically against the *E. coli*, *S. agalactiae*, and *S. epidermidis* strains. Also the released Gentamicin showed suppression of biofilms developed by the Gram-positive bacteria [152].

Tetracycline was loaded in a HA which was calcium deficient in nature and it also acted as the only carrier this material showed a loading capacity of 68% while it also indicated sustained release for a time span of 5 days in phosphate buffered saline of about 88% [153]. Microspheres made of HA was loaded with chlorhexidine and Langmuir–Freudlich adsorption isotherm was adopted in the loading process with a fair uptake [154].

Synthesis of porous composite nanofibers of HA and calcium oxide was done with a higher tetracycline loading capacity. The prolonged release of tetracycline was observed for a time span of 14 days while its antibacterial effect was observed till 7 days [155]. A strontium-incorporated HA, calcium oxide, and calcium carbonate composite mesoporous structured nanofiber was fabricated which did not show the burst release and displayed sustained release and antibacterial effects over 21 days. The loading of the tetracycline was around 0.972 mg/mg which indicated nearly 100% efficiency in loading [156].

Another formulation of tetracycline in injectable and hydrogel having polysaccharide and biodegradability was seen to show proper integrity with carbonate nanospheres and nanoparticles made of HA. The release behavior of tetracycline was observed to be biphasic showing quick initial release of 42.4% drug in the first 24 hours and a slow release of 14.6% drug over a span of 20 days [157].

Electrospun nanofibers of poly-ε-caprolactone, gelatin polymers, and HA composite were prepared and loaded with doxycycline. They showed low drug loading but also showed a biphasic release pattern, with 60% released instantly during the first hour and the rest released gradually over the next 55 hours. It was observed that the percentage of released drug per hour showed statistical similarity strengthening the evidence of slow, sustained release [158]. The effect of these nanofibers was also evaluated in the formation of bone by placing them in the first maxillary molar of male mice of 4 weeks age. For the comparative investigation, a control group was also included. Statistical data showed that the formation of new bone was found to be more after 8 weeks than in that of the controlled group [159].

Fixation pins of titanium dioxide was coated with tobramycin-fused nanoporous HA coating, showed sustained release for 8 days in PBS and it also maintained the concentration of antibiotic over the MIC throughout the entire timeframe [160]. The vancomycin delivered from typical HA and HA-collagen coatings on TiO_2 nanotubes was 92% for HA coatings, 85% for HA-collagen coatings, and 78% [99] for TiO_2 nanotubes [99] after 30 hours [161]. The release of vancomycin from vancomycin-loaded bone-like HA/polyamino-acid (VCBHA/PAA) has a three-phase profile [162] with two release peaks [162]. The initial 48 hours of vancomycin delivery were rapid, proceeded by a constant release phase. Researchers observed that employing VC-BHA/PAA can accomplish a prolonged period of vancomycin administration of 38 days [163] in vitro along with 42 days in vivo [164] in a rabbit model. As per 7-year-long clinical research, a calcium sulfate/HA liquid filler [162], to fill the void of the bone, packed with [162] vancomycin was used to treat 143 osteomyelitis sites on the lower extremity in 125 patients utilizing a percutaneous delivery technique. As a result, no recurrence of osteomyelitis was found in 96.15% of the treated patients, indicating that this approach is both safe and effective [165].

For VC loading and releasing, mesoporous strontium-substituted HA (SrHA) powders [162] having individual submicrometric irregular particles (SrHA-IPs) and nanostructured spherical particles (SrHA-SPs) [162] have been used. Despite the SrHA-SPs being larger than SrHA-IPs (0.88 ± 0.23 μm vs 0.14 ± 0.04 μm) [162], the specific surface areas of SrHA-SPs were still smaller than SrHA-IPs [162]. SrHA-IPs and SrHA-SPs had specific surface areas of 32 and 75 m^2/g, respectively, due to the fact that the SrHA-SP powder was roughly twice as porous as the SrHA-IP powder. As a result, the VC loading efficiencies of SrHA-SP and SrHA-IP powders [99] were 52% and 36%, respectively. Despite the fact that both powders showed consistent drug release profiles over time, the quantity of delivered drug and the rate of dissolution of SrHA-SP powder was greater than that of the SrHA-IP powder. Moreover, as compared to the SrHA-IP powder, the SrHA-SP powder not only released better medication[162], but its overall released quantity steadily grew until 40 hours and then remained stable as of its release rate. Despite this, both delivery devices had similar drug release [99] patterns with no burst or immediate release [99,166].

β-Lactam antibiotics are among the most often used and prescribed broad-spectrum antibiotics. HA nanocrystal surfaces [162] were altered with four monocyclic N-thiol-substituted ß-lactams, and HA loaded with 4 acetoxy-1-(methylthio)-azetidin-2-one [162] demonstrated a good efficacy, continuously delivering antibiotic at 0.4 mol.% a day with about 4.8%–5.8% efficiency of loading [167]. On the AZ31 magnesium alloy (3% aluminum and 1% zinc), a coating of PCL with biocompatibility was successfully fabricated with 2% w/w nano-HA and 1.4% w/w cephalexin. The completed product was submerged in Hank's balanced salt solution [162] for a period of 40 days, and the first day of immersion revealed that 70% of the initial amount of the medication was released [168].

Excipients such as polyvinyl alcohol (PVA) gels that have been refrozen and thawed seem to be good in practice. The initial burst release of 55% was observed in hybrid PVA/HA hydrogels carrying sodium salt of cefuroxime (CFR) within the first 5 hours, followed by a second burst release on day 4 of about 83%, and finally a plateau of roughly 96% over 2 weeks [169]. Another study used the matrix assisted pulsed laser evaporation (MAPLE [99]) method for deposition of bonded slim films depending on HA/poly(lactic-co-glycolic acid) (HA/PLGA [162]) infused to ceftriaxone or CFR, and the biological compatibility of these fine films had been affirmed as per the ability to support the binding and development of the human osteoblasts. The acquired HA/PLGA/antibiotic coatings [99] prevented the preliminary stage of microbiological adherence as well as colonization and biofilm growth on the surfaces of the implant, as per microscopic examination and quantitative measurements. The inherent antibacterial properties of CFX and CFR, which functioned through interrupting the production of the cell wall of the bacteria, were responsible for the suppression of infection by both CFX/HA/PLGA and CFR/HA/PLGA [162,170].

According to a recent study, a stronger affinity for HA than glass-reinforced HA is exhibited by sodium ampicillin. The antibiotic release in this study was divided into three stages, the first of which was characterized by immediate delivery of ampicillin, while the next one was characterized by a slow delivery, and the concluding stage where the remaining drug was released from inner layers of the structure made from HA [171]. Another study discovered a relationship between the quantities of HA in a PLA-HA composite impregnated with clindamycin (CLI) and the rate of degradation in the influence of proteinase K as well as the drug-loading effectiveness of the biocomposite. Furthermore, using ultrasonic waves during the synthesis was beneficial in distributing HA particles within the PLA matrix [99], leading to a component that didn't have a harmful effect to fibroblast cells of human lungs [172].

For controlled CLI delivery, SBA-15, a kind of mesoporous silica, was supplemented by a HA coating. The proportion of CLI which could be fused on this biocompatible SBA-15/HA complex [99] was found to be 23%, and its profile of release in both acidic and physiological pH settings [99] showed a strong immediate delivery in the several initial hours, which was indicative of the concentration of the absorbed drug on the surface of the particles. Around 30 hours later, a cumulative steady release was attained [173,174].

A study of chloramphenicol and metronidazole (MTZ) release from porous HA ceramics with spherical and cylindrical topologies stated that the concentration of medication released is only related to the implant's size and not its shape [175]. The interconnections within pores and porosity, and even the drug placement inside the implant, all impacted the release rates. The maximum capacity for loading of MTZ was double that of CMP. MTZ release was initially modest during the first 4 days, then rose for the following 7 days, then decreased in the last 6 days. In comparison, half of the CMP concentration was delivered throughout the first 6 days, with the remaining medication released over the next 12 days [175]. The electrospinning approach was being used to make therapeutic nanofibers from PVA enhanced using Ag-MTZ/HA sheets in periodontitis therapy, with in vitro and in vivo systems used to assess their properties. Within the produced nanofibers, the medicine had a $96 \pm 2\%$ entrapment efficiency. When comparing normal MTZ and Ag-MTZ [176], the cumulative drug release percentage [99] for a time span of 24 hours suggested a consistent MTZ delivery persisting over hours, showing substantial enhancement of its property. Carbonated HA membrane was another method of MTZ delivery, created by combining varying weight ratios of carbonated HA and bovine-derived gelatin. For different HA contents, the angle of contact of air–water, capacity of MTZ loading, and the content of water were not significantly different [177].

To improve cytocompatibility and antimicrobial properties, HA/polyamide 66 nanoscaffold with a porous nature was designed as well as reinforced using varying quantities of titanium dioxide (TiO_2) and silver (Ag^+) ions as a binary constituent. The composite scaffolds were reported to have antibacterial capabilities inside a rabbit model, as well as encouraging proliferation and cell adhesion of preosteoblasts and boosting the expression of the osteogenic regulator. Moreover, these scaffolds had potent antimicrobial and antiinflammatory activities inside the body, as well as encouraging osteogeneration in osteomyelitis lesions [178].

In vitro antibacterial action of essential oils (EOs) from plants as well as hybrids of plant EOs/HA was undertaken on diverse microbes to study its effects. It was observed that lavender EO inhibited growth of bacteria significantly, with MIC ranging between 0.1 EO wt.% for normal *E. coli* and 0.78 EO wt.% for normal *S. aureus*. These experiments revealed that hybrid entities had better antibacterial action than normal HA, suggesting HA might be employed as a carrier for the administration of low quantities of the antibacterial lavender EO [162] in the future [179]. Additionally, soaking HA scaffolds in lysozyme solutions gave them stability like spongiosa and allowed them to release lysozyme for a minimum of 9 days, displaying bacteriostatic and bactericidal activities over both *Micrococcus luteus* and *Listeria innocua* [180].

Rauschmann et al. in their work tried to develop HA and calcium sulfate nanocrystals as a composite carrier material that is also biodegradable for delivery of antibiotics to bone infection at local sites. A second surgical procedure is required for the removal of polymethacrylate beads [181] used for the local delivery of antibiotics and cytotoxic effects are observed by resorbable calcium sulfate [182–184]. The work aimed to study the antibiotic properties and biocompatibility of a bioresorbable composite consisting of calcium sulfate and a HA in nanoparticulate form (PerOssal). X-Ray, scanning and electron microscopy methods were employed to study the material characteristics of pure calcium sulfate pellets and PerOssal. Soaking of the pellets was done in gentamycin and vancomycin respectively. Over a 10-day period, the release properties of both the antibiotics were investigated from both the materials. Both qualitative and quantitative cytotoxic studies were employed to determine biocompatibility. PerOssal showed specific surface of $106m^2/g$ and for pure calcium sulfate it was $2.2\,m^2/g$. Within 10 days, gentamycin was found to be almost completely eluted from both the carrier materials. PerOssal showed 94.7% while calcium sulfate showed 95.8% elution. In the case of vancomycin, PerOssal showed higher release, i.e., 96.3% while calcium sulfate showed around 74.8%. It was observed that PerOssal shows a higher initial release until the 5th day and lower release afterwards when compared to the calcium sulfate. PerOssal and the nontoxic cell culture medium did not show any significant difference in in vitro cytotoxicity while Calcium sulfate was found to show cytotoxicity in two tests out of four.

In comparison to pure calcium sulfate, substituting it with 50% nanoparticulate HA in the composite material resulted in higher biocompatibility and no in vitro cytotoxicity. In order to prevent the inactivation of the antibiotics, drug loading was done after hardening and sterilizing the pellets. Fast antibiotic loading was observed when the prefabricated pellets were used which resulted in the low setup time and enabled quicker implantation. This composite carrier material possesses the biodegradable properties of calcium sulfate and HA and it is also osteoconductive, so it does not require removal from the site of implant like the polymethacrylate beads.

From the studies, it was concluded that PerOssal shows excellent biocompatibility, sufficient antibiotic release, and good resorption properties and hence can become a new treatment method in osteomyelitis [185].

10.6.1 Genetic carriers

The purpose of this form of therapy is to assure the preservation of therapeutically administrable gene components along with drug targeting within cell nucleus. Another delivery approach for genetic materials is to employ the carrier not just as a passive carrier for delivering the gene to the nucleus, and more like an active agent for treatment, that might contribute to the enhancement in the function achieved by triggering the provided gene inside the targeted cell. When it comes to safety, HA breaks down into Ca and PO_4 ions that cause no hazard to the viability of cell unless they are present in extremely high levels. The pH-sensitivity of HA materials causes them to dissolve in the acid medium of lysosomes and late endosomes [99], allowing nucleotides to be released within the cell prior to being damaged in the proteolytic compartments [162,186]. Furthermore, before reaching the endosome, the stability of HA in vivo is much better than that of polymeric and metallic nanoparticles [187]. Anti-angiogenesis plasmids targeting breast cancer gene therapy were efficiently delivered into MCF-7 cells using HA nanoparticles with amine-functionality. The findings reveal a substantial reduction in vascular endothelial growth factor (VEGF) expression as a consequence of the transfection efficiency. Moreover, the particles did not cause any toxicity in the cells, confirming HA as safe for use as a genetic carrier [188].

The majority of papers describe making HA/genetic material hybrids by combining both of them through coprecipitation [189], and on rare occasions, linking the genes to HA in a postprecipitation procedure. Structures of HA-DNA components are the key aspects in improving and developing genetic delivery methods, and may be adjusted during preparation by adding suitable additives. For example, utilizing PLGA-mPEG [99] as a stabilizing block polymer [99], researchers were able to create 100×25 nm HA nanorods [190], whereas the HA DNA components were created by combining HA with DNA within the presence of calcium chloride. The DNA load-bearing capacity of HA [191] was improved as a result of this method. Another study looked at the expression and transfection of DNA supplied by HA in the following way: initially, 40–60 nm long HA nanorods were made, subsequently they were cross-linked to EGFP-N1 plasmid DNA (pDNA) [162]. After detecting DNA adsorption on HA using electrophoresis, the compound got utilized for in vitro transfection of SGC-7901 cells with around 80% effectiveness. HADNA components were subsequently injected into mice through the tail vein, while the organs such as the liver, brain, and kidney were examined using EGFP fluorescence and TEM to indicate effective delivery of EGFP-N1-pDNA to the abovementioned organs.

Another study looked at the impact of varying Ca/P ratios on transfection effectiveness and size of particles, revealing that adjusting this ratio can allow nanoparticles to pass through several barriers on their journey to the nucleus of cells for transfection [192]. Therefore, a Ca/P molar ratio of about 100–300 was found to be ideal for the binding and condensation of pDNA, resulting in nanoparticles of 25–50 nm in size. The efficiency of transfection was also controlled by the supersaturation ratio [162] throughout coprecipitation of pDNA and HA [162]. In particular, saturation levels of 15 were shown to produce a better association among HA nanomaterials and pDNA enabling transfection than their lower or higher values [162], namely 12.5 and 17.5, respectively [162,193].

As a result, the composite technique might improve loading of nucleic acid and transfection efficiency. Using HA/GO composite materials, researchers transfected cells of breast cancer with the thymidine kinase gene of herpes simplex virus thymidine kinase gene, popularly known as the suicide gene. The tumor cells had a 70% drop in viability, a 22% decrease in apoptosis of cell, and a total suppression of cell growth, whereas normal breast cells had a 90% viability. Because of the biocompatibility of HA and GO, the system had an anticancer impact in cancer cells and acceptable toxicity in healthy cells. The key motivation for employing GO plates was that they considerably enhanced the material's specific surface area, offering significantly extra sites for binding in DNA than HA itself. The use of a large proportion of HA entities in the composite was justified since Ca and PO_4^{3-} strengthen connections between transporter and DNA, thereby also preserving the DNA from damage due to the stabilizing effect of calcium ions. In this scenario, the composite HA/GO method has the additional benefit of facilitating the regulated delivery of DNA [194].

Chowdhury et al. worked on the Ca-Mg nanoprecipitate-based highly efficient gene delivery system for mammalian cell expression. For analysis of structure of gene, regulation, or function, the transfer of desirable genetic sequences into mammalian cells serves as an essential tool [195]. This process is also important for the industrial production of therapeutically important proteins [195] and is very important in strategies of DNA vaccines and gene therapy. Synthetically obtained nonviral systems [195] are highly desired for the abovementioned uses because viral systems have some objectionable limitations like carcinogenicity, immunogenicity, and others. Amongst the most widely utilized nonviral vectors with the prospect for gene therapy application is the coprecipitation of DNA using calcium phosphate based on HA. In current practice, the nonviral techniques are found to be more severely inefficient than the viral techniques. As a result, the research is focusing on this area and the creation of a synthetic vehicle aimed at high-efficiency gene transport accompanied by expression in mammalian cells based on regulated nanoapatite particle growth. Significant inhibition was observed in particle growth when the apatite particles were incorporated with Mg^{2+}. When compared to one of the most extensively used transfection technique, classical calcium phosphate coprecipitation, it led to nanosized particle retention, which has a role in cellular absorption of DNA and subsequent expression ($>$10-fold) [195].

Enhanced cell absorption is also essential, and this may be accomplished by shrinking the HA-DNA complex's structure. In the crystal lattice of HA, for instance, replacing calcium with magnesium hinders crystal development and restricts the size of the particles, resulting in smaller HA-DNA compounds [196]. DNA has an immediate impact on the zeta potential of complexes owing to its increased concentration of phosphates, which are negatively charged, making it comparable to the potential of the cytoplasmic membrane and indicating that lower quantities of DNA [99] inside the HA/DNA complexes would lead to larger cell uptakes [197]. Furthermore, the external energy of nanomaterials may be altered by conjugating various functional groups, which are employed as supporters for cell uptake [198]. A hydrothermal process was employed to create arginine (Arg)-functionalized HA rod-shaped nanoparticles with an average diameter between 50 and 90 nm for a transporter carrying the plasmid of DNA pEGFP-N 1 to a cancer (HeLa cells) cell line and a regular (HAEC cells) cell line. The logic behind the strategy seemed to be that Arg would boost cellular absorption, which was later demonstrated to be true [198]. The approaches for improving HA transfection effectiveness are varied, and one study found that modifying HA particles with glucose improved transfection efficiency (Glc). The association of Glc to glucose transporter 1 (GLUT1) [162], which accelerated the carrier's entry to the cell, was responsible for the enhancing effect. This notion was supported by the finding that inhibiting GLUT1 receptors with the inhibitor STF-31 [162] reduced transfection efficiency. Along with the advantages of carrier entry inside the cell, HA substituted by Glc exhibited greater DNA conservation than pure HA [199].

10.6.2 Anticancer therapeutics

HA has been studied intensely to deliver anticancerous/antitumorous entities in a smart and effective manner to the specific sites and holds the potential to offer efficient clinical and noninvasive treatments. It has also been observed that HA exhibits cytotoxicity against carcinoma cells like melanoma, breast cancer, stomach cancer, hematoma, and colon cancer [200–202]. Nanoparticles of HA have also been demonstrated to have a high selectivity for malignant cells, leaving healthy cells unharmed [203]. HA vehicles have shown an improved ability to target anticancer medicines in this way. In one study, chitosan–lactate–oligosaccharide-layered HA was employed for transporting steroidal analogs, hence improving the targeting efficacy as well as specificity over unadulterated derivatives. The results showed that using this HA carrier caused a two to six times more toxic effect in carcinoma cells than that exhibited by noncancerous cells [204]. A method containing HA nanoparticles treated with chitosan-PLGA and infused with a cancer inhibitor based on androstane showed similar selectivity [205]. Pure HA was found to have the same effects on particular cancer cell lines [206]. The formation of reactive oxygen species (ROS) inside cells [201] and the triggering of apoptosis [200] mediated by the mitochondrion are assumed to be responsible for HA's anticancer effect, as is the suppression of synthesis of protein produced by an elevated amount of endocytosis and the build-up of HA around the endoplasmic reticulum [207].

Several investigations have demonstrated this selectivity and attempted to unravel the mechanism of the HA particle in anticancer activities. In a study the effects of HA on cytotoxicity and apoptosis were demonstrated in ordinary bronchial epithelial cells (16HBE) and a human lung cancer cell line (A549) [99]. This positive selectivity was exhibited by both in vitro and in vivo studies, with carcinoma cells being more vulnerable than normal cells to the action of HA [208]. In relation to normal 16HBE cells, HA raised intracellular [Ca^{2+}]i in A459 cells [99] over a prolonged time span, which was a response that was also observed in healthy and malignant tissues. Also, HA selectively targeted mitochondria in A459 cells [99], causing mitochondrion-responsible death. The reality is that the organelle attacked by HA varies by cell type; for example, in HepG2 cells [209], HA restrains to the nucleus and gathers around the Endoplasmc Reticulum causing interruption of the synthesis of protein in hepatocellular cancerous cells, Bel-7402.

In contrast to previous publications that indicated that HA inhibits cancer cell growth by producing reactive oxygen species (ROS), research postulated a different mechanism: cell cycle inhibition of mitosis. The work showed that HA has a strong harmony toward interacting with ribosomal materials, preventing bonding of mRNA-ribosome, and resulting in reduced synthesis of proteins, thus impacting generation of numerous critical proteinaceous materials for mitosis [207].

The cell toxicity of HA is influenced by its physicochemical qualities, which have an impact on a variety of cellular activities. In HepG2 cells [209], for example, the influence of HA particle size was established in a study. Additional physicochemical features of HA, such as charge of surface, specific surface area, and shape, have clearly influenced biological responses. The association of HA particles with cells, for instance, increased as the specific surface area of HA increased, affecting ROS formation [210]. As a result, nano-HA is thought to be more effective than micro-HA for the induction of apoptosis [211]. It has been demonstrated that HA nanoparticles may cause apoptosis in hepatocellular cancer cells by phagocytosis of fragments less than 100 nm in size [212], and that the anticancer effect of HA grows as the particle size reduces [202,213]. It is worth mentioning that not only could various analytical methods offer wide-ranging data regarding the micromeritics of HA nanoparticles inside body, though opsonization and the hydrodynamic layer that surround the entities also drastically affect the micromeritics, along with the inevitable consequence and also the destiny of the HA nanoparticles inside the body [214]. While several literature findings suggest that HA has repressive effects for tumors, the truth behind this is still unknown. Furthermore, the efficacy of HA particles' anticancer effect is not equivalent to that of traditional chemotherapy medicines. This is indeed a direct result of HA's absence of a particular molecular target, instead relying on a coordinated impact with several cellular targets at the same time. As a result, particle characteristics must be improved in order to demonstrate more prominent antitumor actions. There are two primary methods for improving this activity in general: (1) altering the chemical and morphology of HA; and (2) making composites with HA [215].

A variety of elements can be loaded to crystalline HA, the most common of which are magnesium, zinc, strontium, and silicon. These changes might have a big influence on HA's anticancer properties. For example, in vitro research has confirmed that replacing phosphate by selenite inside HA particles improves not only antimicrobial action and osteoinduction, but also anticancer properties. In another study, needle-shaped HA nanoparticles doped with selenium with an approximate size ranging between 150 and 200 nm were created and they revealed increased tumor-curing abilities in an animal model [216].

When put in mild acidic environments, HA has the benefit of being decomposed into its ionic components [111,217,218]. Because both malignant cells and endosomes have acidic environments, this property is highly useful for the intelligent and environmentally regulated distribution of chemotherapeutic medicines. As a result, the pH gradient of normal cells and cancer cells may be employed as a stimulus for controlled medication release [219]. The pH responsiveness of HA particles was exploited in research to reduce doxorubicin (DOX) leakage in the circulation, decreasing unwanted effects while improving release of the drug in the desired site. Functionalization of HA to folic acid was done for improving particle/cell identification and targeting. The large particle concentration at the tumor region and decreased amounts of aggregation in remote organs showed a significant effectiveness of HA in respect of therapeutic action and targeting, according to their findings [220].

By devising and adopting unique methodologies, several researchers have sought to improve the loading capability of HA with regard to potential anticancer medicines. For instance, one study developed a core/shell particle based on HA with a poly amino acids cap. Poly amino acids increased drug content, and additionally worked like a switch to control the drug's release. The electrostatic association was employed to generate DOX binding to poly amino acid as a model drug. The results showed that medication loading performance and release of drugs have improved [221].

Their complexation with actively targeted ligands, on the other hand, seems to be a more promising method to promote localization of treatment to the tissues of interest. As a result, when HA nanoparticles loaded with chemotherapeutics were complexed with targeting moieties such as bisphosphonates, they showed increased and persistent retention in bone tumor locations [222]. Another possible route is to use external fields to boost passive targeting ability. One of several instances might be the successful utilization of magnetic effect guiding to concentrate magnetic nanoparticles coated with DOX-loaded HA in tumor cells as well as reduce the volume compared to when examined with DOX alone [223].

10.6.3 Bioimaging

Biomedical imaging is a technique for obtaining graphical pictures of sites inside the body of humans in order to identify medical problems and monitor the effectiveness of therapy. Using advanced imaging techniques such as computerized tomography, fluorescence imaging, ultrasound, radio imaging, magnetic resonance imaging, and others, multidimensional data can be visualized, which is considered a breakthrough in diagnostic imaging. This method makes it easier to analyze biological processes like molecule movement across membranes or noticeable alterations in signal transduction. Fluorescence imaging offers a lot of benefits over other methods because of its high sensitivity, safety, and low invasiveness [224–226].

In optical bioimaging, two distinct types of probes have been used: fluorescent dyes and fluorescent nanoparticles, particularly quantum dots, latexes, and silica [227]. HA nanoparticles are a feasible medium for bioimaging purposes, overcoming the durability and autofluorescence concerns present in the earlier technique and also the removal, solubility and toxicity problems encountered with the latter technique [228,229]. Although HA alone has a limited luminosity, supplementing it with lanthanide ions or conjugating it with organic dyes is a standard way to get the optical characteristics needed for in vivo monitoring. As fluorescent probes, HA nanoparticles doped with unique cations like Eu^{3+}, Er^{3+}, Tb^{3+}, and Yb^{3+} have been widely explored [230]. Because of their narrow emission bandwidths, extended fluorescence life span, and strong photochemical tolerance, HA nanoparticles fused with lanthanide can be regarded as good alternatives for fluorescent probes. The amount of the above stated ions has a direct relationship with the intensity of luminescence of these kind of doped systems. Furthermore, the hexagonal shape of HA facilitates constant replacement of calcium ions by multiple rare earth ions without altering the uniformity of the crystal structure. These HA nanoparticles were studied as cell-labeling substances due to their high luminescence under visible light illumination [231,232].

HA nanoparticles are frequently coupled to organic dyes to provide fluorescent characteristics as a substitute to fluorescent metal ions. The creation and use of hollow HA was used in a study to get a larger 8-hydroxyquinoline loading capability than that ordinarily possible (Fig. 10.1). Their findings showed that HA materials made using a ratio of 2:1 ammonium dihydrogen phosphate to calcium carbonate was far more successful in terms of dye entrapment than those made with different ratios [233]. Simple changes in the synthesis technique can frequently result in an ideal coexistence of the material with the dye, as well as a high affinity of their bonding. According to research, after soaking inside SBF, the complex's luminous properties and apatite precipitate over the coating of HA, both of which indicate biological activity, were seen to be preserved. Thermal spraying was used to create this photoluminescent coating, which might be a replacement to supplementing HA with rare-earth metals. It might lead to several novel uses, either alone or in tandem with this ion-doping approach [234,235].

There are other ways based on manufacturing composites apart from the ones described above in order to design fluorescent HA, notably loading with rare-earth elements along with operationalizing using aromatic compounds. The use of metal-free self-catalysis to integrate the surface ligand exchange reaction with the process of atom transfer radical polymerization has been discovered to be a straightforward process for manufacturing fluorescent HA nanoparticles containing human parathyroid hormone. The findings of FT-IR, 1H NMR, XPS, and TGA investigations showed that the fluorescent

Figure 10.1 Fluorescent labels used for luminescent hydroxyapatite nanoparticles [99,230,233].

nanomaterials with superior hydro dispersibility and good fluorescence characteristics were properly manufactured. Poor toxic effect to cells, enhanced absorptive power of cell, and strong biological compatibility were also found in these nanoparticles, indicating that they have a promising future in biological imaging. The benefit of the abovementioned approach is its potential application for changing exteriors of various kinds of nanoparticles, affording them broader applications [236–238]. However, another approach made 2-(methacryloyloxy) ethyl phosphorylcholine polymer nanocomposites with nanoparticle-aggregation-induced dye emission with fluorescent silica [239]. The noncovalent improved Stöber technique was combined with ATRP polymerization to create orange-red fluorescent nanoparticles with silica in this work. Since the pigment was enclosed within the nanoparticles, they were able to avoid the fading of typical organic dyes produced by aggregation [240]. As a consequence, intense fluorescence, good photostability, and mild cytotoxicity were observed throughout and following the treatment while maintaining normal cell shape. Similar results might be accomplished using HA/polymer composites containing photoactive dyes, according to the researchers.

Targeting ligands can be added to the imaging probes, allowing them to bind selectively to the receptors on the specific cell. The imaging sensors would reside at the site of target for a long time and create excellent spatial resolution pictures in this manner. Carbohydrates, vitamins, polymers, proteins, and lipids are examples of targeted components that are connected to nanoprobes employing diverse conjugation strategies, comprising host–guest physical interactions. Furthermore, stabilizing agents and solvents are important in the formation of self-activated fluorescent HA with various structures, such as spheres, cubes, rods, and so on. The most powerful fluorescence is usually linked with electronic states in the nanoparticles' more asymmetric configurations. Biomedical experts will be able to use these versatile nanoparticles to do both screening and therapy simultaneously. In addition to its enormous promise in the area of optical mapping, HA has been investigated as a biological material for other detection approaches, namely SPECT, MRI, and PET [241,242].

Overall, bioimaging gives a valuable set of data which can be utilized to build improved computational models of organs and analyze nanoparticle interactions with them. Anatomical coordinates and biochemical concentrations are two factors that may be used to incorporate imaging data into models. These models, in turn, might lead to new diagnoses and discoveries that can be used to improve bioimaging substances, reducing the gap between theory and experimentation. In this larger context, HA is likely to make a significant impact.

10.7 Safety of hydroxyapatite

Despite the fact that HA's biocompatibility has been documented in numerous research and is considered a dogma according to the vast majority of researchers in the field, there are studies that have given limited toxicity to HA nanoparticle when they implement extremely particular morphologies, dosages, and particle sizes, all of which appear to have a significant impact about how these particles communicate with living organisms and factors [243]. It is frequently forgotten that HA nanoparticles' diversity in terms of toxicity and biocompatibility is also the secret to their prospective application as an efficient therapeutic agent for a variety of disorders, particularly immunological disease, infectious disease and cancer.

Whether HA particles are administered locally, such as to bones or teeth, or systemically, they enter blood vessels [244]. The hemostasis and maintenance of the intravascular space need the organization of endothelium, platelets, and clotting factors, that are altered through the interaction of HA and the abovementioned entities. [245]. Protein corona refers to the bands of bonded or adsorbed proteins that surround nanoparticles as they travel through the circulation, and have a significant impact on particle transport, internalization, and adhesion, as well as the cellular immunological responsiveness to them [246]. Furthermore, the protein corona configuration is influenced by all the above-stated particle properties. The shape, size, and charges of nanoparticles, for example, have a direct impact on protein adsorption and desorption processes [247,248]. According to molecular dynamics data, the process of adsorption of protein to construct the corona takes place via electrostatic interaction [249], suggesting that the character of the corona is influenced not only by the charge, but also the HA particle's surface chemistry and shape which reveal strongly charged phosphate and calcium compounds. Since hundreds of proteins fight to get attached to the exterior to construct a corona, even modest alterations of its shape causes a change in corona structure or the impact orientation of one or more proteins that are present inside, altering the overall bioactivity. The influence of the exterior layer shows that bioactivity and biocompatibility are determined by more than just the particle's internal configuration.

The adherence of proteins in the blood to the particles determines the biological distribution and pharmacokinetics of particles which are parenterally injected, and this adherence is significantly reliant on the shape, content, size, and charge of the particles. Theoretically, adjusting the topography, surface charge distribution, and content of HA from circulation to a specific organ inside the body can therefore regulate its transit. Additionally, the action of the nanoparticle in the bloodstream might affect blood coagulation and hemolysis [250]. The majority of HA hemocompatibility experiments revealed insignificantly low levels of erythrocyte hemolysis, no blood coagulation on HA, and no structural alterations in platelets attached to HA [251–253]. Moreover, other researchers have found that the relationship among HA and the blood is more complicated. For example, researchers looked at HA's hemocompatibility, and the results showed that HA sols prolonged the bleeding as well as the coagulation time inside a rat model, however did not impact on hemolysis observed in rabbit erythrocytes. Furthermore, the effect of HA nanoparticles' typically minute negative surface charge and size on hemolysis was investigated: while the percentage of erythrocyte sedimentation was increased directly in response to the presence of HA entities as well as being directly proportional to the quantity, hemolysis was not spotted, implying outstanding hemocompatibility. Aside from that, the surface charge of HA and the time of incubation had an impact on its association with RBCs. By altering the HA surface, rate of erythrocyte aggregation can be reduced [254].

For better understanding the hazard of HA to blood vessels, researchers evaluated the proinflammatory effects of HA using an in vitro c-culturing setup using endothelial cells of human umbilical vein (HUVECs) and human monocytes (THP-1). The results showed that HA is hazardous to both the cell lines at concentrations of 200 μm/mL and above, while no significant toxicity was observed at subcytotoxic doses. Both HUVECs and monocytes phagocytized HA, but only monocytes produced a proinflammatory action, which triggered inflammatory activation in HUVECs [255]. Using the coprecipitation method at room temperature, a group of scientists created four distinct needle-like HA samples with various sizes spanning from 4 to 10 nm in diameter and around 25–60 nm in length, and evaluated their hemocompatibility using the procedure described previously to see if the ovate particles would express any of the harmful effects seen in asbestos-like systems [256]. The four samples showed less than 5% hemolysis, which indicates a strong compatibility with blood. Hydrophilic elements located exteriorly to needle-shaped HA nanomaterials, like carboxyls and hydroxyls, are said to enable HA's excellent hemocompatibility. Two kinds of HA particles were examined for blood compatibility: HA1, a rod-like structured having low nano dimensions and aspect ratio; and HA2, a broader needle-like having greater aspect ratio. The lethal range of concentration for both the samples was determined, and it was observed that HA1 only lowered cell viability at doses more than 250 μg/mL, whereas in case of HA2 the cell viability was reduced at 50 μg/mL [195]. At 10 and 50 μg/mL [195], HA1 seemed to have no influence on the cell cycle progression, apoptotic index, vWF gene (encoding a glycoprotein that plays a vital importance for coagulation of blood) expression, or the f-actin cytoskeleton [257]. In the equivalent concentration range, HA2 exhibited a detrimental effect on the parameters, which was based on dosing. There was other research that looked at the oxidative stress and genotoxicity reaction of HA upon blood corpuscles, including one that found that at higher concentrations of HA (150–1000 ppm), cell viability reduces but antioxidant capacity rises [258]. The sister chromatid exchange test was used to assess genotoxicity, and there was shown to be a direct relationship with the quantity of HA.

The systemically induced toxic effects of HA indeed is a worry, because the particles, even if locally injected, might spread to other parts of the body. As a result, understanding the place where HA localizes along with physiologic impact on organs is critical. A novel tracking paradigm for long-term (28 days) HA particles of size

~80 nm was developed. The entities were coupled with the radioactive marker ^{125}I, where buildup got monitored following tail injection in a mouse model. The findings showed that HA was concentrated in the kidneys (10% ID/g), liver (35% ID/g), lung (5% ID/g), and spleen (7.5% ID/g), but not in the blood or brain. After 7 days, HA entities were found inside lungs, and following 14 days inside kidneys, the former due to the aggregation of particles and the latter due to HA excretion via urine. The prevalence of HA particles in the liver and spleen following 28 days was due to macrophage endocytosis, since they were cleared from the bloodstream through the reticuloendothelial system (RES). These findings demonstrated that after HA injection, the prolonged actions of HA targeting the spleen and liver must be carefully observed [259]. Furthermore, it should be remembered that HA's intended destination may vary based on the surface characteristics. For example, after being administered intravenously inside a rat model, research looked at the circulation of Ch-PLGA-coated HA, HA coated with chitosan (HA/Ch), and HA. HA entities collected inside the liver, confirming findings generated from previous research. HA/Ch, on other hand, was directed to the liver and spleen, whereas particles of HA/Ch-PLGA got targeted, to lungs. On the other hand, when the HA got eliminated within 24 hours the composite remained much longer inside the organs and blood [260].

Scientists discovered that the toxicity of HA particles was more reliant on the type of the cell than on the particle structure, after studying the influence of the particle structure on the toxicity of cell. They investigated the influence of four distinct HA particle forms (plate-like, sphere-like, rod-like, and needle-like) on RAW264.7 and BEAS-2B cell lines. RAW264.7 cells showed little toxicity, however BEAS-2B cells showed considerable toxicity, only with needle-shaped and plate-shaped entities showing maximum amounts of toxic effect. The IL-6 expression, that was enhanced upon HA availability, was clearly related to BEAS-2B necrosis. Contact to HA increased ROS concentrations, and there was a clear link between ROS formation and toxicity, with rod-shaped HA producing the greatest amount of ROS BEAS-2B [99] cells and needle-shaped HA having a similar impact upon RAW264.7 cells. It is worth noting that ROS generation occurs in tandem with apoptosis, so if the latter was induced by other processes that went unnoticed, it might give the impression that ROS generation is the reason instead of a by-product. The findings also showed that high cellular absorption does not always imply acute toxicity.

Some studies compared two distinct varieties of HA particles, one 30–50 nm in length and 10–20 nm in width [261], and the other 70–90 nm in length and 20–40 nm in width, to see how biocompatibility of HA was affected by particle size. The experiment was carried out on fibroblast cells of mouse (L929), and no substantial inhibition was seen at a dosage of 100 μg/mL [262]. Larger dosages, on the other hand, frequently cause viability decreases, as evidenced by the evaluation of another group of scientists about the lethality on mesenchymal stem cells of bone marrow (BMSCs) with rod-like HA of size below 50 nm. The life span of cells subjected to 1000 g/mL HA was decreased to 60%, and cell spreading was similarly influenced by HA interaction at such large levels [263]. Ultimately, another study examined interleukin-1 alpha concentrations, LDH activity, and viability of MTT cell in human epithelial gingival (HEG) cells to investigate the suitability of HA as per EU cosmetics regulation criteria. Subjected to HA for various points of time, between 10 minutes and 3 hours, at concentrations equivalent to actual exposure, no detrimental effects were seen in HEG cells [264].

Finally, it is indeed worth mentioning that HA, as with most compounds belonging to this group, goes through processes including chemical treatment and mediation by solutions that can affect the interface and bulk phase composition, as well as impact the cells response involving ion-exchange pathways. Such activities, which might hold considerable influence to the cell response to HA, are most frequently overlooked and understudied as factors impacting HA's fate inside the body [265,266].

Therefore, despite the fact that the word "biocompatibility" has long been linked with HA, more study is needed to fully comprehend the impacts of shape, dosage, and particle size on cell lines to assure the safety desired for use in clinical research. Any material, even at the pinnacle of biological properties in the guise of nanoparticles, may become potentially harmful when transformed to nanostructures, and HA should not be an unquestionable alternative to this paradigm [267]. As more applications requiring greater function in communication with pathological cells, like bacterial and neoplastic cells, emerge, the patterns obtained from HA-based nanoparticles will keep evolving to more sophisticated versions, necessitating proper investigation into the biosafety for the abovementioned applications [268]. Lastly, long-term investigations on the biological impacts of HA should be conducted to minimize unintended consequences for exposed species or their offspring.

10.8 Conclusion

HA is the chemical counterpart of biogenic apatite which is the primary constituent of every hard tissues in mammals. It drew interest in the early days of biomaterials science because of the projected biocompatibility based on its chemical and structural similarity to the inorganic constituent of teeth and bone. Furthermore, HA's bioactivity provides it an edge over ceramics which are biocompatible but bioinert in nature like silica, alumina, and titanium, allowing it to interact more closely with the tissues and cells of the bioenvironment. Furthermore, a number of very easy synthesis techniques for HA have been discovered, enabling researchers to create particles with exact shapes and sizes.

Each of these benefits come together to make HA a good choice for implantable medical devices, tissue engineering, gene therapy, medication delivery, and bioimaging, among other biomedical uses. Regardless of the truth that there are several advantages of using HA, its limitations and negative impacts will surely arise and should be thoroughly resolved before a particular clinical usage is chosen. The necessity of the osteoconductive property of HA materials is observed amongst scaffolds for tissue engineering and surgical prosthetics, though the property on its own is insufficient to build a viable and effective tissue alternative. In certain tissues, scaffolds and prostheses that can withstand heavy loads and cyclic stresses are required, and the fragility and poor fracture toughness of HA [162] must be solved through mixing HA with polymeric materials or even different varieties of nanoparticles. Alternatives were presented, however they are largely still in the prototype phase, and it is widely believed that further study is required before the clinical aims can be reached.

HA has become universally acknowledged as a drug carrier that can be loaded with a variety of medicinal entities such as antimicrobials, ligands, nucleic acids, or growth regulators, allowing for targeted medication delivery and precisely regulated rate and site of release of drug, that's the most desired of all drug delivery characteristics. A number of operational procedures may be used to transform HA into a tool for bioimaging, in addition to loading it with therapeutic loads. Because of its endurance to biodegradation and ability to breakdown quickly in specific environments, like the lysosome or even in the presence of cancerous cells, HA has the capability to operate as an intelligent material that responds to environmental stimuli. For its potential to acquire many operational features, HA may be utilized to transport several medications while being tracked and studied in situ, which is one of the requirements for therapeutic and diagnostic uses.

Despite the enormous prospect in a number of medicinal applicability, the majority of them are still awaiting revelation along with proper study. Yet another aspect which investigators must take into account is the possibility that the ultrafine HA particles will cause unwanted adverse effects, such as by crossing through or destroying the supposedly unbreachable and unbreakable biological barriers. These effects should be given a lot of thought, not just because a certain invasiveness found could be turned into a pharmacological efficacy with some creative thinking, but mostly because HA has long been regarded as the epitome of safety in the world of biomaterials, which would be a risky notion that could lead to potential threats for inventors and regulators. In practical terms, particle safety is a result of surface ligands, size, and morphology, as well as other compositional and structural specifications characterizing the particle [269]. As a result, every particle, specifically when it emerges in an innovative structural system, in the nanoparticle field, should be regarded afresh from a safety analysis standpoint before even being permitted to join the clinical stage for application purposes.

Acknowledgments

The authors would like to acknowledge the Department of Science & Technology (DST), Government of India, for their invaluable contributions and support. We are grateful to them for their financial support of this work under Grant Number "DST – IMRCD/ BRICS/Pilot Call 3/NanoDrug/2019".

References

[1] P.N. Kumta, C. Sfeir, D.H. Lee, D. Olton, D. Choi, Nanostructured calcium phosphates for biomedical applications: novel synthesis and characterization, Acta Biomater. 1 (2005) 65–83. Available from: https://doi.org/10.1016/j.actbio.2004.09.008.

[2] T. (Todd) Toyoshima, Biomaterial research in Japan, Biomater. Res. Japan. (2001).

[3] J. Barralet, M. Akao, H. Aoki, H. Aoki, Dissolution Dense Carbonate Apatite Subcutaneously Implanted Wistar Rats., 1999, pp. 10–13.

[4] Biomaterial Research in Japan, 2001.

[5] V. Uskoković, The role of hydroxyl channel in defining selected physicochemical peculiarities exhibited by hydroxyapatite, RSC Adv. (2015).

[6] V. Uskoković, D.P. Uskoković, Nanosized hydroxyapatite and other calcium phosphates: chemistry of formation and application as drug and gene delivery agents, J. Biomed. Mater. Res. - Part. B Appl. Biomater. 96 B (2011) 152–191. Available from: https://doi.org/10.1002/jbm.b.31746.

[7] P. Proussaefs, J. Lozada, G. Valencia, M.D. Rohrer, Histologic evaluation of a hydroxyapatite onlay bone graft retrieved after 9 years: a clinical report, J. Prosthet. Dent. 87 (2002) 481–484. Available from: https://doi.org/10.1067/MPR.2002.122961.

[8] M. Mucalo, Hydroxyapatite (HAp) for Biomedical Applications, Elsevier; Woodhead Publishing, 2015.

[9] S.V. Dorozhkin, Calcium orthophosphates, J. Mater. Sci. 42 (2007) 1061–1095. Available from: https://doi.org/10.1007/s10853-006-1467-8.

[10] S.V. Dorozhkin, Calcium orthophosphates in nature, biology and medicine, Mater. (Basel) 2 (2009) 399–498. Available from: https://doi.org/10.3390/ma2020399.

[11] N.Y. Mostafa, Characterization, thermal stability and sintering of hydroxyapatite powders prepared by different routes, Mater. Chem. Phys. 94 (2005) 333–341. Available from: https://doi.org/10.1016/j.matchemphys.2005.05.011.

[12] A. Afshar, M. Ghorbani, N. Ehsani, M.R. Saeri, C.C. Sorrell, Some important factors in the wet precipitation process of hydroxyapatite, Mater. Des. 24 (2003) 197–202. Available from: https://doi.org/10.1016/S0261-3069(03)00003-7.

[13] M. Letters, The Wet Precipitation Process of Hydroxyapatite, n.d. https://doi.org/10.1016/S0167-577X(03)00266-0.

[14] T. Goto, I.Y. Kim, K. Kikuta, C. Ohtsuki, Hydrothermal synthesis of composites of well-crystallized hydroxyapatite and poly(vinyl alcohol) hydrogel, Mater. Sci. Eng. C. 32 (2012) 397–403. Available from: https://doi.org/10.1016/J.MSEC.2011.11.011.

[15] H.S. Liu, T.S. Chin, L.S. Lai, S.Y. Chiu, K.H. Chung, C.S. Chang, et al., Hydroxyapatite synthesized by a simplified hydrothermal method, Ceram. Int. 23 (1997) 19–25. Available from: https://doi.org/10.1016/0272-8842(95)00135-2.

[16] T. Hattori, Y. Lwadate, Hydrothermal preparation of calcium hydroxyapatite powders, J. Am. Ceram. Soc. 73 (1990) 1803–1805. Available from: https://doi.org/10.1111/j.1151-2916.1990.tb09841.x.

[17] D. Arcos, M. Vallet-Regí, Sol–gel silica-based biomaterials and bone tissue regeneration, Acta Biomater. 6 (2010) 2874–2888. Available from: https://doi.org/10.1016/J.ACTBIO.2010.02.012.

[18] J. Chen, Y. Wang, X. Chen, L. Ren, C. Lai, W. He, et al., A simple sol–gel technique for synthesis of nanostructured hydroxyapatite, tricalcium phosphate and biphasic powders, Mater. Lett. 65 (2011) 1923–1926. Available from: https://doi.org/10.1016/J.MATLET.2011.03.076.

[19] M.P. Bajgai, D.C. Parajuli, S.J. Park, K.H. Chu, H.S. Kang, H.Y. Kim, In vitro bioactivity of sol–gel-derived hydroxyapatite particulate nanofiber modified titanium, J. Mater. Sci. Mater. Med. 21 (2010) 685–694. Available from: https://doi.org/10.1007/s10856-009-3902-2.

[20] A. Sharifah, S. Iis, H. Mohd, R. Singh, Mechanochemical synthesis of nanosized hydroxyapatite powder and its conversion to dense bodies, Mater. Sci. Forum. 694 (2011) 118–122. Available from: https://doi.org/10.4028/http://www.scientific.net/MSF.694.118.

[21] J. Coreño A., O. Coreño A., J.J. Cruz R., C. Rodríguez C., Mechanochemical synthesis of nanocrystalline carbonate-substituted hydroxyapatite, Opt. Mater. (Amst). 27 (2005) 1281–1285. Available from: https://doi.org/10.1016/J.OPTMAT.2004.11.025.

[22] G.C. Koumoulidis, A.P. Katsoulidis, A.K. Ladavos, P.J. Pomonis, C.C. Trapalis, A.T. Sdoukos, et al., Preparation of hydroxyapatite via microemulsion route, J. Colloid Interface Sci. 259 (2003) 254–260. Available from: https://doi.org/10.1016/S0021-9797(02)00233-3.
[23] V. Bayazit, M. Bayazit, E. Bayazit, Evaluation of bioceramic materials in biology and medicine, Dig. J. Nanomater. Biostructures. 5 (2010) 267–278.
[24] S. Ramesh, R. Tolouei, C.Y. Tan, K.L. Aw, W.H. Yeo, I. Sopyan, et al., Sintering of hydroxyapatite ceramic produced by wet chemical method, Adv. Mater. Res. 264–265 (2011) 1856–1861. Available from: https://doi.org/10.4028/http://www.scientific.net/AMR.264-265.1856.
[25] J. Liu, X. Ye, H. Wang, M. Zhu, B. Wang, H. Yan, The influence of pH and temperature on the morphology of hydroxyapatite synthesized by hydrothermal method, Ceram. Int. 29 (2003) 629–633. Available from: https://doi.org/10.1016/S0272-8842(02)00210-9.
[26] D. Liu, K. Savino, M.Z. Yates, Coating of hydroxyapatite films on metal substrates by seeded hydrothermal deposition, Surf. Coat. Technol. 205 (2011) 3975–3986. Available from: https://doi.org/10.1016/J.SURFCOAT.2011.02.008.
[27] C.R. Kothapalli, M. Wei, R.Z. Legeros, M.T. Shaw, Influence of temperature and aging time on HA synthesized by the hydrothermal method, J. Mater. Sci. Mater. Med. 16 (2005) 441–446. Available from: https://doi.org/10.1007/s10856-005-6984-5.
[28] A.A. Chaudhry, S. Haque, S. Kellici, P. Boldrin, I. Rehman, F.A. Khalid, et al., Instant nano-hydroxyapatite: a continuous and rapid hydrothermal synthesis, Chem. Commun. (2006) 2286–2288. Available from: https://doi.org/10.1039/b518102j.
[29] F. Wang, M. Sen Li, Y.P. Lu, Y.X. Qi, Y.X. Liu, Synthesis and microstructure of hydroxyapatite nanofibers synthesized at 37 °C, Mater. Chem. Phys. 95 (2006) 145–149. Available from: https://doi.org/10.1016/J.MATCHEMPHYS.2005.05.034.
[30] J.W. LL Hench, An Introduction to Bioceramics, Advanced S, World Scientific, 1993.
[31] A.R. Boccaccini, M. Erol, W.J. Stark, D. Mohn, Z. Hong, J.F. Mano, Polymer/bioactive glass nanocomposites for biomedical applications: a review, Compos. Sci. Technol. 70 (2010) 1764–1776. Available from: https://doi.org/10.1016/J.COMPSCITECH.2010.06.002.
[32] M. Vallet-Regí, Evolution of bioceramics within the field of biomaterials, Comptes Rendus Chim. 13 (2010) 174–185. Available from: https://doi.org/10.1016/J.CRCI.2009.03.004.
[33] A. Jillavenkatesa, R.A. Condrate, Sol–gel processing of hydroxyapatite, J. Mater. Sci. 33 (1998) 4111–4119. Available from: https://doi.org/10.1023/A:1004436732282.
[34] R.R. Rao, H.N. Roopa, T.S. Kannan, Solid state synthesis and thermal stability of HAP and HAP - β-TCP composite ceramic powders, J. Mater. Sci. Mater. Med. 8 (1997) 511–518. Available from: https://doi.org/10.1023/A:1018586412270.
[35] R. Hynes, Mol. Biol. of Fibronectin 1 (1985) 67–90.
[36] H.K. Varma, Dense Hydroxy Apatite Ceram. Gel Cast. Tech. 29 (1996) 57–61.
[37] R.Z. LeGeros, Calcium phosphates in oral biology and medicine, Monogr. Oral. Sci. (1991) 109–111.
[38] A.A. Mirtchi, J. Lemaitre, N. Terao, Calcium phosphate cements: study of the β-tricalcium phosphate—monocalcium phosphate system, Biomaterials 10 (1989) 475–480. Available from: https://doi.org/10.1016/0142-9612(89)90089-6.
[39] S.H. Kwon, Y.K. Jun, S.H. Hong, H.E. Kim, Synthesis and dissolution behavior of β-TCP and HA/β-TCP composite powders, J. Eur. Ceram. Soc. 23 (2003) 1039–1045. Available from: https://doi.org/10.1016/S0955-2219(02)00263-7.
[40] H. Aoki, Science and medical applications of hydroxyapatite, JAAS 1991 (1991) 123. –123.
[41] F. Driessens, Formation and Stability of Calcium Phosphates in Relation to the Phase Composition of the Mineral in Calcified Tissues, 1st ed., CRC Press, 1983.
[42] M. Vallet-Regí, J.M. González-Calbet, Calcium phosphates as substitution of bone tissues, Prog. Solid. State Chem. 32 (2004) 1–31. Available from: https://doi.org/10.1016/J.PROGSOLIDSTCHEM.2004.07.001.
[43] L.L. Hench, Bioceramics: from concept to clinic, J. Am. Ceram. Soc. 72 (1993) 93–98. J. Am. Ceram. Soc. 74 (1991) 1487–1510.
[44] H. Zhou, J. Lee, Nanoscale hydroxyapatite particles for bone tissue engineering, Acta Biomater. 7 (2011) 2769–2781. Available from: https://doi.org/10.1016/J.ACTBIO.2011.03.019.

[45] F.Z. Cui, Y. Li, J. Ge, Self-assembly of mineralized collagen composites, Mater. Sci. Eng. R. Rep. 57 (2007) 1–27. Available from: https://doi.org/10.1016/J.MSER.2007.04.001.
[46] L. Xia, K. Lin, X. Jiang, Y. Xu, M. Zhang, J. Chang, et al., Enhanced osteogenesis through nano-structured surface design of macroporous hydroxyapatite bioceramic scaffolds via activation of ERK and p38 MAPK signaling pathways, J. Mater. Chem. B 1 (2013) 5403–5416. Available from: https://doi.org/10.1039/c3tb20945h.
[47] K. Lin, L. Xia, J. Gan, Z. Zhang, H. Chen, X. Jiang, et al., Tailoring the nanostructured surfaces of hydroxyapatite bioceramics to promote protein adsorption, osteoblast growth, and osteogenic differentiation, ACS Appl. Mater. Interfaces. 5 (2013) 8008–8017. Available from: https://doi.org/10.1021/am402089w.
[48] C. Zhou, Y. Hong, X. Zhang, Applications of nanostructured calcium phosphate in tissue engineering, Biomater. Sci. 1 (2013) 1012–1028. Available from: https://doi.org/10.1039/c3bm60058k.
[49] Q. Hu, Z. Tan, Y. Liu, J. Tao, Y. Cai, M. Zhang, et al., Effect of crystallinity of calcium phosphate nanoparticles on adhesion, proliferation, and differentiation of bone marrow mesenchymal stem cells, J. Mater. Chem. 17 (2007) 4690–4698. Available from: https://doi.org/10.1039/b710936a.
[50] J.B. Lee, S.H. Lee, S.M. Yu, J.C. Park, J.B. Choi, J.K. Kim, PLGA scaffold incorporated with hydroxyapatite for cartilage regeneration, Surf. Coat. Technol. 202 (2008) 5757–5761. Available from: https://doi.org/10.1016/J.SURFCOAT.2008.06.138.
[51] D. Xue, Q. Zheng, C. Zong, Q. Li, H. Li, S. Qian, et al., Osteochondral repair using porous poly (lactide-co-glycolide)/nano-hydroxyapatite hybrid scaffolds with undifferentiated mesenchymal stem cells in a rat model, J. Biomed. Mater. Res. - Part. A 94 (2010) 259–270. Available from: https://doi.org/10.1002/jbm.a.32691.
[52] I.S. Neira, Y.V. Kolen'Ko, K.P. Kommareddy, I. Manjubala, M. Yoshimura, F. Guitián, Reinforcing of a calcium phosphate cement with hydroxyapatite crystals of various morphologies, ACS Appl. Mater. Interfaces. 2 (2010) 3276–3284. Available from: https://doi.org/10.1021/AM100710B.
[53] E. Piskin, J. Biomater. Sci. Biodegradable Polym. Biomater. (1994) 37–41.
[54] Y. Ji, K. Ghosh, X.Z. Shu, B. Li, J.C. Sokolov, G.D. Prestwich, et al., Electrospun three-dimensional hyaluronic acid nanofibrous scaffolds, Biomaterials 27 (2006) 3782–3792. Available from: https://doi.org/10.1016/J.BIOMATERIALS.2006.02.037.
[55] M. Farokhi, F. Mottaghitalab, S. Samani, M.A. Shokrgozar, S.C. Kundu, R.L. Reis, et al., Silk fibroin/hydroxyapatite composites for bone tissue engineering, Biotechnol. Adv. 36 (2018) 68–91. Available from: https://doi.org/10.1016/J.BIOTECHADV.2017.10.001.
[56] J.Y. Park, C. Yang, I.H. Jung, H.C. Lim, J.S. Lee, U.W. Jung, et al., Regeneration of rabbit calvarial defects using cells-implanted nano-hydroxyapatite coated silk scaffolds, Biomater. Res. 19 (2015) 1–10. Available from: https://doi.org/10.1186/s40824-015-0027-1.
[57] M.H. Kim, B.S. Kim, J. Lee, D. Cho, O.H. Kwon, W.H. Park, Silk fibroin/hydroxyapatite composite hydrogel induced by gamma-ray irradiation for bone tissue engineering, Biomater. Res. 21 (2017) 1–9. Available from: https://doi.org/10.1186/s40824-017-0098-2.
[58] L. Lin, R. Hao, W. Xiong, J. Zhong, Quantitative analyses of the effect of silk fibroin/nano-hydroxyapatite composites on osteogenic differentiation of MG-63 human osteosarcoma cells, J. Biosci. Bioeng. 119 (2015) 591–595. Available from: https://doi.org/10.1016/J.JBIOSC.2014.10.009.
[59] M.K. Ahmed, S.F. Mansour, R. Al-Wafi, S.I. El-dek, V. Uskoković, Tuning the mechanical, microstructural, and cell adhesion properties of electrospun ε-polycaprolactone microfibers by doping selenium-containing carbonated hydroxyapatite as a reinforcing agent with magnesium ions, J. Mater. Sci. 54 (2019) 14524–14544. Available from: https://doi.org/10.1007/s10853-019-03947-1.
[60] P. Yu, R.Y. Bao, X.J. Shi, W. Yang, M.B. Yang, Self-assembled high-strength hydroxyapatite/graphene oxide/chitosan composite hydrogel for bone tissue engineering, Carbohydr. Polym. 155 (2017) 507–515. Available from: https://doi.org/10.1016/J.CARBPOL.2016.09.001.
[61] D.J. Hickey, B. Ercan, L. Sun, T.J. Webster, Adding MgO nanoparticles to hydroxyapatite–PLLA nanocomposites for improved bone tissue engineering applications, Acta Biomater. 14 (2015) 175–184. Available from: https://doi.org/10.1016/J.ACTBIO.2014.12.004.
[62] F. Heidari, M.E. Bahrololoom, D. Vashaee, L. Tayebi, In situ preparation of iron oxide nanoparticles in natural hydroxyapatite/chitosan matrix for bone tissue engineering application, Ceram. Int. 41 (2015) 3094–3100. Available from: https://doi.org/10.1016/J.CERAMINT.2014.10.153.

[63] S. Peng, P. Feng, P. Wu, W. Huang, Y. Yang, W. Guo, et al., Graphene oxide as an interface phase between polyetheretherketone and hydroxyapatite for tissue engineering scaffolds, Sci. Rep. 7 (2017) 1–14. Available from: https://doi.org/10.1038/srep46604.

[64] J.B. Lee, J.E. Kim, M.S. Bae, S.A. Park, D.A. Balikov, H.J. Sung, et al., Development of poly (ε-Caprolactone) scaffold loaded with simvastatin and beta-cyclodextrin modified hydroxyapatite inclusion complex for bone tissue engineering, Polymer (Basel) 8 (2016) 1–10. Available from: https://doi.org/10.3390/polym8020049.

[65] E. Quinlan, A. López-Noriega, E. Thompson, H.M. Kelly, S.A. Cryan, F.J. O'Brien, Development of collagen–hydroxyapatite scaffolds incorporating PLGA and alginate microparticles for the controlled delivery of rhBMP-2 for bone tissue engineering, J. Control. Release. 198 (2015) 71–79. Available from: https://doi.org/10.1016/J.JCONREL.2014.11.021.

[66] S. Kulanthaivel, B. Roy, T. Agarwal, S. Giri, K. Pramanik, K. Pal, et al., Cobalt doped proangiogenic hydroxyapatite for bone tissue engineering application, Mater. Sci. Eng. C. 58 (2016) 648–658. Available from: https://doi.org/10.1016/J.MSEC.2015.08.052.

[67] S. Khalid, S. Calderon-Larrañaga, S. Hawley, M.S. Ali, A. Judge, N. Arden, et al., Comparative anti-fracture effectiveness of different oral anti-osteoporosis therapies based on "real-world" data: a meta-analysis of propensity-matched cohort findings from the uk clinical practice research database and the catalan sidiap database, Clin. Epidemiol. 10 (2018) 1417–1431. Available from: https://doi.org/10.2147/CLEP.S164112.

[68] P.J. Driessens, F.C. De Maeyer, E.A. Fernández, E. Boltong, M.G. Berger, G. Verbeeck, et al., Calcium phosphate cements and their transformation into calcium deficient hydroxyapatite, Bioceramics 9 (1996) 231–234.

[69] S. Ghosh, V. Wu, S. Pernal, V. Uskoković, Self-setting calcium phosphate cements with tunable antibiotic release rates for advanced antimicrobial applications, ACS Appl. Mater. Interfaces. 8 (2016) 7691–7708. Available from: https://doi.org/10.1021/acsami.6b01160.

[70] E.E. Chow LC, Octacalcium Phosphate., Karger Medical and Scientific Publishers, 2001.

[71] J.V. Rau, I.V. Fadeeva, A.S. Fomin, K. Barbaro, E. Galvano, A.P. Ryzhov, et al., Sic Parvis Magna: manganese-substituted tricalcium phosphate and its biophysical properties, ACS Biomater. Sci. Eng. (2019) 6632–6644. Available from: https://doi.org/10.1021/acsbiomaterials.9b01528.

[72] V. Uskoković, V. Graziani, V.M. Wu, I.V. Fadeeva, A.S. Fomin, I.A. Presniakov, et al., Gold is for the mistress, silver for the maid: Enhanced mechanical properties, osteoinduction and antibacterial activity due to iron doping of tricalcium phosphate bone cements, Mater. Sci. Eng. C. 94 (2019) 798–810. Available from: https://doi.org/10.1016/J.MSEC.2018.10.028.

[73] J.V. Rau, V.M. Wu, V. Graziani, I.V. Fadeeva, A.S. Fomin, M. Fosca, et al., The Bone Building Blues: Self-hardening copper-doped calcium phosphate cement and its in vitro assessment against mammalian cells and bacteria, Mater. Sci. Eng. C. 79 (2017) 270–279. Available from: https://doi.org/10.1016/J.MSEC.2017.05.052.

[74] S.V. Dorozhkin, Calcium orthophosphate cements and concretes, Materials (Basel) 2 (2009) 221–291. Available from: https://doi.org/10.3390/ma2010221.

[75] V. Uskoković, S. Ghosh, V.M. Wu, Antimicrobial hydroxyapatite-gelatin-silica composite pastes with tunable setting properties, J. Mater. Chem. B 5 (2017) 6065–6080. Available from: https://doi.org/10.1039/c7tb01794d.

[76] R. Okabayashi, M. Nakamura, T. Okabayashi, Y. Tanaka, A. Nagai, K. Yamashita, Efficacy of polarized hydroxyapatite and silk fibroin composite dressing gel on epidermal recovery from full-thickness skin wounds, J. Biomed. Mater. Res. - Part. B Appl. Biomater. 90 B (2009) 641–646. Available from: https://doi.org/10.1002/jbm.b.31329.

[77] Y. Shin, H. Aoki, N. Yoshiyama, M. Akao, M. Higashikata, Surface properties of hydroxyapatite ceramic as new percutaneous material in skin tissue, J. Mater. Sci. Mater. Med. 3 (1992) 219–221. Available from: https://doi.org/10.1007/BF00713453.

[78] D.Y. Ji, T.F. Kuo, H. Da Wu, J.C. Yang, S.Y. Lee, A novel injectable chitosan/polyglutamate polyelectrolyte complex hydrogel with hydroxyapatite for soft-tissue augmentation, Carbohydr. Polym. 89 (2012) 1123–1130. Available from: https://doi.org/10.1016/J.CARBPOL.2012.03.083.

[79] M. Liu, G. Zhou, W. Song, P. Li, H. Liu, X. Niu, et al., Effect of nano-hydroxyapatite on the axonal guidance growth of rat cortical neurons, Nanoscale. (2012) 3201–3207.
[80] A. Szcześ, L. Hołysz, E. Chibowski, Synthesis of hydroxyapatite for biomedical applications, Adv. Colloid Interface Sci. 249 (2017) 321–330. Available from: https://doi.org/10.1016/J.CIS.2017.04.007.
[81] M. Mazaheri, M. Haghighatzadeh, A.M. Zahedi, S.K. Sadrnezhaad, Effect of a novel sintering process on mechanical properties of hydroxyapatite ceramics, J. Alloy. Compd. 471 (2009) 180–184. Available from: https://doi.org/10.1016/J.JALLCOM.2008.03.066.
[82] L. Zaccaria, S.J. Tharakan, S. Altermatt, Hydroxyapatite ceramic implants for cranioplasty in children: a single-center experience, Child's Nerv. Syst. 33 (2017) 343–348. Available from: https://doi.org/10.1007/s00381-016-3327-4.
[83] A. Haider, S. Haider, S.S. Han, I.K. Kang, Recent advances in the synthesis, functionalization and biomedical applications of hydroxyapatite: a review, RSC Adv. 7 (2017) 7442–7458. Available from: https://doi.org/10.1039/c6ra26124h.
[84] T.J. Levingstone, M. Ardhaoui, K. Benyounis, L. Looney, J.T. Stokes, Plasma sprayed hydroxyapatite coatings: understanding process relationships using design of experiment analysis, Surf. Coat. Technol. 283 (2015) 29–36. Available from: https://doi.org/10.1016/J.SURFCOAT.2015.10.044.
[85] R. Manoj Kumar, K.K. Kuntal, S. Singh, P. Gupta, B. Bhushan, P. Gopinath, et al., Electrophoretic deposition of hydroxyapatite coating on Mg–3Zn alloy for orthopaedic application, Surf. Coat. Technol. 287 (2016) 82–92. Available from: https://doi.org/10.1016/J.SURFCOAT.2015.12.086.
[86] V. Nelea, C. Ristoscu, C. Chiritescu, C. Ghica, I.N. Mihailescu, H. Pelletier, et al., Pulsed laser deposition of hydroxyapatite thin films on Ti-5Al-2.5Fe substrates with and without buffer layers, Appl. Surf. Sci. 168 (2000) 127–131. Available from: https://doi.org/10.1016/S0169-4332(00)00616-4.
[87] J. Hao, S. Kuroda, K. Ohya, S. Bartakova, H. Aoki, S. Kasugai, Enhanced osteoblast and osteoclast responses to a thin film sputtered hydroxyapatite coating, J. Mater. Sci. Mater. Med. 22 (2011) 1489–1499. Available from: https://doi.org/10.1007/s10856-011-4329-0.
[88] G. He, B. Guo, H. Wang, C. Liang, L. Ye, Y. Lin, et al., Surface characterization and osteoblast response to a functionally graded hydroxyapatite/fluoro-hydroxyapatite/titanium oxide coating on titanium surface by sol–gel method, Cell Prolif. 47 (2014) 258–266. Available from: https://doi.org/10.1111/cpr.12105.
[89] S. Singh, R. Manoj Kumar, K.K. Kuntal, P. Gupta, S. Das, R. Jayaganthan, et al., Sol–gel derived hydroxyapatite coating on Mg-3Zn alloy for orthopedic application, JOM 67 (2015) 702–712. Available from: https://doi.org/10.1007/s11837-015-1364-1.
[90] T. Laonapakul, A. Rakngarm Nimkerdphol, Y. Otsuka, Y. Mutoh, Failure behavior of plasma-sprayed HAp coating on commercially pure titanium substrate in simulated body fluid (SBF) under bending load, J. Mech. Behav. Biomed. Mater. 15 (2012) 153–166. Available from: https://doi.org/10.1016/J.JMBBM.2012.05.017.
[91] S. Rammelt, E. Schulze, R. Bernhardt, U. Hanisch, D. Scharnweber, H. Worch, et al., Coating of titanium implants with type-I collagen, J. Orthop. Res. 22 (2004) 1025–1034. Available from: https://doi.org/10.1016/j.orthres.2004.02.011.
[92] M.H. Pham, H.J. Haugen, A. Rinna, J.E. Ellingsen, J.E. Reseland, Hydrofluoric acid treatment of titanium surfaces enhances the proliferation of human gingival fibroblasts, J. Tissue Eng. 10 (2019). Available from: https://doi.org/10.1177/2041731419828950.
[93] H.P. Lim, S.W. Park, K.D. Yun, C. Park, M.K. Ji, G.J. Oh, et al., Hydroxyapatite coating on TiO2 nanotube by sol–gel method for implant applications, J. Nanosci. Nanotechnol. 18 (2018) 1403–1405. Available from: https://doi.org/10.1166/jnn.2018.14868.
[94] Y. Zhang, X. Liu, Z. Li, S. Zhu, X. Yuan, Z. Cui, et al., Nano Ag/ZnO-incorporated hydroxyapatite composite coatings: highly effective infection prevention and excellent osteointegration, ACS Appl. Mater. Interfaces. 10 (2018) 1266–1277. Available from: https://doi.org/10.1021/acsami.7b17351.
[95] N. Murugan, C. Murugan, A.K. Sundramoorthy, In vitro and in vivo characterization of mineralized hydroxyapatite/polycaprolactone-graphene oxide based bioactive multifunctional coating on Ti alloy for bone implant applications, Arab. J. Chem. 11 (2018) 959–969. Available from: https://doi.org/10.1016/J.ARABJC.2018.03.020.

[96] Y. Han, K. Xu, J. Lu, Morphology and composition of hydroxyapatite coatings prepared by hydrothermal treatment on electrodeposited brushite coatings, J. Mater. Sci. Mater. Med. 10 (1999) 243–248. Available from: https://doi.org/10.1023/A:1008914330004.

[97] Y. Han, K. Xu, J. Lu, Z. Wu, The structural characteristics and mechanical behaviors of nonstoichiometric apatite coatings sintered in air atmosphere, J. Biomed. Mater. Res. 45 (1999) 198–203. Available from: https://doi.org/10.1002/(SICI)1097-4636(19990605)45:3 < 198::AID-JBM6 > 3.0.CO;2-Q.

[98] Y. Li, W. Yang, X. Li, X. Zhang, C. Wang, X. Meng, et al., Improving osteointegration and osteogenesis of three-dimensional porous Ti6Al4V scaffolds by polydopamine-assisted biomimetic hydroxyapatite coating, ACS Appl. Mater. Interfaces. 7 (2015) 5715–5724. Available from: https://doi.org/10.1021/acsami.5b00331.

[99] B. Ghiasi, Y. Sefidbakht, S. Mozaffari-Jovin, B. Gharehcheloo, M. Mehrarya, A. Khodadadi, et al., Hydroxyapatite as a biomaterial—a gift that keeps on giving, Drug. Dev. Ind. Pharm. 46 (2020) 1035–1062. Available from: https://doi.org/10.1080/03639045.2020.1776321.

[100] H.S. Alghamdi, R. Bosco, S.K. Both, M. Iafisco, S.C.G. Leeuwenburgh, J.A. Jansen, et al., Synergistic effects of bisphosphonate and calcium phosphate nanoparticles on peri-implant bone responses in osteoporotic rats, Biomaterials 35 (2014) 5482–5490. Available from: https://doi.org/10.1016/J.BIOMATERIALS.2014.03.069.

[101] R. Bosco, M. Iafisco, A. Tampieri, J.A. Jansen, S.C.G. Leeuwenburgh, J.J.J.P. Van Den Beucken, Hydroxyapatite nanocrystals functionalized with alendronate as bioactive components for bone implant coatings to decrease osteoclastic activity, Appl. Surf. Sci. 328 (2015) 516–524. Available from: https://doi.org/10.1016/J.APSUSC.2014.12.072.

[102] K.A. Prosolov, O.A. Belyavskaya, J.V. Rau, O. Prymak, M. Epple, Y.P. Sharkeev, Deposition of polycrystalline zinc substituted hydroxyapatite coatings with a columnar structure by RF magnetron sputtering: role of in-situ substrate heating, J. Phys. Conf. Ser. 1115 (2018) 032077. Available from: https://doi.org/10.1088/1742-6596/1115/3/032077.

[103] K.A. Prosolov, O.A. Belyavskaya, U. Muehle, Y.P. Sharkeev, Thin bioactive Zn substituted hydroxyapatite coating deposited on ultrafine-grained titanium substrate: Structure analysis, Front. Mater. 5 (2018) 1–8. Available from: https://doi.org/10.3389/fmats.2018.00003.

[104] K.A. Prosolov, O.A. Belyavskaya, J. Linders, K. Loza, O. Prymak, C. Mayer, et al., Glancing angle deposition of Zn-doped calcium phosphate coatings by RF magnetron sputtering, Coatings 9 (2019) 1–17. Available from: https://doi.org/10.3390/coatings9040220.

[105] Z.S. Tao, W.S. Zhou, X.W. He, W. Liu, B.L. Bai, Q. Zhou, et al., A comparative study of zinc, magnesium, strontium-incorporated hydroxyapatite-coated titanium implants for osseointegration of osteopenic rats, Mater. Sci. Eng. C. 62 (2016) 226–232. Available from: https://doi.org/10.1016/J.MSEC.2016.01.034.

[106] S. Guimond-Lischer, Q. Ren, O. Braissant, P. Gruner, B. Wampfler, K. Maniura-Weber, Vacuum plasma sprayed coatings using ionic silver doped hydroxyapatite powder to prevent bacterial infection of bone implants, Biointerphases 11 (2016) 011012. Available from: https://doi.org/10.1116/1.4943225.

[107] M.B. Sedelnikova, E.G. Komarova, Y.P. Sharkeev, A.V. Ugodchikova, T.V. Tolkacheva, J.V. Rau, et al., Modification of titanium surface via Ag-, Sr- and Si-containing micro-arc calcium phosphate coating, Bioact. Mater. 4 (2019) 224–235. Available from: https://doi.org/10.1016/J.BIOACTMAT.2019.07.001.

[108] G. Ciobanu, M. Harja, Cerium-doped hydroxyapatite/collagen coatings on titanium for bone implants, Ceram. Int. 45 (2019) 2852–2857. Available from: https://doi.org/10.1016/J.CERAMINT.2018.07.290.

[109] Z.S. Tao, W.S. Zhou, Z. Qiang, K.K. Tu, Z.L. Huang, H.M. Xu, et al., Intermittent administration of human parathyroid hormone (1-34) increases fixation of strontium-doped hydroxyapatite coating titanium implants via electrochemical deposition in ovariectomized rat femur, J. Biomater. Appl. 30 (2016) 952–960. Available from: https://doi.org/10.1177/0885328215610898.

[110] Y. Huang, Z. Xu, X. Zhang, X. Chang, X. Zhang, Y.C. Li, et al., Nanotube-formed Ti substrates coated with silicate/silver co-doped hydroxyapatite as prospective materials for bone implants, J. Alloy. Compd. 697 (2017) 182–199. Available from: https://doi.org/10.1016/J.JALLCOM.2016.12.139.

[111] C.M. Xie, X. Lu, K.F. Wang, F.Z. Meng, O. Jiang, H.P. Zhang, et al., Silver nanoparticles and growth factors incorporated hydroxyapatite coatings on metallic implant surfaces for enhancement of osteoinductivity and antibacterial properties, ACS Appl. Mater. Interfaces. 6 (2014) 8580–8589. Available from: https://doi.org/10.1021/am501428e.

[112] Y. Otsuka, H. Kawaguchi, Y. Mutoh, Cyclic delamination behavior of plasma-sprayed hydroxyapatite coating on Ti–6Al–4V substrates in simulated body fluid, Mater. Sci. Eng. C. 67 (2016) 533–541. Available from: https://doi.org/10.1016/J.MSEC.2016.05.058.

[113] C.P. Charalambous, Calcium phosphate ceramics as hard tissue prosthetics, Class. Pap. Orthop. (2014) 419–421. Available from: https://doi.org/10.1007/978-1-4471-5451-8_106.

[114] R.Z. LeGeros, Calcium phosphate materials in restorative dentistry: a review, Adv. Dent. Res. 2 (1988) 164–180. Available from: https://doi.org/10.1177/08959374880020011101.

[115] R. Rohanizadeh, M. Padrines, J.M. Bouler, D. Couchourel, Y. Fortun, G. Daculsi, Apatite Precipitation after Incubation of Biphasic Calcium-Phosphate Ceramic in Various Solutions: Influence of Seed Species and Proteins, 1998.

[116] G. Daculsi, J.M. Bouler, R.Z. Legeros, Adaptive crystal formation in normal and pathological calcifications in synthetic calcium phosphate and related biomaterials, Int. Rev. Cytol. 172 (1997) 129–191. Available from: https://doi.org/10.1016/S0074-7696(08)62360-8.

[117] G. Daculsi, R.Z. Legeros, E. Nery, K. Lynch, B. Kerebel, Transformation of biphasic calcium phosphate ceramics in vivo: ultrastructural and physicochemical characterization, J. Biomed. Mater. Res. 23 (1989) 883–894. Available from: https://doi.org/10.1002/jbm.820230806.

[118] M. Heughebaert, R.Z. LeGeros, M. Gineste, A. Guilhem, G. Bonel, Physicochemical characterization of deposits associated with HA ceramics implanted in nonosseous sites, J. Biomed. Mater. Res. 22 (1988) 257–268. Available from: https://doi.org/10.1002/jbm.820221406.

[119] P. Ducheyne, S. Radin, L. King, K. Ishikawa, C.S. Kim, In vitro dissolution and precipitation of calcium phosphate phases on various biomaterials correlates with in vivo bioactivity, Bioceramics (1991) 135–144. Available from: https://doi.org/10.1016/B978-0-7506-0269-3.50023-2.

[120] H.K.H.K.K. Hyakuna, T. Yamamuro, Y. Kotoura, M. Oka, T. Nakamura, T. Kitsugi, et al., Surface reactions of calcium phosphate ceramics to various solutions, J. Biomed. Mater. Res. 24 (1990) 471–488. Available from: https://doi.org/10.1002/jbm.820240406.

[121] K. Ishikawa, S. Takagi, L.C. Chow, Y. Ishikawa, E.D. Eanes, K. Asaoka, Behavior of a calcium phosphate cement in simulated blood plasma in vitro, Dent. Mater. 10 (1994) 26–32. Available from: https://doi.org/10.1016/0109-5641(94)90018-3.

[122] G. Daculsi, R.Z. LeGeros, M. Heughebaert, I. Barbieux, Formation of carbonate-apatite crystals after implantation of calcium phosphate ceramics, Calcif. Tissue Int. 46 (1990) 20–27. Available from: https://doi.org/10.1007/BF02555820.

[123] I. Orly, M. Gregoire, J. Menanteau, M. Heughebaert, B. Kerebel, Chemical changes in hydroxyapatite biomaterial under in vivo and in vitro biological conditions, Calcif. Tissue Int. 45 (1989) 20–26. Available from: https://doi.org/10.1007/BF02556656.

[124] R.I. Martin, P.W. Brown, Formation of hydroxyapatite in serum, J. Mater. Sci. Mater. Med. 5 (1994) 96–102. Available from: https://doi.org/10.1007/BF00121697.

[125] A. Boskey, Mineral-matrix interactions in bone and cartilage, Clin. Orthop. Relat. Res. (1992) 244–274.

[126] A.L. Boskey, Matrix proteins and mineralization: an overview, Connect. Tissue Res. 35 (1996) 357–363. Available from: https://doi.org/10.3109/03008209609029212.

[127] T.F.H. Takeuchi, K. Tajima, T. Yamanaka, S. Doi, J. Kanehisa, T. Mori, et al., Doi, The mechanism of dental plaque and dental calculus formation by fibronectin: inhibition of seeded crystal growth of hydroxyapatite in metastable calcium phosphate solution by fibronectin, Nippon. Shishubyo Gakkai Kaishi 30 (1988) 110–113.

[128] D. Hanein, B. Geiger, L. Addadi, Fibronectin adsorption to surfaces of hydrated crystals. An analysis of the importance of bound water in protein-substrate interactions, Langmuir 9 (1993) 1058–1065. Available from: https://doi.org/10.1021/la00028a030.

[129] M.B. Contri, F. Boraldi, F. Taparelli, A. De Paepe, I.P. Ronchetti, Matrix proteins with high affinity for calcium ions are associated with mineralization within the elastic fibers of pseudoxanthoma elasticum dermis, Am. J. Pathol. 148 (1996) 569–577.

[130] C.A. van Blitterswijk, J.J. Grote, W. Kuÿpers, C.J.G. Blok-van Hoek, W.T. Daems, Bioreactions at the tissue/hydroxyapatite interface, Biomaterials 6 (1985) 243–251. Available from: https://doi.org/10.1016/0142-9612(85)90020-1.

[131] R. Holmes, H. Hagler, Porous hydroxyapatite as a bone graft substitute in maxillary augmentation: an histometric study, J. Cranio-Maxillofacial Surg. 16 (1988) 199–205. Available from: https://doi.org/10.1016/S1010-5182(88)80047-7.

[132] D.R. Hardy, P. Frayssinet, G. Bonel, T. Authom, S.A.L. Naelou, P.E. Delincé, Two-year outcome of hydroxyapatite-coated prostheses: two femoral prostheses retrieved at autopsy, Acta Orthop. 65 (1994) 253–257. Available from: https://doi.org/10.3109/17453679408995449.

[133] K. Shimazaki, V. Mooney, Comparative study of porous hydroxyapatite and tricalcium phosphate as bone substitute, J. Orthop. Res. 3 (1985) 301–310. Available from: https://doi.org/10.1002/jor.1100030306.

[134] M. Akao, H. Aoki, K. Kato, A. Sato, Dense polycrystalline β-tricalcium phosphate Prosthet applications, , J. Mater. Sci. 17 (1982) 343–346. Available from: https://doi.org/10.1007/BF00591468.

[135] M.B. Nair, H.K. Varma, A. John, Triphasic ceramic coated hydroxyapatite as a niche for goat stem cell-derived osteoblasts for bone regeneration and repair, J. Mater. Sci. Mater. Med. 20 (2009) 809–812. Available from: https://doi.org/10.1007/s10856-008-3598-8.

[136] T.L. Vassilev, Aluminium phosphate but not calcium phosphate stimulates the specific IgE response in guinea pigs to tetanus toxoid, Allergy 33 (1978) 155–159. Available from: https://doi.org/10.1111/j.1398-9995.1978.tb01527.x.

[137] B.B. Cogne, M. Ballet, J.J. Schmitt C, Total and IgE antibody levels following booster immunization with aluminum absorbed and nonabsorbed tetanus toxoid in humans, Ann. Allergy 54 (1985) 148–151.

[138] N. Goto, H. Kate, Local tissue irritating effects and adjuvant activities of calcium phosphate and aluminium hydroxide with different physical properties, 15 (1997) 1364–1371.

[139] T.T. Glant, K. Mikecz, A. Arzoumanian, A.R. Poole, Proteoglycan-induced arthritis in BALB/c mice. Clinical features and histopathology, Arthritis Rheum. 30 (1987) 201–212. Available from: http://www.ncbi.nlm.nih.gov/pubmed/3827960.

[140] B.Y.D.E. Trentham, A.S. Townes, A.H. Kang, Autoimmunity to type II collagen an experimental model of arthritis, 146, (1978) 10–11.

[141] P.H. Wooley, J.R. Seibold, J.D. Whalen, J.M. Chapdelaine, Pristane-induced arthritis. the immunologic and genetic features of an experimental murine model of autoimmune disease, Arthritis Rheum. 32 (1989) 1022–1030. Available from: https://doi.org/10.1002/anr.1780320812.

[142] G.M. Cannon, G.W. Woods, M.L. Clayton, F, Induction of arthritis in DA rats by incomplete freund's adjuvant, J. Rheumatol. 20 (1993) 7–11.

[143] S. Kleinau, H. Erlandsson, R. Holmdahl, L. Klareskog, Adjuvant oils induce arthritis in the DA rat. I. Characterization of the disease and evidence for an immunological involvement, J. Autoimmun. 4 (1991) 871–880. Available from: https://doi.org/10.1016/0896-8411(91)90050-M.

[144] M.M. Griffiths, J.J. Langone, M.M. Lightfoote, Biomater. Granulomas 304 (1996) 295–304.

[145] V. Uskoković, Nanostructured platforms for the sustained and local delivery of antibiotics in the treatment of osteomyelitis, Crit. Rev. Ther. Drug. Carr. Syst. 32 (2015) 1–59. Available from: https://doi.org/10.1615/CritRevTherDrugCarrierSyst.2014010920.

[146] M.V. Ciocilteu, A.G. Mocanu, A. Mocanu, C. Ducu, O.E. Nicolaescu, V.C. Manda, et al., Hydroxyapatite-ciprofloxacin delivery system: synthesis, characterisation and antibacterial activity, Acta Pharm. 68 (2018) 129–144. Available from: https://doi.org/10.2478/acph-2018-0011.

[147] M.U. Munir, A. Ihsan, Y. Sarwar, S.Z. Bajwa, K. Bano, B. Tehseen, et al., Hollow mesoporous hydroxyapatite nanostructures; smart nanocarriers with high drug loading and controlled releasing features, Int. J. Pharm. 544 (2018) 112–120. Available from: https://doi.org/10.1016/J.IJPHARM.2018.04.029.

[148] G.S. Kumar, R. Govindan, E.K. Girija, In situ synthesis, characterization and in vitro studies of ciprofloxacin loaded hydroxyapatite nanoparticles for the treatment of osteomyelitis, J. Mater. Chem. B. 2 (2014) 5052–5060. Available from: https://doi.org/10.1039/c4tb00339j.

[149] V.C.A. Martins, G. Goissis, Nonstoichiometric hydroxyapatite-anionic collagen composite as support for the double sustained release of gentamicin and norfloxacin/ciprofloxacin, Artif. Organs. 24 (2000) 224–230. Available from: https://doi.org/10.1046/j.1525-1594.2000.06517.x.

[150] Z.C. Xiong, Z.Y. Yang, Y.J. Zhu, F.F. Chen, Y.G. Zhang, R.L. Yang, Ultralong hydroxyapatite nanowires-based paper co-loaded with silver nanoparticles and antibiotic for long-term antibacterial benefit, ACS Appl. Mater. Interfaces. 9 (2017) 22212−22222. Available from: https://doi.org/10.1021/acsami.7b05208.
[151] K. Sangeetha, M. Ashok, E.K. Girija, G. Vidhya, G. Vasugi, Strontium and ciprofloxacin modified hydroxyapatites as functional grafts for bone prostheses, Ceram. Int. 44 (2018) 13782−13789. Available from: https://doi.org/10.1016/J.CERAMINT.2018.04.221.
[152] M.E. Butini, S. Cabric, A. Trampuz, M. Di Luca, In vitro anti-biofilm activity of a biphasic gentamicin-loaded calcium sulfate/hydroxyapatite bone graft substitute, Colloids Surf. B Biointerfaces 161 (2018) 252−260. Available from: https://doi.org/10.1016/J.COLSURFB.2017.10.050.
[153] K. Madhumathi, T.S. Sampath Kumar, Regenerative potential and anti-bacterial activity of tetracycline loaded apatitic nanocarriers for the treatment of periodontitis, Biomed. Mater. 9 (2014) 035002. Available from: https://doi.org/10.1088/1748-6041/9/3/035002.
[154] C.A. Soriano-Souza, A.L. Rossi, E. Mavropoulos, M.A. Hausen, M.N. Tanaka, M.D. Calasans-Maia, et al., Chlorhexidine-loaded hydroxyapatite microspheres as an antimicrobial delivery system and its effect on in vivo osteo-conductive properties, J. Mater. Sci. Mater. Med. 26 (2015). Available from: https://doi.org/10.1007/s10856-015-5505-4.
[155] S.W. Tsai, S.S. Huang, W.X. Yu, Y.W. Hsu, F.Y. Hsu, Fabrication and characteristics of porous hydroxyapatite-CaO composite nanofibers for biomedical applications, Nanomaterials 8 (2018). Available from: https://doi.org/10.3390/nano8080570.
[156] S.W. Tsai, W.X. Yu, P.A. Hwang, S.S. Huang, H.M. Lin, Y.W. Hsu, et al., Fabrication and characterization of strontium-substituted hydroxyapatite-CaO-$CaCO_3$ nanofibers with a mesoporous structure as drug delivery carriers, Pharmaceutics 10 (2018). Available from: https://doi.org/10.3390/pharmaceutics10040179.
[157] B. Ren, X. Chen, S. Du, Y. Ma, H. Chen, G. Yuan, et al., Injectable polysaccharide hydrogel embedded with hydroxyapatite and calcium carbonate for drug delivery and bone tissue engineering, Int. J. Biol. Macromol. 118 (2018) 1257−1266. Available from: https://doi.org/10.1016/J.IJBIOMAC.2018.06.200.
[158] R. Ramírez-Agudelo, K. Scheuermann, A. Gala-García, A.P.F. Monteiro, A.D. Pinzón-García, M.E. Cortés, et al., Hybrid nanofibers based on poly-caprolactone/gelatin/hydroxyapatite nanoparticles-loaded Doxycycline: Effective anti-tumoral and antibacterial activity, Mater. Sci. Eng. C. 83 (2018) 25−34. Available from: https://doi.org/10.1016/J.MSEC.2017.08.012.
[159] L. Ding, P. Zhang, X. Wang, J. Hao, K. Aoki, S. Kuroda, et al., Effect of doxycycline-treated hydroxyapatite surface on bone apposition: a histomophometric study in murine maxillae, Dent. Mater. J. 37 (2018) 130−138. Available from: https://doi.org/10.4012/dmj.2017-007.
[160] J.H. Sörensen, M. Lilja, T.C. Sörensen, M. Astrand, P. Procter, S. Fuchs, et al., Biomechanical and antibacterial properties of Tobramycin loaded hydroxyapatite coated fixation pins, J. Biomed. Mater. Res. - Part. B Appl. Biomater. 102 (2014) 1381−1392. Available from: https://doi.org/10.1002/jbm.b.33117.
[161] D. Ionita, D. Bajenaru-Georgescu, G. Totea, A. Mazare, P. Schmuki, I. Demetrescu, Activity of vancomycin release from bioinspired coatings of hydroxyapatite or TiO2 nanotubes, Int. J. Pharm. 517 (2017) 296−302. Available from: https://doi.org/10.1016/J.IJPHARM.2016.11.062.
[162] http://www.tandfonline.com, (n.d.).
[163] B. Zihayat, A. Khodadadi, M. Torabi, M. Mehdipour, M. Basiri, M. Asadi-Shekarri, Wound healing activity of sheep's bladder extracellular matrix in diabetic rats, Biomed. Eng. - Appl. Basis Commun. 30 (2018) 1−8. Available from: https://doi.org/10.4015/S1016237218500151.
[164] Z. Cao, D. Jiang, L. Yan, J. Wu, In vitro and in vivo drug release and antibacterial properties of the novel vancomycin-loaded bone-like hydroxyapatite/poly amino acid scaffold, Int. J. Nanomed. 12 (2017) 1841−1851. Available from: https://doi.org/10.2147/IJN.S122864.
[165] J.C. Karr, Lower-extremity osteomyelitis treatment using calcium sulfate/hydroxyapatite bone void filler with antibiotics seven-year retrospective study, J. Am. Podiatr. Med. Assoc. 108 (2018) 210−214. Available from: https://doi.org/10.7547/16-096.
[166] R.P. Singh, G. Singh, H. Singh, Sub-micrometric mesoporous strontium substituted hydroxyapatite particles for sustained delivery of vancomycin drug, J. Aust. Ceram. Soc. 55 (2019) 405−414. Available from: https://doi.org/10.1007/s41779-018-0248-6.

[167] D. Giacomini, P. Torricelli, G.A. Gentilomi, E. Boanini, M. Gazzano, F. Bonvicini, et al., Monocyclic β-lactams loaded on hydroxyapatite: new biomaterials with enhanced antibacterial activity against resistant strains, Sci. Rep. 7 (2017) 1–12. Available from: https://doi.org/10.1038/s41598-017-02943-2.

[168] A. Zomorodian, C. Santos, M.J. Carmezim, T.M.E. Silva, J.C.S. Fernandes, M.F. Montemor, "In-vitro" corrosion behaviour of the magnesium alloy with Al and Zn (AZ31) protected with a biodegradable polycaprolactone coating loaded with hydroxyapatite and cephalexin, Electrochim. Acta 179 (2015) 431–440. Available from: https://doi.org/10.1016/J.ELECTACTA.2015.04.013.

[169] T. Wan, G.K. Stylios, M. Giannoudi, P.V. Giannoudis, Investigating a new drug delivery nano composite membrane system based on PVA/PCL and PVA/HA(PEG) for the controlled release of biopharmaceuticals for bone infections, Injury 46 (2015) S39–S43. Available from: https://doi.org/10.1016/S0020-1383(15)30053-X.

[170] D. RĂdulescu, V. Grumezescu, E. Andronescu, A.M. Holban, A.M. Grumezescu, G. Socol, et al., Biocompatible cephalosporin-hydroxyapatite-poly(lactic-co-glycolic acid)-coatings fabricated by MAPLE technique for the prevention of bone implant associated infections, Appl. Surf. Sci. 374 (2016) 387–396. Available from: https://doi.org/10.1016/J.APSUSC.2016.02.072.

[171] A.C. Queiroz, J.D. Santos, F.J. Monteiro, I.R. Gibson, J.C. Knowles, Adsorption and release studies of sodium ampicillin from hydroxyapatite and glass-reinforced hydroxyapatite composites, Biomaterials 22 (2001) 1393–1400. Available from: https://doi.org/10.1016/S0142-9612(00)00296-9.

[172] J. Pradid, W. Keawwatana, U. Boonyang, S. Tangbunsuk, Biological properties and enzymatic degradation studies of clindamycin-loaded PLA/HAp microspheres prepared from crocodile bones, Polym. Bull. 74 (2017) 5181–5194. Available from: https://doi.org/10.1007/s00289-017-2006-2.

[173] G. Gonzalez, A. Sagarzazu, A. Cordova, M.E. Gomes, J. Salas, L. Contreras, et al., Comparative study of two silica mesoporous materials (SBA-16 and SBA-15) modified with a hydroxyapatite layer for clindamycin controlled delivery, Microporous Mesoporous Mater. 256 (2018) 251–265. Available from: https://doi.org/10.1016/J.MICROMESO.2017.07.021.

[174] W. Nie, C. Peng, X. Zhou, L. Chen, W. Wang, Y. Zhang, et al., Three-dimensional porous scaffold by self-assembly of reduced graphene oxide and nano-hydroxyapatite composites for bone tissue engineering, Carbon N. Y. 116 (2017) 325–337. Available from: https://doi.org/10.1016/J.CARBON.2017.02.013.

[175] O. Zamoume, S. Thibault, G. Regnié, M.O. Mecherri, M. Fiallo, P. Sharrock, Macroporous calcium phosphate ceramic implants for sustained drug delivery, Mater. Sci. Eng. C. 31 (2011) 1352–1356. Available from: https://doi.org/10.1016/J.MSEC.2011.04.020.

[176] A. Deepak, A.K. Goyal, G. Rath, Development and characterization of novel medicated nanofiber for the treatment of periodontitis, AAPS PharmSciTech 19 (2018) 3687–3697. Available from: https://doi.org/10.1208/s12249-018-1177-y.

[177] R. Ardhani, O.A. Setyaningsih, I.D. Hafiyyah, et al., Preparation of carbonated apatite membrane as metronidazole delivery system for periodontal application, Key Eng. Mater. 696 (2016) 250–258. Available from: https://doi.org/10.4028/http://www.scientific.net/KEM.696.250.

[178] K. Deshmukh, M.M. Shaik, S.R. Ramanan, M. Kowshik, Self-activated fluorescent hydroxyapatite nanoparticles: a promising agent for bioimaging and biolabeling, ACS Biomater. Sci. Eng. 2 (2016) 1257–1264. Available from: https://doi.org/10.1021/acsbiomaterials.6b00169.

[179] D. Predoi, S.L. Iconaru, N. Buton, M.L. Badea, L. Marutescu, Antimicrobial activity of new materials based on lavender and basil essential oils and hydroxyapatite, Nanomaterials 8 (2018). Available from: https://doi.org/10.3390/nano8050291.

[180] B. Mueller, L. Treccani, K. Rezwan, Antibacterial active open-porous hydroxyapatite/lysozyme scaffolds suitable as bone graft and depot for localised drug delivery, J. Biomater. Appl. 31 (2017) 1123–1134. Available from: https://doi.org/10.1177/0885328216688074.

[181] K. K., Gentamicin-PMMA-beads in treating bone and soft tissue infections (author's transl), Zentralbl. Chir. 104 (1979) 934–942.

[182] A. Coetzee, Regeneration of bone in the presence of calcium sulfate, Arch. Otolaryngol. 106 (1980) 405–409.

[183] G.H. Lee, J.G. Khoury, J. Bell, J.A. Buckwalter, Adverse reactions to osteoset bone graft substitute, IOWA Orthop. J. 22 (2002) 35–39.

[184] H.N. Robinson, D. Alk, D. Sandbank, J. Farber R, Inflammatory reactions associated with a calcium sulfate bone substitute, Ann. Transplant. 4 (1999) 91–97.

[185] M.A. Rauschmann, T.A. Wichelhaus, V. Stirnal, E. Dingeldein, L. Zichner, R. Schnettler, et al., Nanocrystalline hydroxyapatite and calcium sulfate as biodegradable composite carrier material for local delivery of antibiotics in bone infections, Biomaterials 26 (2005) 2677–2684. Available from: https://doi.org/10.1016/j.biomaterials.2004.06.045.

[186] S. Bisht, G. Bhakta, S. Mitra, A. Maitra, pDNA loaded calcium phosphate nanoparticles: highly efficient non-viral vector for gene delivery, Int. J. Pharm. 288 (2005) 157–168. Available from: https://doi.org/10.1016/J.IJPHARM.2004.07.035.

[187] D. Lee, K. Upadhye, P.N. Kumta, Nano-sized calcium phosphate (CaP) carriers for non-viral gene deilvery, Mater. Sci. Eng. B. 177 (2012) 289–302. Available from: https://doi.org/10.1016/J.MSEB.2011.11.001.

[188] H. Xu, D. Wei, X. Gai, Y. Jiang, Amine functionalised hydroxyapatite nanoparticles for anti-angiogenesis gene therapy of breast cancer, Micro Nano Lett. 11 (2016) 416–419. Available from: https://doi.org/10.1049/mnl.2016.0078.

[189] T. Welzel, I. Radtke, W. Meyer-Zaika, R. Heumann, M. Epple, Transfection of cells with custom-made calcium phosphate nanoparticles coated with DNA, J. Mater. Chem. 14 (2004) 2213–2217. Available from: https://doi.org/10.1039/b401644k.

[190] G.J. Wu, L.Z. Zhou, K.W. Wang, F. Chen, Y. Sun, Y.R. Duan, et al., Hydroxylapatite nanorods: An efficient and promising carrier for gene transfection, J. Colloid Interface Sci. 345 (2010) 427–432. Available from: https://doi.org/10.1016/J.JCIS.2010.01.048.

[191] S.H. Zhu, B.Y. Huang, K.C. Zhou, S.P. Huang, F. Liu, Y.M. Li, et al., Hydroxyapatite nanoparticles as a novel gene carrier, J. Nanopart. Res. 6 (2004) 307–311. Available from: https://doi.org/10.1023/b:nano.0000034721.06473.23.

[192] D. Olton, J. Li, M.E. Wilson, T. Rogers, J. Close, L. Huang, et al., Nanostructured calcium phosphates (NanoCaPs) for non-viral gene delivery: Influence of the synthesis parameters on transfection efficiency, Biomaterials 28 (2007) 1267–1279. Available from: https://doi.org/10.1016/J.BIOMATERIALS.2006.10.026.

[193] M.A. Khan, V.M. Wu, S. Ghosh, V. Uskoković, Gene delivery using calcium phosphate nanoparticles: optimization of the transfection process and the effects of citrate and poly(l-lysine) as additives, J. Colloid Interface Sci. 471 (2016) 48–58. Available from: https://doi.org/10.1016/J.JCIS.2016.03.007.

[194] T.Y. Cheang, Y.Y. Lei, Z.Q. Zhang, H.Y. Zhou, R.Y. Ye, Y. Lin, et al., Graphene oxide–hydroxyapatite nanocomposites effectively deliver HSV-TK suicide gene to inhibit human breast cancer growth, J. Biomater. Appl. 33 (2018) 216–226. Available from: https://doi.org/10.1177/0885328218788242.

[195] E.H. Chowdhury, M. Kunou, M. Nagaoka, A.K. Kundu, T. Hoshiba, T. Akaike, High-efficiency gene delivery for expression in mammalian cells by nanoprecipitates of Ca – Mg phosphate, 341 (2004) 77–82. https://doi.org/10.1016/j.gene.2004.07.015.

[196] A. Hanifi, M.H. Fathi, H.M.M. Sadeghi, J. Varshosaz, Mg^{2+}substituted calcium phosphate nano particles synthesis for non viral gene delivery application, J. Mater. Sci. Mater. Med. 21 (2010) 2393–2401. Available from: https://doi.org/10.1007/s10856-010-4088-3.

[197] J. Klesing, S. Chernousova, M. Epple, Freeze-dried cationic calcium phosphate nanorods as versatile carriers of nucleic acids (DNA, siRNA), J. Mater. Chem. 22 (2012) 199–204. Available from: https://doi.org/10.1039/c1jm13502c.

[198] G.H. Wang, Y.Z. Zhao, J. Tan, S.H. Zhu, K.C. Zhou, Arginine functionalized hydroxyapatite nanoparticles and its bioactivity for gene delivery, Trans. Nonferrous Met. Soc. China. 25 (2015) 490–496. Available from: https://doi.org/10.1016/S1003-6326(15)63629-9.

[199] H. Komuro, T. Sasano, N. Horiuchi, K. Yamashita, A. Nagai, The effect of glucose modification of hydroxyapatite nanoparticles on gene delivery, J. Biomed. Mater. Res. - Part. A 107 (2019) 61–66. Available from: https://doi.org/10.1002/jbm.a.36523.

[200] X. Chen, C. Deng, S. Tang, M. Zhang, Mitochondria-dependent apoptosis induced by nanoscale hydroxyapatite in human gastric cancer SGC-7901 cells, Biol. Pharm. Bull. 30 (2007) 128–132. Available from: https://doi.org/10.1248/bpb.30.128.

[201] J. Xu, P. Xu, Z. Li, J. Huang, Z. Yang, Oxidative stress and apoptosis induced by hydroxyapatite nanoparticles in C6 cells, J. Biomed. Mater. Res. - Part. A. 100 A (2012) 738–745. Available from: https://doi.org/10.1002/jbm.a.33270.

[202] B. Li, B. Guo, H. Fan, X. Zhang, Preparation of nano-hydroxyapatite particles with different morphology and their response to highly malignant melanoma cells in vitro, Appl. Surf. Sci. 255 (2008) 357–360. Available from: https://doi.org/10.1016/J.APSUSC.2008.06.114.

[203] V. Uskoković, M.A. Iyer, V.M. Wu, One ion to rule them all: the combined antibacterial, osteoinductive and anticancer properties of selenite-incorporated hydroxyapatite, J. Mater. Chem. B. 5 (2017) 1430–1445. Available from: https://doi.org/10.1039/c6tb03387c.

[204] N.L. Ignjatović, M. Sakač, I. Kuzminac, V. Kojić, S. Marković, D. Vasiljević-Radović, et al., Chitosan oligosaccharide lactate coated hydroxyapatite nanoparticles as a vehicle for the delivery of steroid drugs and the targeting of breast cancer cells, J. Mater. Chem. B. 6 (2018) 6957–6968. Available from: https://doi.org/10.1039/c8tb01995a.

[205] N.L. Ignjatović, K.M. Penov-Gaši, V.M. Wu, J.J. Ajduković, V.V. Kojić, D. Vasiljević-Radović, et al., Selective anticancer activity of hydroxyapatite/chitosan-poly(d,l)-lactide-co-glycolide particles loaded with an androstane-based cancer inhibitor, Colloids Surf. B Biointerfaces. 148 (2016) 629–639. Available from: https://doi.org/10.1016/J.COLSURFB.2016.09.041.

[206] A. Adamiano, V.M. Wu, F. Carella, G. Lamura, F. Canepa, A. Tampieri, et al., Magnetic calcium phosphates nanocomposites for the intracellular hyperthermia of cancers of bone and brain, Nanomedicine 14 (2019) 1267–1289. Available from: https://doi.org/10.2217/nnm-2018-0372.

[207] Y. Han, S. Li, X. Cao, L. Yuan, Y. Wang, Y. Yin, et al., Different inhibitory effect and mechanism of hydroxyapatite nanoparticles on normal cells and cancer cells in vitro and in vivo, Sci. Rep. 4 (2014) 1–8. Available from: https://doi.org/10.1038/srep07134.

[208] Y. Sun, Y. Chen, X. Ma, Y. Yuan, C. Liu, J. Kohn, et al., Mitochondria-targeted hydroxyapatite nanoparticles for selective growth inhibition of lung cancer in vitro and in vivo, ACS Appl. Mater. Interfaces. 8 (2016) 25680–25690. Available from: https://doi.org/10.1021/acsami.6b06094.

[209] Y. Yuan, C. Liu, J. Qian, J. Wang, Y. Zhang, Size-mediated cytotoxicity and apoptosis of hydroxyapatite nanoparticles in human hepatoma HepG2 cells, Biomaterials 31 (2010) 730–740. Available from: https://doi.org/10.1016/J.BIOMATERIALS.2009.09.088.

[210] X. Zhao, B.C. Heng, S. Xiong, J. Guo, T.T.Y. Tan, F.Y.C. Boey, et al., In vitro assessment of cellular responses to rod-shaped hydroxyapatite nanoparticles of varying lengths and surface areas, Nanotoxicology 5 (2011) 182–194. Available from: https://doi.org/10.3109/17435390.2010.503943.

[211] G. Li, L. Ye, J. Pan, M. Long, Z. Zhao, H. Yang, et al., Antitumoural hydroxyapatite nanoparticles-mediated hepatoma-targeted trans-arterial embolization gene therapy: In vitro and in vivo studies, Liver Int. 32 (2012) 998–1007. Available from: https://doi.org/10.1111/j.1478-3231.2012.02761.x.

[212] I.W. Bauer, S.P. Li, Y.C. Han, L. Yuan, M.Z. Yin, Internalization of hydroxyapatite nanoparticles in liver cancer cells, J. Mater. Sci. Mater. Med. 19 (2008) 1091–1095. Available from: https://doi.org/10.1007/s10856-007-3124-4.

[213] W. Jiang, B.Y.S. Kim, J.T. Rutka, W.C.W. Chan, Nanoparticle-mediated cellular response is size-dependent, Nat. Nanotechnol. 3 (2008) 145–150. Available from: https://doi.org/10.1038/nnano.2008.30.

[214] J. Gómez-Morales, M. Iafisco, J.M. Delgado-López, S. Sarda, C. Drouet, Progress on the preparation of nanocrystalline apatites and surface characterization: overview of fundamental and applied aspects, Prog. Cryst. Growth Charact. Mater. 59 (2013) 1–46. Available from: https://doi.org/10.1016/j.pcrysgrow.2012.11.001.

[215] V.M. Wu, S. Tang, V. Uskoković, Calcium phosphate nanoparticles as intrinsic inorganic antimicrobials: the antibacterial effect, ACS Appl. Mater. Interfaces. 10 (2018) 34013–34028. Available from: https://doi.org/10.1021/acsami.8b12784.

[216] W. Yanhua, H. Hao, Y. Li, S. Zhang, Selenium-substituted hydroxyapatite nanoparticles and their in vivo antitumor effect on hepatocellular carcinoma, Colloids Surf. B Biointerfaces. 140 (2016) 297–306. Available from: https://doi.org/10.1016/J.COLSURFB.2015.12.056.
[217] H. Xiong, S. Du, J. Ni, J. Zhou, J. Yao, Mitochondria and nuclei dual-targeted heterogeneous hydroxyapatite nanoparticles for enhancing therapeutic efficacy of doxorubicin, Biomaterials 94 (2016) 70–83. Available from: https://doi.org/10.1016/J.BIOMATERIALS.2016.04.004.
[218] B. Kundu, D. Ghosh, M.K. Sinha, P.S. Sen, V.K. Balla, N. Das, et al., Doxorubicin-intercalated nano-hydroxyapatite drug-delivery system for liver cancer: an animal model, Ceram. Int. 39 (2013) 9557–9566. Available from: https://doi.org/10.1016/J.CERAMINT.2013.05.074.
[219] L.E. Gerweck, Tumor pH: implications for treatment and novel drug design, Semin. Radiat. Oncol. 8 (1998) 176–182. Available from: https://doi.org/10.1016/S1053-4296(98)80043-X.
[220] W. Sun, J. Fan, S. Wang, Y. Kang, J. Du, X. Peng, Biodegradable drug-loaded hydroxyapatite nanotherapeutic agent for targeted drug release in tumors, ACS Appl. Mater. Interfaces. 10 (2018) 7832–7840. Available from: https://doi.org/10.1021/acsami.7b19281.
[221] D. Li, X. Huang, Y. Wu, J. Li, W. Cheng, J. He, et al., Preparation of pH-responsive mesoporous hydroxyapatite nanoparticles for intracellular controlled release of an anticancer drug, Biomater. Sci. 4 (2016) 272–280. Available from: https://doi.org/10.1039/c5bm00228a.
[222] H. Maeda, K. Tsukigawa, J. Fang, A. Retrospective, 30 years after discovery of the enhanced permeability and retention effect of solid tumors: next-generation chemotherapeutics and photodynamic therapy—problems, solutions, and prospects, Microcirculation 23 (2016) 173–182. Available from: https://doi.org/10.1111/micc.12228.
[223] R. Rastegar, H. Akbari Javar, M. Khoobi, P. Dehghan Kelishadi, G. Hossein Yousefi, M. Doosti, et al., Evaluation of a novel biocompatible magnetic nanomedicine based on beta-cyclodextrin, loaded doxorubicin-curcumin for overcoming chemoresistance in breast cancer, Artif. Cells, Nanomed. Biotechnol. 46 (2018) 207–216. Available from: https://doi.org/10.1080/21691401.2018.1453829.
[224] A.R. Kherlopian, T. Song, Q. Duan, M.A. Neimark, M.J. Po, J.K. Gohagan, et al., A review of imaging techniques for systems biology, BMC Syst. Biol. 2 (2008) 1–18. Available from: https://doi.org/10.1186/1752-0509-2-74.
[225] P. Sharma, S. Brown, G. Walter, S. Santra, B. Moudgil, Nanoparticles for bioimaging, Adv. Colloid Interface Sci. 123–126 (2006) 471–485. Available from: https://doi.org/10.1016/J.CIS.2006.05.026.
[226] V. Ntziachristos, Fluorescence molecular imaging, Annu. Rev. Biomed. Eng. 8 (2006) 1–33. Available from: https://doi.org/10.1146/annurev.bioeng.8.061505.095831.
[227] H. Liu, F. Chen, P. Xi, B. Chen, L. Huang, J. Cheng, et al., Biocompatible fluorescent hydroxyapatite: synthesis and live cell imaging applications, J. Phys. Chem. C. 115 (2011) 18538–18544. Available from: https://doi.org/10.1021/jp206843w.
[228] S.S. Syamchand, G. Sony, Multifunctional hydroxyapatite nanoparticles for drug delivery and multimodal molecular imaging, Microchim. Acta. 182 (2015) 1567–1589. Available from: https://doi.org/10.1007/s00604-015-1504-x.
[229] V. Uskoković, S. Tang, M.G. Nikolić, S. Marković, V.M. Wu, Calcium phosphate nanoparticles as intrinsic inorganic antimicrobials: in search of the key particle property, Biointerphases 14 (2019) 031001. Available from: https://doi.org/10.1116/1.5090396.
[230] R. Jagannathan, M. Kottaisamy, Eu^{3+} luminescence: a spectral probe in M_5 $(PO_4)_3$ X apatites (M = Ca or Sr; X = F^-, Cl^-, Br^- or OH^-), J. Phys. Condens. Matter. 7 (1995) 8453–8466. Available from: https://doi.org/10.1088/0953-8984/7/44/014.
[231] A. Doat, M. Fanjul, F. Pellé, E. Hollande, A. Lebugle, Europium-doped bioapatite: a new photostable biological probe, internalizable by human cells, Biomaterials 24 (2003) 3365–3371. Available from: https://doi.org/10.1016/S0142-9612(03)00169-8.
[232] A. Doat, F. Pellé, N. Gardant, A. Lebugle, Synthesis of luminescent bioapatite nanoparticles for utilization as a biological probe, J. Solid. State Chem. 177 (2004) 1179–1187. Available from: https://doi.org/10.1016/J.JSSC.2003.10.023.
[233] D.R.K. Weerasuriya, W.P.S.L. Wijesinghe, R.M.G. Rajapakse, Encapsulation of anticancer drug copper bis(8-hydroxyquinoline) in hydroxyapatite for pH-sensitive targeted delivery and slow release, Mater. Sci. Eng. C. 71 (2017) 206–213. Available from: https://doi.org/10.1016/J.MSEC.2016.10.010.

[234] T. Matsuya, Y. Otsuka, M. Tagaya, S. Motozuka, K. Ohnuma, Y. Mutoh, Formation of stacked luminescent complex of 8-hydroxyquinoline molecules on hydroxyapatite coating by using cold isostatic pressing, Mater. Sci. Eng. C. 58 (2016) 127–132. Available from: https://doi.org/10.1016/J.MSEC.2015.08.020.
[235] I.A. Neacsu, A.E. Stoica, B.S. Vasile, E. Andronescu, Luminescent hydroxyapatite doped with rare earth elements for biomedical applications, Nanomaterials 9 (2019). Available from: https://doi.org/10.3390/nano9020239.
[236] Z. Li, Z. Liu, M. Yin, X. Yang, Q. Yuan, J. Ren, et al., Aptamer-capped multifunctional mesoporous strontium hydroxyapatite nanovehicle for cancer-cell-responsive drug delivery and imaging, Biomacromolecules 13 (2012) 4257–4263. Available from: https://doi.org/10.1021/bm301563q.
[237] V. Stanić, S. Dimitrijević, D.G. Antonović, B.M. Jokić, S.P. Zec, S.T. Tanasković, et al., Synthesis of fluorine substituted hydroxyapatite nanopowders and application of the central composite design for determination of its antimicrobial effects, Appl. Surf. Sci. 290 (2014) 346–352. Available from: https://doi.org/10.1016/J.APSUSC.2013.11.081.
[238] Y.H. Yang, C.H. Liu, Y.H. Liang, F.H. Lin, K.C.W. Wu, Hollow mesoporous hydroxyapatite nanoparticles (hmHANPs) with enhanced drug loading and pH-responsive release properties for intracellular drug delivery, J. Mater. Chem. B. 1 (2013) 2447–2450. Available from: https://doi.org/10.1039/c3tb20365d.
[239] L. Mao, X. Liu, M. Liu, L. Huang, D. Xu, R. Jiang, et al., Surface grafting of zwitterionic polymers onto dye doped AIE-active luminescent silica nanoparticles through surface-initiated ATRP for biological imaging applications, Appl. Surf. Sci. 419 (2017) 188–196. Available from: https://doi.org/10.1016/J.APSUSC.2017.05.041.
[240] J.D. Byrne, T. Betancourt, L. Brannon-Peppas, Active targeting schemes for nanoparticle systems in cancer therapeutics, Adv. Drug. Deliv. Rev. 60 (2008) 1615–1626. Available from: https://doi.org/10.1016/J.ADDR.2008.08.005.
[241] B. Sandhöfer, M. Meckel, J.M. Delgado-López, T. Patrício, A. Tampieri, F. Rösch, et al., Synthesis and preliminary in vivo evaluation of well-dispersed biomimetic nanocrystalline apatites labeled with positron emission tomographic imaging agents, ACS Appl. Mater. Interfaces. 7 (2015) 10623–10633. Available from: https://doi.org/10.1021/acsami.5b02624.
[242] A. Adamiano, M. Iafisco, M. Sandri, M. Basini, P. Arosio, T. Canu, et al., On the use of superparamagnetic hydroxyapatite nanoparticles as an agent for magnetic and nuclear in vivo imaging, Acta Biomater. 73 (2018) 458–469. Available from: https://doi.org/10.1016/J.ACTBIO.2018.04.040.
[243] G. Tomoaia, L.B. Pop, I. Petean, M. Tomoaia-Cotisel, Significance of surface structure on orthopedic materials, Mater. Plast. 49 (2012) 48–54.
[244] F. Jung, S. Braune, Thrombogenicity and hemocompatibility of biomaterials, Biointerphases 11 (2016) 029601. Available from: https://doi.org/10.1116/1.4938557.
[245] M. Hoffman, D.M.M. Iii, J.P. Riddel, B.E. Aouizerat, C. Miaskowski, D.P. Lillicrap, et al., A cell-based model of hemostasis "for internal educational purposes only. Not for dissemination"For Internal Educational Purposes Only. Not for Dissemination," Arterioscler. Thromb. Vasc. Biol. 24 (2003) 797–811.
[246] Y. Yan, K.T. Gause, M.M.J. Kamphuis, C.S. Ang, N.M. O'Brien-Simpson, J.C. Lenzo, et al., Differential roles of the protein corona in the cellular uptake of nanoporous polymer particles by monocyte and macrophage cell lines, ACS Nano 7 (2013) 10960–10970. Available from: https://doi.org/10.1021/nn404481f.
[247] X. Chen, Q. Wang, J. Shen, H. Pan, T. Wu, Adsorption of leucine-rich amelogenin protein on hydroxyapatite (001) surface through -COO- claws, J. Phys. Chem. C. 111 (2007) 1284–1290. Available from: https://doi.org/10.1021/jp0646630.
[248] X. Dong, Q. Wang, T. Wu, H. Pan, Understanding adsorption-desorption dynamics of BMP-2 on hydroxyapatite (001) surface, Biophys. J. 93 (2007) 750–759. Available from: https://doi.org/10.1529/BIOPHYSJ.106.103168.
[249] J.W. Shen, T. Wu, Q. Wang, H.H. Pan, Molecular simulation of protein adsorption and desorption on hydroxyapatite surfaces, Biomaterials 29 (2008) 513–532. Available from: https://doi.org/10.1016/J.BIOMATERIALS.2007.10.016.

[250] A. Mayer, M. Vadon, B. Rinner, A. Novak, R. Wintersteiger, E. Fröhlich, The role of nanoparticle size in hemocompatibility, Toxicology 258 (2009) 139–147. Available from: https://doi.org/10.1016/J.TOX.2009.01.015.

[251] W. Zhang, H. Liu, W. Yang, C. Liu, M. Xie, R. Guo, et al., Hydroxyapatite/silk fibroin composite biomimetic scaffold for dental pulp repair, Bioinspired, Biomim. Nanobiomater. 8 (2019) 231–238. Available from: https://doi.org/10.1680/jbibn.18.00050.

[252] A. Bhattacharjee, A. Gupta, M. Verma, P.A. Murugan, P. Sengupta, S. Matheshwaran, et al., Site-specific antibacterial efficacy and cyto/hemo-compatibility of zinc substituted hydroxyapatite, Ceram. Int. 45 (2019) 12225–12233. Available from: https://doi.org/10.1016/J.CERAMINT.2019.03.132.

[253] C.H. Ooi, Y.P. Ling, W.Z. Abdullah, A.Z. Mustafa, S.Y. Pung, F.Y. Yeoh, Physicochemical evaluation and in vitro hemocompatibility study on nanoporous hydroxyapatite, J. Mater. Sci. Mater. Med. 30 (2019). Available from: https://doi.org/10.1007/s10856-019-6247-5.

[254] Y. Han, X. Wang, H. Dai, S. Li, Nanosize and surface charge effects of hydroxyapatite nanoparticles on red blood cell suspensions, ACS Appl. Mater. Interfaces. 4 (2012) 4616–4622. Available from: https://doi.org/10.1021/am300992x.

[255] X. Liu, J. Sun, Potential proinflammatory effects of hydroxyapatite nanoparticles on endothelial cells in a monocyte-endothelial cell coculture model, Int. J. Nanomed. 9 (2014) 1261–1273. Available from: https://doi.org/10.2147/IJN.S56298.

[256] Y. Katanasaka, T. Ida, T. Asai, K. Shimizu, F. Koizumi, N. Maeda, et al., Antiangiogenic cancer therapy using tumor vasculature-targeted liposomes encapsulating 3-(3,5-dimethyl-1H-pyrrol-2-ylmethylene)-1,3-dihydro-indol-2-one, SU5416, Cancer Lett. 270 (2008) 260–268. Available from: https://doi.org/10.1016/J.CANLET.2008.05.009.

[257] C. Santos, S. Turiel, P. Sousa Gomes, E. Costa, A. Santos-Silva, P. Quadros, et al., Vascular biosafety of commercial hydroxyapatite particles: discrepancy between blood compatibility assays and endothelial cell behavior, J. Nanobiotechnol. 16 (2018) 1–15. Available from: https://doi.org/10.1186/s12951-018-0357-y.

[258] H. Turkez, M.I. Yousef, E. Sönmez, B. Togar, F. Bakan, P. Sozio, et al., Evaluation of cytotoxic, oxidative stress and genotoxic responses of hydroxyapatite nanoparticles on human blood cells, J. Appl. Toxicol. 34 (2014) 373–379. Available from: https://doi.org/10.1002/jat.2958.

[259] X.G. Sun J, Tissue distribution of intravenously administrated hydroxyapatite nanoparticles labeled with 125I, J. Nanosci. Nanotechnol. 11 (2011) 10996–11000. Available from: https://doi.org/10.1166/jnn.2011.3956.

[260] N. Ignjatović, S. Vranješ Djurić, Ž. Mitić, D. Janković, D. Uskoković, Investigating an organ-targeting platform based on hydroxyapatite nanoparticles using a novel in situ method of radioactive 125Iodine labeling, Mater. Sci. Eng. C. 43 (2014) 439–446. Available from: https://doi.org/10.1016/J.MSEC.2014.07.046.

[261] http://www.scribd.com, (n.d.).

[262] T. Ding, Y. Xue, H. Lu, Z. Huang, J. Sun, Effect of particle size of hydroxyapatite nanoparticles on its biocompatibility, IEEE Trans. Nanobiosci. 11 (2012) 336–340. Available from: https://doi.org/10.1109/TNB.2012.2190418.

[263] S. Syama, P.J. Sreekanth, H.K. Varma, P.V. Mohanan, Zinc oxide nanoparticles induced oxidative stress in mouse bone marrow mesenchymal stem cells, Toxicol. Mech. Methods. 24 (2014) 644–653. Available from: https://doi.org/10.3109/15376516.2014.956914.

[264] J.M. Ramis, C.C. Coelho, A. Córdoba, P.A. Quadros, M. Monjo, Safety assessment of nano-hydroxyapatite as an oral care ingredient according to the EU cosmetics regulation, Cosmetics 5 (2018) 1–13. Available from: https://doi.org/10.3390/COSMETICS5030053.

[265] J. Gustavsson, M.P. Ginebra, E. Engel, J. Planell, Ion reactivity of calcium-deficient hydroxyapatite in standard cell culture media, Acta Biomater. 7 (2011) 4242–4252. Available from: https://doi.org/10.1016/J.ACTBIO.2011.07.016.

[266] J. Gustavsson, M.P. Ginebra, J. Planell, E. Engel, Osteoblast-like cellular response to dynamic changes in the ionic extracellular environment produced by calcium-deficient hydroxyapatite, J. Mater. Sci. Mater. Med. 23 (2012) 2509–2520. Available from: https://doi.org/10.1007/s10856-012-4705-4.

[267] C.F. Jones, D.W. Grainger, In vitro assessments of nanomaterial toxicity, Adv. Drug. Deliv. Rev. 61 (2009) 438–456. Available from: https://doi.org/10.1016/J.ADDR.2009.03.005.
[268] K.S. Kang, J.E. Trosko, Stem cells in toxicology: fundamental biology and practical considerations, Toxicol. Sci. 120 (2011). Available from: https://doi.org/10.1093/toxsci/kfq370.
[269] V. Uskoković, Nanotechnologies: what we do not know, Technol. Soc. 29 (2007) 43–61. Available from: https://doi.org/10.1016/J.TECHSOC.2006.10.005.

CHAPTER 11

Nanoceramics: surfaces and coatings for consumer goods

Ramesh K. Guduru[1], Uttakantha Dixit[1], Bhooma Bhagat[2] and Kalisadan Mukherjee[2]
[1]Department of Mechanical Engineering, School of Technology, Pandit Deendayal Energy University, Gandhinagar, Gujarat, India
[2]Department of Chemistry, School of Technology, Pandit Deendayal Energy University, Gandhinagar, Gujarat, India

11.1 Introduction

Ceramic materials consisting at least one component in the nanometer range (1−100 nm) are termed as nanoceramics. They are usually made up of nanosized/nanostructured particles, fibers, tubes, sheets, rods, and thin films etc. [1]. Ceramics are inorganic nonmetallic solids made of metal or nonmetal compounds. When compared with the metals and polymers, they exhibit high chemical stability due to their inertness. Ceramic compounds under different external environments can be shaped and hardened in the presence of high temperatures. The ceramics are divided into oxide ceramics, nonoxide ceramics and composites. The advancements in ceramics are taking place at a high pace and advanced ceramics like Si_3N_4 and SiC are used for mechanical and tribological applications [2]. Some of the advanced ceramics include (1) functional ceramics: dielectrics and piezoelectrics; (2) structural ceramics: oxides, nitrides, carbides, and borides, etc.; (3) ceramic coatings and fire resistant glazing and glasses for optoelectronics [3]; and (4) bioceramics: hydroxyapatite and alumina. Oxide ceramics are a group of ceramics containing not more than 15% silica with little or no glass phase. These types of ceramics help to withstand oxidation and provide strength to the product. They are usually made through chemical process and then heated to form the ceramic. Some of the oxide ceramics are used in mechanical seals (Al_2O_3, Si_3N_4, SiC), prosthetic devices (Al_2O_3, $Al_2O_3-ZrO_2$), and ball bearings (Si_3N_4) etc. [4]. Nonoxide ceramics are covalently bonded with high temperature resistance. They are well suited to extreme environments, for example, TiN, TiC, Si_3N_4, SiC offer high heat resistance and corrosion resistance, and silicon-based materials show low oxygen permeability. Thus, ceramics offers a whole wide spectrum of properties for different applications ranging from structural to high temperature, tribology, biological/biomedical, electronics, and sensing. They are also used extensively for coating applications. Table 11.1 shows a list of ceramics and their properties for various applications. Table 11.2 lists the

Industrial Applications of Nanoceramics
DOI: https://doi.org/10.1016/B978-0-323-88654-3.00012-3

Table 11.1 Various ceramic compounds/coatings and their characteristics [5].

Ceramic coatings	Characteristics
Aluminum oxide	Economy wear and corrosion protection
Chromium oxide	Economy wear and corrosion protection
Zirconium oxide	Thermal barrier protection and rub resistance for compressor and turbine blade tips
Zircoat and zircoania coatings	Thermal barrier protection and solid particle erosion resistance; accommodates expansion and contraction of underlying metal substrates

Table 11.2 Different approaches/techniques used for synthesis of ceramics [2].

Ceramic powder synthesis methods			
Solid phase reactant	**Liquid phase reactant**	**Gas phase reactant**	**Miscellaneous synthesis methods**
• Solid state reactant • Thermal decompositions of a solid • Oxidation or reduction of a solid	• Precipitation • Solution heating and cooling • Evaporation salting-out • Chemical reaction with insoluble product Hydrothermal SynthesisForced insolubility • Dissolution reprecipitation	• Evaporative Condensation • Gas phase reaction with solid product Thermal decomposition • Oxidation or reduction reactions • Combination reaction with a solid product	• Solvent removal • Spray drying • Freeze drying • Spray roasting • Sol–gel synthesis • Melt solidification

typical techniques used for the synthesis of various ceramic materials, including powders, films and coatings. More information on the synthesis and applications of different types of ceramics can be found in the review articles [1].

Lately, nanoceramics have gotten a lot of attention from scientists and industry because of their remarkable properties when compared to the bulk-scale ceramics. Nanoceramics display superior mechanical properties like superplasticity, enhanced machinability, sturdiness, strength, and bioactivity, because of their fine grain size and controllable crystallinity [3]. Nanoceramic coatings exhibit superior thermal resistivity and hydrophobicity, along with other characteristics like repulsion toward dry particles and oil. They make a surface scratch safe, improve hardness, and make it impervious to microorganisms [4]. A list of typical applications for nanoceramic coatings in the consumer markets and technological fields is given below.

11.2 Applications of nanoceramic coatings based on the end users in many applications

Due to its excellent qualities, such as high hardness, wear resistance, corrosion resistance, and thermal stability, nanoceramic coatings are used in a variety of industries and end-user sectors. Following are some specific uses for nanoceramic coatings adapted to various end-user industries:

- Aerospace: gas nozzles, thermal insulators, space mirrors and nose cones [6].
- Automotive: diesel port liners, manifold insulation, catalyst support systems, flow separator housings, regenerator cores, and turbine nozzles.
- Electrical: connector housings, heater and resistor supports, washers, inserts & stand- offs, instrument and appliance insulators, coil forms and bobbins.
- Electronics: wafer chucks, insulators, vacuum tube structures, microwave housings, arc barriers, X-ray equipment, and physical vapor deposition (PVD) applications.
- Heat treating: brazing and carburizing fixtures, induction heating tubes, furnace and tooling insulation, guides, kiln furniture, welding jigs, hot forming dies.
- Metallurgical: molten metal crucibles, nozzles, troughs, liners, transfer rollers, structural parts, filters, thermocouple sheaths, permanent molds.
- Petrochemical: high-temperature corrosion and wear-resistant components.
- Plastics: hot die parts for thermoplastic forming equipment.

As this chapter focuses more on nanoceramic coatings and their commercial products, more emphasis is given to the currently available nanoceramic products along with their characteristics and applications in different segments of the consumer as well as technological markets. Following is an overview of nanoceramic-based consumer/technological products currently existing in the global markets and these are expected to increase in number and range as the technological advancements and innovations progress in terms of the synthesis and tailoring of the properties of nanoceramic coatings. The following list is categorized based on the applications of various nanoceramic coatings along with their brand or trademark names.

11.2.1 Nanoceramic products for biomedical applications

11.2.1.1 Nanoceramic protect

It is the most advanced ceramic protection system in the world [7]. It offers a full range of surface protection coating products: Hard 9H Gold, Hard 9H Red, Hard 6H Blue, and Hard 9HEVO + .

11.2.1.2 Nanoceramic protect—Evo clean for hands

The product consists of Hard 9HEvo + ceramic protection with silver particles which intertwine with the structure of the coating forming a cross link [8]. The product has self-cleaning properties toward microorganisms and unpleasant odors.

Properties:

- Made of quaternary ammonium salts as they are odorless, colorless, and nonflammable.
- Contains glycerin and betaine to moisturize the hands.

11.2.1.3 Nanoceramic protect—Evo clean for surfaces

These products are used for disinfecting surfaces. At low concentration they show excellent bactericidal properties [9].

Properties:

- Stable in a wide range of pH and temperature, they are characterized by quick action and have detergent properties to remove dirt from the surface.
- Effective as preservatives in industries from cosmetics to pharmaceuticals.

11.2.2 Medical–biocidal–Innoclean IPS

The product can be used for the disinfection of surfaces such as smartphones, tablets, keyboards, countertops, and desks [10].

They have bactericidal, fungicidal, and virucidal effects.

11.2.2.1 Nanoceramic protect-fabric

Impregnation of fabric upholstery, alcantara, and various types of fabrics [11]. Effectively hydrophobicizes textile surfaces (becomes waterproof). It protects against the ingress of water, juice, coffee, mud, dirt, UV radiation. Extends the life of materials. The product can be applied to a fabric (not to suede or leather).

11.2.2.2 Nanoceramic Protect N°01—Glass Cleaner

Nanoceramic Protect N°01—Glass Cleaner is specially formulated to give a streak-free and smear-free finish. It has been designed to remove the most stubborn dirt, leaving glass surfaces clean and clear [12]. It can be used around the home and is suitable for cleaning windows, mirrors, ceramics, chrome, laminate, and plastic surfaces.

11.2.2.3 Nanoceramic Protect N°03—Descaling Gel

Nanoceramic Protect N°01—Descaling Gel is a product in the form of a gel for descaling household ceramics and tapware [13]. Excellent removal of scale and persistent dirt gel with mint fragrance, safe for all surfaces, biodegradable & ecological product.

11.2.2.4 Nanoceramic Protect N°05–APC

Nanoceramic Protect APCN°05—Universal cleaning liquid. A product created for cleaning hard surfaces (varnished, glass, chrome, ceramic, stone). Also perfect for cleaning leather, material upholstery, carpets, plastics—for use in industry and household. It is safe to use, efficient and effective, with a pleasant fragrance. It has a very high ability to dissolve dirt, fat, and oil.

11.2.2.5 Nanoceramic Protect N°10—Quick Detailer

Nanoceramic Protect N°10—Quick Detailer is created for the care of hard surfaces (varnished, glass, chrome, ceramic, stone). Easily and quickly removes all forms of dirt. Does not leave streaks and does not scratch the surface. The emulsion formula picks up dirt particles from the cleaned surface, leaving it clean, protected, with a high gloss appearance. It strengthens protective coatings. The product with a pleasant aroma is an emulsion based on polymers that contains avocado oils and modified organic oils. The product masks smaller scratches, and is completely biodegradable.

11.2.3 Lumitester PD-30

Lumitester PD30 is designed for rapid hygiene monitoring, by measuring the bioluminescence generated during the enzymatic degradation of ATP and AMP. As it tests the degree of contamination of a tested surface, the company decided to use it to demonstrate the effectiveness of their products [14]. These tests use adenosine triphosphate (ATP) and adenosine monophosphate (AMP), which are the sources of energy for organisms, as indicators of contamination; not only microbes but also contamination itself can be detected.

11.2.3.1 Ceramic Queen (superhydrophobic nanoceramic coating for car)

Ceramic Queen is a water-based self-ceramic coating [15]. It is designed in a way so that users can quickly apply a quick coating on their car. There is no need for buffing, the users spray the coat on the car and wipe it off with water. It can be used with car shampoo, and thus is very popular among car wash shops. It has superhydrophobic water repellency, great gloss, and antiscratch properties.

Ceramic Queen lasts for 3–6 months and can be used on any part of the exterior of the car including metal, wheels, plastic, and glass. It has a very high solid content density and thus can be mixed up to a 1:15 ratio depending on the method of application

11.2.4 G3 antimicrobial bioglass

11.2.4.1 Class III medical device

The G3 patented bioglass (ES2361695) is phosphorus-free and consists of a SiO_2-Al_2O_3-Na_2O-B_2O_3-CaO composition that has demonstrated to be very effective against Gram + and Gram − bacteria, fungi, and the corresponding biofilms [5]. Also, previous studies have demonstrated that G3 completely avoids biofilm formation of the most common bacteria present in the human mouth.

Finally, preclinical tests performed with coated ceramic abutments in beagle dogs showed complete inhibition of biofilm formation and a reduction in bone loss >70%.

Compared to other marketed bioglass, G3 shows higher biocompatibility and osseointegration when implanted in contact with bone.

11.2.5 Nanova hygiene

11.2.5.1 Nanova hygiene multisurface self-protecting nanocoating

It is a water-borne thin film coating that keeps surfaces free from microbes [6]. The active compound in this product contains the combination of metal nanoparticles and nonmigratory cationic preservative, which is extremely effective in deactivation of enveloped RNA virus and bacterial DNA genome. It is easy to use and has no cytotoxicity. This effective self-protecting nanocoating remains active on the surface for a minimum of 90 days if not damaged mechanically. It provides a water-borne thin film protection against viruses and bacteria.

11.2.5.2 Nanova hygiene vegetable and fruit cleaner

It is formulated with food-grade eco-friendly and biodegradable ingredients to clean the vegetables and fruits in just 10 minutes by safely removing germs. It is easy to use, nontoxic, does not alter any taste or smell in the vegetables and fruits, and keeps them fresh for a longer period of time.

Properties:

- Safe and easy to use
- Enriched with neem extracts
- Food-grade and human safe
- No preservative, detergent, soap, chlorine and alcohol
- Treated fruits and vegetables maintain their original taste and flavor
- Eco-friendly and biodegradable ingredients

11.2.6 Advanced Nanotech Lab

The nanotechnology-based antimicrobial product Hydraguard plus can dominate and replace disinfectants in times of pandemics [16].

11.2.6.1 Hydraguard plus

It is a durable antimicrobial coating system that is nontoxic, transparent, nanoleaching, self-curing system, and its rare nonmigrating technology works 24/7 to constrain the growth of bacteria, fungi, and algae on surfaces like glass, metal, plastic, stone, ceramic, etc. The nanospikes that are developed on the surface help to break down the cell wall of microorganisms, which are thus rendered harmless. Hydraguard Plus has been tested for physical abrasion and also with weather meter testing from renowned laboratories in India, and astonishing results were delivered. This means that even after 90 days, it does not deplete its antiviral potency. It can kill 99.99% of bacteria and viruses like Ebola, Ms2, and many more for 90 days, making disinfectants obsolete.

11.2.6.2 Food-grade hydrophobic coating

In daily life, there can be a significant (up to 15%) waste of liquid food products as residue in food containers and also in the manufacturing plant as the fillers of the food are stained due to liquid flow. This waste occurs because liquids (e.g., soup, liquid margarine, and honey) remain adhered to the surface. One strategy to reduce the waste is to employ hydrophobic coatings (i.e., coatings that are extremely repellent to water), which have tremendous practical applications, including self-cleaning and drag reduction. Hydrophobic coatings are fabricated through a combination of surface texture (e.g., microscale texture or nanoscale texture or hierarchical texture) and chemical modification that imparts low solid surface energy. They can reduce liquid waste by lowering the adhesion between the container and aqueous liquid. As per the FDA, any coating that comes in contact with food must be compliant as per food safety standards.

Properties:

- The abrasion-resistant, durable coating is food safe and can prevent stains and make the surface self-cleaning or easy to clean. Cleaning can be done with a damp cloth without using strong chemicals.
- Cures at atmospheric temperature and has easy application via wipe or spray.

11.2.6.3 Hydraplas—self-cleaning for plastic

A one of its kind nanocoating for plastic surfaces like acrylonitrile butadiene styrene (ABS), polycarbonate, etc. Once coated, the water forms tiny droplets and rolls off the surface and does not spread to a uniform film even though plastics are already nonabsorbent for liquids. The coating creates a thin invisible barrier that prevents dirt and other foreign matter from adhering to the surface. Most of the dirt along with the water is effectively removed from the surface immediately, which means that the cleaning intervals are longer and the cleaning of the plastic surface is much easier and less time-consuming.

11.2.6.4 Protect your plastic and plexiglas surfaces: Hydraplas

It is a brilliant nanotechnology-based coating that creates a permanent, easy-to-clean silica coating over the plastic and Perspex surfaces of your bathroom. Not only will it protect them from damage from limescale, but it will also make cleaning them a breeze. It greatly improves bathroom hygiene as Hydraplas nanocoating always allows you to maintain a cleaner and safer home environment; it prevents the growth of harmful molds and bacteria on the surfaces of your bathroom—one of the greatest health hazards that any household faces. It saves time, money, and the ecosystem. The surfaces protected by the nanocoating do not require scrubbing or expensive detergents, because they simply do not get as dirty. This way, you can save both the time you would need to clean your bathroom and the money you'd spend on cleaners while being kind to the planet.

11.2.7 Nano4-hygiene Life

Nano4-hygiene Life is a product that provides full antibacterial protection for hospitals, schools, public transport, homes, and every place where bacterial deposition through human tactile contact is possible.

Studies on sanitization shows that bacterias can survive for more than 30 days on walls, surfaces, and medical supplies. It has also been confirmed that all common commercial disinfectants fail to fully eliminate pathogens. Their short-term effect is followed by the reappearance of bacteria in just a few hours, and so the disinfected surfaces have to be resanitized. In contrast, Nano4-Hygiene Life provides lasting antibacterial protection for 3 years.

Protection against common everyday issues:

- Safe antibacterial functionality
- Prevents body odor: bacteria convert sweat into odor causing substances
- Permanent fungicidal function
- Deprives house dust mites of food
- Fungal infection: athlete's foot
- Prevents mold growth on glossy and porous surfaces: Releases toxins allergic reactions, respiratory or immune system responses
- Prevents microscratches and reduces damage by improving the friction coefficient
- Facilitates cleaning and removal of limescale, soot, grease, dust, etc.
- Promotes comfort, well-being and safety
- Free from halogens (especially fluorine, persistent, bioaccumulative and toxic (PBT) or very persistent and very bioaccumulative (vPvB))

Effects:

- Nano4-Hygiene Life provides a residual self-sanitizing activity against athlete's foot fungus (*Trichophyton mentagrophytes*) on treated socks.
- Nano4-Hygiene Life prevents 99.9% of the growth of athlete's foot fungus on the sock
- Nano4-Hygiene Life technology facilitates the removal of dirt and biological deposits as well as soap residues and inorganic dirt.
- By forming an ultrathin layer on glass, Nano4-Hygiene Life protects, among other things, against microscratches. Under mechanical abrasion, the protective layer is

abraded before the substrate is damaged. A unique glide function ensures that any abrasive contact materials glide over the treated surface and leave fewer traces.
- Surfaces that are protected become evidently harder. This is acheived by a two-step method.
  The surface is first cleaned and polymerized (step 1).
- Subsequently, the modified silica delivers a safe and permanent antimicrobial function (step 2).
- A nonmigrating antimicrobial glass layer is thus created.

11.2.8 Nanosono Ltd. (NSC Ltd.)

It is in the key position to offer a cost-effective solution to a soon-to-be standardized market [8]. In the near future, medical devices and fabrics that may potentially cause infection will require antibacterial capabilities for non transmitting any bacteria or virus when close to the patient. The most effective way to protect the patients and the staff is by avoiding infection in the first place, preferably by eradicating the bacteria from the devices and the equipment that comes in contact with the patients and the hospital staff. In addition, we make sure the patients' environment, such as operating rooms, is as sterile as possible. The company is also focused on lowering the risk of infection by making "carrier" materials within the patients' direct environment such as robes, linens, etc.

11.2.9 Nanoceramic products in automobile industry

Nanotechnology includes the production and use of physical, chemical, and biological entities at nuclear or subatomic scale and furthermore the coordination of the subsequent nanostructures into bigger frameworks. In this manner, nanotechnology manages the enormous arrangement of materials and items that depend on a change in their physical properties as their sizes are so small [9]. Nanotechnology guarantees a leap forwards in fields such as materials and manufacturing. Nanoparticles, for instance, take advantage of their huge surface area to volume ratio, so their optical properties become a function of the particles' diameter. When fused into a bulk material, these can emphatically impact the mechanical properties such stiffness or elasticity [10]. For instance, conventional polymers can be supported by nanoparticles, prompting novel materials to be utilized as lightweight substitutes for metals. Such improved materials will enable a weight reduction together with an increase in durability and enhanced functionality [10].

At nanoscale, the surface area to volume proportion increases and therefore, more atoms are exposed to the surface hence, making surface properties more predominant than the properties of the bulk material [11]. These upgraded properties help with resistance against oxidation, mechanical abrasions, corrosion, and high temperature

[12]. Nanoparticles are very small, smaller than the wavelength of the visible light, and subsequently no reflection happens from these particles. The dispersion impact makes longer wavelengths be deflected less than the shorter wavelengths by nanoparticles causing a color effect. Controlling the size of the nanoparticles serves to achieve the desired wavelength region for planned applications, such as windshield, light reflection and scattering, clear glass, and so forth [10]. At the point when particles are decreased to the nanometer range, their energy levels become discrete, that is, they get quantized. Charge carriers are ready to move freely in bulk materials, yet when its size is reduced, scattering at the boundaries will occur. Subsequently the electronic properties of the particles are influenced. By reducing the size of the particles, ferromagnetism disappears and moves to super paramagnetism due to the huge surface energy [4].

This section attempts to discuss the applications of nanotechnology in automotive sector and bring some examples of application of nanoceramics in each set of products being used in car industries.

11.2.10 Exterior applications

It has been discovered that the incorporation of nanoparticles, for example, Al_2O_3, SiO_2, ZrO_2, and TiO_2 into a clearcoat network could improve scratch resistance. Ceramic nanoparticles have been found to be suitable hardening materials to significantly improve clearcoat hardness and hence scratch resistance [13]. However, the improvement can't be effectively attained when the particles are poorly dispersed. The inorganic fillers do not have an intrinsic affinity to organic phase and this leads to phase separation. The particles (>100 nm) deteriorate clear coat properties particularly the optical clearness [17]. Attempts have been done to take care of this issue by surface modification of fillers with organosilanes to deliver them hydrophobically and consequently improve their dispersibility into the polymeric matrix [18]. The surface modification does not exclusively impact dispersibility but can result in stronger physical/chemical interfacial adhesion between particles and the matrix. Different factors might be compelling for the impacts of nanofillers on the scratch resistance of a clearcoat, such as the particles chemistry, size, shape, and surface change. It has been shown that nanoparticles could improve clearcoat properties in an unexpected way.

Nanovarnishes at the moment are presenting scratch resistance and maintaining paint brilliance for an extended time [19]. This property is feasible because of embedded ceramic debris within the final varnish layer within the nanometer range. Conventional paints encompasses the types of binders with less mobility, in comparision of which nanopaints consists binders that provide excessive elasticity and inorganic debris with excessive strength. In the usual process, increasing the cross links in the coating's binder can increase their scratch resistance, but because they are less flexible,

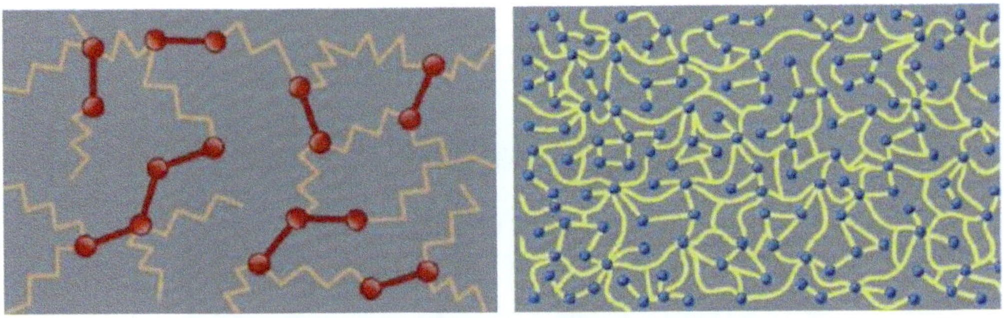

Figure 11.1 (A) Conventional paint with binder; (B) nanoceramic paints. Source: *Reproduced by permission J. Mathew, J. Joy, S.C. George, Potential applications of nanotechnology in transportation: a review, J. King Saud. University-Science 31.4 (2019) 586–594, of Copyright Elsevier, 2019.*

they have very low impact resistance [20]. The tightly packed nanodebris make the paint scratch-resistant. Fig. 11.1A and B shows the conventional and nanopaint structure [21].

11.2.11 Wear resistant nanocoatings for engines

Scratch and wear are criteria parameters which will be considered for the metal parts utilized in auto engines. Electrodeposited hard chrome and microstructure ceramic coatings are the most utilized sorts of protective coatings for motor parts [22]. The ceramic coatings are frequently applied on metal parts utilizing thermal spray. In plasma splashing, the coating powder supported with ceramic particles is injected into a plasma stream following by heating and acceleration toward the metal substrate [23]. The ceramic quickly cools and produces a coating layer over the substrate. However, there are restrictions for the utilization of microstructure ceramic and electrodeposited chrome coatings. Chrome coatings incorporate hazardous materials impacting the climate and are likewise costly. The conventional microstructure ceramic coatings are more affordable than chrome coating but are brittle and show low adhesion to the substrate. Attempts have been made to discover other substitutions. Nanostructure ceramic coatings have been used to improve metal parts of engines against abrasion and wear, as shown in Figs. 11.2 and 11.3. Decreasing the size of materials' microstructure, like grain size, particle size, or layer thickness, can essentially modify its properties [12].

Recently, new nanoceramic composites (alumina–titanium ceramic coatings) have shown excellent wear resistance [24]. However, there are issues with spraying nanoparticles because of their low mass and poor fluidity. One approach to tackle this issue is agglomerating nanoparticles into micrometer-sized aggregates. However, to acquire the initial nanostructure after spraying, the process should be carefully controlled. Alumina, alumina titanium, cemented tungsten carbides, or zirconia powders are

Figure 11.2 Ceramic-coated engine body structure.

Figure 11.3 Engine cylinders coated with nanoceramic coating. Source: *Reproduced by permission S. Du, C. Liu, D. Huang, A shearlet-based separation method of 3D engineering surface using high definition metrology, Precis. Eng. 40 (2015) 55–73, of Copyright Elsevier, 2015.*

examples of the nanostructured coatings obtained from agglomerated particles plasma spraying [25]. Both wear and abrasion resistance of the metal substrate will be extensively improved after spraying with the nanostructured materials.

These sources of friction could decrease engine life and increase oil consumption. Coatings could reduce friction and result in lower oil use. Some of these coatings are Ni–Mo–MoS_2, Ni–BN, graphite–Ni, and so on. Recently, nanostructured materials have been used to improve the friction properties of piston rings [26]. Zirconium ceramic coatings can change surface properties. Nanosize zirconium powder can be scattered in a mineral oil. The nanoparticles can arrive at working surface of the engine when the cylinder moves. The nanosize zirconium help ceramic particles better bond to the metal surface. Heat generated during engine operation would be sufficient to fix ceramic powder attached to the engine surface [25]. Subsequent to restoring, a ceramic coating produces a hard and smooth surface at various parts of the engine including cylinder walls, piston rings, piston top, valve tops and bearing surfaces. The nanosize zirconium particles can likewise increase fuel economy, power output, and oil burning and reduce noise, vibration of engine, and pollution discharge [10].

Nanotechnology may be utilized in decreasing the friction, thereby increasing fuel efficiency. By making use of a nanocrystalline coating on a cylinder wall, friction and abrasion may be decreased along with fuel consumption. Coating substances with imbedded nanocrystals with a length of 60–130 nm comprising iron carbide and boride allow surfaces with low friction properties [27].

11.2.12 Baltic Nano Technologies

JSC "Baltic Nano Technologies" is a high-tech company that creates and manufactures nanotechnology-based surface protection products. The company was founded in 2007. The company introduced a new innovative products Hendlex to the market. The products are designed to protect various surfaces from destructive environmental impacts. The main goal is to create and bring innovative surface protection products to the market, consulting with their customers about nanotechnology, and providing professional services to a wide range of clients [28]. Some of the products are:

11.2.12.1 Plastic renewer: nanocoating for plastic surfaces

The product cleans and protects plastic parts at the same time [29]. Renews plastic, refreshes color, and protects the surface from UV. It applies an additional layer, which provides an "easy to clean" effect. The product is suitable for interior plastic and vinyl parts.

11.2.12.2 Nanoscrub: cleaner and/or nanocoatings remover

It is an effective nanoscrub, used to prepare surface before using nanocoatings. The scrub removes nanocoatings from glass, plastic, painted surfaces, and car bodies. The product is water based, user-friendly, and odorless.

11.2.12.3 Wax remover: wax cleaner

It is used to remove wax from surfaces [30]. The cleaner effectively cleans the surface of wax and its residues from surfaces without damaging it.

11.2.12.4 Paint prepare: cleaner for surface preparation

It is a concentrated cleaner, used to prepare the surface before using nanocoatings [31]. The cleaner removes oil and cleans the surface. It is suitable for all nonwetting surfaces (e.g., glass, plastic, car body, etc.).

11.2.12.5 Sprayable ceramic: sprayable nanoceramic coating for cars, motorcycles, aircrafts, ships

It is a highly advanced, easy to use sprayable ceramic coating that provides protection for automotive paint, rims, and plastic or metal trims [32]. Coated surfaces have outstanding hydrophobic properties, strong chemical resistance, exceptional gloss, and deep shine.

11.2.13 Cori Coatings

Cori Coating specializes in paintwork recovery, paint protection, and premium maintenance care for both the exterior and interior of vehicles. In addition to paint protection, Cori Coating also provides upholstery coating which helps prevent staining from food, soil, and other impurities of daily use, which helps to keep the car clean and extends the life of the interior. Probably the most significant benefit of a ceramic coating is that it repels water. By creating a hydrophobic layer, the coating causes the water to bead on the surface and slide off more easily. This is known as the self-cleaning effect. A ceramic coasting is a liquid polymer which is applied by hand to the outer surface of a vehicle. The coating creates a bond with the vehicles factory paint through a chemical reaction and this creates a protective layer that also shines. The basic chemical composition of a ceramic coating is silicon dioxide (SiO_2). Ceramic coatings may also be referred to as quartz coating.

11.2.14 Ceramic Pro (nanoceramic coatings)

Using highly advanced nanotechnology, Ceramic Pro have developed a full range of ceramic coating solutions, designed to solve a wide range of problems. With qualities such as 9H hardness, being permanent and highly hydrophobic, as well as resistant to chemicals, oxidation, corrosion, UV-rays, and graffiti, the products are trusted by professionals globally. The versatility of Ceramic Pro coatings makes them suitable for a variety of items, such as luxury vehicles, art, jewelery, high-end apparel, furniture, and interior elements made of rare natural materials [33].

11.2.14.1 Ceramic Pro top coat [33]

Top Coat allows the surface to stay cleaner longer as dirt and grime will not stick to it. The superhydrophobic effect of the coating will cause water to bead up and roll off the surface along with any dirt and grime. Top Coat is usually applied over 9H in the layering process. Though not as durable as 9H, it can also be applied on its own. Top Coat also enhances gloss depth on all painted surfaces for that wet look that is often hard to attain with synthetic waxes.

- Superhydrophobic effect
- Weather and UV resistance
- High gloss finish

- Antigraffiti
- Advanced chemical resistance
- Oxidation and corrosion resistance

11.2.14.2 Ceramic pro sport

Commonly used on surfaces previously treated with CP 9 H or CP Top Coat in order to refresh their hydrophobic and UV-resistant properties but can also be used as a 6–12 months stand-alone coating [33].

- Easy to apply
- Super hydrophobic effect
- Self-cleaning effect
- Chemical resistance
- UV resistance
- Prevents water spotting
- High gloss finish

11.2.14.3 Ceramic Pro strong [33]

Strong is an industrial product and will work great on painted surfaces where protection is more important than a flawless finish. Strong has similar characteristics to 9H but creates a thicker layer once applied.

- Scratch resistance (Above 9H)
- Super hydrophobic effect
- Weather and UV resistance
- Thermal resistance (up to 2200°C)
- Antigraffiti
- Advanced chemical resistance
- Oxidation and corrosion resistant

11.2.15 Kavaca (paint protection film)

11.2.15.1 Kavaca paint protection films

Kavaca paint protection films (PPFs) are top-quality paint protection films. The products feature unique qualities such as almost instant recovery from minor scratches without the use of heat, improved gloss, and advanced nonyellowing effect. Their visual qualities are improved by the use of nanoceramic technology that grants the PPFs the shine and resistance to UV, achieved through many years of research in the field of nanotechnology. The range of products includes films for body, lights, and windows, and they are available in different colors, and in matt and gloss finishes [34].

11.2.16 IGL coatings

11.2.16.1 EcocoatKenzo

Synthesized using the latest nanotechnology and graphene nanoplatelets, the new and improved graphene-reinforced EcocoatKenzo has increased durability of up to 5 years and higher resistance to water spots. EcocoatKenzo's unmatched clarity and gloss radiates a rich and natural glow reminiscence of bespoke wax and its thick 10H hardness layer protect 12 times longer than waxes [35].

11.2.16.2 Ecoclean Precoat

It is a nanotechnology based intense cleaner designed to dissolve oil particles and remove polish residue to prepare the paintwork for IGL Coatings Ecocoat series. Bonds of waxes, oils, and polishes are weakened by precoat and are wiped from the surface easily, creating the possibility to acquire the perfect ecocoat finish [36].

11.2.17 Ceramtec

11.2.17.1 Ceramaseal isolators

Hermetic, vacuum-sealed Ceramaseal products are built to endure extreme conditions. They stand up to ultrahigh vacuum (UHV) environments, temperatures ranging from −269°C to +450°C, pressures in excess of 1700 bar, and corrosive or caustic environments, while maintaining an unsurpassed level of reliability and performance [37].

11.2.17.2 Ceramaseal multipin connector

These products are known around the globe for their reliability. For ceramic-to-metal and glass-to-ceramic components and glass-sealing technology CeramTec offers both standard and custom designs for use in vacuum, ultrahigh vacuum, high-pressure, cryogenic, and high-temperature applications [38].

11.2.18 Irancar care

11.2.18.1 Nanoceramic car body spray

- Super hydrophobic [39]
- High gloss
- Fast and easy
- It repels rain, dust and pollution
- Keeps surfaces cleaner for longer
- Excellent protection against UV rays
- Can be used for all colors
- Can be used for ceramic surface

11.2.19 Nanoceramic 9H

11.2.19.1 Solar glass shield

Solar Glass Shield is a super hydrophobic nanocoating [40]. It repels water and dust. The solution is based on nanotechnology. It forms a thin layer on the surface of solar panels. It prevents the accumulation of a soiling layer and dust on the surface of panels. This coating will make the solar panel easy to clean and will reduce the cleaning frequency of the solar plant. It can reduce the cleaning cycle by as much as 40%. It helps in saving water required for cleaning the panels.

11.2.20 Nanoproceramic

11.2.20.1 Nano Pro soft fabric SiO_3 ceramic coating (cleaner for car interior)

Nano Pro soft fabric SiO_3 is a specially manufactured ceramic coating for cars and vehicles [41]. This is one of the top products for car interior protection.

- Chemically engineered for soft automotive interior seats.
- Perfect for suede and absorbent material.
- It gives ultimate protection to soft materials.
- Manufactured with a water-based SiO_3 formulation.

11.2.20.2 Nano Pro ceramic plus 10H radical

Nano Pro Ceramic Plus 10H radical is a ceramic gloss to protect paint [42]. This 10H glass coating leaves a vehicle's body with a gloss finish. The 10H radical is one of the strongest, shatterproof and flexible ceramic coatings. It comes with a 5-year warranty. It can be applied outside, using hands.

- Delivers a high gloss finish.
- 2.5 times stronger than traditional 9H.
- Registered with CarFax.
- It gives you unsurpassed protection and shines.

11.2.20.3 Nano Pro ceramic soap

This concentrated SiO_2 ceramic soap is a ready-to-use ceramic foam soap [43]. It gives a high gloss finish. It can be used to wash cars and is an advanced car wash solution. It requires 1.5 oz per bucket of water.

- Leaves a high gloss finish.
- Creates a ceramic adhesion.
- Sufficient for five vehicles at a time.
- A permanent ceramic solution.

11.2.20.4 Vinyl protectant for cars

Ceramic coating vinyl wrap is manufactured for convertible tops and soft canvas tops [44]. It is a spray application and the best vinyl protectant for cars. It comes with a warranty of 18 months. In addition, it efficiently minimizes the potential of interior leaking.

- Perfect for automotive convertible tops.
- It is hydrophobic.
- Reduces interior leaking.
- It is quick to spray.

11.2.20.5 9H Nanoceramic coating

Nano Pro "Hybrid" 9 H automotive ceramic coating is a powerful and perfect 9 H nano-ceramic coating [45]. It gives a scratch-proof body. It is best used for hydrophobic car paints and leaves a 5-star shine paint protection effect. It can be applied by hand and needs no external application. It is registered with CarFax and comes with 3-year protection.

- 85% cure time in 4 hours
- Excellent automotive ceramic coating
- Registered with CarFax
- Powerful and flexible

11.2.21 Nanoceramic Protect south

11.2.21.1 Nanoceramic Protect—Hard 9H

The ceramic protective coating of Nanoceramic Protect Hard 9 H was tested and examined many times [46]. Those tests were performed by research units and detailing experts and customers. The high 9H hardness and flexibility of the preparation guarantees the endurance of the 9H protection coating even in the extremely difficult atmospheric conditions after the application of one layer.

11.2.21.2 Nanoceramic Protect rims and tires

Ceramic plastic protection [47]

- Deep and rich color
- Perfect plastic conservation
- High hydrophobicity

11.2.21.3 Nanoceramic Protect glass

Nanoceramic Protect Glass is a product, thanks to which car windows are perfectly protected and the driver gains an incomparable improvement of visibility both during day and night [47]. Nanomolecules in the Nanoceramic Protect glass preparation form a durable ceramic protective covering on the glass surface. That layer repels water molecular (hydrophobic effect), causing their immediate and automatic removal from the car glass while driving.

11.2.21.4 Nanoceramic Protect plastic

Nanoceramic Protect Plastic is a high-quality product that finds use in a maintenance of car elements made from gum or plastics [48]. The ceramic coating on the surface of protected components acts like a specific cover from a negative influence of external factors.

Advantages of using Nanoceramic Protect Plastic:

- High hydrophobicity
- Plastic conservation
- Resistant to low temperatures
- Ceramic protection covering based on a nanotechnology
- Perfect hydrophobicity (water molecules repulsion)
- High durability on water
- Better resistance of tires
- Ease of maintaining plastic (impurities do not stick to the covering)
- Protection from environmental factors

11.2.21.5 Nanoceramic Protect Leather

Nanoceramic Protect Leather makes the leather upholstery more resistant (against UV radiation and other harmful external factors) and nourished and also brings a rich color to it. The ceramic coating, which crystallizes after application, enables clean and protected upholstery [49].

Laying the protection covering counteracts against the creation of awkward impurities—the residue does not stick to the upholstery, and liquids do not absorb into the upholstery structure (the hydrophobicity effect), making impurities easy to remove [50].

11.2.22 BossautoInnova, S.A

11.2.22.1 Nanoskin nanoceramic

It increases the strength of gloss paint and protects against oxidation, UV rays, acidic rain, tree sap, and even insects [51]. It is applicable for the protection treatment of painted surfaces and gel coats. Nanoceramic also removes small-sized defects (holograms) with its abrasive action. It is applicable by orbital polisher or air sander with a soft polishing foam pad and is excellent for all kinds of paint and gel coats. Use a white/black foam pad for a high glossy finish without haze, opacity, and holograms. It increases the paint's life by 40%–50% and reduces dirt accumulation by 70%–80%.

- Long-lasting protection
- Very high glossy
- Hologram free
- Low consumption
- Doesn't stain the plastics (*)

- Antidust
- UV absorber

11.2.23 System X ceramic protection

11.2.23.1 System X Diamond SS

System X Diamond SS is a superslick, thin, flexible, permanent 9 H + ceramic with previously unachievable gloss, slickness, and hydrophobic properties [52]. This top-of-the-line self-cleaning coating is slick to the touch and imparts a deep, wet gloss to paintwork. Diamond's unparalleled paint protection against environmental factors will keep your paintwork looking better than new as long as you own it. The film thickness ranges from 0.8 to 1 μm. The chemical resistance quality is excellent. The temperature resistance is >760°C [52].

11.2.23.2 System X max

System X max is a first-of-its-kind ceramic technology that is more durable and environmentally resistant than ever before [52]. X-Max features a stronger, tougher shell up to four times thicker, two times stronger, and two times more durable than previous coatings. Max's increased strength makes it more resistant to harmful water spotting.

11.2.24 Nanoident Carline series

11.2.24.1 Nanoident Supraperl SSH

Nanoident Supraperl SSH nanocoating and sealing spray for automotive glass offers an easy to clean effect and very high rain pearling, easy application, excellent durability to windshield wiper abrasions, and immediate pearl function without speed [53]. Free visibility in the rain from 50 km/hr Stable up to 20,000 km with drying time of 1–2 minutes.

11.2.24.2 The Nanoident Aluprotec SSH

The Nanoident Aluprotec SSH coating or sealing has been specially developed to protect coated aluminum wheels against dirt, pollution, and brake dust. The nanoproduct can be very easily applied.

11.2.25 Pearl nanoceramic coatings

11.2.25.1 Pearl Nano Cerami-Coat Quartz 9H

Pearl Nano Cerami-Coat Quartz 9H contains unique UV screens which proactively lock out harmful atmospheric degradation and sun bleaching giving the treated surfaces the ultimate in long-life protection, even in the harshest extremes of the hot desert and inclement polar climate conditions [54].

Nano Cerami-Coat Quartz 9H delivers the very latest in super hydrophobic technology (lotus-leaf effect) ensuring water, dirt, and other surface contaminants are repelled, leaving treated surfaces cleaner for longer without the need to ever wax again throughout ownership.

11.2.25.2 Pearl Nano Cerami-Glass Quartz 20/20

Pearl Nano Cerami-Glass Quartz 20/20 is a technologically advanced nanoengineered, transparent protective ceramic coating, formulated to chemically bond at a nanoscopic level to glass windscreens and windows [54].

Containing unique UV screens which proactively lock out harmful atmospheric degradation and sun-bleaching, it gives the treated surfaces the ultimate in long-life protection, even in the harshest extremes of the hot desert and inclement polar climate conditions.

11.2.25.3 Pearl Nano Cerami-Rims Quartz HD [55]

Pearl Nano Cerami-Rims Quartz HD is a technically advanced nanoengineered, transparent protective ceramic coating, formulated to chemically bond at a nanoscopic level with alloy wheels, contours, spokes, brake calipers, and exhaust pipe tips. It withstands high temperatures in excess of 200°C, making it the perfect solution for hard-working wheels.

11.2.26 Nanoceramic protect automotive products

11.2.26.1 Nanoceramic protect remover

It is a chemical, high-quality degreasing alcohol-based preparation. This primer is necessary before applying ceramic protective coatings for car paint, chrome, and glass. After application of Nanoceramic protect remover, the applied preparations adhere firmly to the cleaned surface, which makes them last longer on the surface. The product should be used before all car paint and glass protection.

11.2.26.2 Nanoceramic Protect Soft

Nanoceramic Protect Soft is a high-quality product for protecting car paint surfaces from the harmful influence of external and environmental causes. With an easy and fast (about 30 minutes) method of preparation and application, it allows an excellent effect with a minimum workload. Thanks to the nanoceramic structure, the covering gained protects the paint from permanent impurities and environmental factors, which have a negative influence on the condition of the paint finish. The product's molecular structure creates an invisible matrix and will add a ~20%–30% increase in gloss to the finish. The advantage of this matrix structure allows savings of both time and money. Nanoceramic Protect Soft stays on the paint's surface for up to 12 months, however it is recommended to refresh the surface after 6 months. Application of NCP

Soft will provide protection for up to 12 months, and offer UV radiation resistance, protection from environmental factors, high hydrophobicity (repels water, dirt, debris), ease for cleaning and maintaining, high paint glow, and nourished color.

11.2.26.3 Nanoceramic Protect Sealant

Nanoceramic Protect Sealant is a unique synthetic wax made from a combination of the highest quality polymers, without the addition of silicone. It masks smaller scratches, discolorations, and mists, leaving a protective layer that lasts up to 6 months. Nanoceramic Protect Sealant facilitates the car washing process and ensures protection against: UV rays, road salt, and insect adhesion.

11.2.26.4 Ceramic Glass Protection "invisible wiper"

Use of Nanoceramic Protect Glass "invisible wiper" gives:

- Car glass. conservation
- Creation of a ceramic protection covering based on a nanotechnology.
- Better visibility through glasses.
- Perfect hydrophobicity (water molecular repulsion).
- Resistance to low temperatures, ease to clean frozen glasses.
- High durability against chemical influence.
- Better resistance to mechanical damages.
- Ease of keeping glasses clean (impurities do not stick to the covering).
- Protection from environmental factors.
- Very good visibility even in difficult atmosphere conditions.

11.2.27 Autotriz Coatings

11.2.27.1 V1—3D Matrix Coating

A revolutionary coating that utilizes three-dimensional (3D) matrix nanostructure technology to produce extreme clarity, amazing reflection, and gloss on paint [56].

11.2.27.2 V2—Hybrid Coating

The latest formulation that combines the Pure Glass Coating and the unique 3D Matrix Coating. It is a blend of inorganic materials that creates a hard protective barrier on the paint surface.

11.2.27.3 V3—Plastic Coating

It is a special transparent protective layers that protects plastic and trim parts. It offers excellent resistance and does not crack, bubble, peel, or discolor at temperatures up to 700°C.

11.2.27.4 V4—Wheel Coating

The toughest protective barrier that bonds strongly on metal, aluminum, or chrome wheels. It has extreme heat resistance and contaminant resistance to efficiently repel brake dust, road grime, and dirt.

11.2.27.5 V5—Glass Coating

A nanotechnology fluorine-based protection for glass surfaces provides extreme durability and super water repellent properties to allow water to flow off windscreens at speeds as low as 60 km/h.

11.2.27.6 V6—Leather Coating

A nongreasy leather nanocoating that is formulated by the most advanced acrylic latex emulsion with an extra UV barrier. It protects against permanent staining from food and drinks to preserve it from longer time period.

11.2.28 KubeBond Series

11.2.28.1 KubeBond NanoX

The coating forms a ceramic that enables strong protection and creates a surface that stays clean for a longer period. Just one application transforms all the paintwork into a shiny surface that is resistant to all weather conditions [57].

11.2.28.2 KubeBond Nu Interior

KubeBond Nu Interior provides the car's interior plastic and rubber surfaces with a strong protective layer because of its ceramic molecular net [57]. It leaves interior surfaces looking brand new and shiny, and makes them just as protected as every other surface in the car that uses the KubeBond series

11.2.28.3 KubeBond RimShield

KubeBond RimShield is developed for rims, exhaust system, and brake calipers [57]. A spray application achieves a permanent 9H hardness nanoceramic coating and forms an excellent resistance layer against brake dust, damaging substances, and heat.

11.2.29 Jade Ceramics

11.2.29.1 Jade Ice PRO

Jade Ice PRO is a clear, nanocrystalline ceramic coating that protects automobile substrates from weather, chemicals, physical abrasion, and UV rays, while repelling water, dirt, and other contaminants via a self-cleaning effect. Jade Ice PRO applies an ultra-hard, inert, durable coating that is several microns thick and imparts a deep, reflective shine [58].

11.2.29.2 Jade Quartz PRO

Jade Quartz PRO is a clear, nanocrystalline ceramic coating that protects automobile substrates from weather, chemicals, physical abrasion, and UV rays while repelling water, dirt, and other contaminants via a self-cleaning effect [59]. Jade Quartz PRO applies an ultrahard, inert, durable coating that is several microns thick and imparts a deep, reflective shine.

11.2.29.3 CUPÌDO

It is an amazing and unique nanocomposite sealant [60]. The formula acts as a bridge between traditional sealants and nanoceramic products. Cupido generates a polymeric matrix that has a definite structure compared to the flexible elastomer which can anchor and properly adhere to the painted surface adding protecting against harsh abrasion caused by certain car detailing steps like washing, waxing, polishing, and others. This nanoceramic product also fights against both alkaline and acid chemical aggression. In short, it generates a chemical as well as physical barrier. Use of Cupido is also recommended for protection of superhydrophobic nanofluoropolymer, which is generally subject to the phenomenon of water spotting.

Features of Cupido:

- Easy to use
- Long lasting and durable
- Amazing beading and sheeting effect
- Can be used over chrome, paint and glass surface
- Silky smooth touch
- Delivers high brightness
- Unbeatable protection with highly glossy finish
- Doesn't contain any alcohol or petroleum distillates.

11.2.30 Ethos

Ethos car care is the US-based company that focuses on high-quality and locally made ingredients to make products that deliver better quality or life [61]. The company makes its formulas from scratch and built the products from the ground up. Other ceramic paint protection products include the Ethos ceramic wax pro-aerospace coating protection.

11.2.31 Carpro

As a leader in car detailing and ceramic coating products, Carpro was established in 2009 [62]. From the start, it has used nanotechnology to achieve advances in both fields and set itself apart from the competition. Besides the company's flagship quartz, there's also the Carpro reload spray sealant and the Carpro flyby forte 15 mL full kit.

11.2.32 Nasiol

In 2008, the company was established in Turkey to work on nanospinning techniques [63]. In 2012, it changed its focus to protective nanocoatings and changed its name to Artekya. One of its top products is the NasiolZR5

11.2.32.1 Nanoskin Nanoceramic

Nanoskin Nanoceramic Synthetic Sealant is a combination of ultrafine nanoabrasives and protective ceramic coating designed to create a brilliant shine with outstanding protection in one single step. Extremely fine polishing agents remove oxidation, wash marring, swirl marks, and stains.

11.2.32.2 Hydrax for automotive

Hydrax is an extremely-lasting, 9H-certified water repellent 9H ceramic coating for exterior surfaces, such as cars and aircraft. It is based on liquid glass coating technology, which simply means SiO_2-based technology.

Properties:

- Easy to apply solutions saves application cost of hydrophobic coatings
- UV resistant
- 9H ceramic coating resists minor scratches

11.2.32.3 Hydra for Infrastructure

Even though nanocoatings are suitable for almost any substrate, not every coating adheres to all substrates. Each hydrophobic coating caters specially to a different surface, be it glass or metal or plastic or ceramic tile

- Designed for harsh environments
- Durable water repellent coatings catering to market demands
- Improved visibility
- Improved visibility for glasses with hydrophobic coatings
- Stain-resistant

11.2.33 Modesta nanoceramic coating

Modesta is the most exclusive automotive surface protection and nanoceramic coating in the world and only available to select detailers [64]. It is the hardest coating in the world, and the most difficult nanoceramic coating to install correctly. Modesta was designed specifically for those with discerning tastes for its exceedingly rich and glossy finish, as well as its ability to resist the elements and protect like no other.

11.2.34 Nanoslic stencil coating

Nanoslic offers the world's most advanced coating for solder paste stencils, integrating surface preparation, coating, and curing to provide unparalleled printing improvements, transfer efficiency, and ease-of-cleaning [65]. Nanoslic coatings use a primary binder polymer that is ceramic, meaning it is primarily composed of an inorganic material. On a weight basis Nanoslic coatings are largely composed of silica; other materials are added for specific formulations and applications. Nanoslic materials are inherently capable of maintaining properties at temperatures well beyond nonceramic polymers. Nanoslic coatings are resistant to most solvents and are unaffected at a wide range of pH. Most formulations achieve 9H hardness. By incorporating specific functional groups into the polymer, various properties can be achieved including:

- Hydrophobicity
- Oleophobicity
- Ambient temperature curing
- Heat-induced crosslinking
- Flexibility
- Outdoor weathering
- Conductivity
- Improved temperature resistance
- Reduced coefficient of friction

11.2.34.1 Lat 26° Nanoceramic Coating

Lat 26° Nanoceramic Coating is specifically formulated for marine paints and gelcoats, resulting in a deep gloss that lasts! Two coats will provide a years' worth of protection [66]. It offers incredible repellence against exhaust, dirt, and saltwater, and great protection against yellowing and oxidation.

11.2.35 Wr15 Nano Coat

It is formulated for easy-to-clean surface protection of ceramics and glass [67]. It forms an invisible nanocoating on the surface and offers resistance to water, pollutions, dust, UV, as well as being antibacterial, anticorrosive, and protects from acids, alkalis, saline weather, water-based chemicals, mold, algae, and fungus growth. It allows exterior and interior use. The treatment is stable and cannot be removed except by harsh cleaners or reagents, such as abrasive cleaners. The treatment provides a hydrophobic surface that allows simple rinsing away of soils and sludge from the surface with no residual surface staining. The treated surface does not change the surface, is highly repellent to oils and water, and durable to alkaline cleaners and detergents.

11.2.36 Vetro Sol Car coating products

Vetro Sol is a manufacturer of ceramic nanocoating chemicals for all surfaces for the automotive industry and car detailing industry. Ceramic nanocoating chemical is uses as a protective coating for various surfaces in the detailing industry and the car care industry.

11.2.36.1 Vetro sol Cera Glare

This coating is a unique ceramic molecular compound formula (nanoceramics), that forms transparent, very stable silicone layers, thus making them an ideal basis for functional coating systems that are anticorrosive, scratch-resistant, dirt-repellent, UV protective, and hydrophobic. Cera Glare is used on painted metal surfaces on:

- Trains
- Buses
- Metal facades and shutters
- Containers, trailers, trucks, heavy transportation vehicles
- Car body paint and under body coating
- Pipeline coating
- Anticorrosive coating

The coating is also used on leather, rims, dashboards etc. Vetro sol Cera Glare is a low viscosity, methyl-substituted organic silicone dioxide-based coating technology. It is used as a binder in high-performance coatings, which can be cured between room temperature and 250°C.

By using Vetro sol Cera Glare both transparent and pigmented coatings with the following properties can be prepared: corrosion protection, antitarnish, easy-to-clean, thermal barrier, high temperature and chemical resistance

11.2.36.2 Veteo Shield Super Hydrophobic Nano Coating

It works on any glass, anywhere, anytime supergreen, and reduces by 50% on average the energy requirements and carbon emissions for routine washing of architectural glass. It offers new ways of creating real added value, both tangible and intangible, at each step of the glass supply chain;

11.2.36.3 Fabric Shield

Vetro Sol Fabric Shield is an unique super hydrophobic coating for fabric. This is a sprayable coating. It will give you excellent water repellence, dirt repellence, and oil repellence on fabrics.

11.2.36.4 Cera Glare

This coating is used on leather, rims, dashboards, etc. of cars. This product can be used for almost all types of fabric to protect them from oil, water, dirt, mud, etc. It protects and gives time to clean the surface before any damage to high-value suits, sofas, carpets, curtains, pillows, shoes, or fabrics that you want to protect from getting dirty and damaged because of water, dirt, mud, bird droppings, oil, etc.

11.2.37 9H Liquid Diamond Coating

- Silicon dioxide base coating
- Binder for coating systems
- Both room-temperature and heat curable
- Excellent adhesion to most substrates
- Outstanding weather resistance
- High UV and chemical resistance
- Very good high temperature stability

11.2.38 Carzspa Crystalshield Ceramic Coating [68]

Crystalshield Ceramic Coatings contain upwards of 80% high-grade SiO_2, which hardens in contact with air giving the surface a smooth and protective layer. They will add additional protection to your vehicle's exterior and give it a shiny, glossy finish that even brand-new cars do not have. It contains nanoceramic particles that enter the pores of the car's clearcoat. When exposed to air, these nanobased ceramic particles begin to crystallize, forming a hard layer of glass over the clear coat. Since these particles interlock with the surface molecules of the car's paint, they create a semipermanent layer of protection which will protect the car's paint from dirt, debris, UV rays, and chemical stains for years.

11.2.39 Flamingo Nanoceramic Coating

Specifications

- 10 H hardness for scratch resistance [69]
- Super lyophobic coating on surface
- High gloss
- Strong weatherability
- UV protection and corrosion prevention
- Premium protective coating film

11.2.40 Taciturn Caretech OPC Private Limited

The product range includes 9H nanoceramic car coating kit, 9H gold nanoceramic car coating medium, 9H gold nanoceramic car coating deluxe, and marble/granite nanocoating liquid [70]. The nanotech silicone makes the vehicle exterior hydrophobic and scratch resistant.

11.2.41 Gold coating

Taciturn 9H Gold Nanoceramic coating is a silicone-based liquid. It can be used to coat the exterior of cars/motorcycles/automobiles. It uses the latest nanotechnology to create a crystal-like layer/coating on top of your vehicle, making your vehicle shiny, glossy, and scratch-resistant. It also provides protection against UV radiation, oxidation, and fire/heat. This product uses nanotech and is far more effective and long-lasting than wax coating/waxing or sealant. Once it is applied there is no need for waxing, as this coating is very durable and will protect the shiny look of your paint, and allow your vehicle to look brand new for a longer time. This product, once applied, will make the exterior paint of your vehicle hydrophobic, i.e., it adds a coating which is water and oil repellent.

11.2.42 Nanodiamond4-Carpaint

Upon applying Nanodiamond4-Carpaint on the car's surface, a multilayer coating forms and within minutes the thickness of the car paint surface can be substantially increased [71]. The application produces an additional superprotective invisible coating, a kind of armor which shields your car from scratches at a resistance scale from 9H to 12H.

Product qualities:

- Scratch resistance (optimum scale 9H–12H)
- 10-year permanent protection
- Car paint protection against fire
- Antigrafffiti
- Self-cleaning
- Super waterproofing hydrophobic factors
- Protection from UV radiation
- Extreme weather conditions protection
- Advanced chemical (acid) resistance
- 100% invisible sealing protection
- Frost-proofing and tolerance
- High protection against ice
- Shiny glossy finishing
- Extreme tolerance to high temperatures (up to 1200°C)
- Anticorrosion protection
- Antioxidation protection
- Extreme resistance on chemical corrosion effect
- Shiny paint coverage
- The coating cannot be affected by machine washing and can be removed only by painting procedure.

- 100% environment-friendly
- Money saving by reducing up to 90% the use of chemical detergents

11.2.43 Nos2 Research and Development Private Limited

Arcus Pro 12 H Micro NanoTemper Glass ceramic coating is a revolutionary product that creates a superior protective layer to keep auto surfaces strong and shiny, just like a diamond [72]. All it takes is a one-time application followed up with minimal maintenance. Arcus Pro 12H is a product of Choose Nanotech's long experience in semiconductor industrial coating. It complies with industrial testing criteria. It contains a cubic ceramic molecular matrix structure that offers excellent resistance to scratches and chemicals. The surface coated by Arcus Pro 12H adds a hard cubic ceramic layer, giving the best protection from damaging contaminants and harsh chemicals.

11.2.44 Nikavi H_2O Guard & Gloss 100 mL—nanocoating agent spray

- H_2O Guard & Gloss can be used both wet and dry because of its unique bonding technology [73]. Derived from a successful liquid paint sealant, this water-activated paint sealer adds a protective layer to extend the life of current waxes and/or sealants.
- The fastest way to seal your paint! not only does H_2O Guard & Gloss give you ceramic coating like results in a few minutes, but you also have 3–6 month protection applied as a stand-alone water-activated spray sealant.
- Safe for all exterior surfaces, including trim! It can darken plastics, add depth to paint, and protect shine in minutes. Nikavi H_2O Guard & Gloss uses the most advanced chemical acrylic technology to provide unmatched beading capability.
- Rain protector: a perfect polish for application on cars, bikes, scooters (plastic, rubber, vinyl, rexene, surface etc.), powder coated accessories, car and bike headlights, and taillights.

11.2.45 N-WET rain and water repellent nanocoating

N-WET is a hydrophobic water-repellent nanocoating, that is 500 times thinner than a human hair 100% natural and eco-friendly and effective for up to 12 months depending on the type of coating on any glass surface [74].

11.2.46 Super Ceramic Coating

This AntiScratch Ceramic Car Formula works as a strong protective layer above your paint. The Super Ceramic Coating acts as a strong protective layer above paintwork, protecting it from UV rays and environmental contaminants. With just a thin coat of the coating, it forms an extremely hard shell on the car's surface that makes it easy to clean the car without worrying about scratches. What's even better is that it also creates a beautiful shine on the car's exterior! Super Ceramic Coating protects from color fading by adding a glass layer (SiO_2) above the paint surface. The coating lasts 2 years and the hydrophobic properties will help to clean your vehicle easily. The dirt and debris will not stick to the paint after coating. The easiest way to understand ceramics coatings is to think of them as a second layer of skin, or a sacrificial layer of protection over your vehicle paint.

11.3 Applications of nanoceramics in textiles/fashion industry

The use of nanotechnology in the textile industry is increasing rapidly due to its unique and valuable properties. There is much potential for useful applications of nanotechnology in the cotton and other textile industries. Its application can economically extend the properties and values of textile and processed products [75]. Nanotechnology makes fabrics multifunctional, allowing us to produce fabrics with special features, such as antibacterial activity, UV resistance, ease of cleaning, waterproofing, stain resistance, and odor resistance [76]. The future success of nanotechnology in textile applications lies in the area where new principles are incorporated into durable multifunctional fiber systems without compromising fiber-specific properties such as the ability to be worked, flexibility, and more.

This study focuses on the application of nanotechnology in the textile industry, which focuses on improving various properties of textiles [77]. If nanocrystalline epiezoceramic particles are incorporated into fabrics, the finished fabric can convert exerted mechanical forces into electrical signals enabling the monitoring of bodily functions such as heart rhythm and pulse if they are worn next to the skin [78]. Nanoparticles, which include metal oxides and ceramics, are utilized in fabric to alter surface properties and provide fabric features. Because of nanosize debris are transparent, and do not have color, shows less brightness of the fabric substrates, preventing nanoparticles from aggregation that's the important thing to acquire a favored performance [79]. The fabric which has been handled with TiO_2 and MgO nanoparticles is changed by the active carbons that are used as chemical and biological shielding materials.

The photocatalytic activity of TiO_2 and MgO nanoparticles can break dangerous and toxic chemicals and biological agents [80] These nanoparticles may be preengineered to stick to fabric substrates through the use of spray coating or electroplating methods. Finishing with nanoparticles can convert fabrics into sensor-based materials [81]. If nanocrystalline piezoceramic particles are incorporated into fabrics, the finished fabric can convert exerted mechanical forces into electrical signals enabling the monitoring of bodily functions such as heart rhythm and pulse if they are worn next to skin [78].

Ceramic powder has a high melting point and low thermal conductivity because air is trapped in the internal pores. Ceramic powder is widely used as a heat insulator.

The application of ceramics to fibers has already been studied [17]. The use of ceramics has been shown to slightly improve the insulation of fibers compared to fibers without the addition of ceramics. The production of textiles containing ceramic powder is complicated by the difficulty of introducing the ceramic powder into the textiles. However, ceramic powders such as silica (SiO_2) and Zirconia (ZrO_2) can be selected due to their high emissivity and ability to form a uniform layer of tissue in the infrared region [76]. Insulation performance characteristics of various fiber materials have been studied [82]. Spacer fabrics are used because they are known as three-dimensional (3D) fabrics. They are manufactured in various thicknesses and are thicker than weaves and knits, so they can be used for insulation applications.

11.4 Ceramic Pro Textile

Ceramic Pro Textile comes from the high-tech industry of the photoelectron semiconductor assembly which is reliable and an inorganic compound. Ceramic Pro Textile dramatically reduces the surface energy of textiles or suede, so that when liquids come into contact with it, they form beads and simply roll off while keeping the textile substrate completely dry. This enables the fabric, suede, or tissue surface to be free from the water/dust and all other liquids, without affecting the look or feel [83].

- Super hydrophobic effect—140°
- Excellent wear resistance
- Ceramic Pro Textile can only be applied by professional detailers.
- Excellent durability
- Repels spills and stains
- Keeps fibers clean

11.5 Armatex Ceramic Coated Fabrics

Armatex Ceramic Coated Fabrics consist of a cured, high-temperature ceramic coating, applied to a variety of bases, creating a high-temperature fabric. The ceramic coating enhances strength, adds dimensional stability, and reduces the fraying of the fabric in fabrication. These heat-resistant fabrics also have improved weather resistance due to the ceramic coating. Typical uses for Armatex Ceramic Coated Fabrics include removable insulation pads and covers, protection against molten metal splash and welding splatter, and thermal insulation seals and barriers. The Armatex product line includes.

11.5.1 Armatex Coated Fabrics and Textiles

Armatex SF Silicone Coated Fabrics are typically used for insulation blankets and pads for the insulation industry, for welding curtains and blankets in any industry where welding takes place, as insulation for the power generation industry, for cargo covers on airplanes, on aircraft jet-way bridges, and the list goes on [84].

Armatex also coats textiles including round and square braided ropes, tapes, and tadpole gaskets. Coated textiles are often used in sealing applications for door seals, furnace seals, oven seals, and gas fireplaces, where airtight seals are required. Armatex coated ropes are often used as seals in primary aluminum smelters, where the sealing applications require a product that can survive under very extreme conditions.

11.5.2 Thermopak fabricated seals, curtains, and blankets

- Better strength (tensile and tear) after thermal exposure than typical robot protective fabrics [85]
- Very flexible—flexibility virtually unaffected by temperature
- Excellent high/low-temperature performance capability
- Excellent resistance to a wide range of fluids and chemicals
- High thermal conductivity
- Manufactured for a wide variety of robots
- Excellent ozone and UV resistance
- Excellent electrical insulating properties
- Excellent weatherability
- Maximum value

11.5.3 Hytex high-temperature-resistant textiles, such as ropes, tapes, sleeving, and fabrics

Hytex 400 Kevlar Textiles are high-strength, lightweight, industrial textiles excellent for use in aerospace, industrial, and commercial applications. As an organic, aramid fiber, Kevlar Industrial Textiles have high chemical resistance, moderate thermal resistance, and excellent dimensional stability.

11.5.4 Siltex silica fabrics and silica textiles

Silica fabric and textiles are high-temperature resistant materials that can be used constantly in the operating temperature of 1800°F to 982°C. This product are high-temperature resistant fabric formulated to withstand thermal expansion and contraction under fluctuating heat conditions.

11.6 Nanoident Carline Series

11.6.1 NanoidentPermatexS4

The NanoidentPermatexS4 nanocoating or sealant has been specifically developed to protect textile seats against dirt and stains. The product is water based and can be easily applied. To preclean the seats there is a multipurpose cleaner. NanoidentPermatexS4 contains no toxic ingredients.

11.7 Nanoceramic Protect

11.7.1 Nanoceramic Protect Leather Set

Laying the protection covering counteracts against creation of awkward impurities—the residue does not stick to the upholster, and liquids do not absorb into the upholster structure (the hydrophobicity effect), which make impurities easy to remove. Unusually effective and durable reparation, devoted to leather car upholstery, which require specific protection. The ceramic coating, which crystallizes after an application, causes gain of a clean and maximal protected upholstery.

11.7.2 Nanoceramic Protect Fabric Protector

Nanoceramic Protect Fabric is the product, based on latest nanotechnology coating technique which makes it suitable for all types of fabrics. Convertible roofs, car upholstery, floor mats, carpets and upholstered furniture are some of the examples for this product. This can be employed in convertible roof tops as they keep fading over a period of time due to the constant exposure to sunlight and other weathering conditions. This fabric brings an inexpensive repair to roof top damages. The regular use of the Nanoceramic Protect Fabric guarantees better resistance to humidity and counteracts against repeated grime of roof.

11.8 Taciturn Nano Silver antimicrobial coating textile

Nano Textile antimicrobial coating is the best available coating for all fabric/textile materials [86]. It makes all fabric material antimicrobial and safe from bacteria/fungi/virus after coating for up to 30 washes.

11.9 Future prospects

As listed above, nanoceramics are found to have extensive applications in various segments of consumer markets ranging from automobile engineering products to biomedical and textiles. Large-scale synthesis of nanoceramics with recent innovations can enable them to become versatile materials for a vast range of applications extending to sensors and high-end thermal shields. Even though the composition of a glass coating and a ceramic automobile coating may be the same, there are changes in their atomic structure that lead to some structural variations. For instance, only heating a glass covering can transform it into a ceramic coating. As a result, the erratic structure of glass is transformed into the more stable, ordered structure of ceramic [87]. Although nanoceramics exhibit improved mechanical, hydrophobic, antimicrobial, and osseointegration in biomedical applications, the structure-dependent electrical, thermal, and optical properties with reduced optical transparency might limit some of their applications in the electronic/optical-related applications. On the other hand, tuning the band structures of nanoceramics with different dopants with controllable electronic properties might enable their suitability for selective sensing, thermal, and specific electronic applications. On top of these, determining the short/long-term negative impacts of nanoceramics on human lives and the environment is of utmost importance when they are implemented in powder or any loose form for practical applications, and this is where more research is required to confirm their friendly nature to the humans and environment with minimal to no impact. Nevertheless, the current global market projections clearly reflect a significant growth of nanoceramics in the upcoming decades in both the developed and developing nations.

References

[1] D. Eliche-Quesada, L. Pérez-Villarejo, P. José Sánchez-Soto, Introductory Chapter: Ceramic materials - synthesis, characterization, applications and recycling, Ceramic Materials - Synthesis, Characterization, Applications and RecyclingCeramic Materials - Synthesis, Characterization, Applications and Recycling (2019) 1–5. Available from: https://doi.org/10.5772/intechopen.84710.

[2] T.A. Ring, Ceramic powder synthesis, Fundamentals of Ceramic Powder Processing and Synthesis, 81–83, 1996. Available from: https://doi.org/10.1016/b978-012588930-8/50004-8.

[3] A.K. Mishra, Sol-gel based nanoceramic materials: preparation, properties and applications, Sol-Gel Based Nanoceramic Materials: Preparation, Properties and Applications., 1–2s97, 2016. Available from: https://doi.org/10.1007/978-3-319-49512-5.

[4] O. Van Der Biest, Nanoceramics: Issues and opportunities, Int. J. Appl. Ceram. Technol. 10 (4) (2013) 565–576. Available from: https://doi.org/10.1111/ijac.12074.

[5] "G3 Bioglass has been awarded the Solar Impulse Efficient Solution Label." https://nanoker.com/our-g3-bioglass-has-been-awarded-the-solar-impulse-efficient-solution-label/ (accessed Aug. 07, 2021).

[6] "Top Nanotechnology Company in India for Nanotechnology Manufacturing - Nanova Care Coat."

[7] "Nanodiamond4-Carprotect_ by NANO4LIFE."

[8] S. Du, C. Liu, D. Huang, A shearlet-based separation method of 3D engineering surface using high definition metrology, Precis. Eng. 40 (2015) 55–73.

[9] C. Contado, Nanomaterials in consumer products: a challenging analytical problem, Front. Chem. 3 (2015). Available from: https://doi.org/10.3389/fchem.2015.00048.

[10] M. Mohseni, B. Ramezanzadeh, H. Yari, M. Moazzami, The role of nanotechnology in automotive industries, New Advances in Vehicular Technology and Automotive Engineering, 2012, , pp. 3–54. 10.5772/49939.

[11] J. Mathew, J. Joy, S.C. George, Potential applications of nanotechnology in transportation: a review, J. King Saud. Univ. - Sci. 31 (4) (2019) 586–594. Available from: https://doi.org/10.1016/j.jksus.2018.03.015.

[12] V.D.N. Rao, D.M. Kabat, H.A. Cikanek, C.A. Fucinari, G. Wuest, Material systems for cylinder bore applications – plasma spray technology, SAE Tech. Pap. 412 (1997). Available from: https://doi.org/10.4271/970023.

[13] "Nanomaterials in consumer products_ a challenging analytical problem_ Enhanced Reader."

[14] "Lumitester PD30_ Hygiene monitoring, cell counting_ Test systems_ Measuring Instruments_ Labware_ Carl Roth - International.."

[15] "Ceramic Queen(Superhydrophobic Nanoceramic Coating for car) | tradekorea." https://www.tradekorea.com/product/detail/P760700/Ceramic-Queen(Superhydrophobic-Nanoceramic-Coating-for-car).html# (accessed Aug. 07, 2021).
[16] "Nanotechnology Based Antimicrobial_ Advanced Nanotech Lab.."
[17] R.A. McIntyre, Common nano-materials and their use in real world applications, Sci. Prog. 95 (1) (2012) 1−22. Available from: https://doi.org/10.3184/003685012X13294715456431.
[18] A.S.H. Makhlouf, E. Guerrero, Toward smart and durable nanoceramic and nanobioceramic coatings and their medical and industrial applications, Advances in Smart Coatings and Thin Films for Future Industrial and Biomedical Engineering Applications, Elsevier, 2019, pp. 383−403.
[19] "Fact Sheet Use of nanomaterials in coatings 1. Description of application 1.1 Products and purpose of using nanomaterials 1."
[20] "Nanoceramic in consumer products."
[21] J. Mathew, J. Joy, S.C. George, Potential applications of nanotechnology in transportation: a review, J. King Saud. Univ.-Sci. (2019) 586−594. 31.4.
[22] K.M. Batoo, Nanoceramics and Their Applications, Advanced Structured Materials, 118, Springer Verlag, 2019, pp. 143−158.
[23] A. Majumdar, S. Jana, Glass and Glass-Ceramic Coatings, Versatile Materials for Industrial and Engineering Applications, 2001.
[24] S.K. Ameta, A.K. Rai, D. Hiran, R. Ameta, S.C. Ameta, Use of nanomaterials in food science, Biogenic Nano-Particles their Use Agro-Ecosystems, Springer Singapore, 2020, pp. 457−488.
[25] "Ceramic Coating Applications and Research Fields for Internal Combustion Engines | IntechOpen." https://www.intechopen.com/chapters/29756 (accessed Aug. 06, 2021).
[26] M. Ciniviz, M.S. Salman, E. Canlı, H. Köse, Ö. Solmaz, 7 Ceramic Coating Applications and Research Fields for Internal Combustion Engines. [Online]. Available: http://www.intechopen.com.
[27] M.E. Vance, T. Kuiken, E.P. Vejerano, S.P. McGinnis, M.F. Hochella, D.R. Hull, Nanotechnology in the real world: redeveloping the nanomaterial consumer products inventory, Beilstein J. Nanotechnol. 6 (1) (2015) 1769−1780. Available from: https://doi.org/10.3762/bjnano.6.181.
[28] "Car & motorcycle - Hendlex." https://hendlex.com/car-motorcycle/ (accessed Aug. 06, 2021).
[29] "PLASTIC RENEWER - Hendlex." https://hendlex.com/my-product/plastic-renewer/ (accessed Aug. 06, 2021).
[30] "WAX REMOVER - Hendlex." https://hendlex.com/my-product/wax-remover/ (accessed Aug. 06, 2021).
[31] "PAINT PREPARE - Hendlex." https://hendlex.com/my-product/paint-prepare/ (accessed Aug. 06, 2021).
[32] "SPRAYABLE CERAMIC DC60 - Hendlex." https://hendlex.com/my-product/sprayable-ceramic-dc60/ (accessed Aug. 06, 2021).
[33] "Ceramic Pro Nano Technology Surface Protection Products." https://ceramicpro.com/products/ (accessed Aug. 06, 2021).
[34] "Ceramic Pro Kavaca Advanced Paint Protection Films." https://ceramicpro.com/kavaca/paint-protection-film/ (accessed Aug. 06, 2021).
[35] "ecocoat kenzo - IGL Coatings." https://iglcoatings.com/products/ecocoat-kenzo (accessed Aug. 06, 2021).
[36] "ecoclean precoat - IGL Coatings." https://iglcoatings.com/products/ecoclean-precoat (accessed Aug. 06, 2021).
[37] "Ceramaseal Isolators." https://www.ceramtec.com/ceramaseal/isolators/ (accessed Aug. 06, 2021).
[38] "Ceramaseal Multipin Connectors." https://www.ceramtec.com/ceramaseal/multi-pin-connectors/ (accessed Aug. 06, 2021).
[39] "Nanoceramic Spray Car Car Body |." https://irancar.care/store/automotive/carpro-reload/ (accessed Aug. 06, 2021).
[40] "NANOCERAMIC 9H - Products." https://www.nanoceramic-9h.com/products (accessed Aug. 06, 2021).
[41] "Buy Best Car Interior Coating - Top Ceramic Coating Manufecturer in Florida." https://nanopro-ceramic.com/product/nano-pro-soft-fabric-sio3-ceramic-coating/ (accessed Aug. 06, 2021).

[42] "Buy Best 10h Nanoceramic Coating - Top Ceramic Coating Company in Florida." https://nanoproceramic.com/product/nanopro-ceramic-plus-10h-radical/ (accessed Aug. 06, 2021).
[43] "Shop for Best Ceramic soap from top Ceramic Coating Company." https://nanoproceramic.com/product/nanopro-ceramic-soap/ (accessed Aug. 06, 2021).
[44] "Best Vinyl Protectant for Cars - Buy Vinyl Coating for Cars." https://nanoproceramic.com/product/best-vinyl-protectant-for-cars/ (accessed Aug. 06, 2021).
[45] "Best 9h Scratch-proof Ceramic Coating (Hydrophobic) - Buy Best Ceramic Coating For Cars." https://nanoproceramic.com/product/nano-pro-hybrid-9h-automotive-ceramic-coating/ (accessed Aug. 06, 2021).
[46] "Ceramic Paint Protection | Nanoceramicprotect." https://www.ncpsouth.com/ceramic-paint-protection (accessed Aug. 06, 2021).
[47] "Ceramic Coatings – Nanoceramic Protect." https://www.nanoceramicprotect.com/ceramic-coatings/ (accessed Aug. 06, 2021).
[48] "Ceramic Glass | Nanoceramic Protect South | United States." https://www.ncpsouth.com/copy-of-ceramic-rims-and-tire (accessed Aug. 06, 2021).
[49] "Nanoceramic Protect – Robertson Racing LLC." https://nanoceramicprotectnorthamerica.com/nanoceramic-protect/ (accessed Aug. 06, 2021).
[50] "Products – Nanoceramic Protect." https://www.nanoceramicprotect.com/products/ (accessed Aug. 06, 2021).
[51] "NANOSKIN NANOCERAMIC." https://www.bossauto.com/en/nanoskin/nanoskin-nanoceramic/products (accessed Aug. 06, 2021).
[52] "Ceramic Paint Protection - System X Ceramic Protection Products." https://www.systemx.com/products/ (accessed Aug. 06, 2021).
[53] "Automotive | German Coating Technology." https://gctnano.wordpress.com/nanobionic-products/automotive-and-marine/ (accessed Aug. 06, 2021).
[54] "Pearl Nano Cerami-Coat Quartz 9H Super-Hydrophobic Nano Coatings." https://pearlnanocoatings.com/pearl-nano-cerami-coat/ (accessed Aug. 06, 2021).
[55] "Pearl Nano Cerami-Rims Quartz HD Super-Hydrophobic Nano Coatings." https://pearlnanocoatings.com/pearl-nano-cerami-rims/ (accessed Aug. 06, 2021).
[56] "Automotive | AUTOTRIZ- The Future of Nanoceramic Coating." https://www.auto-triz.com/automotive (accessed Aug. 06, 2021).
[57] "KubeBond | Products." https://www.kubebond.com/products/ (accessed Aug. 06, 2021).
[58] "Jade Ceramic – The Ultimate in Ceramic Coatings." https://jade-ceramic.com/#!/portfolio (accessed Aug. 06, 2021).
[59] "Jade Quartz PRO – Jade Ceramic." https://jade-ceramic.com/project/jade-quartz-pro/ (accessed Aug. 06, 2021).
[60] "Cupido | Best Ceramic Coating for Cars in all Nanoceramic products." https://www.manmachineworks.com/cupido.html (accessed Aug. 07, 2021).
[61] "Professional Car Detailing Products - Ethos Car Care_ Ethos Car Care.."
[62] "CarPro India briSngs you Cquartz Finest, Reload, IronX, Forte, ceramic coating India, glass coating India, paint protection cars India, ceramic coating cars India.."
[63] "Hydrophobic Nanoceramic Coatings Manufacturer." https://www.nasiol.com/ (accessed Aug. 07, 2021).
[64] "MODESTA - Advanced Liquid Glass & Protective Coatings.."
[65] "NanoSlic Ceramic Coating Technology - NanoSlic Coatings.."
[66] "Nanoceramic Coating_ Premium Detailing Products.."
[67] E. Cleaning, S. Protection, F.O.R. Ceramics, WR15 NANO COAT is formulated for EASY CLEANING SURFACE PROTECTION FOR, 15.
[68] "Ceramic Coating for Cars – 9h & 10h Crystal Shield Ceramic Coating.."
[69] "Flamingo Car Care Tech Co.."
[70] "Taciturn Caretech OPC Private Limited, New Delhi - Manufacturer of Air Purifier and 9h nanoceramic coating.."
[71] "Nanodiamond.."

[72] "Car Engine Carbon Cleaning Machine and Engine Decarbonizing Machine Manufacturer_ Nos2 Research & Development Private Limited, Coimbatore.."
[73] "NIKAVI H2O Guard & Gloss 100mL - Nano Coating Agent Spray.."
[74] "N-WET Rain and Water Repellent Nano Coating - Car Windshield Glass Hydrophobic Coating, Waterproof Rainproof Coat Spray for Rear view Mirror Side Windows Windshield - 50 mL_ Buy N-WET Rain and Water Repellent.."
[75] K.M. Batoo, August. Nanoceramics and Their Applications, 118, Springer, Singapore, 2019.
[76] M. Hosseini, A.S. Hamdy Makhlouf, Industrial Applications for Intelligent Polymers and Coatings, 2016.
[77] A. Kiani, M. Rahmani, S. Manickam, B. Tan, Nanoceramics: Synthesis, Characterization, and Applications, 2014, 2014.
[78] M. Ziaei, M. Ghane, Thermal insulation property of spacer fabrics integrated by ceramic powder impregnated fabrics, J. Ind. Text. 43 (1) (2013) 20–33. Available from: https://doi.org/10.1177/1528083712446384.
[79] K. Koo, J. Choe, Y. Park, The application of PCMMcs and SiC by commercially irect dual-complex coating on textile polymer, Appl. Surf. Sci. 255 (20) (2009) 8313–8318. Available from: https://doi.org/10.1016/j.apsusc.2009.05.039.
[80] K. Patra, Application of nanotechnology in textile engineering: an overview, J. Eng. Technol. Res. 5 (5) (2013) 104–111. Available from: https://doi.org/10.5897/jetr2013.0309.
[81] M.H. Shim, C.H. Park, H.S. Shim, Effect of ceramics on the physical and thermo- physiological performance of warm-up suit, Text. Res. J. 79 (17) (2009) 1557–1564. Available from: https://doi.org/10.1177/0040517508095605.
[82] G. Barroso, Q. Li, R.K. Bordia, G. Motz, Polymeric and ceramic silicon-based coatings—a review, J. Mater. Chem. A 7 (5) (2019) 1936–1963.
[83] "Ceramic Pro Textile – Ceramic Pro India." https://ceramicpro.co.in/ceramic-pro-textile/ (accessed Aug. 06, 2021).
[84] "ARMATEX | Refractory Coating | Silicone Coated Fabric | Mid-Mountain Materials." https://mid-mountain.com/armatex/ (accessed Aug. 06, 2021).
[85] "THERMOPAK Robot Covers." https://mid-mountain.com/thermopak-robot-covers-2/ (accessed Aug. 06, 2021).
[86] "Nano textile coatings – Taciturn Nano Silver antimicrobial coating textile Manufacturer from New Delhi.."
[87] Cori Coatings, Available from: http://www.cori-coatings.be/ceramic-coating/why-to-use-it/. (accessed 06.08.21).

CHAPTER 12

Applications of nanoceramics to promote environmental sustainability

Anjali Goyal, Areeba Khalid and Lovleen Gupta
Department of Environmental Engineering, Delhi Technological University, Delhi, India

12.1 Introduction

With the growing cognizance of the ill effects of anthropogenic activities, more and more people are shifting toward sustainable solutions to their daily problems, due to which the environmental sector has seen rapid growth in the last decade. As per the United Nations WCED: "Environmental sustainability is about acting in a way that ensures future generations have the natural resources available to live an equal, if not better, way of life as current generations." It is one of the global challenges faced by today's generation as it is under threat due to the tremendous acceleration in environmental pollution as a result of significant escalation in new pollutants, urbanization, civilization, climate change, and industrialization. Nanoceramics is one of the advanced nanomaterials that provides various methods to combat environmental pollution and aid in environmental sustainability.

Nanoceramics refers to ceramic materials in the nanometer range, that is, 1−100 nm, with high adsorption capacity, antimicrobial actions, high strength, hydrophilicity, wear resistance, biocompatibility, as well as distinct surface qualities, for example, superplasticity, machinability, hardness, strength, and bioactivity. Nanoceramics are easily available, which makes their use possible to control environmental pollution and attain sustainability. Some examples of nanoceramics are TiO_2, SiO_2, Al_2O_3, ZrO_2, and ZiO_2.

Nanoceramics are immensely popular for use in various industries such as biomedical industry, textiles industry, cosmetics industry, environmental industry, nanoceramics in energy industry, construction industry (the ceramic industry), aircraft and engine industries, and the electronics and electrical industry. This chapter will make you aware of various applications of nanoceramics in the environmental industry with the aim of promoting environmental sustainability.

12.2 Application of nanoceramics in the environmental industry

The environmental industry is a collection of big and small-scale setups that aim to achieve environmental sustainability, by organizing and working toward managing and

Industrial Applications of Nanoceramics
DOI: https://doi.org/10.1016/B978-0-323-88654-3.00015-9

protecting the natural resources and helping to heal the harm done to them. They introduce technologies that, for instance, have a low carbon footprint, and can help in improving the air quality, water quality, or soil quality. The industry comprises various sectors to provide products and services that include cleaner technologies, pollution control services, waste management, greenhouse gas management, generation associated with environmental protection, and remediation of contaminated property. In the environmental industry, the following are the applications of nanoceramics which contribute to environmental sustainability:

1. Water treatment and remediation
2. Sensing devices
3. Dye-sensitized solar cells
4. Air purification and remediation
5. Antibacterial materials

12.2.1 Water treatment and remediation

Water treatment and remediation is a technique that enhances the quality of raw water by eliminating pollutants and undesired components or reduces their concentration to make the water more suitable for any given end-use, without any long-term or short-term risk of any severe health effects. The end use may be the industrial supply of water, drinking water, irrigation, river flow maintenance, water recreation, and many other uses. This remediation is essential to living a healthy human life. Nanoceramics can treat wastewater using three techniques:

1. Membrane processes
2. Desalination
3. Adsorbent technique

12.2.1.1 Membrane processes

Nanoceramics have proven to be extremely helpful in water remediation as they possess diverse advantageous properties, such as thermal and chemical stability, high robustness, small pore size, more efficient membrane modules, high porosity, higher flux, and large hydrophilic surface. These properties prove to be of great benefit for large-scale water purification. The only disadvantage of nanoceramics is that they are brittle in nature, which makes it susceptible to cracking. The following are some of the examples of nanoceramic membranes that are used in water remediation and treatment:

1. Alumina membrane
2. Zirconia membrane
3. Titania membrane
4. Silica membrane
5. Zeolite membrane
6. Ceramic membrane

12.2.1.1.1 Ceramic membrane

Ceramic is a popular and essential material for membrane fabrication. Ceramic membranes have recently gained immense popularity due to their eccentric yet amazing properties. They are classified into two categories: inorganic and organic materials, as shown in Fig. 12.1. Ceramic and polymeric membranes are the two most conventionally used membranes in the field of wastewater treatment. Ceramic materials for wastewater treatment membranes are chosen based on their unique physical and chemical features, such as hydrophilicity, porosity and pore size, chemical and thermal stabilities, mechanical strength, cost-effectiveness, and other water treatment benefits. Five major alternatives for manufacturing ceramic membranes for water and wastewater treatment are alumina, zirconia, titania, silica, and zeolite [1].

Ceramic membranes demonstrate greater technical leverage, allowing for narrow and well-defined pore size distributions, higher porosity, better separation, and a higher flux; higher thermal, mechanical, and chemical stability, as well as longer membrane lifetimes; and higher hydrophilicity, high fluxes at low pressures, and less fouling [2–4] when compared to polymeric membranes. Ceramic membranes have a limited market in commercial applications and scientific work at the moment [1]. Ceramic membranes are used in microfiltration (MF), ultrafiltration (UF), nanofiltration (NF), and reverse osmosis (RO), as indicated in the schematic diagram in Fig. 12.2. Composite-type ceramic membranes have also been generated, which are made by an amalgamation of more than one type of ceramic.

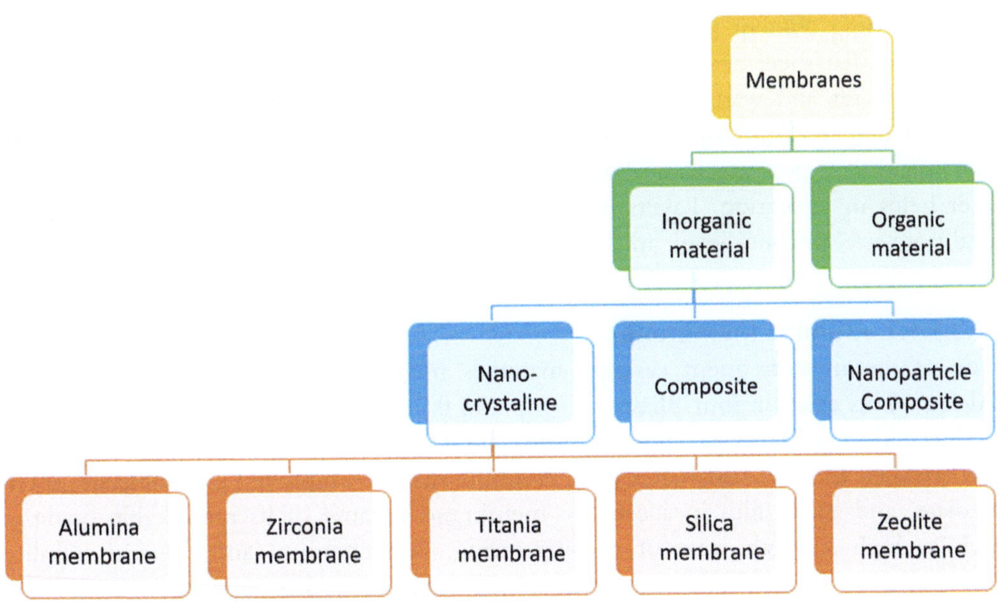

Figure 12.1 Classification of membranes.

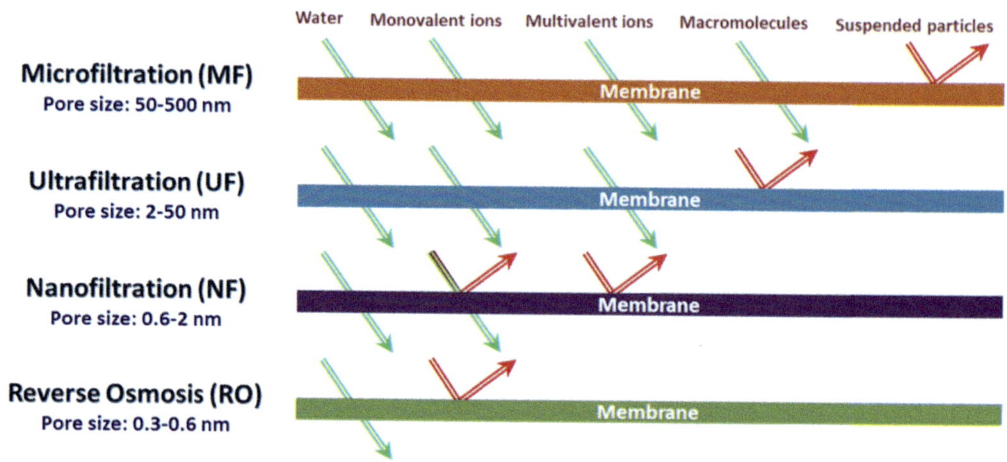

Figure 12.2 Schematic diagrams of MF, UF, NF, and RO. *MF*, Microfiltration; *NF*, nanofiltration; *RO*, reverse osmosis; *UF*, ultrafiltration. Source: *From Z. He, Z. Lyu, Q. Gu, L. Zhang, J. Wang, Ceramic-based membranes for water and wastewater treatment, Colloids Surf. A: Physicochem. Eng. Asp. 578 (2019) 123513. https://doi.org/10.1016/j.colsurfa.2019.05.074.*

Ceramic membranes are utilized in water treatment in two different configurations: tubular and flat-sheet. When compared to flat-sheet membranes, tubular membranes have a higher surface-to-volume ratio which makes them more feasible for water treatment processes. Removal of natural organic matter (NOM) from drinking water sources [5]; treatment of oil-field wastewater [6]; cleaning of dyeing effluent [7]; reuse of car-washing effluent [8]; and desalination [9] are just some of the applications for tubular and flat-sheet ceramic membranes. Membranes made of ceramic for the treatment of water and wastewater often have a microstructure that is not symmetrical, with a denser dynamic top layer on a more permeable and bigger molecular substrate. The membrane gets mechanical support from the substrate and the denser dynamic layer helps in separation. Layers in the middle are sometimes used in modifying the total pore size, porosity level, and toughness [1].

12.2.1.1.2 Alumina membrane

One of the most frequent ceramic materials used to make membranes is alumina (Al_2O_3). This exists in four-phase: α, γ, η, and θ; its thermodynamically stable phase is α-alumina (corundum). α and γ alumina are two phases of this ceramic membrane that are used to treat wastewater [10]. Alumina may be employed as a substrate, intermediate, and active film in a ceramic membrane because of its remarkable qualities, such as high strength, ease of manufacturing, and chemical and thermal stability. According to the nanoscale range, the alumina membrane has applications in UF and NF of wastewater.

Alumina membrane is extensively used to remove contaminants from wastewater. Zou et al. [10] reported upon the use of alumina UF membrane to treat dye wastewater. According to research results, a cosintering-produced bilayer α-alumina UF membrane with a molecular weight cutoff (MWCO) of around 10 kDa with water permeance of 70 $L/m^2/h/bar$ had good filtration performance. In addition, most of the treated dye wastewater had a rejection rate of greater than 99%. High-flux solvent-resistant ceramic NF membranes are presently being researched. To treat the water from industry, the alumina membrane's water permeance shall be more than 6 $L/m^2/h/bar$ [11,12]. However, Wang et al., in one of their studies, [13], have prepared γ-Al_2O_3/α-Al_2O_3 hollow fiber NF membranes with different MWCO, and pure water permeance of 17.4 $L/m^2/h/bar$ was measured with a pore size of 1.61 nm, which could be used to treat the water from industries.

12.2.1.1.3 Zirconia membrane

Another typical ceramic material used in wastewater treatment membranes is zirconia (ZrO_2). Zirconia has different structures at different temperatures: monoclinic crystal structures at room temperature, and tetragonal and cubic phases at higher temperatures. Among the five major ceramic membranes which we will explore in this chapter, zirconia has the highest hydrophilicity and thermal resistance [3,14,15]. High hydrophilicity results in excellent flux and less fouling during wastewater treatment, while membranes with high thermal stability are better suited to liquid-phase applications in severe environments [3]. All of the benefits listed above make the zirconia membrane suitable for use as the upper surface material in UF and NF wastewater treatment. The treatment of high-salinity wastewater is essential to protect the environment, and desalination can also be done using the zirconia membrane.

12.2.1.1.4 Titania membrane

Rutile, anatase, and brookite are the three minerals that make up titania (TiO_2). Titania is the most chemically resistant of the five ceramic materials used to make membranes [3]. Titania has a one-of-a-kind ability to mineralize organic molecules when exposed to UV light, hence its membranes can perform photocatalytic oxidation and filtration side by side. Titania has been utilized in wastewater treatment as a support, medium, and active layer for UF and NF, as well as for photocatalytic properties [16–21]. This membrane has water treatment capabilities due to its unique photocatalytic feature. It possesses various other unique properties such as high permeability and antifouling and antibacterial qualities at the same time. Due to its great chemical stability compared to other ceramic materials, a titanium membrane may be utilized to treat wastewater for corrosive media [3]. On the other hand, titania membranes encounter challenges such as reduced surface area and anatase structural instability at high temperatures.

12.2.1.1.5 Silica membrane

Silica (SiO_2) membrane is used for wastewater treatment. The pores size of a silica membrane can be reduced to 3–5 Å, which is the diameter range of a water molecule and hydrated salt ions [4,22]. Due to this reason, silica membranes can be used to treat wastewater via desalination. Because of its strong affinity for water, one of the downsides of the silica membrane is that it has an unstable structure when it comes into contact with it, and so its hydrostability must be improved [4,22].

12.2.1.1.6 Zeolite membrane

Zeolites are hydrated aluminosilicates that are crystalline and include cations from groups I and II. The pore structure of this membrane is unique, with two types of pores: uniform subnanometer zeolite pores and intercrystal micropores with a specific size distribution [23]. Both the interparticle pores and the intraparticle pores transfer water. The zeolite membrane's high hydrophilicity can also help to prevent fouling [24]. Zeolite membrane can be utilized to treat wastewater and for desalination using the process of UF, RP, and pervaporation [23,25,26]. A more hydrophilic zeolite membrane surpassed the alumina membrane in terms of fouling resistance. However, obtaining detect-free zeolite membranes of the desired thickness remains a difficulty [4]. Hence, the zeolite membrane needs to be improved for wastewater treatment by working on process innovation and optimization.

12.2.1.1.7 Removing cations

Nanoceramics can be used to remove cations from wastewater. Based on a size-exclusion mechanism, a zirconium-based metal–organic framework (Zr-MOF) membrane demonstrates multivalent ion retention (98.0% for Mg, 86.3% for Ca^{2+}, and 99.3% for Al^{3+}). It also displays good permeability and chemical stability in saline water [27]. The NaCl and NaSO retention of a thin-film nanocomposite membrane comprising porous MCM-41 silica nanoparticles synthesized via in situ interfacial polymerization for water filtration is around 97.9% + 0.3% and 98.5% + 0.2%, respectively. Increased hydrophilicity, toughness, zeta potential, and permeate water flux come from the addition of MCM-41 silica nanoparticles [28,29]. The pore size of the cation-rejection membrane is primarily responsible for repelling cations.

12.2.1.1.8 Removing nanoparticles

Nanoparticles may readily enter the human body while drinking water because of their nanoscale size. This may result in severe health issues and thus the removal of nanoparticles becomes an important task. The use of nanocrystalline silicon as a dynamic etching layer in the development of nanoporous silicon membranes has been shown to be effective in the removal of gold (Au) nanoparticles from water. The membrane can remove up to 80% of gold nanoparticles [28].

12.2.1.1.9 Removing biological contaminants

Biological contaminants can be a source of many waterborne diseases; they can include contaminants like algae, bacteria, and viruses. An alkaline hydrothermal synthesis and hot press technique were used to produce a layered structure TiO_2 nanowire UF membrane [19]. The diameter of these nanowires is 20 nm and they have shown successful results in removing polyethylene glycol, polyethylene oxide, HA, and *Escherichia coli* from water [30].

12.2.1.2 Desalination of water

Desalination is the technique of removing mineral (salts) components from saline water. Due to the increasing population and changing climate, water scarcity has become a major issue of concern. "An estimated 55 million people globally are affected by droughts every year, and they are the most serious hazard to livestock and crops in nearly every part of the world," says the World Health Organization. Desalination can be a step toward solving this problem by ridding the water of its salts and making it suitable for drinking and agricultural purposes. It can affirm water security for future generations. Desalination membranes are also used in the treatment of effluents from industries and the reusing of municipal wastewater. Nanoceramic and ceramic-based nanocomposite membranes have shown great application in the process of desalination of water. Techniques like NF, UF, and RO are employed to achieve this aim.

Great inherent properties are shown by zeolite for the desalination of water. For example, when using zeolite membranes to RO concentrated solutions having various cations, a significant rejection rate is seen [23]. Yet it is a challenge to attain zeolite membranes that are defect-free with certain thicknesses [4].

A superporous and superhydrophobic carbon nanotube (CNT) network structure was made by researchers (in Fig. 12.3, you can see more experimental details) in which especially structured ceramics with high thermal and mechanical stabilities were introduced. This combination resulted in the formation of two distinct CNT membrane structures: made of partially covered CNTs and fully covered CNTs (FC-CNTs). FC-CNT in comparison with existing hydrophobic membranes shows higher and more stable water flux as well as high salt rejection properties, which is a result of its amazing antiwetting characteristics and superhydrophobic and superporous structure [31].

The UF process shows great application in the desalination of water due to its small pore size. But to decrease the cost further, MF can also prove to be a great option, as a MF membrane does not need trans-membrane pressure and has higher flux when compared to UF [32,33]. An alumina MF membrane with a nano-TiO_2 coating distributed on its grain has shown to increase the hydrophilic character of its surface. Moreover, the performance of this membrane is superior to that of the unmodified membrane [34].

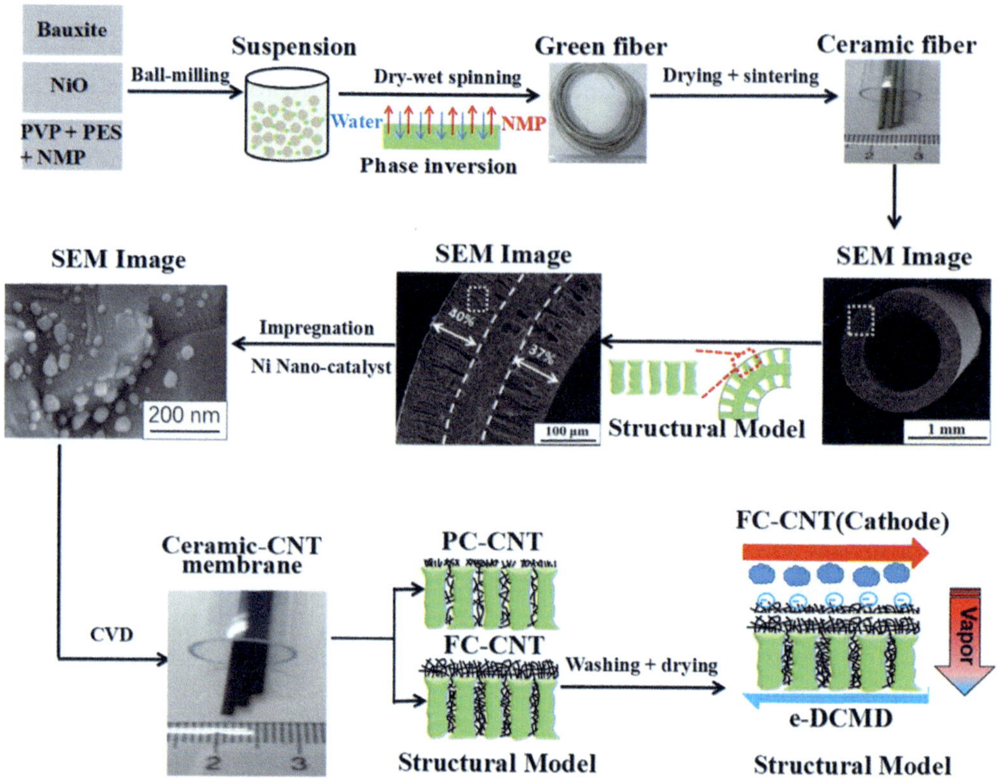

Figure 12.3 An illustration of the ceramic-based CNT membrane production and DCMD process, including essential processes, sample photos, and simplified structural models. *CNT*, Carbon nanotubes; *DCMD*, Direct-contact membrane distillation. Source: *From Y. Dong, L. Ma, C.Y. Tang, F. Yang, X. Quan, D. Jassby, M.J. Zaworotko, M.D. Guiver, Stable superhydrophobic ceramic-based carbon nanotube composite desalination membranes, Nano Lett. 18 (9) (2018) 5514–5521. https://doi.org/10.1021/acs.nanolett.8b01907.*

12.2.1.2.1 Solar desalination technology

For a solar evaporator, a CNFA (CNT@SiO_2 nanofibrous aerogel) with a vertically oriented vessel and permeable vessel walls was created. CNFA has an extremely low density and porosity due to its highly linked cellular architecture. It comprises numerous paths for water and salt transportation ascribed to its huge number of vessels and pores, which constitute 99.7% of its total volume. Since the density of hydrous CNFA is less than water, it floats on the surface of the water during solar evaporation. Additionally, it also withstands tremendous compression and automatically rebounds to its initial state when the pressure is removed. It shows robust mechanical strength due to the presence of SiO_2 nanoparticles and collapsible vessel walls. Owing to its properties it may be anticipated that CNFA possesses a huge potential in the process of desalination of water [35].

12.2.1.3 Adsorbent

Adsorption is one of the very efficient techniques used for the elimination of a variety of contaminants from wastewater. It emerged as a promising technique for removing nondegradable organic components in wastewater at low concentrations. The adsorption capacity for pollutants is influenced by the following factors: interaction between contaminant and adsorbent, pollutant concentration, surface area of adsorbent, particle size, pH, temperature, and time of contact. Previously, activated carbon (AC) was the only efficient adsorbent used for the removal of pollutants, but nowadays ceramics and ceramic nanocomposites have come to light as adsorbents in the areas where traditionally AC was used. Some of the major advantages of these ceramic adsorbents are as follows: because of their low cost, increased activity, and extended life period, the overall operation cost is lower [36]. Detoxification of polluted and wastewater using different anionic contaminants can be undertaken using various adsorbents, such as various phases of aluminum oxides (γ-Al_2O_3, α-Al_2O_3), oxyhydroxides, and hydroxides. For the removal of arsenates, gibbsite (c-$AL(OH)_3$) was employed [37] while sulfates and selenates can be removed using activated alumina [38]. Aluminum oxides have a high adsorption capacity because of their textural properties, like pore volume, area of the surface, and pore size, thus making them suitable and significant materials for catalytic and anion adsorption processes [39]. There is a need to adsorb fluoride ions from the groundwater as its concentration is exceeding the WHO permissible limits day by day, and for this amorphous Al_2O_3 (Al_2O_3/CNTs) synthesized on CNTs was recently used to adsorb these fluoride ions from water. However, at an equilibrium fluoride concentration of 12 mg/L in the pH range of 5–9, the adsorption capacity of the ceramics composite was found to be 4 × than that of c-Al_2O_3 and 13.5 × that of AC [40]. Fabrication of $Fe(OH)_3$ through in situ precipitation on thermally activated Al_2O_3 powder allows arsenic removal from drinking water with low-cost adsorbents. This manufactured ceramic material demonstrated a pH-dependent binding capacity, as well as improved removal capacity and selectivity for the removal of arsenite ions and arsenate [41].

Pansini et al. investigated the usage of metal–ceramic nanocomposites made by a patented process of thermal modification of cation exchanged zeolites in a reducing environment; these nanocomposites have the potential to be used to remove agrochemicals from water [42]. One of the most often used ceramic materials is porous γ-alumina (γ-Al_2O_3). To expand the application possibilities, various attempts have been undertaken to create multifunctional γ-Al_2O_3 ceramic composite nanoparticles, followed by magnetization and seeded polymerization with epoxide functional methacrylate (Glycidyl Methacrylate). Bristy et al. demonstrated a four-step method for producing epoxide functional γ-Al_2O_3 ceramic nanocomposites particles known as γ-Al_2O_3/Fe_3O_4/SiO_2/PGMA. The multifunctional γ-Al_2O_3 ceramic nanocomposite particles generated are useful in biotechnology, catalysis, and as pollutant adsorbents [43].

12.2.2 Sensing devices

Industrial activities are increasing continuously, and due to the rapid population growth and industrial development, our surroundings become more polluted day by day at an increasing rate. As a result, to avoid environmental disasters, it has become important to monitor and control such pollutants. Traditional analytic instruments are expensive, time-consuming, and rarely employed in real time in the field.

Ceramic sensor technology is becoming popular as a cost-effective alternative to conventional sensor solutions as the need arises for unique physical, chemical, and biological detection systems (2017). To monitor the changes in ambient gas concentration, gas sensors are used by detecting the gases and differentiating the smell. To choose effective solid-state sensors from ceramic or ceramic nanocomposites, the following are the desirable characteristics: the correlation between electrophysical, electronic, catalytic activity, thermodynamic surface, and structural characteristics. Semiconductors like TiO_2, SnO_2, ZnO, and others have played a significant role in the environmental industry, particularly in gas sensing. Their distinct optical properties, chemical stability, powerful oxidation capabilities, affordability, non-toxic nature, and ample surface area have greatly benefited this field [44]. Nowadays, the use of ceramics in graphene-based nanocomposites has also gained much attention around the globe. The few-layered graphene with the introduction of ceramic materials has allowed the development of a composite that produces astoundingly high charge carrier characteristics and electrochemical performance [45]. Several ceramic-reduced graphene oxides (rGO) composites like WO_3/rGO (2019), NiO/rGO [46], SnO_2 nanoparticles-rGO [47], In_2O_3-rGO [48], SnO_2 quantum wire-rGO [49,50], ZnO-rGO [51,52], Cu_2O-rGO [53], and many more are known to be used as sensors for various gases such as NO_2, H_2S, NH_3, CH_4, etc.

At ambient temperature, Feng et al. demonstrated that rGO-enveloped Co_3O_4 composite nanofibers showed an exceptional sensitivity toward varied concentrations of NH_3 ranging from 5 to 100 ppm [54]. According to Galstyan et al., the nanostructured ZnO is a potential chemical gas sensor with very high and reversible responses toward the following gases: NO_2, H_2, and CH_4 [55]. Jeevitha et al. [56] reported that the porous and conductive 5% rGO/WO_3 nanocomposite sensor has shown enhanced sensing performance toward NH_3, and in the presence of other potentially interfering gases, it is very selective for NH_3. At room temperature, it also demonstrates exceptional stability and repeatability. This excellent sensing property of rGO/WO_3 nanocomposites is because of the increased area of the surface and pore size as compared to pure WO_3 nanospheres (2019). Zhang et al. studied the sensing capacities of NiO/rGO nanocomposite film made by hydrothermal synthesis. The results showed that the sensor is very highly responsive, and takes less recovery time for the detection of CH_4 [46]. According to Zhang et al., the SnO_2-rGO nanocomposites show an excellent sensing capacity toward NO_2 at low operating temperatures. The addition of

SnO_2 to the rGO matrix improved the sensing capabilities for graphene-based sensing materials dramatically [47].

Apart from the ceramic oxides, carbon-based materials, for example, graphene, and CNTs, after fabrication with the electrodes are also utilized to detect a variety of organic contaminants.

12.2.3 Dye-sensitized solar cells

In recent years nanoceramics have been increasingly used in dye-sensitized solar cells (DSSCs) to increase their efficiency along with the surety of their long-term durability. Fig. 12.4 shows a schematic illustration of a DSSC. TiO_2 solar cells developed by Gratzel et al. are quite popular because of their ease of production and low cost [57]. They have been shown to have achieved an energy conversion efficiency of 10% [58].

TiO_2 is a nanoceramic that is extensively utilized in DSSCs to ensure long-term stability. When hydrophobic complexes are used as charge transfer photosensitizers in solar cells based on TiO_2, they exhibit amazing stability with regard to water-induced desorption [59,60]. An integral element in the DSSCs is their rate of electron transfer, which defines the overall efficiency of the cells. Dark current is produced when electrons are injected into the conduction band along with an oxidized redox mediator, which thereby decreases the efficiency of charge collection. This problem can be solved with the utilization of TiO_2 films equipped with hydrophobic sensitizers that reduce the formation of dark currents [59]. The utilization of single-crystalline TiO_2 nanotubes as semiconductors also makes highly efficient DSSCs. The electron transfer rate of single-crystalline TiO_2 nanotubes is very high compared to TiO_2 films composed of nanoparticles. It also displays a double short circuit current density in comparison with titania nanoparticles. Titania nanotube cells demonstrate a conversion rate of light to the electricity of around 5% [61].

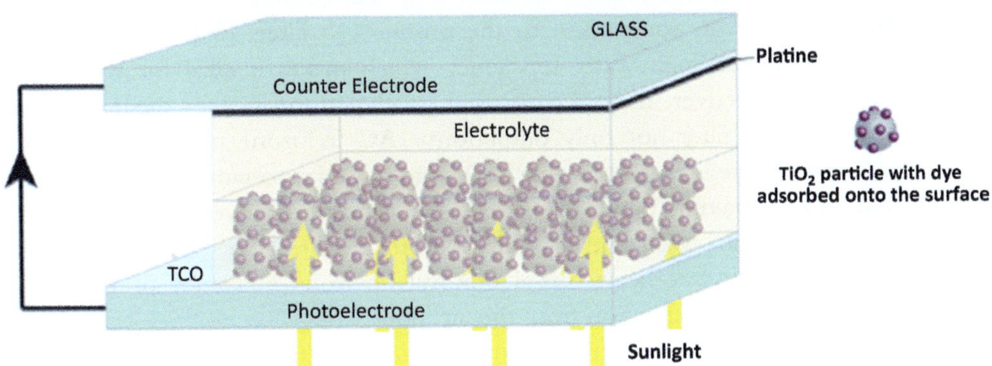

Figure 12.4 Schematic representation of dye-sensitized solar cells. Source: *From M.K. Nazeeruddin, E. Baranoff, M. Grätzel, Dye-sensitized solar cells: a brief overview, Solar Energy 85 (6) (2011) 1172–1178. https://doi.org/10.1016/j.solener.2011.01.018.*

Nanoceramic crystalline semiconductor oxides have additionally demonstrated great adsorption capacity of dye molecules, owing to their large surface area. This has helped in increasing the adsorption of the incident sunlight. ZnO and Fe_2O_3 are ceramic nanocrystalline semiconducting material oxides that showcase many physical properties ideal for their utilization in the DSSCs. They are economical, have great electrical conductivity, exceptional substrate cohesion, and chemical stability [62,63]. However, the technique used for the preparation of these semiconductors greatly affects the performance of the solar cells. There are many techniques present, but among these techniques, sol−gel is the most preferred method [64]. To enhance the properties of solar cells, either by increasing the surface area or by using hierarchical and multidimensional structures as well as by increasing the pathways, ZnO nanorods or nanowires can be used instead of nanoparticles (2012). Moreover, Zn doping of DSSCs made of TiO_2 nanorods photoanodes and $CuCrO_2$ nanoparticles has led to appreciable enhancement in its energy conversion efficiency. The utilization of $CuCrO_2$:Zn nanoparticles in DSSCs leads to a 39% increase in efficiency, relative to the undoped ones [65].

12.2.4 Air pollution

Air pollution has become an issue of great concern. It has caused many detrimental effects like climate change, global warming, and has become the root of many health problems including the common cough, Alzheimer's, bronchitis, and even cancer. Nanoceramics have shown favorable hope in solving this problem by cleaning the polluted air and thereby countering the contribution of industrial pollution. CNT fiber filters are used to filter out aerosol particles from the air with great efficiency. Nanoceramics assist in improving the filter quality of these HEPA (high-efficiency particulate air) CNT filters. The HEPA CNT filters with quartz fiber growth increase the filter specific area of the filter by 12 times more than the pristine QF filter [66]. The pressure drop across the filter increased by about 50% in comparison to the pristine QF filters [66]. This CNT/QF composite filter exhibited high water repellency which makes it ideal for working in humid conditions. Moreover, they also showed good performance in the removal of ozone. The CNT/QF filter not only outperforms AC in ozone conversion but also shows a better pressure resistance according to Yang et al. Nanoceramic composite filters of Al_2O_3 and SiO_2-containing CNTs have been shown to remove yeast cells along with various heavy metals from water and particulates from the air. They exhibit yeast filtration to an efficiency of 98%, heavy metal removal up to an efficiency of 100%, and exceedingly good particulate filtration from the air [66].

Quartz fiber filters also show unique properties in comparison to other filter types. Their structure consists of a combination of fine and coarse layout that creates a fibrous filter that is both efficient and permeable. They show great air permeability, hydrophilicity, stability when exposed to high temperature, and high chemical resistance. During

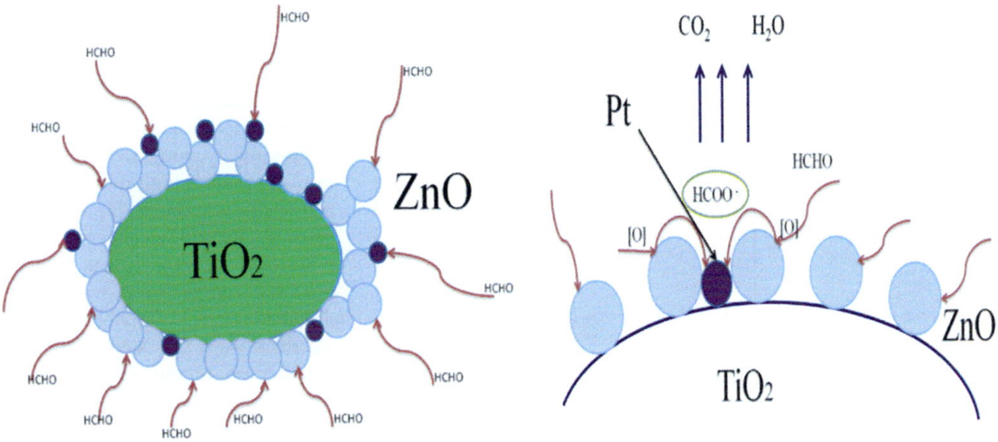

Figure 12.5 Schematic of degradation of HCHO over Pt/TiO_2-ZnO_2 catalyst. *HCHO*, formaldehyde. Source: *From C. Nie, L. Liu, R. He, Pt/TiO_2-ZnO in a circuit photo-electro-catalytically removed HCHO for outstanding indoor air purification, Sep. Purif. Technol. 206 (2018) 316–323. https://doi.org/10.1016/j.seppur.2018.06.024.*

the thermal analysis method, quartz fiber filters are used as a sampler to measure the mass of particulate matter and organic carbon present in the air [67]. Another nanoceramic, TiO_2 is used in combination with Zn and Pt as a cathode in the photoelectrocatalytic oxidation process (you can see the reaction in Fig. 12.5) for the removal of HCHO [68]. Recently, the prevention of infectious disease transmission has emerged as an urgent need worldwide due to the current health crisis. TiO_2 has been used in a fluidized bed reactor that contains Ag@TiO_2-Cu^{2+} perlite supported catalyst and a visible–UVA LED lamp. These materials when combined exhibit excellent antibacterial activity. This photocatalytic system has enhanced the capacity for air disinfection and has emerged in the field of indoor air filtration technologies as a prospective candidate.

12.2.5 Antibacterial materials

In water there are chemical as well as biological contaminants that make the water impure and unhealthy for consumption. The presence of viruses and other microbes that thrive in water bodies, as well as NOM and toxic biological contaminants, are the main causes leading to biological contamination. Among all the microorganisms, nowadays cyanobacterial toxins are drawing much attention due to their potential carcinogenic threats, so there is a need to remove these toxins from the water.

Ceramic and ceramic nanocomposite adsorbents can efficiently remove biological contaminants and can be applied on a large scale. Nanomaterials such as Ag, ZnO, and TiO_2 have the charge capacity and due to this can disinfect the waterborne disease-causing microbes [69]. Ohira along with others explored the effects of crystalline sizes of MgO

Table 12.1 Application of various nanoceramics materials to promote environmental sustainability.

S. no.	Applications	Examples of materials	Some important properties
1.	Water treatment and remediation	Al_2O_3 membrane, ZrO_2 membrane, TiO_2 membrane, SiO_2 membrane, zeolite membrane, Zr-MOF, MCM-41 silica	Strong antimicrobial action, hydrophilicity, roughness, low human toxicity, high mechanical and chemical stability, high permeability and selectivity, photocatalytic activity, zeta potential, and permeate water flow
2.	Desalination	FC-CNT, nano-TiO_2, Al_2O_3, SiO_2, CNFA	High rejection rate, hydrophobic, higher and stable water flux, antiwetting characteristics, porous, small pore size, robust mechanical strength, chemical stability
3.	Adsorbent	γ-Al_2O_3, α-Al_2O_3, oxyhydroxides, Al_2O_3/CNTs, γ-$Al_2O_3/Fe_3O_4/SiO_2$/PGMA, hydroxides, c-AL $(OH)_3$	High specific surface area, small pore size, longer life span, adsorption sites that are easily accessible, selective and more adsorption sites, and a short intra-particle diffusion distance
4.	Sensing devices	TiO_2, SnO_2, ZnO, WO_3/rGO, NiO/rGO, SnO_2 nanoparticles-rGO, In_2O_3-rGO, SnO_2 quantum wire-rGO, ZnO-rGO, Cu_2O-rGO, CNTs	Good chemical inertness, strong oxidizing power, low cost, non-toxicity, larger surface and, unique optical properties, good sensitivity, high selectivity, stability and repeatability, increased pore size, and surface area
5.	Dye-sensitized solar cells	TiO_2, ZnO, Fe_2O_3, $CuCrO_2$, $CuCrO_2$:Zn	Photocatalytic activity under solar irradiation, good electrical conductivity, substrate adherence, hydrophobic, ease of fabrication, low human toxicity, high energy conversion efficiency, high stability and selectivity, low cost
6.	Air purification and remediation	CNT/QF, Al_2O_3, SiO_2, TiO_2, Ag@TiO_2-Cu^{2+}, Pt/TiO_2-ZnO_2 catalyst	Hydrophobic, excellent air permeability, hydrophilicity, high-temperature stability, high chemical resistance, and antibacterial activity
7.	Antibacterial material	ZnO, TiO_2, MgO, $ZnTiO_3$, Al_2O_3, zeolite	Strong antibacterial activity, disinfectant, low toxicity and cost, high charge capacity, chemical stability, and ease of use are all advantages

CNTs, Carbon nanotubes; *FC-CNT*, fully covered carbon nanotube; *rGO*, reduced graphene oxides; *Zr-MOF*, zirconium-based metal–organic framework.

and ZnO on antibacterial activity. In his study, he found that the crystalline-sized powder shows excellent antibacterial properties with regard to *Staphylococcus aureus* and *E. coli* [70].

A ceramic, nanocrystalline $ZnTiO_3$ was produced using urea as fuel through solution combustion. This was then evaluated with regard to its antibacterial activity, and a remarkable effect was observed on the following strains: Gram-negative *Klebsiella aerogenes, Pseudomonas desmolyticum, E. coli*; and Gram-positive *S. aureus* [71]. Sabbani along with others invented biocidal surfaces with the deposition of Ag nanoparticles on zeolite films which were grown on a micropatterned α-Al_2O_3 containing a large surface area which resulted in the absolute elimination of *E. coli* within 2 h [72].

12.3 Conclusion

Table 12.1 summarizes the various materials that we have mentioned in this chapter for the particular application of nanoceramics to promote environmental sustainability, and some of the important properties that contribute to their respective applications.

References

[1] Z. He, Z. Lyu, Q. Gu, L. Zhang, J. Wang, Ceramic-based membranes for water and wastewater treatment, Colloids Surf. A: Physicochem. Eng. Asp. 578 (2019) 123513. Available from: https://doi.org/10.1016/j.colsurfa.2019.05.074.

[2] M. Padaki, R. Surya Murali, M.S. Abdullah, N. Misdan, A. Moslehyani, M.A. Kassim, et al., Membrane technology enhancement in oil-water separation. A review, Desalination 357 (2015) 197−207. Available from: https://doi.org/10.1016/j.desal.2014.11.023.

[3] B. Hofs, J. Ogier, D. Vries, E.F. Beerendonk, E.R. Cornelissen, Comparison of ceramic and polymeric membrane permeability and fouling using surface water, Sep. Purif. Technol. 79 (3) (2011) 365−374. Available from: https://doi.org/10.1016/j.seppur.2011.03.025.

[4] A. Kayvani Fard, G. McKay, A. Buekenhoudt, H. Al Sulaiti, F. Motmans, M. Khraisheh, et al., Inorganic membranes: preparation and application for water treatment and desalination, Materials 11 (1) (2018) 74. Available from: https://doi.org/10.3390/ma11010074.

[5] S. Lee, J. Cho, Comparison of ceramic and polymeric membranes for natural organic matter (NOM) removal, Desalination 160 (3) (2004) 223−232. Available from: https://doi.org/10.1016/S0011-9164(04)90025-2.

[6] M. Ebrahimi, K.S. Ashaghi, L. Engel, D. Willershausen, P. Mund, P. Bolduan, et al., Characterization and application of different ceramic membranes for the oil-field produced water treatment, Desalination 245 (1−3) (2009) 533−540. Available from: https://doi.org/10.1016/j.desal.2009.02.017.

[7] L. Xu, W. Li, S. Lu, Z. Wang, Q. Zhu, Y. Ling, Treating dyeing waste water by ceramic membrane in crossflow microfiltration, Desalination 149 (1−3) (2002) 199−203. Available from: https://doi.org/10.1016/S0011-9164(02)00759-2.

[8] S. Moazzem, J. Wills, L. Fan, F. Roddick, V. Jegatheesan, Performance of ceramic ultrafiltration and reverse osmosis membranes in treating car wash wastewater for reuse, Environ. Sci. Pollut. Res. 25 (9) (2018) 8654−8668. Available from: https://doi.org/10.1007/s11356-017-1121-9, http://www.springerlink.com/content/0944-1344.

[9] A. Larbot, L. Gazagnes, S. Krajewski, M. Bukowska, W. Kujawski, Water desalination using ceramic membrane distillation, Desalination 168 (1−3) (2004) 367−372. Available from: https://doi.org/10.1016/j.desal.2004.07.021.

[10] D. Zou, M. Qiu, X. Chen, Y. Fan, One-step preparation of high-performance bilayer α-alumina ultrafiltration membranes via co-sintering process, J. Membr. Sci. 524 (2017) 141–150. Available from: https://doi.org/10.1016/j.memsci.2016.11.025, http://www.elsevier.com/locate/memsci.
[11] S.R. Chowdhury, R. Schmuhl, K. Keizer, J.E. Ten Elshof, D.H.A. Blank, Pore size and surface chemistry effects on the transport of hydrophobic and hydrophilic solvents through mesoporous γ-alumina and silica MCM-48, J. Membr. Sci. 225 (1–2) (2003) 177–186. Available from: https://doi.org/10.1016/j.memsci.2003.07.018, http://www.elsevier.com/locate/memsci.
[12] Z. Sadeghian, F. Zamani, S.N. Ashrafizadeh, Removal of oily hydrocarbon contaminants from wastewater by γ-alumina nanofiltration membranes, Desalin. Water Treat. 20 (1–3) (2010) 80–85. Available from: https://doi.org/10.5004/dwt.2010.1154, http://www.deswater.com/readpaper.php?id = 824.
[13] Z. Wang, Y.M. Wei, Z.L. Xu, Y. Cao, Z.Q. Dong, X.L. Shi, Preparation, characterization and solvent resistance of γ-Al_2O_3/α-Al_2O_3 inorganic hollow fiber nanofiltration membrane, J. Membr. Sci. 503 (2016) 69–80. Available from: https://doi.org/10.1016/j.memsci.2015.12.039, http://www.elsevier.com/locate/memsci.
[14] Y. Zhu, D. Wang, L. Jiang, J. Jin, Recent progress in developing advanced membranes for emulsified oil/water separation, NPG Asia Mater. 6 (5) (2014). Available from: https://doi.org/10.1038/am.2014.23, http://www.natureasia.com/asia-materials/.
[15] X. Da, X. Chen, B. Sun, J. Wen, M. Qiu, Y. Fan, Preparation of zirconia nanofiltration membranes through an aqueous sol-gel process modified by glycerol for the treatment of wastewater with high salinity, J. Membr. Sci. 504 (2016) 29–39. Available from: https://doi.org/10.1016/j.memsci.2015.12.068, http://www.elsevier.com/locate/memsci.
[16] A.W.C. Van Den Berg, L. Gora, J.C. Jansen, M. Makkee, T. Maschmeyer, Zeolite A membranes synthesized on a UV-irradiated TiO_2 coated metal support: the high pervaporation performance, J. Membr. Sci. 224 (1–2) (2003) 29–37. Available from: https://doi.org/10.1016/S0376-7388(03)00345-4, http://www.elsevier.com/locate/memsci.
[17] M. Ebrahimi, D. Willershausen, K.S. Ashaghi, L. Engel, L. Placido, P. Mund, et al., Investigations on the use of different ceramic membranes for efficient oil-field produced water treatment, Desalination 250 (3) (2010) 991–996. Available from: https://doi.org/10.1016/j.desal.2009.09.088.
[18] T. Van Gestel, C. Vandecasteele, A. Buekenhoudt, C. Dotremont, J. Luyten, R. Leysen, et al., Alumina and titania multilayer membranes for nanofiltration: preparation, characterization and chemical stability, J. Membr. Sci. (2002) 73–89. Available from: https://doi.org/10.1016/s0376-7388(02)00053-4.
[19] X. Zhang, T. Zhang, J. Ng, D.D. Sun, High-performance multifunctional TiO_2 nanowire ultrafiltration membrane with a hierarchical layer structure for water treatment, Adv. Funct. Mater. 19 (23) (2009) 3731–3736. Available from: https://doi.org/10.1002/adfm.200901435Singapore, http://www3.interscience.wiley.com/cgi-bin/fulltext/122680215/PDFSTART.
[20] K.A. Manjumol, S. Sankar, B.N. Nair, M. Midhun, P.A. Mohamed, K.G.K. Warrier, A novel approach to formulate high flux multifunctional ultrafiltration membranes from photocatalytic titania composite precursors on multi-channel tubular substrates, RSC Adv. 6 (63) (2016) 58813–58822. Available from: https://doi.org/10.1039/c6ra14117j, http://pubs.rsc.org/en/journals/journalissues.
[21] M.A. Anderson, M.J. Gieselmann, Q. Xu, Titania and alumina ceramic membranes, J. Membr. Sci. 39 (3) (1988) 243–258. Available from: https://doi.org/10.1016/S0376-7388(00)80932-1.
[22] M. Elma, C. Yacou, J.C. Diniz da Costa, D.K. Wang, Performance and long term stability of mesoporous silica membranes for desalination, Membranes 3 (3) (2013) 136–150. Available from: https://doi.org/10.3390/membranes3030136, http://www.mdpi.com/2077-0375/3/3/136/pdf.
[23] M. Kazemimoghadam, New nanopore zeolite membranes for water treatment, Desalination 251 (1–3) (2010) 176–180. Available from: https://doi.org/10.1016/j.desal.2009.11.036.
[24] J. Cui, X. Zhang, H. Liu, S. Liu, K.L. Yeung, Preparation and application of zeolite/ceramic microfiltration membranes for treatment of oil contaminated water, J. Membr. Sci. 325 (1) (2008) 420–426. Available from: https://doi.org/10.1016/j.memsci.2008.08.015.

[25] H. Ahn, H. Lee, S.B. Lee, Y. Lee, Dehydration of TFEA/water mixture through hydrophilic zeolite membrane by pervaporation, J. Membr. Sci 291 (1–2) (2007) 46–52. Available from: https://doi.org/10.1016/j.memsci.2006.12.033.

[26] A.K. Basumatary, R.V. Kumar, K. Pakshirajan, G. Pugazhenthi, Removal of trivalent metal ions from aqueous solution via cross-flow ultrafiltration system using zeolite membranes, J. Water Reuse Desalin 7 (1) (2017) 66–76. Available from: https://doi.org/10.2166/wrd.2016.211, http://jwrd.iwaponline.com/content/ppiwajwrd/7/1/66.full.pdf.

[27] X. Liu, N.K. Demir, Z. Wu, K. Li, Highly water-stable zirconium metal-organic framework UiO-66 membranes supported on alumina hollow fibers for desalination, J. Am. Chem. Soc. 137 (22) (2015) 6999–7002. Available from: https://doi.org/10.1021/jacs.5b02276, http://pubs.acs.org/journal/jacsat.

[28] J.P.S. DesOrmeaux, J.D. Winans, S.E. Wayson, T.R. Gaborski, T.S. Khire, C.C. Striemer, et al., Nanoporous silicon nitride membranes fabricated from porous nanocrystalline silicon templates, Nanoscale 6 (18) (2014) 10798–10805. Available from: https://doi.org/10.1039/c4nr03070b, http://www.rsc.org/publishing/journals/NR/Index.asp.

[29] Z. Wang, A. Wu, L. Colombi Ciacchi, G. Wei. Recent advances in nanoporous membranes for water purification. Nanomaterials (Basel). 8 (2) (2018) 65. Available from: https://doi.org/10.3390/nano8020065.

[30] X. Chen, M. Qiu, H. Ding, K. Fu, Y. Fan, A reduced graphene oxide nanofiltration membrane intercalated by well-dispersed carbon nanotubes for drinking water purification, Nanoscale 8 (10) (2016) 5696–5705. Available from: https://doi.org/10.1039/c5nr08697c, http://www.rsc.org/publishing/journals/NR/Index.asp.

[31] Y. Dong, L. Ma, C.Y. Tang, F. Yang, X. Quan, D. Jassby, et al., Stable superhydrophobic ceramic-based carbon nanotube composite desalination membranes, Nano Lett. 18 (9) (2018) 5514–5521. Available from: https://doi.org/10.1021/acs.nanolett.8b01907.

[32] X.S. Yi, S.L. Yu, W.X. Shi, S. Wang, N. Sun, L.M. Jin, et al., Estimation of fouling stages in separation of oil/water emulsion using nano-particles Al_2O_3/TiO_2 modified PVDF UF membranes, Desalination 319 (2013) 38–46. Available from: https://doi.org/10.1016/j.desal.2013.03.031.

[33] D. Vasanth, G. Pugazhenthi, R. Uppaluri, Cross-flow microfiltration of oil-in-water emulsions using low cost ceramic membranes, Desalination 320 (2013) 86–95. Available from: https://doi.org/10.1016/j.desal.2013.04.018.

[34] Q. Chang, J.E. Zhou, Y. Wang, J. Liang, X. Zhang, S. Cerneaux, et al., Application of ceramic microfiltration membrane modified by nano-TiO_2 coating in separation of a stable oil-in-water emulsion, J. Membr. Sci. 456 (2014) 128–133. Available from: https://doi.org/10.1016/j.memsci.2014.01.029.

[35] X. Dong, L. Cao, Y. Si, B. Ding, H. Deng, Cellular structured CNTs@SiO_2 nanofibrous aerogels with vertically aligned vessels for salt-resistant solar desalination, Adv. Mater. 32 (2020) e1908269. https://onlinelibrary.wiley.com/doi/abs/10.1002/adma.201908269.

[36] G.Z. Kyzas, J. Fu, K.A. Matis, The change from past to future for adsorbent materials in treatment of dyeing wastewaters, Materials 6 (11) (2013) 5131–5158. Available from: https://doi.org/10.3390/ma6115131Greece, http://www.mdpi.com/1996-1944/6/11/5131/pdf.

[37] M.M. Miedaner, R. Weerasooriya, H.J. Tobschall, Chapter 17: 1-pK modeling strategies for the adsorption of some trace elements onto gibbsite. Interface Science and Technology, Elsevier, Germany, 2006, pp. 469–490. Available from: https://doi.org/10.1016/S1573-4285(06)80061-X. http://www.elsevier.com/wps/find/bookdescription.cws_home/BS_3710/description#description.

[38] H. Wijnja, C.P. Schulthess, Vibrational spectroscopy study of selenate and sulfate adsorption mechanisms on Fe and Al (Hydr)oxide surfaces, J. Colloid Interface Sci. 229 (1) (2000) 286–297. Available from: https://doi.org/10.1006/jcis.2000.6960, http://www.elsevier.com/inca/publications/store/6/2/2/8/6/1/index.htt.

[39] E. Kumar, A. Bhatnagar, W. Hogland, M. Marques, M. Sillanpää, Interaction of anionic pollutants with Al-based adsorbents in aqueous media—a review, Chem. Eng. J. 241 (2014) 443–456. Available from: https://doi.org/10.1016/j.cej.2013.10.065.

[40] Y.H. Li, S. Wang, A. Cao, D. Zhao, X. Zhang, C. Xu, et al., Adsorption of fluoride from water by amorphous alumina supported on carbon nanotubes, Chem. Phys. Lett. 350 (5–6) (2001) 412–416. Available from: https://doi.org/10.1016/S0009-2614(01)01351-3.

[41] J. Hlavay, K. Polyák, Determination of surface properties of iron hydroxide-coated alumina adsorbent prepared for removal of arsenic from drinking water, J. Colloid Interface Sci. 284 (1) (2005) 71–77. Available from: https://doi.org/10.1016/j.jcis.2004.10.032.

[42] M. Pansini, F. Sannino, A. Marocco, P. Allia, P. Tiberto, G. Barrera, et al., Novel process to prepare magnetic metal-ceramic nanocomposites from zeolite precursor and their use as adsorbent of agrochemicals from water, J. Environ. Chem. Eng. 6 (1) (2018) 527–538. Available from: https://doi.org/10.1016/j.jece.2017.12.030.

[43] S.S. Bristy, M.A. Rahman, K. Tauer, H. Minami, H. Ahmad, Preparation and characterization of magnetic γ-Al_2O_3 ceramic nanocomposite particles with variable Fe_3O_4 content and modification with epoxide functional polymer, Ceram. Int. 44 (4) (2018) 3951–3959. Available from: https://doi.org/10.1016/j.ceramint.2017.11.187.

[44] S. Das, V. Jayaraman, SnO_2: a comprehensive review on structures and gas sensors, Prog. Mater. Sci. 66 (2014) 112–255. Available from: https://doi.org/10.1016/j.pmatsci.2014.06.003.

[45] V. Dhand, K.Y. Rhee, H. Ju Kim, D. Ho Jung, A comprehensive review of graphene nanocomposites: RESEARCH status and trends, J. Nanomater. (2013)(2013). Available from: https://doi.org/10.1155/2013/763953.

[46] D. Zhang, H. Chang, P. Li, R. Liu, Characterization of nickel oxide decorated-reduced graphene oxide nanocomposite and its sensing properties toward methane gas detection, J. Mater. Sci.: Mater. Electron. 27 (4) (2016) 3723–3730. Available from: https://doi.org/10.1007/s10854-015-4214-6.

[47] H. Zhang, J. Feng, T. Fei, S. Liu, T. Zhang, SnO_2 nanoparticles-reduced graphene oxide nanocomposites for NO_2 sensing at low operating temperature, Sens. Actuators B: Chem. 190 (2014) 472–478. Available from: https://doi.org/10.1016/j.snb.2013.08.067.

[48] R.S. Andre, L.A. Mercante, H.M. Murilo, et al., Correa, enhanced and selective ammonia detection using In_2O_3/reduced graphene oxide hybrid nanofibers, Appl. Surf. Sci. 473 (2016) 1205–1212.

[49] Z. Song, Z. Wei, B. Wang, Z. Luo, S. Xu, W. Zhang, et al., Sensitive room-temperature H_2S gas sensors employing SnO_2 quantum wire/reduced graphene oxide nanocomposites, Chem. Mater. 28 (4) (2016) 1205–1212. Available from: https://doi.org/10.1021/acs.chemmater.5b04850.

[50] Z. Song, J. Liu, Q. Liu, H. Yu, W. Zhang, Y. Wang, et al., Enhanced H_2S gas sensing properties based on SnO_2 quantum wire/reduced graphene oxide nanocomposites: equilibrium and kinetics modeling, Sens. Actuators B: Chem. 249 (2017) 632–638. Available from: https://doi.org/10.1016/j.snb.2017.04.023.

[51] N. Kumar, A. Srivastava, H.S.G. Patel, G. Bipin Das, Facile synthesis of ZnO–reduced graphene oxide nanocomposites for NO_2 gas sensing applications, Eur. J. Inorg. Chem. 2015 (2015) 1912–1923. https://chemistry-europe.onlinelibrary.wiley.com/doi/abs/10.1002/ejic.201403172.

[52] F. Meng, Y. Chang, W. Qin, Z. Yuan, J. Zhao, J. Zhang, et al., ZnO-reduced graphene oxide composites sensitized with graphitic carbon nitride nanosheets for ethanol sensing, ACS Appl. Nano Mater. 2 (5) (2019) 2734–2742. Available from: https://doi.org/10.1021/acsanm.9b00257, https://pubs.acs.org/journal/aanmf6.

[53] Q. He, J. Liu, X. Liu, G. Li, P. Deng, J. Liang, Preparation of Cu_2O-reduced graphene nanocomposite modified electrodes toward ultrasensitive dopamine detection, Sensors 18 (1) (2018) 199. Available from: https://doi.org/10.3390/s18010199.

[54] Q. Feng, X. Li, J. Wang, A.M. Gaskov, Reduced graphene oxide (rGO) encapsulated Co_3O_4 composite nanofibers for highly selective ammonia sensors, Sens. Actuators B: Chem. 72 (10) (2016) 864–870. Available from: https://doi.org/10.1016/j.snb.2015.09.021.

[55] V. Galstyan, E. Comini, C. Baratto, G. Faglia, G. Sberveglieri, Nanostructured ZnO chemical gas sensors, Ceram. Int. 41 (10) (2015) 14239–14244. Available from: https://doi.org/10.1016/j.ceramint.2015.07.052.

[56] G. Jeevitha, R. Abhinayaa, D. Mangalaraj, N. Ponpandian, P. Meena, V. Mounasamy, et al., Porous reduced graphene oxide (rGO)/WO_3 nanocomposites for the enhanced detection of NH_3 at room temperature, Nanoscale Adv. 1 (2019) 1799–1811. 10.1039/C9NA00048H.

[57] D. Kuang, P. Comte, S.M. Zakeeruddin, D.P. Hagberg, K.M. Karlsson, L. Sun, et al., Stable dye-sensitized solar cells based on organic chromophores and ionic liquid electrolyte, Sol. Energy 85 (6) (2011) 1189–1194. Available from: https://doi.org/10.1016/j.solener.2011.02.025.

[58] Michael Grätzel, Dye-sensitized solar cell, J. Photochem. Photobiol. C: Photochem. Rev. 4 (2003) 145–153.

[59] M.K. Nazeeruddin, E. Baranoff, M. Grätzel, Dye-sensitized solar cells: a brief overview, Sol. Energy 85 (6) (2011) 1172–1178. Available from: https://doi.org/10.1016/j.solener.2011.01.018.

[60] M.K. Nazeeruddin, P. Péchy, T. Renouard, S.M. Zakeeruddin, R. Humphry-Baker, P. Comte, et al., Engineering of efficient panchromatic sensitizers for nanocrystalline TiO_2-based solar cells, J. Am. Chem. Soc. 123 (8) (2001) 1613–1624. Available from: https://doi.org/10.1021/ja003299u.

[61] M. Adachi, J. Electrochem. Soc. 150 (2003)

[62] N. Özer, F. Tepehan, Optical and electrochemical characteristics of sol gel deposited iron oxide films, Sol. Energy Mater. Sol. Cell 56 (2) (1999) 141–152. Available from: https://doi.org/10.1016/s0927-0248(98)00152-4, http://www.sciencedirect.com/science/journal/09270248/100.

[63] A.Y. Oral, M.H.G. Aslan, E.L.A. Mens-ur, E. Bas-aran, Preparation of-axis orientation zinc-oxide thin films and the study of their micro-structure and optical properties, Sol. Energy Mater. Sol. Cell 82 (2004) 543–552.

[64] S.M. Reda, Synthesis of ZnO and Fe_2O_3 nanoparticles by sol–gel method and their application in dye-sensitized solar cells, Mater. Sci. Semicon. Proc. 13 (5–6) (2010) 417–425. Available from: https://doi.org/10.1016/j.mssp.2011.09.007.

[65] M. Asemi, M. Ghanaatshoar, Conductivity improvement of $CuCrO_2$ nanoparticles by Zn doping and their application in solid-state dye-sensitized solar cells, Ceram. Int. 42 (6) (2016) 6664–6672. Available from: https://doi.org/10.1016/j.ceramint.2016.01.017.

[66] R. Zhang, F. Wei, High-efficiency particulate air filters based on carbon nanotubes, Nanotube Superfiber Materials: Science, Manufacturing, Commercialization, Elsevier, China, 2019, pp. 643–666. Available from: https://doi.org/10.1016/B978-0-12–812667-7.00026-4. http://www.sciencedirect.com/science/book/9780128126677.

[67] C.C. Ho, F. Kang, G.M. Chang, S.J. You, Y.F. Wang, Application of recycled lanthanum-doped TiO_2 immobilized on commercial air filter for visible-light photocatalytic degradation of acetone and NO, Appl. Surf. Sci. 465 (2019) 31–40. Available from: https://doi.org/10.1016/j.apsusc.2018.09.136.

[68] C. Nie, L. Liu, R. He, Pt/TiO_2-ZnO in a circuit Photo-electro-catalytically removed HCHO for outstanding indoor air purification, Sep. Purif. Technol. 206 (2018) 316–323. Available from: https://doi.org/10.1016/j.seppur.2018.06.024.

[69] M.T. Amin, A.A. Alazba, U. Manzoor, A review of removal of pollutants from water/wastewater using different types of nanomaterials, Adv. Mater. Sci. Eng. 2014 (2014). Available from: http://www.hindawi.com/journals/amse/0.10.1155/2014/825910.

[70] T. Ohira, O. Yamamoto, Correlation between antibacterial activity and crystallite size on ceramics, Chem. Eng. Sci. 68 (1) (2012) 355–361. Available from: https://doi.org/10.1016/j.ces.2011.09.043.

[71] R.S. Raveendra, P.A. Prashanth, R. Hari Krishna, N.P. Bhagya, B.M. Nagabhushana, H. Raja Naika, et al., Synthesis, structural characterization of nano $ZnTiO_3$ ceramic: an effective azo dye adsorbent and antibacterial agent, J. Asian Ceram. Soc 2 (4) (2014) 357–365. Available from: https://doi.org/10.1016/j.jascer.2014.07.008, https://www.tandfonline.com/loi/tace20.

[72] S. Sabbani, D. Gallego-Perez, A. Nagy, W. James Waldman, D. Hansford, P.K. Dutta, Synthesis of silver-zeolite films on micropatterned porous alumina and its application as an antimicrobial substrate, Microporous Mesoporous Mater. 135 (1–3) (2010) 131–136. Available from: https://doi.org/10.1016/j.micromeso.2010.06.020.

CHAPTER 13

Nanoceramics in the energy storage industry

S. Kishore Babu[1], J. John Donald Raj[1], Deepankara Vrushabhadas Shastri[1], Shilpa Rajkumar[1] and Kantha D. Arunachalam[2]
[1]Department of Physics and Nanotechnology, SRM Institute of Science and Technology, Kattankulathur, Tamil Nadu, India
[2]Faculty of Sciences, Marwadi University, Rajkot, Gujarat, India

13.1 Introduction

Despite the rapidly expanding need for more efficient and cost-effective energy storage technologies and devices, there is worldwide pressure to install energy-efficient and environment-friendly innovations focused on slowing the rate of global climate change. Even though lithium batteries are currently the most widely used conventional energy sources, they have some inherent limitations arising from their electrochemical storage mechanism. Additionally, there are some physical limits to their performance/energy storage densities, controlled by the components used for the electrodes, which are evaluated by the material properties used for the anode and cathode. These also have a limited life span and significant servicing expenses, which make them uneconomical. Research and development of new energy sources storage systems, such as supercapacitors, has been driven by this, and they are currently garnering global interest from scholars and industrial investors alike. In this area, a breakthrough is now taking place, which involves the development of novel hybrid systems as well as new kinds of materials with extremely high permittivity, among other things. This project was concerned with the field of improved nanoscale materials and its application to new generations of energy storage devices and the considerable influence that this would have on the world economy as a result. The main objective was to build novel methods to generate new ceramic composite nanoscale materials for energy storage devices with high-performance ceramic-solid states. These materials include complicated nanoceramic multilayered materials as well as other types of complex materials [1—3].

13.2 Electrochemical energy storage devices

Several energy storage devices, such as fuel cells, batteries, supercapacitors, and conventional capacitors, are now employed to minimize the demand for robust and cost-efficient energy storage systems. The performance of different power-storing devices was evaluated

Industrial Applications of Nanoceramics
DOI: https://doi.org/10.1016/B978-0-323-88654-3.00018-4

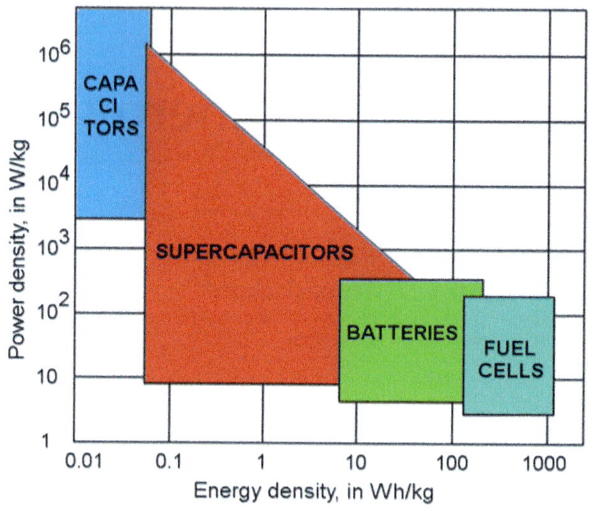

Figure 13.1 Ragone chart showing energy density versus power density for different types of energy-storage devices.

using a Ragone chart, such that the energy (in Wh/kg) values are compared to the power (in W/kg) values. A Ragone diagram is initially utilized to compare battery performance. However, any energy storage device can also be compared, as shown in Fig. 13.1.

13.3 Batteries

One or more electrochemical cells comprise the electric battery, which converts stored chemical energy into electrical energy. Every cell includes a positive and a negative terminus cathode and anode. Electrolytes enable ions to pass between electrodes and terminals, allowing the battery to run a current. Because of their high energy densities, high coulombic efficiency, low self-discharge characteristics, and various chemical potentials accessible with multiple electrode designs, lithium-ion batteries (LIBs) are widely used as power sources in a wide range of applications [4–7]. Over the last decade, there has been an increasing trend toward next-generation LIBs with high charge capacities/power densities created for electric cars (EVs), hybrid electric vehicles, aerospace applications, and autonomous electric gadgets (e.g., hybrid solar batteries). However, conventional LIBs shown in Fig. 13.2 (e.g., $LiCoO_2$), which are primarily based on the Li^+ intercalation mechanism, cannot provide the large charge capacities ($>$ 200 mAh/g) necessary for the specific applications. As a result, high-energy-density battery technologies such as Ni-rich TMO oxides, high voltage spinel, Li-sulfur, Li-air, organic electrode, Li-CO_2, and solid-state batteries have recently piqued the interest of researchers. They profit not just from the increased charging capacity but also the lower cost. Furthermore, they are less hazardous than the extensively used LIBs (e.g., $LiCoO_2$) [8–11].

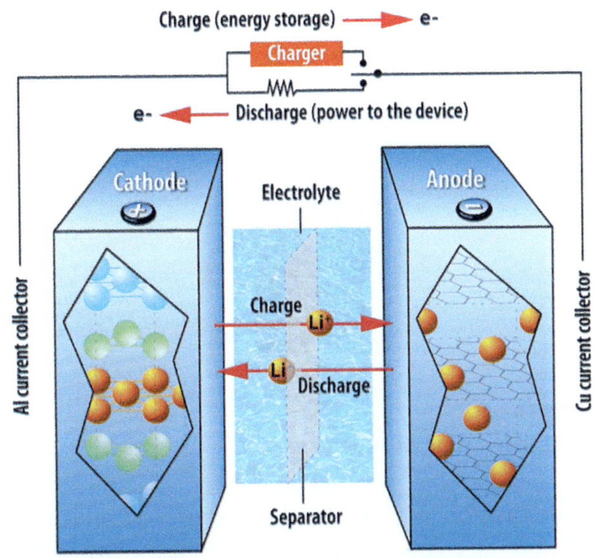

Figure 13.2 Graphical representation of conventional Li-ion battery.

13.4 Capacitors

A capacitor is a device that stores energy in an electric field formed between two conductors with equal but opposite electric charges. Capacitors are critical components of electrical circuits because they keep energy in the microfarad range and aid in filtering. Capacitors have two primary functions: one is to charge or discharge electricity, and the other is to stop the flow of direct current. This is used to describe filters that extract or eliminate specific frequencies. The energy stored in a capacitor is proportional to the voltage between the plates; indeed, the capacitor's capacity to retain energy due to voltage (potential difference between the two leads) leads in an attempt to maintain a constant voltage. When the voltage across a capacitor increases, it pulls power from the rest of the circuit and is said to be charging; however, when the voltage across a capacitor decreases, it acts as a power source, supplying current to the rest of the circuit. A capacitor is considered to be discharging while it is in this state.

13.5 Supercapacitor

Based on the device architecture, supercapacitors, also known as ultracapacitors or electrochemical capacitors, are classified as symmetric, asymmetric, or hybrid supercapacitors. Supercapacitors are categorized into three varieties based on the charge storage method of electrode materials: electrical double-layer capacitors (EDLCs), pseudocapacitors (PCs), and battery-like supercapacitors shown in Fig. 13.3.

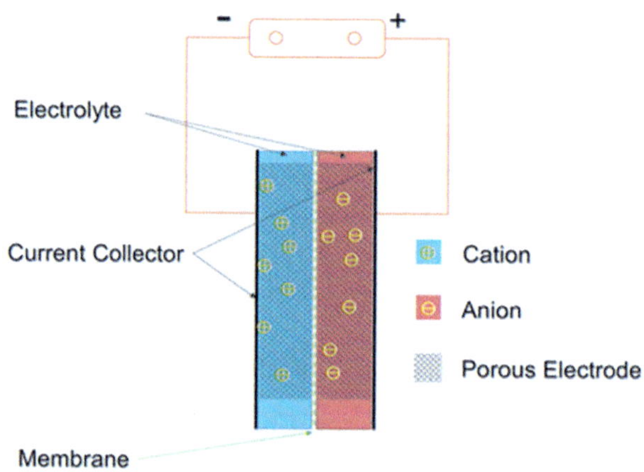

Figure 13.3 Schematic representation of supercapacitor.

The adsorption and desorption of ions in the electric double layer generated at the electrode–electrolyte interface are the key ways EDLCs store charges [12]. Because the entire process is purely physical and does not include any chemical processes, EDLCs have a high power density [13]. The low capacitance, on the other hand, limits its range of uses. Because of their high conductivity and vast surface area, carbon-based materials, such as activated carbon, graphene, and carbon nanotubes, are commonly employed as electrode materials for EDLCs. Due to Faraday redox processes happening at or near the surface of active materials, PCs frequently have higher specific capacitance than EDLCs [14]. Poor conductivity and long-term cycling stability are also disadvantaging electrode materials for PCs, and these concerns are still deadly drawbacks. As Thierry Brousse points out, electrode materials like MnO_2 can be constructed as "pseudocapacitive" materials since they have similar electrochemical behavior to capacitive carbon electrodes [15–17]. There are two ways of different capacitors and batteries: the amount of storage and the speed with which the power (charge) is delivered. The batteries may store higher loads than ordinary capacitors. However, they are unable to provide energy very quickly as a significant negative to the batteries. On the other hand, the capacitors offer energy (charge) very fast (fast discharge); however, the principal disadvantage of condensers is that the battery storage fails to store a considerable quantity of demand. The use of supercapacitors can solve these two drawbacks (large charge storage and rapid charge delivery). Supercapacitors work like regular capacitors, however, supercapacitors vary in two ways: massive electrodes and extremely thin dielectric (the distance of separation among electrodes is very tiny) compared with regular condensers are available in supercapacitors. This allows greater capacity (more extensive load shop) than conventional condensers to be achieved. Supercapacitors are mainly classified into three types: (1) double layer capacitors, (2) PCs, and (3) hybrid capacitors shown in Fig. 13.4.

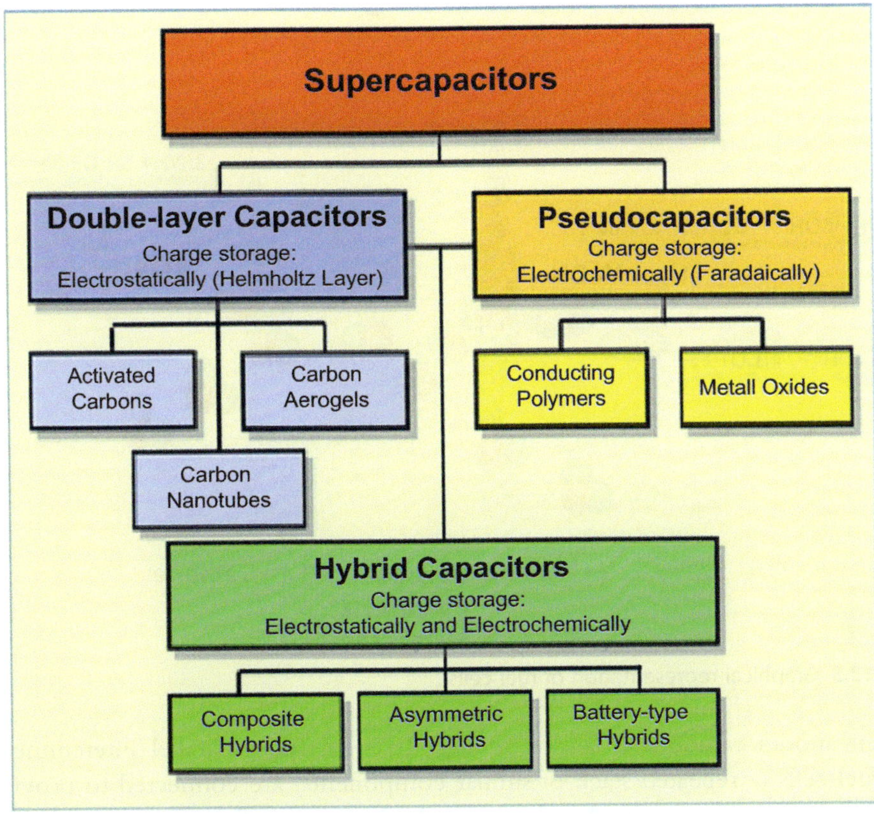

Figure 13.4 Three types of supercapacitors.

13.6 Fuel cells

The fuel cell is an electrical (galvanic) instrument that converts the free electrical energy of a chemical reaction into electric energy (electricity), heat and water/steam by-products when hydrogen and air are indeed the reactants, and, in some types of a fuel cell, carbon dioxide and lower remaining forms of hydrocarbons may also be used depending upon fossil fuel. In this mechanism, there is no ignition, and so no NO^{-2} is produced. Sulfur is poison in the fuel cells, so that the energy must be extracted before it is fed to any fuel cell; hence SO^2 is not produced. On-demand, a fuel cell constantly generates power while the fuel and oxidizer are supplied. As a reference, the primary or secondary cell is also an electrochemical device producer and needs to bring away once the battery has been depleted. An electrical energy storage device is a rechargeable or secondary battery that produces or uses electricity with a reversible reaction and has finite lives. The fuel cell components are the anode, anode catalysts, electrolytes, catalyst, cathode, bipolar/interconnections, and gas screens/prevents

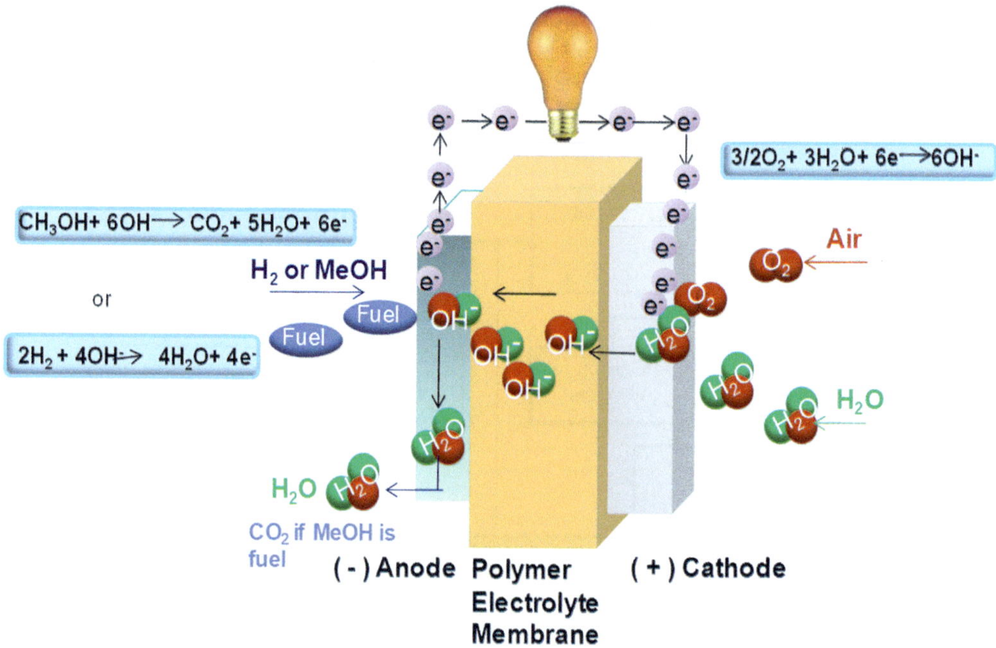

Figure 13.5 Graphical representation of fuel cells.

between anodes/cathode, as shown in Fig. 13.5. In series/parallel interconnections, these fuel cells (a repeated stack of similar components) are connected to provide the desired tension and current. Anode and cathode consist of porous gaseous diffusion layers, commonly composed of highly conductive electrons (and theoretically conducive of zero protons) such as thin absorbent layers of graphite. Platinum in low-temperature fuel cells, and nickel and other elements, depending on the fuel cell type are among the most popular catalysts. The electrolyte consists of such material as a proton conductivity and an electron conductivity that is theoretically zero. Depending on the fuel cell type, the load carriers vary (from the anode to the cathode or vice versa). The bipolar platforms gather the electrical current, distribute reactive gases, and split them into a fuel cell stack [18–20].

13.7 Synthesis of ceramic oxides for energy storage devices

For chemists, material scientists, and engineers, the connection between processing solids and the required structure, compositions, and qualities is still challenging. The diffusion controls the formation of solids through the ceramic method by reactants and products of atoms and ionic species. It, therefore, requires the repeated smear, pelletization, and carbonization of reactants (oxides or carbonates) at high temperatures for more extended periods (those are soft chemical routes). The current job of the solid-state is to improve the material reactivity to speed up the kinetic reaction since the speed of a solid-state response depends upon ionic conductivity or diffusion [21]. The temperature generates a sound state diffusion and latrine vibrations. Organized synthesis and controlling structure are prioritized in modern materials since they can customize the range of attributes needed for demanding applications. Solid-state reaction, sol–gel method, combustion method, mechanical alloying, plasma processing, and vapor deposition all can play a role in complex materials' controlled structure and composition. Multifunctional oxides are prepared via a solid-state reaction using ceramic powder, also known as ceramic-powder-based processing. Although this method has drawbacks, such as a high-temperature reaction, a restricted level of chemical homogeneity, and poor sintering capability, it is currently being used by many industries. Therefore, various chemical processing techniques have been developed to process powders with a more uniform composition, better reactivity, and good sintering ability for the past several years. Mechanical alloying and mechanochemical procedures have lately been employed to generate interactions across organisms. Recent projects using the base energy storage of supercapacitors have required novel high capacitance materials to speed up the reaction kinetics and reduce processing time [22–24].

13.8 Solid-state reaction for nanoceramics

Polycrystalline material is frequently synthesized using solid-state reactions. Usually, a very high temperature is required for the reaction to occur. Reaction parameters such as temperature, pressure, and the environment of the response affect the chemical and morphological properties of the reagents, including reactivity, surface area, and free energy change. In the solid-state reaction method, you also benefit from a simplified production process and high output. There is a great deal of essential information in solid-state science, which includes mineralogy, quantum physics, and even design for advanced materials. In addition to investigating solid-state reactions, a portion of it is devoted to studying homogeneous and heterogeneous reactions. It is essential to distinguish between responses in the solid state and reactions between solids. The entire array of solid reactions, from breakdown to catalytic activity, is encompassed by the first. The only requirement is that just one of the reactants be solid. Reaction products are not always solids, and solid-solid reactions only occur between solids. In all three of these reactions, no solvents are used. At room temperature, solids will not react with one other. They must be heated to a higher temperature, such as 1000°C–1500°C. Firm feasibility and solid-state reaction rate depend on the elements outlined above: the reaction circumstances, the structural features of the reactants, the surface area of the solids, and the reactivity of the solids. Conventional solid-state reactions were previously done in a conventional oven and a conventional melting process. The difference between the oven approach and the melting technique is that a high temperature was employed to promote reactions in the former. Still, in the latter, the reactants melt and then interact in liquid form, leading to a pasty paste that hardens into a solid. The response utilizes the rate of diffusion, which is temperature-dependent. Since the diffusion rate is faster at higher temperatures, it's easier for a new crystalline solid to develop [25].

13.9 Sol–gel method for nanoceramics

Sol–gel solution processing can make a wide range of advanced materials, including ceramic and organic–inorganic hybrids. The building blocks of this concept (gel) are a colloidal suspension (sol) and gelation of the sol to create a network in a continuous liquid phase are the building blocks of this concept (gel). These colloids are most often synthesized using a metal or metalloid element and several reactive ligands surrounding it. To generate a dispersible oxide, the initial material is treated into a sol. The particles are created from the liquid from the sol, and the liquid/sol separation causes them to expand and change shape. With continued drying and calcination, dense materials are produced. Producing nanomaterials using the sol–gel process is effective. Chemists widely used them for the manufacture of oxide materials. With the abundance of sol–gel derived compounds, there are countless applications. The sol–gel process can produce ultrafine and homogenous ceramic powder, the protective covering, refractory ceramic, thin films, and fibers [26].

13.10 Combustion synthesis method for nanoceramics

The most appealing method for synthesizing advanced materials is combustion synthesis, which allows for synthesizing powders and near-net form products of ceramics, functionally graded materials intermetallics and composites. The basic process to achieve combustion synthesis of ceramic oxide is to combine powdered metal with metal oxide mixtures in a specific ratio and then compact the mix. In an oxidizing atmosphere, the combustion of ceramic oxide produces ceramic oxide compounds. The appearance of metal powder and an oxidizing agent are synonyms for the proposition that the metal powder is the fuel and the oxidizing agent is oxygen. To obtain the final product, both ingredients promote combustion and induce phase transition. However, to allow for proper reaction, the process must be carried out under sufficient oxygen pressure. This operating point can cause the metal to melt, which may lead to coalescence and grain development of an as-synthesized product, which results from the combustion process caused by the heat generated by the metal at temperatures of approximately 1400°C–1600°C [27].

13.11 Volume combustion synthesis for nanoceramics

Until a reaction happens throughout the volume, the whole temperature is increased uniformly in a regulated manner. It is commonly known as the thermal explosion model, better suited for weaker exothermic reactions requiring annealing before ignition. An explosion is a nondestructive phenomenon. It only refers to the increase in temperature once a response has begun. While it may be easier to think about volume combustion synthesis (VCS) as a means of synthesis, it may be more accurate to describe it as a manner of synthesis [28].

13.12 Mechanical milling and ball milling for nanoceramics

Vibrational, attritor, universal, and planetary mills are all employed in mechanical milling. In these mills, impact or shear fracture mode operations can be made, or both operations can be done in tandem. This procedure gives the material being milled the necessary energy. Since it is usually needed to develop new surfaces, like grinding and crushing, it is typically used to fracture and create new characters. In the energy conversion process, this energy can also be used to form flaws such as dislocations and generate heat due to friction. Involving the milling process to change the physical characteristics of the material to make nanocrystalline particles by varying the parameters can complete the plastic deformation and dynamic recovery of the material to produce nanocrystalline particles. It is because the larger polycrystals have randomly oriented, high-angle grain boundaries. Milling parameters are essential in determining the fracture mode and ultimate particle size and morphology, but the qualities of the milled material are also significant. Smaller particles need lesser milling power to permit the equivalent reduction in size than larger particles, so it is important to choose an appropriate particle size initially. A simplistic interpretation of the Griffith theory of brittle fracture, as it pertains to internal faults, is that particles smaller than a critical length are more likely to have a flaw of that length, therefore being weaker. The phrase "reactive friction" or "mechanical chemistry-synthesis" was extended to activate high-energy ball milling. A range of element powder compounds, including oxides, nitrides, chalcogenides and semimetals, have been synthesized. Ball-reactive milling methods are unique in their solid-state, as reactions that ordinarily need high temperatures can be active at ambient temperature because of energy accumulating as stress defects or by decreasing the size of the kernel. These flaws cut down the diffusion path from bulk to short circuit diffusion and ultimately minimize the chemical reaction activation energy. Ball-frying can decrease the activation energy required by metallic iron to reduce cuprous oxide from 575 to 9 kJ/mol [29,30].

References

[1] R. Shilpa, R. Mudachathi, Increasing the free spectral range of Fabry-Perot cavities using dissimilar 1D photonic crystals, Mater. Lett. 299 (2021). Available from: https://doi.org/10.1016/j.matlet.2021.130089, http://www.journals.elsevier.com/materials-letters/.

[2] T.E. Balaji, H. Tanaya Das, T. Maiyalagan, Recent trends in bimetallic oxides and their composites as electrode materials for supercapacitor applications, ChemElectroChem. 8 (10) (2021) 1723–1746. Available from: https://doi.org/10.1002/celc.202100098, http://onlinelibrary.wiley.com/journal/10.1002/(ISSN)2196-0216.
[3] T.R. Roy, J. John Donald Raj, A. Sen, Inelastic tunnel transport and nanoscale junction thermoelectricity with varying electrode topology, Adv. Theory Simul. 4 (6) (2021). Available from: http://www.advtheorysimul.com.10.1002/adts.202100054.
[4] M. Armand, J.M. Tarascon, Building better batteries, Nature 451 (7179) (2008) 652–657. Available from: https://doi.org/10.1038/451652a.
[5] J.M. Tarascon, M. Armand, Issues and challenges facing rechargeable lithium batteries, Nature 414 (6861) (2001) 359–367. Available from: https://doi.org/10.1038/35104644.
[6] J.B. Goodenough, K.S. Park, The Li-ion rechargeable battery: a perspective, J. Am. Chem. Soc. 135 (4) (2013) 1167–1176. Available from: https://doi.org/10.1021/ja3091438.
[7] J.B. Goodenough, Y. Kim, Challenges for rechargeable Li batteries, Chem. Mater. 22 (3) (2010) 587–603. Available from: https://doi.org/10.1021/cm901452z.
[8] D. Larcher, J.-M. Tarascon, Towards greener and more sustainable batteries for electrical energy storage, Nat. Chem. (2015) 19–29. Available from: https://doi.org/10.1038/nchem.2085.
[9] M.A. Pellow, C.J.M. Emmott, C.J. Barnhart, S.M. Benson, Hydrogen or batteries for grid storage? A net energy analysis, Energy Environ. Sci. 8 (7) (2015) 1938–1952. Available from: https://doi.org/10.1039/c4ee04041d, http://www.rsc.org/Publishing/Journals/EE/About.asp.
[10] A. Bilich, K. Langham, R. Geyer, L. Goyal, J. Hansen, A. Krishnan, et al., Life cycle assessment of solar photovoltaic microgrid systems in off-grid communities, Environ. Sci. Technol. 51 (2) (2017) 1043–1052. Available from: https://doi.org/10.1021/acs.est.6b05455, http://pubs.acs.org/journal/esthag.
[11] M. Arbabzadeh, J.X. Johnson, G.A. Keoleian, P.G. Rasmussen, L.T. Thompson, Twelve principles for green energy storage in grid applications, Environ. Sci. Technol. 50 (2) (2016) 1046–1055. Available from: https://doi.org/10.1021/acs.est.5b03867, http://pubs.acs.org/journal/esthag.
[12] M. Jayachandran, S. Kishore Babu, T. Maiyalagan, M.R. Kannan, R. Goutham Kumar, Y. Sheeba Sherlin, et al., Effect of various aqueous electrolytes on the electrochemical performance of porous NiO nanocrystals as electrode material for supercapacitor applications, Mater. Lett. 302 (2021). Available from: https://doi.org/10.1016/j.matlet.2021.130415, http://www.journals.elsevier.com/materials-letters/.
[13] S. Rajkumar, E. Elanthamilan, T.E. Balaji, A. Sathiyan, N.E. Jafneel, J.P. Merlin, Recovery of copper oxide nanoparticles from waste SIM cards for supercapacitor electrode material, J. Alloys Compd. 849 (2020). Available from: https://doi.org/10.1016/j.jallcom.2020.156582, https://www.journals.elsevier.com/journal-of-alloys-and-compounds.
[14] S. Kishore Babu, M. Jayachandran, T. Maiyalagan, T. Vijayakumar, B. Gunasekaran, Metal-organic framework (MOF-5) incorporated on $NiCo_2O_4$ as electrode material for supercapacitor application, Mater. Lett. 302 (2021). Available from: https://doi.org/10.1016/j.matlet.2021.130338, http://www.journals.elsevier.com/materials-letters/.
[15] J. Wang, S.Q. Zhang, J. Electrochem. Soc. (D75) (2001).
[16] E.A.F.B. Frackowiak, Carbon 39 (2001).
[17] C.S. Du, J. Yeh, Nanotechnology 16 (2005).
[18] R.H. Thring, Fuel Cells for Automotive Applications, 2004.
[19] G. Hoogers, Fuel Cell Technology Handbook, 2003.
[20] J. Larminie, A. Dicks, Fuel Cell Systems Explained, 2003.
[21] B.R. Bloor, M.C. Flemings, Mahajan, The Encyclopedia of Advanced Materials, 1994.
[22] L.Z. Pei, W.Y. Yin, J.F. Wang, J. Chen, C.G. Fan, Q.F. Zhang, Low temperature synthesis of magnesium oxide and spinel powders by a sol-gel process, Mater. Res. (2010) 339–343. Available from: https://doi.org/10.1590/S1516–14392010000300010.
[23] A. Goldstein, A. Goldenberg, Y. Yeshurun, M. Hefetz, Transparent $MgAl_2O_4$ spinel from a powder prepared by flame spray pyrolysis, J. Am. Ceram. Soc. 91 (12) (2008) 4141–4144. Available from: https://doi.org/10.1111/j.1551-2916.2008.02788.x.

[24] V. Montouillout, D. Massiot, A. Douy, J.P. Coutures, Characterization of $MgAl_2O_4$ precursor powders prepared by aqueous route, J. Am. Ceram. Soc. 82 (12) (1999) 3299–3304. Available from: https://doi.org/10.1111/j.1151-2916.1999.tb02243.x.
[25] A. Buekenhoudt, F. Snijkers, Comprehensive Membrane Science and Engineering, 2010.
[26] E. Yilmaz, M. Soylak, Handbook of Nanomaterials in Analytical Chemistry, n.d.
[27] D. Biljana, S., Nikola I. Ilic, Magnetic, Ferroelectric, and Multiferroic Metal Oxides, 2018.
[28] A. Varma, A.S. Mukasyan, A.S. Rogachev, K.V. Manukyan, Solution combustion synthesis of nanoscale materials, Chem. Rev. 116 (23) (2016) 14493–14586. Available from: https://doi.org/10.1021/acs.chemrev.6b00279, http://pubs.acs.org/journal/chreay.
[29] A. Calka, A. Chowdhury, K. Konstantinov, Rapid synthesis of functional oxides by electric discharge assisted mechanical milling method, J. Alloys Compd. (n.d.).
[30] C. Andrzej, A. Chowdhury, High Di-Electric Constant Nano-Structure Ceramics Synthesis using Novel Electric Discharge Assisted Mechanical Milling and Magneto Ball Milling and its Properties, n.d.

CHAPTER 14

Application of nanoceramics in energy industries: present developments and future scopes

Priyanka Gogoi, Diganta Sarma and Kalyanjyoti Deori
Department of Chemistry, Dibrugarh University, Dibrugarh, Assam, India

14.1 Introduction

Ceramics comes from a Greek word that means "pottery" and it includes all inorganic nonmetallic materials [1]. Heating, which provides the hard and strong mechanical properties to these materials, plays a key role during its synthesis [2]. All ceramic materials can be divided broadly into two categories: traditional ceramics and advanced ceramics. Advanced ceramics, which can also be recognized as engineering ceramics, can be further divided into three groups: oxide ceramics, nonoxide ceramics, and composite ceramics. Alumina (Al_2O_3) and zirconia (ZrO_2) are typical oxides that belongs to oxide ceramics, whereas nonoxide ceramics include borides, carbides, silicides, etc. Ceramics are solids that can withstand high temperature and exhibit excellent hardness. High melting points, low thermal and electrical conductivity, high elastic modulus, strength, and brittleness are prevalent in ceramics [3]. However, conventional bulk ceramic materials are brittle in nature, which limits their application at elevated temperature [4]. Composite ceramics were developed in order to resolve the well-documented problems by reinforcing different phases to one another to introduce mechanisms within, such as crack deflection, load transfer, and crack blunting [5]. Most ceramic materials can withstand high temperature, although brittleness is the most challenging technical barrier which limits their practical utility. Despite the disadvantages, ceramic materials play an increasingly important role in the production of various industrially important compounds and advanced materials with excellent properties and unique characteristics.

14.2 Advantages of nanoceramics

Nanomaterial is commonly defined as matter approximately 1−100 nm in size at least along one dimension. Nanoparticles may occur in different shapes, for example, rods, fibers, wires, tubes, cubes, spheres, or in sheets like graphene. Nanoceramics is an

Industrial Applications of Nanoceramics
DOI: https://doi.org/10.1016/B978-0-323-88654-3.00019-6

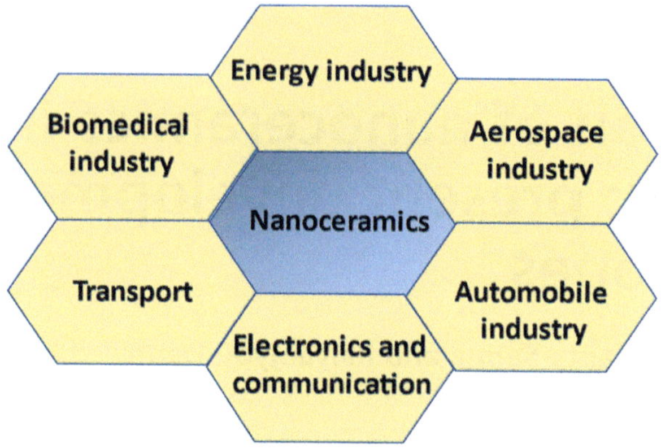

Figure 14.1 Application of nanoceramics in various fields.

important class of products of nanotechnology first discovered in the early 1980s. Their structure is different from regular ceramics. It is a subcategory of nanocomposites, composed of a metal and nonmetallic elements but considered to be nonmetallic [6]. Nanoceramics consist of a homogeneous structure of repeating elements, unlike nanocomposites which are composed of a heterogeneous mixture. Due to the smaller size of the constituent particles, nanoceramics have unique mechanical and surface characteristics compared to bulk ceramics. Superplasticity, thermal strength, toughness, machinability, and bioactivity are among the properties that make them potential candidates for many emerging technologies. Hence, the field of application of nanoceramics is larger than that for bulk ceramics materials. Also, they possess some distinctive properties which make them very useful for coatings. On many materials, nanoceramics been applied as protective layers. These are regarded as hard coatings and durable at elevated temperature [7]. Health and biomedical, electronics and communications, energy industries, space technologies, automotive industries, transport, construction industries are the various fields that have extensive applications of nanoceramics. Among its countless applications in various fields, this chapter mainly focuses on its vital applications in energy industries. Fig. 14.1 shows a schematic of nanoceramics used in various fields.

14.3 Applications of nanoceramics in energy industries

The use of energy has been considered to be the building block for the growth of human society. In this era of modern society, energy sectors have become one of the utmost essential needs not only for transportation and communication but also in agriculture, power generation, and waste collection. Demand for energy is increasing worldwide continuously and according to the forecast of the International Energy

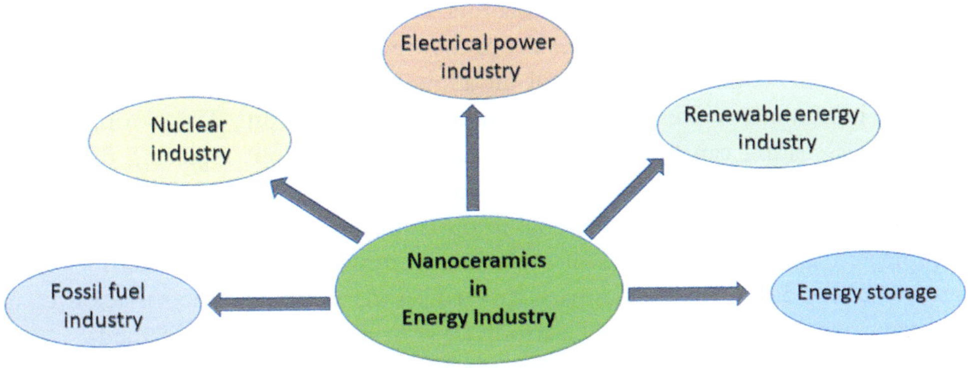

Figure 14.2 Application of nanoceramics in various energy industries.

Agency, the growth will be up to 18,000 MTOE (million-ton oil equivalent) by 2030. Energy industries basically comprise electrical power industries, nuclear power industries, renewable energy industries, and most importantly fossil fuel industries, which include petroleum industries, refineries, coal industries, etc. [8]. Nanoceramics have become an integral part of most of the energy industries. From its discovery until today, it has gained much attention from scientists and researchers. In consequence, the market size of nanoceramics is increasing expeditiously owing to the growing demand from the energy industries. Developing the concept of sustainable and renewable energy and their increasing need will increase its market size in coming years [9]. Highly durable, wear-resistant, and user-friendly nanoceramics have been employed in various energy industries (Fig. 14.2).

14.3.1 Nanoceramics in fossil fuel industry

The demand for crude oils to satisfy energy needs is increasing day by day. Yet, much of today's technology is inclined toward generating renewable and sustainable energy harvesting pathways as pollution has been increasing tremendously in every part of the world. A number of nanotechnology-based materials have been introduced in drilling, enhanced oil recovery, completion, and refineries to meet the immediate demands [10]. Nanocomposites [11,12], nanosensors [13,14], and coating nanoparticles [15,16] are amongst them. Nanoceramics is one of the promising materials for this purpose. Nanoceramics have profound applications in petrochemical industries. Nanoceramics have been widely used as nanocoatings to modify the surface of the materials in order to get enhanced wear and corrosion-resistant properties in the oil and gas industries [10]. The plasma spray method is frequently applied to obtain a ceramic layer coating on a surface, which is very simple in concept, although difficult in practice. One of the very good examples of these is Al_2O_3—13 TiO_2 nanoceramic coating, which exhibits very high wear resistance, and is being successfully utilized as drill bits for ultradeepwater drilling [17]. In the fossil fuel industries, the drill bit operates under extremely harsh conditions, which results in the frequent replacement of bit. Applying this nanoceramic coating over the drill bit allows it to last longer in extreme conditions and it minimizes the operational and maintenance cost. This strong layer of nanoceramics coating can be repaired after application without replacing the bit. For fabrication of this nanoceramic material as coating, the plasma spray method is being applied by injecting the coating material into plasma stream. It is heated in the stream and projected towards the surface of substrate, as shown in Fig. 14.3. With a velocity range of 1200–1500 m/s, the plasma is sprayed on the substrate. The surface is cooled down after being coated and finally a strong layer of nanocoating is obtained.

Compared to a conventional coating, the hardness of Al_2O_3–13 TiO_2 is remarkable which can be tested using a "cup-test." The conventional coating deforms and reveals typical cracking while a coated coupon is forced onto a 1-inch diameter steel ball. Deformation also takes place in nanostructured one but no macroscopic fracture is seen. Another benefit obtained from nanostructured ceramics coating is that it is less expensive to apply. Grinding and polishing consists of around 40% of the total cost of coating. Nanoceramics coating takes around half the time that is required for common ceramics coating [18]. In addition, greater longevity and reliability are other advantages of the material. Conventional ceramic coating lacks adequate bond strength, and

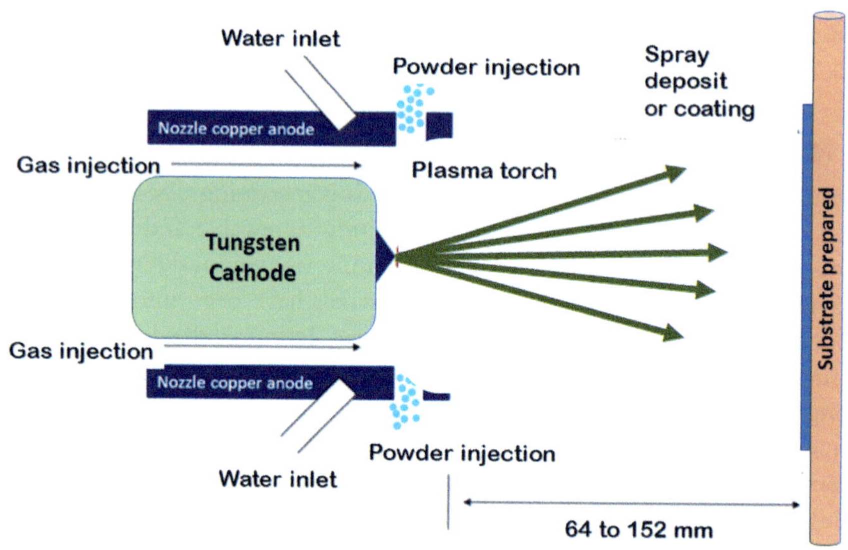

Figure 14.3 Nanocoating by aluminum oxide (Al_2O_3) and titanium oxide (TiO_2). Source: *From Q. Jin, L. Zhao, B. Cui, J. Wang, H. Ma, R. Zhang, Y. Liu, X. Zhang, Enhanced energy storage properties in lead-free $BaTiO_3$@Na0.5K0.5NbO_3 nano-ceramics with nanodomains: via a core-shell structural design, J. Mater. Chem. C 8(15) (2020) 5248–5258. https://doi.org/10.1039/d0tc00179a.*

hence is not suitable for application over metallic surfaces. This limitation is overcome by nanostructured ceramics materials due to the considerable toughness. Al_2O_3−13 TiO_2 can be applied over a titanium surface to reduce the corrosion that occurs near submarine steel structures. Another important part of the oil and gas industries is the transmission pipes through which crude oil and gas can be transferred from one place to another. Nanoceramics coating inhibits the corrosion that takes place inside the pipes by protecting the surface from air and moisture. Acrylic resin water-based nanocomposites along with a nanometer tunnel are applied to the inner surface of pipes and it shows excellent corrosion resistance properties, hence reduces the maintenance cost [19]. Other nanoceramics materials, such as tungsten carbide-cobalt, chrome oxide, and zirconia, are under development and industrialists are hopeful to find a wide range of applications from these in near future.

14.3.2 Nanoceramics in nuclear power industry

Nuclear power industries are one of the strongest clean energy sources that is widely expanded to generate more than half of the global electricity than any other source. The major challenges that nuclear power systems face are related to the corrosive environment, high temperature, and radiation damage to the material during operation [20]. In order to increase power conversion efficiencies, a high operating temperature is necessary, for which coolant is used. But the problem related to most of the coolants is that they are extremely corrosive and the corrosive nature is enhanced by high temperature and radiation [21]. Ceramics, in this regard, stand as favorable materials that can withstand high temperature and also are inert to most of the corrosive environments. Ceramics are extensively employed as nuclear fuel and waste disposal materials [22]. Oxide ceramics of uranium and plutonium have acted as genuine fuel pellets for water-cooled as well as liquid metal-cooled reactors. Oxides of these two radioactive elements have some very unique characteristics that make them useful to withstand the extreme high temperature in nuclear reactors. The high melting point and highly variable oxygen-to-metal ratio are among the characteristics that provide outstanding resistance to radiation damage. Besides one drawback of these oxide ceramics, which is low thermal conductivity, other advantages include low fabrication cost, no swelling during burnup, and inertness to applied coolant materials [23]. The inherent brittleness of these ceramics materials somehow prevents their widespread utilization. For toughness, radiation tolerance, and chemical compatibility, nanoceramics and their applications in nuclear power plants are worth highlighting. The development of nanostructured ceramics for next-generation nuclear power plant is now an ongoing field of research. Oxide nanoceramics are a promising material due to their environment-friendly nature, especially with light water, liquid helium, and liquid metals [24]. Al_2O_3 nanoceramic thin film has been synthesized for protecting steels in heavy liquid metal (HLM) at very high temperature [25]. HLMs, for example, lead-bismuth eutectic, are used as coolant for next-generation nuclear reactors. But corrosion by HLM of the steel structures of the reactors is the major drawback of the system. The Al_2O_3 nanoceramic coating grown by pulsed laser deposition (PLD) successfully protects steel in the HLM environment even at high temperatures. As-fabricated coated specimen of steel has been examined in the HLM environment at 550°C for 500 hours [25].

Fig. 14.4A indicates the coated face of a sample before exposure and Fig. 14.4B and C indicate the uncoated and coated face, respectively, after exposure of the sample. The uncoated face is visibly colored and corroded but the coated face was safely protected from corrosion after exposure. Ferré and group further studied the mechanical response of the as-synthesized Al_2O_3 nanoceramics and found a homogeneous dispersion of it in an amorphous matrix initially [26]. As irradiation takes place, crystallization is induced progressively followed by grain growth. Crystallization induces hardness while grain growth introduces softening to the material.

Besides all the advantages, many challenges are also associated with these industries. Fuel storage and disposal is a big question unless proper care is taken, since these wastes contain radioactive isotopes with prolonged half-lives. In spite of relevant measures, some unfortunate accident still take place. In 2011, the Fukushima Daiichi nuclear power plant accident was caused by a 9.0 magnitude earthquake. It occurred due to a lack of protection in primary fuel cladding of the reactor. In order to prevent these types of unfortunate accidents in light-water nuclear reactors, a EU-funded project is working on developing a new ceramic material. The primary aim of the project is to develop stronger oxide thin films and a new accident-tolerant fuel for modern nuclear light-water reactors. Scientists are putting efforts to create more efficient nanoceramic materials which will be useful for the next-generation nuclear reactors.

Figure 14.4 Fabricated coated specimen: (A) coated face before exposure, (B) uncoated face, and (C) coated face after exposure. The uncoated face is corroded while the coated face is protected from exposure [27]. Source: *From García Ferré, Advanced Al_2O_3 coatings for high temperature operation of steels in heavy liquid metals: a preliminary study, Corros. Sci. 77, 375–378. https://doi.org/10.1016/j.corsci.2013.07.039.*

14.3.3 Nanoceramics in electrical power industry

Since1882, from the day the distribution of electrical power started, it has served as one of the most important needs to human society, covering the production, transmission, and distribution of electric power [28]. The application of nanotechnologies in various electronics, sensors, and photovoltaics is considered to be a key factor to achieve rapid economic growth. Nanoceramics, amongst many nanomaterials, stand out as very promising candidates for the electronics and electrical power industries. The special features of nanoceramics in electrical power industries include mechanical strength at high temperature, high corrosion resistance, and thermal shock resistance. In recent years, nanoceramics have become a promising alternative that can replace piezoelectric materials, which convert an electric stimulus into mechanical energy [29,30]. These materials have the ability to provide very high electrical conductivity that has been achieved never before by conventional ceramic materials. The electrical properties of these nanoceramics have been investigated with the help of FTIR spectroscopy. David et al. have studied the electrical properties of polycrystalline ZnO thin-film nanoceramics by FTIR measurement and observed the effect of thermal heating on electrical conductivity [31]. In another work, to convert insulating nanoceramics to conductive composites, single-walled carbon nanotubes (SWCNTs) have been employed and they show a significant increase in electrical conductivity through the nanoceramic material [32]. The coating composite improves the fracture toughness of the material and widens the potential use of these materials.

14.3.4 Nanoceramics in renewable energy industry

The 21st century requires carrying out of almost all daily needed industrial processes through green and sustainable routes in order to meet the environmental norms. The extensive usage of fossil fuels has been identified as the primary cause of environmental pollutions. Renewable energy resources, also regarded as clean sources of energy, play a key role in achieving future energy needs [33]. Solar energy, wind energy, hydrothermal, biomass, geothermal, and tidal energy belong to renewable energy sources. Nanoceramics have gained considerable attention for their broad range of applications in renewable energy industries. Solar energy (photovoltaics), as a clean source for power generation, is increasing the demand for nanoceramics powder as it is widely used to produce photovoltaic cells. Bismuth ferrite $BiFeO_3$ (BFO) nanoceramics, in this context, were first successfully synthesized in a pure form by Venkataraman's group using a low-temperature sol—gel synthesis route [34]. Analytical data suggest a rhombohedrally distorted single-phase perovskite structure of BFO nanoceramics along with a small amount of $Bi_2Fe_4O_9$ as a secondary phase which could be advantageous in increasing the optical absorption of the material. The presence of $Bi_2Fe_4O_9$ makes BFO more efficient for application in photovoltaic devices. However, there are a few limitations of BFO nanoceramics that prevent it from wide application; hence Bi-site substituted BFO nanoceramics materials have been introduced [35–38]. In this regard, single-phase $Bi_{1-x}La_xFeO_3$ nanorods were prepared by Delfard et al. at 200°C which showed enhanced optoelectronic properties [39]. Along with this developing field, glass-ceramics, that is, glass containing nanometer to micron sized crystals, is another rapidly developing field in the area of modern optics and photonics [40]. The idea of applying these glass-ceramic materials in thermophotovoltaics is not new [41]. However, the field is still under development with the search for new nanoceramic materials that might possess some distinct benefits over glass-ceramics [42]. For solar cell fabrication and processing, thin-film nanoceramics are a very promising material [43]. Thin films of manganese dioxide (MnO_2), cadmium sulfide (CdS), and tantalum carbide (TaC) nanoceramics were synthesized which have high absorption in the UV–vis range; hence they can be used in optoelectronic devices [44]. Results show that with increasing concentration of TaC nanoparticles, the optical conductivity of MnO_2-CdS nanocomposite increases.

Another renewable energy source is the conversion of wind power into a useful form of energy and it has attracted considerable attention in recent times [45]. Wind blades are the most important component of wind turbines that should satisfy some specifications [46]. Low weight, high strength, high stability, high stiffness, and wide availability are amongst the requirements [47]. Natural materials such as wood, canvas, and, later, aluminum and its alloys were utilized as blade material, though these require frequent maintenance, which decreases the efficiency to capture wind energy. Reducing the weight of the blades is one of the major ways to improve efficiency. In order to enhance the working efficiency, ceramic nanomaterials-based lightweight components are being used [48]. Nanoceramic composite materials, such as Al_2O_3-SiC and Si_3N_4-SiC, are notable examples that provide considerable hardness and strength to the blades that are of relevance for application in wind energy industries [49]. A significant drawback with bulk ceramic materials is their inability to enhance fracture toughness and low durability, which has been amended in many technologically important nanoceramics/nanocomposites. Future materials development related to wind energy must emphasize the development of lightweight composites with high fracture toughness and high elastic modulus.

With the rapid growth of industrialization, the presence of carcinogenic heavy metals in wastewater leads to water pollution [50]. Nanoceramics contribute characteristic properties toward wastewater management. For the removal of lead particles, hydroxyapatite nanoceramic material (HAp) ($Ca_{10}(PO_4)_6(OH)_2$) has been successfully employed [51]. The abovementioned material was synthesized by a wet chemical method and further characterized with the help of XRD, TGA, UV–vis, and AFM. After application to Pb-containing water it showed excellent results with regard to the removal of heavy metals [52]. Hence, hydroxyapatite nanoceramic material is an inexpensive technique to capture toxic metal impurities that damage health.

14.3.5 Application of nanoceramics in energy storage

People have found different ways to store energy since the discovery of electricity. Over the past decade, the fabrication of efficient and long-lasting modern energy storage devices has been a growing competition amongst the researchers. Batteries show effective performance in storing energy, although supercapacitors are considered as the most prominent device for energy storage due to their fast charge–discharge capacity [53]. Bulk ceramics dielectric capacitors are used in many energy storage devices as they typically hold a large dielectric permittivity, however, due to their poor flexibility and high specific mass, they show low charge storage performance, and hence are unable to meet the requirements [54]. Nanoceramics, with enhanced electrochemical properties, can fulfill most of the requisite parameters to overcome the shortcomings of bulk ceramics [55]. At present, electrochemical supercapacitors are not capable of being recognized as high-energy storage dielectric material due to their high self-discharge. Hence, the development of a new hybrid material was carried out; nanostructured ABO_3 perovskite ceramic composites achieve high dielectric permittivity [56]. The lead-free perovskite nanostructured ceramics, $BaTiO_3$, is an example that is mainly utilized as a dielectric material for multilayered ceramic capacitors. Apart from 1D perovskite nanocrystals, metal oxide nanoceramics, such as $NiCo_2O_4$, Fe_3O_3, and $CoMoO_4$, were fabricated to make flexible supercapacitor devices that show efficient charge storage capacity. Interestingly, these nanoceramic fabricated supercapacitors show an increase in specific capacitance when operational temperature rises from 10°C to 70°C. Furthermore, efforts have been made to synthesize toxic/heavy metal-free nanoceramic materials using greener routes. Recently, lead-free novel environment-friendly BaTiO3@Na0.5K0.5NbO3(BT@KNN) nanoceramics have been designed with a core–shell structure, as shown in Fig. 14.5 [57]. BT particles with approximate

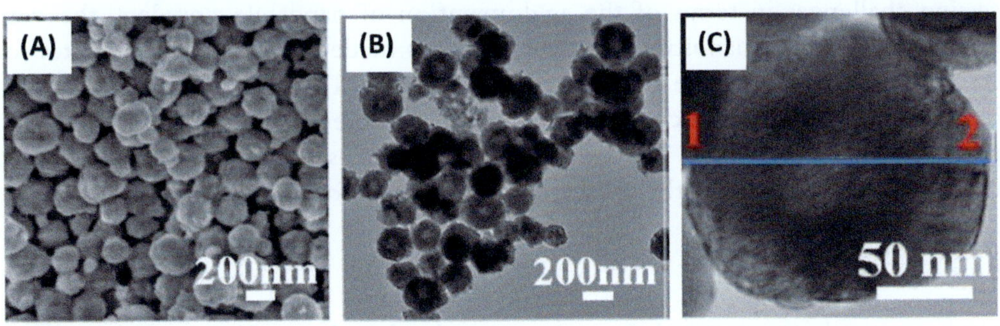

Figure 14.5 (A) SEM, (B) TEM image of BT@KNN particles, and (C) TEM of a single BT@KNN particle. *SEM*, scanning electron microscope; *TEM*, transmission electron microscope. Source: *From Q. Jin, L. Zhao, B. Cui, J. Wang, H. Ma, R. Zhang, Y. Liu, X. Zhang, Enhanced energy storage properties in lead-free BaTiO3@Na0.5K0.5NbO3 nano-ceramics with nanodomains: via a core-shell structural design, J. Mater. Chem. C 8(15) (2020) 5248–5258. https://doi.org/10.1039/d0tc00179a.*

size of 200 nm with a rhombohedral and tetragonal nanodomain have been prepared using a coprecipitation method [58]. As-synthesized nanoceramics show excellent energy storage performance due to their enhanced dielectric characteristics and temperature stability.

Some other toxic element-free hybrid nanoceramic materials have also been synthesized similar to BT@KNN, such as $Na_{0.5}Bi_{0.5}TiO_3$, $K_{0.5}Na_{0.5}NbO_3$, which is widely used in multilayered ceramic capacitors. This research will encourage further development of more efficient lead-free energy storage ceramics in the near future.

14.4 Future scope

Owing to the growing demand as well as improvements in the quality of nanoceramic materials their market growth has increased, especially in the energy industries. Nanoceramic coating is a primary component not only within energy and aerospace industries, but also in the biomedical sector [59]. According to reports, within the next year, 2024, nanoceramic coatings are expected to grow over $2 billion, which gives a positive indication of rapid growth of the industry. Researchers and industry are finding modern nanoceramic materials that can be used in other renewable energy industries, like geothermal, tidal, and biomass industries, so as to provide more suitable and user-friendly applications in near future. A new type of nanostructured ceramics that is expected to be observed soon is nanoglass ceramic technology, which will be a promising material for energy industries [60]. Nanoceramics are expected to play a key role in next-generation electric cars, flexible lighting systems, lightweight batteries, and supercapacitors. Although nanoceramics and nanoceramic composites possess some processing challenges in terms of the agglomeration of nanoscale powders, the very high surface area of nanostructured ceramics has made them useful candidates for catalytic application in the fuel conversion process. The development of energy industries is the key to success for a country. Nanotechnology development is a positive sign for mankind. The fundamental as well as applied research has to be further improved to make the synthesized material fit for commercial application.

14.5 Conclusion

Nanoceramics have integrated themselves into the daily lives of human beings. Their wide range of applications in automobile, aerospace, biomedical, and energy industries have made the lives of people easier and smarter. The chapter in this book covers most of the applications of nanoceramics in the energy industries. High upfront cost and low awareness about the material benefits and utilization amongst industrialists may provide a negative impact on the growing field of nanoceramics. The present ongoing research focuses on improving the processing and exploring more industrial applications. Today's world demands most of the industrial processes to be sustainable and in this regard nanoceramic materials have been recognized and widely applied.

References

[1] A.N. Samant, N.B. Dahotre, Laser machining of structural ceramics—a review, J. Eur. Ceram. Soc. 29 (6) (2009) 969–993. Available from: https://doi.org/10.1016/j.jeurceramsoc.2008.11.010.

[2] H.T. Ting, K.A. Abou-El-Hossein, H.B. Chua, Review of micromachining of ceramics by etching, Trans. Nonferrous Met. Soc. China (Engl. Ed.) 19 (1) (2009) s1–s16. Available from: https://doi.org/10.1016/S1003-6326(10)60237-3.

[3] L. Ma, K.E. Hendrickson, S. Wei, L.A. Archer, Nanomaterials: science and applications in the lithium-sulfur battery, Nano Today 10 (3) (2015) 315–338. Available from: https://doi.org/10.1016/j.nantod.2015.04.011, http://www.elsevier.com/wps/find/journaldescription.cws_home/706735/description#description.

[4] I.P. Tuersley, A. Jawaid, I.R. Pashby, Review: Various methods of machining advanced ceramic materials, J. Mater. Process. Tech. 42 (4) (1994) 377–390. Available from: https://doi.org/10.1016/0924-0136(94)90144-9.

[5] A.G. Evans, Engineering property requirements for high performance ceramics, Mater. Sci. Eng. 71 (C) (1985) 3–21. Available from: https://doi.org/10.1016/0025-5416(85)90202-2.

[6] A. Karatutlu, A. Barhoum, A. Sapelkin, Liquid-phase synthesis of nanoparticles and nanostructured materials, Emerging Applications of Nanoparticles and Architectural Nanostructures: Current Prospects and Future Trends, Elsevier Inc, Turkey, 2018, pp. 1–28. Available from: https://doi.org/10.1016/B978-0-323-51254-1.00001-4, http://www.sciencedirect.com/science/book/9780323512541.

[7] P. Fauchais, A. Vardelle, M. Vardelle, Thermally sprayed nanoceramic and nanocomposite coatings, Handbook of Nanoceramic and Nanocomposite Coatings and Materials, Elsevier Inc, France, 2015, pp. 225–256. Available from: https://doi.org/10.1016/B978-0-12-799947-0.00010-9, http://www.sciencedirect.com/science/book/9780127999470.

[8] R. Fouquet, P.J.G. Pearson, Seven centuries of energy services: the price and use of light in the United Kingdom (1300–2000), Energy J. 27 (1) (2006) 138–178.

[9] X.H. Wang, I.W. Chen, X.Y. Deng, Y.D. Wang, L.T. Li, New progress in development of ferroelectric and piezoelectric nanoceramics, J. Adv. Ceram. 4 (1) (2015). Available from: https://doi.org/10.1007/s40145-015-0132-6, http://www.springer.com/materials/special + types/journal/40145.

[10] Z. Zhe, A. Yuxiu, Nanotechnology for the oil and gas industry—an overview of recent progress, Nanotechnol. Rev. 7 (4) (2018) 341–353. Available from: https://doi.org/10.1515/ntrev-2018-0061, http://www.degruyter.com/view/j/ntrev?rskey = tED3So&result = 1.

[11] S. Kango, S. Kalia, A. Celli, J. Njuguna, Y. Habibi, R. Kumar, Surface modification of inorganic nanoparticles for development of organic-inorganic nanocomposites–a review, Prog. Polym. Sci. 38 (8) (2013) 1232–1261. Available from: https://doi.org/10.1016/j.progpolymsci.2013.02.003.

[12] S.M. Sajadi, K. Kolo, M. Pirouei, S.A. Mahmud, J.A. Ali, S.M. Hamad, Natural iron ore as a novel substrate for the biosynthesis of bioactive-stable ZnO@CuO@iron ore NCs: a magnetically recyclable and reusable superior nanocatalyst for the degradation of organic dyes, reduction of Cr(vi) and adsorption of crude oil aromatic compounds, including PAHs, RSC Adv 8 (62) (2018) 35557–35570. Available from: https://doi.org/10.1039/c8ra06028b, http://pubs.rsc.org/en/journals/journal/ra.

[13] J.H. Kim, D.A. Heller, H. Jin, P.W. Barone, C. Song, J. Zhang, et al., The rational design of nitric oxide selectivity in single-walled carbon nanotube near-infrared fluorescence sensors for biological detection, Nat. Chem. 1 (6) (2009) 473–481.

[14] M. Khalil, B.M. Jan, C.W. Tong, M.A. Berawi, Advanced nanomaterials in oil and gas industry: design, application and challenges, Appl. Energy 191 (2017) 287–310. Available from: https://doi.org/10.1016/j.apenergy.2017.01.074, http://www.elsevier.com/inca/publications/store/4/0/5/8/9/1/index.htt.
[15] C. Glaus, R. Rossin, M.J. Welch, G. Bao, In vivo evaluation of 64 Cu-labeled magnetic nanoparticles as a dual-modality PET/MR imaging agent, Bioconjugate Chem. 21 (4) (2010) 715–722.
[16] M. Ates, S. Caliskan, E. Özten, Supercapacitor study of reduced graphene oxide/Zn nanoparticle/polycarbazole electrode active materials and equivalent circuit models, J. Solid. State Electrochem. 22 (10) (2018) 3261–3271. Available from: https://doi.org/10.1007/s10008-018-4039-3, http://link.springer-ny.com/link/service/journals/10008/index.htm.
[17] S. Sengupta, A. Kumar, Nano-ceramic coating—a means of enhancing bit life and reducing drill string trips. Int. Petrol. Technol. Conf. 2013 1–9. Available from: https://doi.org/10.2523/IPTC-16474-MS.
[18] L.L. Shaw, D. Goberman, R. Ren, M. Gell, S. Jiang, Y. Wang, et al., Elsevier Sequoia SA United States The dependency of microstructure and properties of nanostructured coatings on plasma spray conditions, Surf. Coat. Technol. 130 (1) (2000) 1–8. Available from: https://doi.org/10.1016/S0257-8972(00)00673-3.
[19] E. Noveiri, S. Torfi, Nano coating application for corrosion reduction in oil and gas transmission pipe: a case study in south of Iran, in: International Conference on Advanced Materials Engineering IPCSIT 15, 2012, pp. 104–110.
[20] S. Chu, A. Majumdar, Opportunities and challenges for a sustainable energy future, Nature. 488 (7411) (2012) 294–303. Available from: https://doi.org/10.1038/nature11475.
[21] S.J. Zinkle, G.S. Was, Materials challenges in nuclear energy, Acta Mater. 61 (3) (2013) 735–758. Available from: https://doi.org/10.1016/j.actamat.2012.11.004.
[22] H. Tsai, Carbide and nitride nuclear fuels, Encycl. Mater. Sci. Eng. (1986) 493–495.
[23] S.J. Zinkle, J.T. Busby, Structural materials for fission & fusion energy, Mater. Today 12 (11) (2009) 12–19. Available from: https://doi.org/10.1016/S1369-7021(09)70294-9.
[24] S.J. Zinkle, K.A. Terrani, J.C. Gehin, L.J. Ott, L.L. Snead, Accident tolerant fuels for LWRs: a perspective, J. Nucl. Mater. 448 (1–3) (2014) 374–379. Available from: https://doi.org/10.1016/j.jnucmat.2013.12.005.
[25] F. García Ferré, M. Ormellese, F. Di Fonzo, M.G. Beghi, Advanced Al_2O_3 coatings for high temperature operation of steels in heavy liquid metals: a preliminary study, Corros. Sci. 77 (2013) 375–378. Available from: https://doi.org/10.1016/j.corsci.2013.07.039.
[26] F.G. Ferre, A. Mairov, L. Ceseracciu, Y. Serruys, P. Trocellier, C. Baumier, et al., Radiation endurance in Al_2O_3 nanoceramics, Sci. Rep. 6 (2016) 1–9.
[27] F. Ferre, M. Ormellese, F. Di Fonzo, M.G. Beghi, Advanced Al_2O_3 coatings for high temperature operation of steels in heavy liquid metals: a preliminary study, Corros. Sci. 77 (2013) 375–378.
[28] R.C. Armstrong, C. Wolfram, K.P. De Jong, R. Gross, N.S. Lewis, B. Boardman, et al., The frontiers of energy, Nat. Energy 1 (1) (2016). http://www.nature.com/nenergy/0.10.1038/nenergy.
[29] R.N. Torah, S.P. Beeby, N.M. White, Improving the piezoelectric properties of thick-film PZT: the influence of paste composition, powder milling process and electrode material, Sens. Actuators, A: Phys. 110 (1–3) (2004) 378–384. Available from: https://doi.org/10.1016/j.sna.2003.09.027. United Kingdom.
[30] Y. Saito, H. Takao, T. Tani, T. Nonoyama, K. Takatori, T. Homma, et al., Lead-free piezoceramics, Nature. 432 (7013) (2004) 84–87. Available from: https://doi.org/10.1038/nature03028.
[31] C. David, F. Paumier, B.P. Tinkham, D. Eyidi, M. Marteau, P. Guerin, Modification of electrical properties induced by annealing of ZnO: B thin films deposited by chemical vapour deposition: kinetic investigation of evolution author links opens overlay panel, Mater. Sci. Eng. B 178 (2013) 698–702.
[32] G.D. Zhan, J.D. Kuntz, J.E. Garay, A.K. Mukherjee, Electrical properties of nanoceramics reinforced with ropes of single-walled carbon nanotubes, Appl. Phys. Lett. 83 (6) (2003) 1228–1230. Available from: https://doi.org/10.1063/1.1600511.
[33] E. Omar, A.R. Haitham, B. Frede, Renewable energy resources: current status, future prospects and their enabling technology, Renew. Sustain. Energy Rev. (2014) 748–764. Available from: https://doi.org/10.1016/j.rser.2014.07.113.

[34] W.A. Wani, K. Ramaswamy, S. Kundu, B. Harihara Venkataraman, Influence of thermal treatment on the physical properties of bismuth ferrite nanoceramics for promising multifarious device applications, AIP Conf. Proc. (2020). Available from: http://scitation.aip.org/content/aip/proceeding/aipcp2269.

[35] F. Yan, M.O. Lai, L. Lu, T.J. Zhu, Variation of leakage mechanism and potential barrier in la and Ru co-doped $BiFeO_3$ thin films, J. Phys. D: Appl. Phys. 44 (43) (2011). Available from: https://doi.org/10.1088/0022-3727/44/43/435302, http://iopscience.iop.org/0022-3727/44/43/435302/pdf/0022-3727_44_43_435302.pdf. Singapore.

[36] S. Rajmohan, V. Jeseentharani, A. Manikandan, J. Pragasam, Co-precipitation synthesis method, characterizations and humidity sensing applications of perovskite-type mixed oxide La1-xCoxVO$_3$-δ nanocomposites, Nanosci. Nanotechnol. Lett 8 (5) (2016) 393–398. Available from: https://doi.org/10.1166/nnl.2016.2148, http://docserver.ingentaconnect.com/deliver/connect/asp/19414900/v8n5/s5.pdf?expires = 1474658914&id = 88698305&titleid = 72010039&accname = Elsevier + BV&checksum = C6349EF7CC71ED5284E596C0155108B0.

[37] A.T. Ravichandran, J. Srinivas, R. Karthick, A. Manikandan, A. Baykal, Facile combustion synthesis, structural, morphological, optical and antibacterial studies of $Bi1-xAlxFeO_3$ $(0.0 \leq x \leq 0.15)$ nanoparticles, Ceram. Int. 44 (11) (2018) 13247–13252. Available from: https://doi.org/10.1016/j.ceramint.2018.04.153.

[38] M.H. Basiri, H. Shokrollahi, G. Isapour, Effects of la content on the magnetic, electric and structural properties of $BiFeO_3$, J. Magnet. Mag. Mater. 354 (2014) 184–189. Available from: https://doi.org/10.1016/j.jmmm.2013.10.020.

[39] N.B. Delfard, H. Maleki, A.M. Badizi, M. Taraz, Enhanced structural, optical, and multiferroic properties of rod-like bismuth iron oxide nanoceramics by dopant lanthanum, J. Supercond. Nov. Magnet. 33 (4) (2020) 1207–1214. Available from: https://doi.org/10.1007/s10948-019-05294-3, http://www.springer.com/east/home/generic/search/results?SGWID = 5-40109-70-35500892-0.

[40] T. Berthier, V.M. Fokin, E.D. Zanotto, New large grain, highly crystalline, transparent glass-ceramics, J. Non-Crystalline Solids. 354 (15–16) (2008) 1721–1730. Available from: https://doi.org/10.1016/j.jnoncrysol.2007.08.052.

[41] R.E. Nelson, A brief history of thermophotovoltaic development, Semicond. Sci. Technol. 18 (2003) 141–143.

[42] V.M. Orera, R.I. Merino, Ceramics with photonic and optical applications, Boletin de. la. Soc. Espanola de Ceramica y. Vidrio 54 (1) (2015) 1–10. Available from: https://doi.org/10.1016/j.bsecv.2015.02.002, http://ac.els-cdn.com/S0366317515000035/1-s2.0-S0366317515000035-main.pdf?_tid = 11e3f0a6-5d-11-11e5-9419-00000aab0f01&acdnat = 1442476547_86e4ee2c6ae3515adae922051ec861e0.

[43] D. Patidar, R. Sharma, N. Jain, T.P. Sharma, N.S. Saxena, Optical properties of CdS sintered film, Bull. Mater. Sci. 29 (1) (2006) 21–24. Available from: https://doi.org/10.1007/BF02709350, http://www.ias.ac.in/matersci/bmsfeb2006/21.pdf. India.

[44] M.C.N. Martínez, M.P. Kobyla, P. Mazierski, J. Wółkiewicz, G. Trykowski, A. Malankowska, et al., Self-organized TiO_2-MnO_2 nanotube arrays for efficient photocatalytic degradation of toluene, Molecules. 22 (564) (2017) 1–14.

[45] O.T. Thomsen, Sandwich materials, Renew. Sust. Energy Rev. 53 (2009) 209–224.

[46] F. Findik, K. Turan, Materials selection for lighter wagon design with a weighted property index method, Mater. Des. 37 (2012) 470–477. Available from: https://doi.org/10.1016/j.matdes.2012.01.016.

[47] P. Brøndsted, H. Lilholt, A. Lystrup, Composite materials for wind power turbine blades, Annu. Rev. Mater. Res. 35 (2005) 505–538. Available from: https://doi.org/10.1146/annurev.matsci.35.100303.110641.

[48] A. Mukhopadhyay, B. Basu, Consolidation-microstructure-property relationships in bulk nanoceramics and ceramic nanocomposites: a review, Int. Mater. Rev. 52 (5) (2007) 257–288. Available from: https://doi.org/10.1179/174328007X160281.

[49] K. Niihara, New design concept of structural ceramics, Ceram. nanocomposites. Nippon. Seramikkusu Kyokai Gakujutsu Ronbunshi/Journal Ceram. Soc. Jpn. 99 (1154) (1991) 974–982. Available from: https://doi.org/10.2109/jcersj.99.974.

[50] U. Wingenfelder, C. Hansen, G. Furrer, R. Schulin, Removal of heavy metals from mine waters by natural zeolites, Environ. Sci. Technol. 39 (12) (2005) 4606–4613. Available from: https://doi.org/10.1021/es048482s.
[51] V.N. Narwade, M.P. Mahabole, K.A. Bogle, R.S. Khairnar, Waste water treatment by nanoceramics: removal of lead particles, Int. J. Eng. Sci. Innov. Technol. (IJESIT) 3 (3) (2014) 324–329.
[52] M.P. Mahabole, R.C. Aiyer, C.V. Ramakrishna, B. Sreedhar, R.S. Khairnar, Synthesis, characterization and gas sensing property of hydroxyapatite ceramic, Bull. Mater. Sci. 28 (6) (2005) 535–545. Available from: https://doi.org/10.1007/BF02706339, http://www.ias.ac.in/matersci/bmsoct2005/535.pdf. India.
[53] R. Gupta, Nano-ceramics for energy storage applications, J. Mater. Sci. Eng. 4 (4) (2015) 38–49.
[54] Z. Yang, H. Du, S. Qu, Y. Hou, H. Ma, J. Wang, et al., Significantly enhanced recoverable energy storage density in potassium-sodium niobate-based lead free ceramics, J. Mater. Chem. A. 4 (36) (2016) 13778–13785. Available from: https://doi.org/10.1039/c6ta04107h, http://pubs.rsc.org/en/journals/journalissues/ta.
[55] Z. Haibo, M.M. Ali, X. Bing, A. Malik, L. Kai, Z. Yiwei, et al., Polymer matrix nanocomposites with 1D ceramic nanofillers for energy storage capacitor applications, ACS Appl. Mater. Interfaces (2019) 1–37. Available from: https://doi.org/10.1021/acsami.9b15005.
[56] T.J. Park, Y.B. Mao, S.S. Wong, Synthesis and characterization of multiferroic $BiFeO_3$ nanotubes, Chem. Commun. (2004) 2708–2709.
[57] Q. Jin, L. Zhao, B. Cui, J. Wang, H. Ma, R. Zhang, et al., Enhanced energy storage properties in lead-free $BaTiO_3$@Na0.5K0.5NbO_3 nano-ceramics with nanodomains: Via a core-shell structural design, J. Mater. Chem. C. 8 (15) (2020) 5248–5258. Available from: https://doi.org/10.1039/d0tc00179a, http://pubs.rsc.org/en/journals/journal/tc.
[58] X.T. Zhang, B. Cui, J. Wang, Q. Jin, The effect of a barium titanate xerogel precursor on the grain size and densification of fine-grained $BaTiO_3$ ceramics, Ceram. Int. 45 (2019) 10626–10632.
[59] A.S.H. Makhlouf, E. Guerrero, Toward smart and durable nanoceramic and nanobioceramic coatings and their medical and industrial applications, Advances in Smart Coatings and Thin Films for Future Industrial and Biomedical Engineering Applications, Elsevier, United States, 2019, pp. 383–403. Available from: https://doi.org/10.1016/B978-0-12-849870-5.00021-5, http://www.sciencedirect.com/science/book/9780128498705.
[60] O. Van Der Biest, Nanoceramics: issues and opportunities, Int. J. Appl. Ceram. Technol. 10 (4) (2013) 565–576. Available from: https://doi.org/10.1111/ijac.12074.

CHAPTER 15

Nanoceramics in advanced materials industry for renewable energy and storage

Arpan Kar*, Sowmik Maji*, Somoprova Halder*, Subhasis Roy* and Bhaskar Chandra Das*
Department of Chemical Engineering, University of Calcutta, Kolkata, West Bengal, India

15.1 Introduction

Ever since discovering fossil fuels, their consumption has increased steadily, and now, we are in a threatening position of its decline. Fossil fuels are said to provide almost 85% of the world's total energy demand. It is estimated that the rate of petroleum consumption is 10^5 times of its natural production. Such a huge consumption rate has led to the estimation that the world's fossil fuel reserve would be depleted by 2050 [1,2]. So, it is high time that we search for sustainable energy sources. The various such energy sources available are solar, wind, fuel cells, geothermal, biomass, and tidal energy. The good news is that due to extensive research on renewable energy and with the development of modern technology, it is assumed that 50% of the world's energy would be supplied from renewable sources by 2040.

Moreover, renewable energy comes with the additional advantage of low CO_2 release. The concentration of CO_2 in the air has increased considerably over the past 150 years with current value of 385 ppm and is expected to cross 550 ppm this century [3,4]. So renewable energy will play a vital role in boosting air quality. It is also estimated that the total energy that could be conceivably extracted only from wind energy is 72 terawatts, bearing in mind that the total human energy consumption is 16 terawatts, as of 2006.

Despite the various advantages of renewable energy, the challenge that we are facing is the efficiency of its extraction. In this regard, a special field called Nanotechnology has proven to be beneficial. It has been observed that substances, when reduced to their nanoscale (typically 1−100 nm) show various enhanced properties like increased physical, chemical, electrical, mechanical, optical, and magnetic properties. Using these enhanced properties, it has been possible to extract energy quite efficiently. The various areas that nanotechnology has penetrated are discussed in the following section.

* All authors contributed equally.

Industrial Applications of Nanoceramics
DOI: https://doi.org/10.1016/B978-0-323-88654-3.00020-2

15.2 Synthesis of nanostructures for industrial applications

Nanoparticles are the building blocks of the technologically improved future. The process of designing and creating devices on the nanoscale is termed nanofabrication. The increasing trend and ability to miniaturize the size of devices opens many doors for the development of new ways to capture, store, and transfer energy. Improvements in the efficiency of nanofabrication technologies are critical and are continuously being worked upon by scientists all over the world with an aim to solve many energy-related problems that the world is currently facing. Various methods are employed to prepare nanoparticles having a different range of chemical compositions. However, the production of pure, nonaggregated nanoparticles in bulk is still a challenging task.

15.3 Plasma-based and flame hydrolysis method

This is a high-temperature method. Material is sprayed over the flame to transform its ions. Then they are either rapidly cooled the superheated vapor (in the gas-to-particle process) or the solvent of the slurry containing suspend droplets is evaporated (in the droplets-to-particle process). Here ultrafine and spherical particles are obtained [5–7].

15.4 Chemical vapor deposition

At first, the material is vaporized, and after that, condensation process under vacuum condition, either the same material or new material is obtained.

15.5 Sol–gel processing

Here nanoparticles are synthesized at lower temperatures, and it also provides to achieve highly pure nanomaterial [6]. There are main routes:

1. inorganic route (metal salts in aqueous solution)
2. organic route (metal alkoxides in organic solvents)

After hydrolysis of raw materials, polymerization occurs to form a cluster of small particles. Then colloidal suspension of oxide particles is stabilized by electric double layer or steric repulsion. The viscosity of the solution increases as the gelling process proceeds.

15.6 Laser pyrolysis

This technique is used to synthesis a large variety of oxide and nonoxide nanoparticles.

In this gas- and vapor-phase synthesis method, when a sufficient degree of supersaturation is reached in the vapor phase, nanoparticles formation starts suddenly. At high temperatures, coalescence is faster than coagulation, and spherical particles are formed. This process has the disadvantage that at low temperature, agglomeration formation takes place as the coalescence process slowed down and nonspherical particles are formed [8,9].

15.7 Nanotechnology for energy devices

Nanotechnology has been showing great potential in boosting the renewable energy industries. It has been employed to increase efficiency in the energy generation of photovoltaic cells, windmills, hydrogen fuel cells, geothermal power generators, and biodiesel. Owing to their size in the nanometer range, which comprises of only some hundreds of atoms put together, their band structure largely differs from the bulk of the same material to a large extent. Due to a high surface to volume ratio, coupled with preferentially exposed facets in differently engineered nanostructures, they have an extremely high and efficient catalytic property. That is why, silver shows antimicrobial properties, platinum and gold acts as a catalyst while aluminum becomes combustible [10]. Suh et al. [11] have given a detailed discussion on nanoparticles in their review.

An advanced form of solar cells is thin and flexible, cheaper, easy to produce, and requires less material. The idea is that they can be incorporated into your briefcase, which will charge any laptop or cell phone kept inside it. Also, these transparent solar panels can be used in the windows of the skyscrapers to generate power.

There is a lot of heat generated in a car that is wasted. So, car companies have thought of a way to convert this heat into useful electricity. Sheets of nanotubes are used to make thermocells that generate electricity when one side of the cell is at a different temperature than the other. By wrapping these cells around the hot parts of cars, electricity can be generated from the heat.

Light bulbs made of plasmonic cavities—a nanoengineered polymer matrix—are highly efficient as well as shatter-proof. A very innovative idea of using nanocrystals in filaments of bulbs that can convert infrared radiation into visible light thus saving energy and electricity bills.

Windmills' efficiency has been enhanced by using borocarburized steel to make the components of the gearbox, which provides great abrasive resistance. Also, lighter and longer blades are made using epoxy-containing carbon nanotube (CNT), increasing the amount of electricity generated.

Geothermal energy extraction has also become efficient by nanotechnology. In countries like Iceland, geothermal energy has been exploited extensively. The source of geothermal energy lies deep down below the Earth's surface, about 1500 m below. The heat there is extracted by injecting cold fluids through pipes. This heat from the fluid is then used to generate electricity. Nanoparticles can improve the heat retaining capacity and thus making it possible to extract heat near the earth's surface at a lower temperature.

Batteries and capacitors are used to store energy. Using nanoparticles has also increased the efficiency of these. Graphite electrodes are generally used in batteries. But it is now being replaced by CNT for its high surface area, good electrical conductivity, and linear geometry. The result is batteries that are smaller and lighter are made. Also, nanobatteries that can be charged 60 times faster than the existing ones have been developed. The energy storage capacity of the capacitor is proportional to the electrode surface area, millions of nanoelectrodes are used to get more surface area and thus higher energy. For hydrogen fuel cells, nanoblades have been used to store large volumes of hydrogen [12].

The conventional aluminum-made transmission wires have drawbacks because of their high resistance and the fact that a considerable part of electrical energy is converted into heat. The increased temperature of the wires results in their sagging and reduces the distance between any successive pole. CNT having high conductivity, flexibility, and resistance to heat sag, are a better candidate for such a purpose.

Another field where nanotechnology is emerging as a potential inducer is bioenergy. Bioenergy is the energy that is stored in plants and animals. Potential sources are biomass, vegetable oils, biodiesel, etc. Various researches have been done and are still going on to make this energy a primary source and replace conventional fossil fuels.

Other applications of nanotechnology include the use of single-walled CNTs (SWCNTs) in liquid chromatography. Andre et al. [13–15] concluded that using SWCNT in the silica column instead of C18 results in the reduction of HETP value from 11.5 to 7.1 μm. Sandron et al. [14] revealed through their study that, when used as stationary phase, coating mercaptopropylsilica (Si-RSH) with AgNP greatly enhances the efficiency.

One-dimensional (1D) nanostructures have shown promise to increase energy density, charge transport, and cycling life of energy storage systems. These nanostructures are mainly used in batteries or other electrodes because of their shorter bi-continuous ion and electron transport pathways, which results in higher battery performance [16].

Additionally, 1D nanostructures are capable of enhancing charge storage by double layering and can also be used in supercapacitors because of their fast pseudocapacitive surface redox kinetics. Research is being carried on to develop novel designs, and the controllable synthesis of these materials will be developed much more in-depth.

2D nanomaterials have a layered morphology and form sheet-like nanostructures. The interlayer space can be manipulated for nonlayered materials, called 2D nanofluidic channels. The introduction of porosity in 1D and 2D nanomaterials has attracted a lot of research to be used for energy storage and catalytic applications through facile charge and mass transport.

In the following sections, some important areas where nanotechnology has been exploited extensively have been discussed in detail.

15.7.1 Developing nanostructured electrodes in fuel cells

The idea of a "*Hydrogen Economy*" is emerging to fight the alarms about global warming and the expected future disappearance of oil resources [17]. Hydrogen is gaining importance due to no carbon dioxide emission and is considered one of the most promising energy carriers [18,19]. A fuel cell is nothing but an electrochemical device, which produces electrical energy directly from chemical energy. The fuel cell that uses hydrogen and polymer as the electrolyte is known as a polymer electrolyte membrane fuel cell (PEMFC). Among the different types of fuel cells, PEMFCs have the capability to become a sustainable power source for transportation due to their lower operating temperatures (60°C–80°C) and higher power density. In PEMFCs, the only by-product is water, and therefore these devices can produce energy with zero carbon dioxide emissions [20]. Besides the environmental advantages, high efficiency and their malleability to use renewable fuel sources are unique features of PEMFCs [20].

Generally, two electrodes (anode and cathode) and electrode catalyst layers separated by a proton-conducting polymer electrolyte membrane constitute a PEMFC membrane electrode assembly. Backing layers of a porous carbon gas diffusion layer support the electrodes, which ensures effective diffusion of the gases to the catalyst (Figs. 15.1 and 15.2).

The reactant gases hydrogen (H_2) and oxygen (O_2) are introduced to the cathode and anode of the PEMFC, respectively, resulting in the hydrogen oxidation reaction to produce protons (H^+) at the anode. A polymer electrolyte generally transfers these produced protons from the anode to the cathode side, and electrons are transferred through an external circuit, where these protons and reactant (O_2 gas) [reduction reaction (ORR)] participate in the chemical reaction to produce energy along with water as the only by-product. The individual electrochemical reactions occurring at the cathode and anode are

$$\text{At anode: } 2H_2 \rightarrow 4H^+ + 4e^- (\text{HOR})$$

$$\text{At cathode: } O_2 + 4H^+ + 4e^- \rightarrow 2H_2O(\text{ORR})$$

$$\text{Overall cell reaction: } 2H_2 + O_2 \rightarrow 2H_2O$$

However, despite its potential to become a source of sustainable and green energy, commercialization of fuel cell technology is yet to realize. The main impasse to its mass production is the "cost factor" [22].

In the PEMFC system, a significant amount of Pt, forming cathode and anode, is the main reason behind its high cost [17]. Reducing the size of Pt to nanoscale will result in more surface area with more active catalytic sites [23]. With the size around 3 nm of Pt nanoparticles the best electrochemical stability can be achieved [24]. But during the sol–gel process, Pt tends to agglomerate and form clusters. Chen and Xing [25] used poly-(vinylpyrrolidone) (PVP) to prevent aggregation and attained high dispersion.

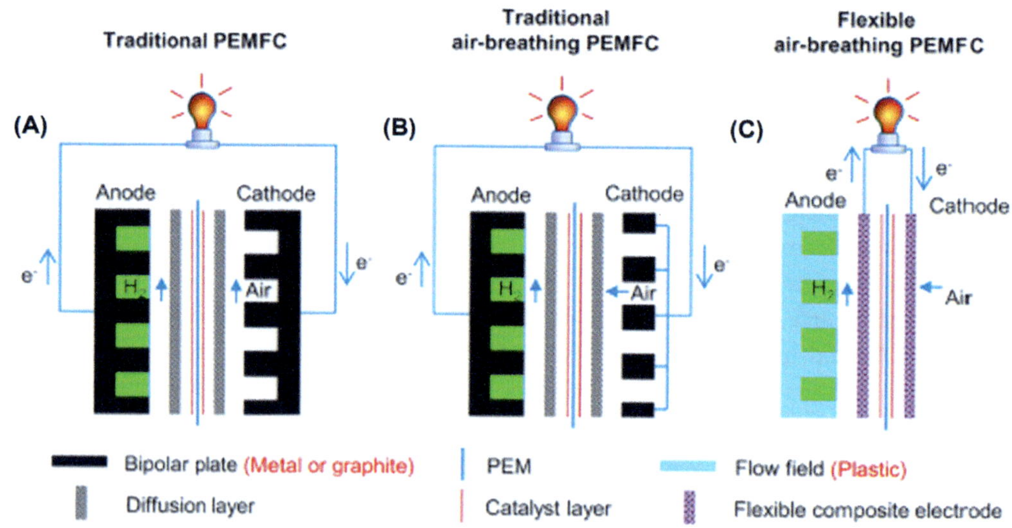

Figure 15.1 (A) Structure of conventional PEMFC. Air or O_2 is pumped into the PEMFC. (B) Schematic structure of traditional air-breathing PEMFC. Air naturally diffuses into PEMFC. (C) Structure of flexible air-breathing PEMFC. Air naturally diffuses into PEMFC [21]. Source: *Reprinted (adapted) with permission from F. Ning, X. He, Y. Shen, H. Jin, Q. Li, D. Li, S. Li, Y. Zhan, Y. Du, J. Jiang, H. Yang, X. Zhou, Flexible and lightweight fuel cell with high specific power density, ACS Nano 11 (6) (2017) 5982–5991. Copyright (2017) American Chemical Society.*

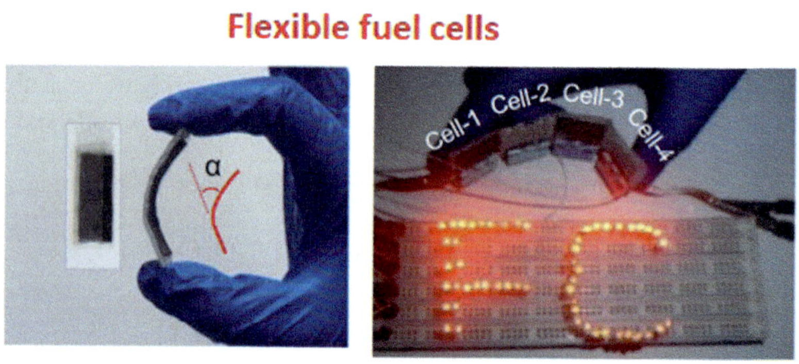

Figure 15.2 Flexible fuel cell. Source: *Reprinted (adapted) with permission from F. Ning, X. He, Y. Shen, H. Jin, Q. Li, D. Li, S. Li, Y. Zhan, Y. Du, J. Jiang, H. Yang, X. Zhou, Flexible and lightweight fuel cell with high specific power density, ACS Nano 11 (6) (2017) 5982–5991. Copyright (2017) American Chemical Society.*

Due to higher surface areas compared to low aspect ratio nanostructured material, metallic nanomaterials like nanorods [26] and nanowires [27] have shown superior electrochemical activities.

Another essential step for generating electrical current in PEMFC is the oxygen reduction reaction at the cathode. But because of the presence of carbon monoxide

(CO) in the feed gas, Pt suffers from deactivation above 100°C as CO gets absorbed [28]. Fe–phenanthroline ($C_{12}H_8N_2$) complexes show an activity comparable to commercial Pt catalysts and an alternative catalyst [29]. Because of higher active surface area, higher stability during electrochemical reactions, and enhanced conductivity with substrate materials, three dimensional (3D) nanostructured materials can be used to achieve the appreciable catalytic performance of electrocatalyst materials which help to shrink the overpotential required for facile ORR reaction in fuel cells [30].

The cathode and anode used in PEMFC are generally made up of platinized carbon cloth as it serve both as electrocatalyst and gas diffusion electrodes. The CNTs can be considered as promising supporting materials to make electrocatalyst as they enhance the efficiency because of their high chemical stability and good electrical conductivity besides their high surface area for deposition of metal catalyst nanoparticles which acts as electrocatalyst [31].

Another constraint for Pt/C catalyst is low CO tolerance which can be moderately overcome by depositing a thin layer of Ru over the surface of the electrode, which can act as a CO filter. However, nanocrystalline PtMo/C catalyst has shown better performance than aforesaid catalyst [32].

Nonnoble metal catalysts like graphite carbon nitrides (g-C_3N_4) blended with high surface area catalyst supporting materials like carbon black [33] or mesoporous carbon (CMK-3) [34] are seen as an alternative due to their improved catalytic performance and stability compared to noble metal catalysts because of their high cost and scarcity. Vertically aligned nitrogen-containing CNTs with their higher electrocatalytic activity, long-term durability, and greater resistance to the fuel crossover compared to the Pt for ORR [35], can be considered in developing Pt free electrocatalyst material [36].

High proton conductivity, sufficient mechanical strength, and stability, optimum chemical and electrochemical stability, low H_2 and O_2 crossover, and reasonable production cost are the main criteria for PEMFC membranes [37]. The management of water content and reactant crossover difficulties can be conquered by combining polymer electrolyte membranes with nanometer-size particles of Pt (by preventing reactant crossover by recombining H_2 and O_2 to create water) and metal oxides (by absorbing water produced at Pt particle) [38].

15.7.2 Enhancement of lithium-ion batteries through nanostructured electrodes

An example of a rechargeable battery is a lithium-ion battery or Li-ion battery. The lithium-ion battery is regarded as a promising energy storage device because of its high power and energy density. These batteries are the main parts for portable and smart electronics, hybrid electric vehicles, telecommunication, transportation, and renewable energy sectors because of their high energy density and longer lifespan.

15.7.2.1 Components

There are three primary functional components of a lithium-ion battery.

1. Negative electrode
2. Positive electrode
3. Electrolyte (nonaqueous, because the aqueous electrolyte is not stable at low potential)

These three components take part in electrochemical reactions. In these batteries, lithium ions move from the negative electrode through an electrolyte to the positive electrode during discharge, and while charging, the reverse incident happens. In a conventional lithium-ion battery, metal oxides such as layered $LiCoO_2$ are placed as positive electrodes and graphitic carbon as the negative electrode.

15.7.2.2 Cell reaction

The half-reaction at positive electrode is:

$$LiCoO_2 \leftrightarrow Li_{1-n}CoO_2 + nLi^+ + ne^-$$

The half-reaction at negative electrode is:

$$nLi^+ + ne^- + C \leftrightarrow Li_yC$$

The energy output of a lithium-ion cell is nothing but the voltage times the charge.

15.7.2.3 Disadvantages

Though these batteries are successfully used rigorously all over the world, these batteries have a few shortcomings. Low electronic conductivity, low Li^+ transfer efficiency, rate capability, volume expansion/contraction during repeated cycling processes, and loss of active materials due to the corrosion during electrolytes limit these devices' performance.

To enhance performance in terms of high energy/power density, good mechanical/thermal stability, environmental friendliness, and low clean energy storage cost, and hybrid electric vehicles usage, many plausible ways have been proposed, including doping cations or anions to optimize the crystalline lattice of active materials, enhance electron/Li^+ ion conductivity and reactivity by reducing particle size [39]. However, these techniques have some limitations, as nanoparticle often leads to agglomeration of particles results in overall reduced surface area, while the advantages of composite structures are only limited to the surface or interface.

The characteristic time for lithium ions to diffuse through an electrode material varies with the square of the characteristic diffusion length. The diffusion coefficient is increased, or the diffusion length is decreased to decrease the diffusion time [40].

15.7.3 Nanostructured materials as a cathode for lithium-ion batteries

Nanotechnologies have created great candidates for cathode electrodes with better-conducting pathways, faster charge transport, and good strain relaxation [41]. The most widely experimented nanostructured cathode materials include layered $LiMO_2$, olivine-type $LiMPO_4$, and spinel-type LiM_2O_4, where M is one or more transition metal (Co, Ni, Mn, etc.) elements. Among them, $LiCoO_2$ [42] and $LiNiO_2$ [43] transition metal oxides with layered structure $LiFePO_4$ [44] with olivine structure and $LiMn_2O_4$ [45] with 3D spinel structure are the most successful nanocathode materials due to their chemical and structural stability, high lithium-ion capacity and favorable electrical properties like high lithium ions and electrons conductivity (Fig. 15.3).

$LiCoO_2$: $LiCoO_2$ is the most widely used transition metal oxide in commercial lithium-ion batteries (LIBs) as the cathode. They have high rate capability. But because of agglomeration and increase in contact resistance, they have issues like instability and capacity fading. With a protective layer of several materials such as aluminum phosphates ($AlPO_4$) and magnesium oxide (MgO) [47] this problem can be solved, which shows stable capacitance even after 40 cycles.

$LiMn_2O_4$: These compound gained popularity because of exhibition higher rate performance due to 3D pathway in spinel framework, increased abundance, low cost, low toxicity, and better safety, as compared to $LiMn_2O_4$ [48]. These nanostructures have disadvantages of capacity fading due to Jahn–Teller distortion, dissolution of Mn into the electrolyte, loss of crystallinity, and development of micro-strain. However, these problems are somewhat solved by constructing nanocomposites of $LiMn_2O_4$ with CNTs and graphene nanosheets where large discharge capacities can be achieved [49].

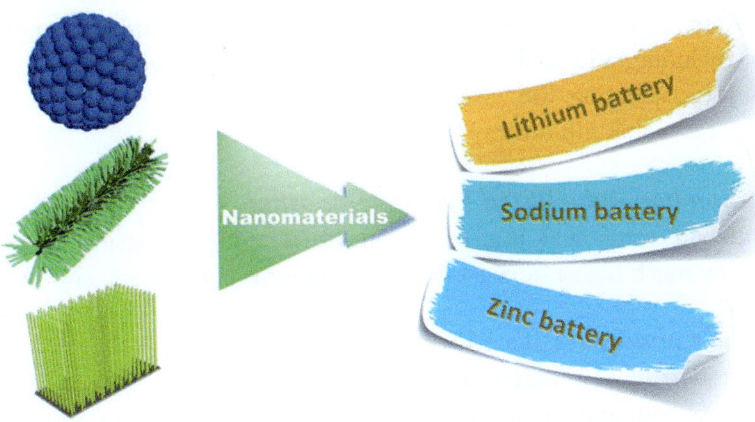

Figure 15.3 Different types of batteries [46]. Source: *Reprinted (adapted) with permission from Y. Zhang, N. Liu, Nanostructured electrode materials for high-energy rechargeable Li, Na and Zn batteries, Chem. Mater. 29 (22) (2017) 9589–9604. Copyright (2017) American Chemical Society.*

$LiFePO_4$: Because of its high theoretical capacity (170 mAh/g), low cost, safety, good thermal stability, and eco-friendly nature $LiFeO_4$ has been considered as another candidate for cathode [44]. This compound has been successfully implemented commercially in an electric vehicle as a cathode and stationary storage application because of strong overcharge tolerance [50]. At the nanoscale, it has improved electrochemical performance due to shorter diffusion length and increased effective reactive areas.

V_2O_5: For Li-ion intercalation applications, vanadium oxide usage has many advantages such as low cost, easy synthesis, abundant source, and high energy densities. The increased surface area of V_2O_5 composite electrode can overcome slow electrochemical kinetics [51]. Also, defects in V_2O_5 nanoroll can enhance electrochemical performance [52].

15.7.4 Nanostructured materials as an anode for lithium-ion batteries

Graphite (theoretical capacitance of 372 mAh/g) is preferred as the anode because of high electronic (in-plane) conductivity due to delocalized π-bonds, low expansion lithium insertion, and suitable structure for lithium-ion intercalation and diffusion. It also has petite pulverization due to its mechanical and chemical stability [53].

Significant improvements for low theoretical capacitance and low lithium insertion capacity are essential, which can be achieved by hollow nanotubes due to:

1. Increment of electrode/electrolyte contact area due to the high surface area.
2. Shorter lithium transport distance due to reduced size.
3. Extra space for Li^+ storage.
4. Prolongation of cycling life due to void space in hollow structure [54].

15.7.4.1 Carbon nanotubes

Pure CNTs are considered a strong contender for anode material due to hollow structure, low density, high conductivity, high rigidity, and high tensile strength [54]. The short CNTs, including a lot of lateral defects, enable easier intercalation and deintercalation of Li ions, affecting the electrochemical performance of lithium storage and enhancing the reversible lithium capacity [55].

15.7.4.2 Carbon nanotube-based hybrid composites

CNT-based hybrid composites (e.g., transition metal oxides, inorganic material like Si, metals, etc.) show much higher storage capacity and prevent pulverization and crumbling [54].

15.7.4.3 Carbon nanotubes/silicon

Silicon is considered as a potential replacement due to its theoretical capacity of 4200 mAh/g, low cost, and abundance [56]. But it suffers from uncompromising volume change during charging and discharging, which results in pulverization and crumbling and leads to electrode capacity loss. In addition, silicon has lower conductivity. By introducing CNTs in Si, the hybrid composites have shown better electronic conduction performance and structural buffering [57]. Multiwalled CNTs with Si nanoparticles have shown the high specific capacity and superior capacity retention, besides high rate capability [58].

15.7.4.4 Carbon nanotubes/oxide

Transition metal oxides are preferred due to high capacity for anode while they have showed poor cycle stability and low cycle efficiency. By incorporating CNTs into metals have improves electrochemical performance due to highly activated surface area and also prevent volume contraction/expansion and aggregation of nanoparticles [59]. CNTs/SnO_2 composite has high theoretical values of discharge and charge capacities of SnO_2 (1494 and 782 mAh/g, respectively) [59]. CNTs/MnO_2 composites are utilized due to high storage capacity, low cost, eco-friendliness and natural abundance [60]. Other transition metal oxides, for example, Co_3O_4 [61], NiO [62], TiO_2 [63], MoO_2 [64], Fe_3O_4 [65], V_2O_5 [66] etc., have drawn attention for CNT hybrid anode material. MWCNTs/NiO composites have shown superior electrochemical performance with much higher lithium storage capacity [62]. CNT/Fe_3O_4 composites have shown higher capacity retention, higher reversible capacity, and higher rate capability [65].

15.7.4.5 Carbon nanotubes/metal

Although Sn anode has the high reversible capacity (approximately 994 mAh/g from $Li_{4.4}Sn$), there is an inevitable occurrence of significant volume expansion (259%) results in crumbling, which leads to poor cycle performance. By introducing CNTs in Sn anode, this significant volume expansion problem somewhat might have been solved [67]. SWCNT/Sn nanoparticle composites have shown improved anode durability and high reversible capacity [68]. Intermetallic Sn can be seen as a candidate for anode due to relatively small volume expansion, high specific capacities, and good cyclability [69]. CNT-encapsulated (7.6 wt.% of CNT) Sn−Sb nanorods can maintain high specific capacity [70].

15.7.4.6 TiO_2 nanotubes

TiO_2 is seen as a replacement for anode material due to its high working voltage and structural stability during lithium insertion/extraction because of small volume expansion. However, it shows the disadvantage of poor rate capability, which can be overcome by incorporating conducting/active materials into TiO_2 and reducing its size [71,72].

15.7.5 Nanoceramics for development of efficient solar cells

Currently, 80% of the global energy demand is met by fossil fuels. The remaining fossil fuels will become extinct in the next 100 years. This energy supply problem can be overcome with a most abundant carbon-neutral energy source, sunlight, if technology can harvest it efficiently and cost-effectively. Energy from sunlight can be transformed in three different ways for utilization of society: solar electricity, solar fuel, and solar thermal.

The solar cells. Their efficiency can be achieved up to 25% [73]. Si as a material has the advantages of high abundance, nontoxicity, and long-term stability. But the cost of making ultrapure crystalline Si is very high and is also an indirect band gap semiconductor. A thick layer of Si is required to attain maximum absorption of sunlight, which makes the whole device more costly. To overcome this problem, Schottky diodes are used over Si p−n junction diode. In Schottky diode, the semiconductor has a thin layer of indium doped tin oxide to increase the absorbance [74]. Due to the high cost of indium and being less abundance, graphene is used as a replacement.

In 2007, the CNTs−Si heterojunction concept is developed to overcome the problem of Si p−n junction [75]. But this heterojunction forms both Schottky and p−n junctions and has low conductivity. Hence, graphene is used in place of CNTs. In the graphene-Si heterojunction concept, multiple layers of graphene sheets, overlapped and interconnected, are used to resolve the problem. It will increase conductivity and also acts as an antireflecting layer [76]. In graphene-Si composite, performance can be lowered due to the high density of defects present at Si and graphene heterojunction. With p-type doping working function of graphene can be increased for enhancing efficiency [77].

In Cu(In, Ga)Se_2 (CIGS) based solar cell, In_2S_3 can be used as buffer layer effectively in place of CdS [78] to obtain Cd free buffer layer for an environmental reason despite the fact CIGS with CdS buffer layer has highest conversion efficiency among thin-film solar cell [79]. In_2Se_3 can be utilized in low-cost photovoltaics (PV) applications due to optical and electrical properties, high absorption coefficiency, and optimum band gap energy for solar energy conversion [80].

After crystalline Si-based solar cells and thin-film solar cells, third generation solar cells like dye-sensitized solar cells and polymer-based solar cells are cheaper. Dye-sensitized solar cells are consist of two things: electrodes and electrolytes. The anode is coated with wide band gap semiconductors like TiO_2, and the cathode is coated with the layer of Pt on a glass having transparent fluorine-doped tin oxide coating. The void between two electrodes is filled with redox couple of I^-/I^{3-} [81]. The liquid electrolyte provides a high conductivity pathway, which is the main reason behind the high conversion efficiency [63,82]. Semiconductors are dipped in the dye solution. Upon exposure of light, charge separation occurs by electron injection from dye to TiO_2 at semiconductor junction.

The conversion efficiency of dye-sensitized solar cells ranges between 6% and 13%. Dye-sensitized solar cells have achieved the highest conversion efficiency using liquid electrolytes. But the main problem with this system is leakage [63]. Solid polymer is used instead of liquid one where polyether plays the dominant role [83]. Due to high conductivity, good thermal stability, biodegradability, etc., the biopolymer-salt complex can also be considered a good replacement [84].

Due to low cost, easily processable, and flexible renewable energy sources, organic semiconducting material is used in solar cells. But the main problem of organic semiconductor-based solar cells is the low electron mobility [85]. The transport of electrons has to rely on another material that will provide the interface for electron transfer [85]. Combining n-type inorganic semiconductors with conjugated polymers can solve this problem. These polymer-inorganic nanocomposites involve both solution processing of polymer semiconductors and high electron mobility of inorganic semiconductors [86].

15.7.6 Nanoceramics in wind energy generation

The second most popular renewable energy source after solar energy that comes to mind is wind energy. A wind energy system comprises a wind turbine that converts the kinetic energy of flowing wind into electrical energy. This electrical energy is then used for various purposes like electricity for households or hydrogen production for fuel cells by water splitting. Besides solar energy, wind energy can be generated daily since it depends only on wind flow. So, in windy countries like United States, Germany, Sweden, Denmark, wind power is already giving a tough competition to other conventional energy sources. Current estimation claims that more than 23 billion kWh of cheap and clean electricity are being produced worldwide. India's power production from wind energy is estimated to be around 1000 MW. Sweden has built an offshore wind park with a potential of 500 MW and has a production capacity of 2 TWh annually, serving 250,000 homes with electricity.

One of the issues that a wind turbine faces is working under severe operating conditions such as high loading, unsteady/interrupted operation, system vibration, which led to wearing and failure of the turbine gearbox components. Carburizing steel to enhance hardness is a general practice. But recent studies have shown that boronizing provides much better hardness compared to carburization [87]. The dense iron boride (FeB and Fe_2B) layer significantly enhances the wear resistance making it a suitable surface treatment for the wind turbine gearbox industry.

According to Greco et al. [88], an optical micrograph study of the wear track on the carburized and borocarburized surface reveals significant abrasive wear on the carburized surface while the wear track of the borocarburized surface is smoother, indicating polishing wear.

15.7.7 Application of nanoceramics in biodiesel and biogas utilization

With the depletion of fossil fuel deposits, there is a growing inclination toward searching for renewable and cheap energy sources. Fuels produced from biological sources are sustainable sources. Bioenergy in biomass, biofuels, biogas, vegetable oil, and biodiesel are being studied and exploited and are hoping to replace the conventional energy sources. Biomass and biogas are the cheapest and widely practiced bioenergy sources. Sweden, during 2008 has nearly 15,000 cars and hundreds of buses and trucks running in biogas [89].

Mahmood and Hussain [90] produce biofuels like bioethanol, biodiesel, and charcoal from spent tea using cobalt nanoparticles. They concluded that this process would be beneficial for tea-producing countries. Makam and Goh [91] experimented with the effects of magnetic SWCNTs by incorporating iron oxide on SWCNT on biofuel production. They used immobilized amyloglucosidase enzyme, which was adsorbed on SWCNTs. They concluded immobilized enzymes show greater stability and efficiency and retain their activity when stored at 4°C for at least a month.

Biodiesel is methyl or ethyl esters of fatty acids. It is considered renewable since it is produced from plant (vegetable oils) and animal (fats) sources, making its production inexhaustible. Production from animal fats is rare. Although biodiesel produces carbon dioxide when combusted, it is considered carbon-neutral because the plants that are used to make biodiesel utilizes atmospheric CO_2 as they grow and thus form a cycle. Further biodiesel contains less carbon and more oxygen and hydrogen than fossil fuel diesel and thus reduces the emission of particulates from unburnt carbon. Biodiesel is produced mainly by (1) blending of vegetable oil with diesel oil, (2) microemulsion, (3) pyrolysis, and (4) transesterification. Among these, transesterification is preferred chiefly [92].

Owing to the threat of depleting fossil fuels, research on biodiesel and its efficiency improvement has emerged as a very attractive field. Various attempts have been made to realize the vision of making biodiesel the primary fuel and fossil fuel diesel secondary. But the issue is with its cost of production, which is higher than conventional diesel. To mitigate this problem nanocatalysts are now being used in the synthesis of biodiesel. Chen and Guo [93], in their review, have explained the use of dispersed sulfated zirconia nanoparticles as high-efficiency catalysts for bis(indolyl)methane and biodiesel production. Sulfated zirconia is produced using PVPas surfactant. Sajith and Sobhan [94] experimented on the effects of cerium oxide nanoparticles as fuel additives on biodiesel. It was observed that the addition of cerium oxide to biodiesel had increased the fuel's flash and fire points (Fig. 15.4) thus making its handling safer. Also, there was a significant reduction of both hydrocarbon (Fig. 15.5) and NO_x (Fig. 15.6) emissions. Qiu et al. [95] studied the use of solid heterogeneous nanocatalyst in the production of biodiesel. Powdered zirconia (ZnO_2) was impregnated in potassium bitartrate ($C_4H_4O_6HK$) followed by calcination at elevated temperature which showed enhanced catalytic activity for transesterification of soyabean oil and methanol. The highest yield reached up to 98.03%, and the properties were close to that commercial diesel fuel. Zhang et al. [96] performed a study on transesterification of soyabean oil and methanol using a carbon-based monolithic nanocatalyst by catalytic distillation. He concluded that the transesterification of the two is not efficient enough in the traditional way. But the simultaneous reaction and the separation of the catalytic distillation column can break the equilibrium limit of the reaction and give a higher yield. Wen et al. [97] used KF/CaO as a nanocatalyst to produce biodiesel from Chinese

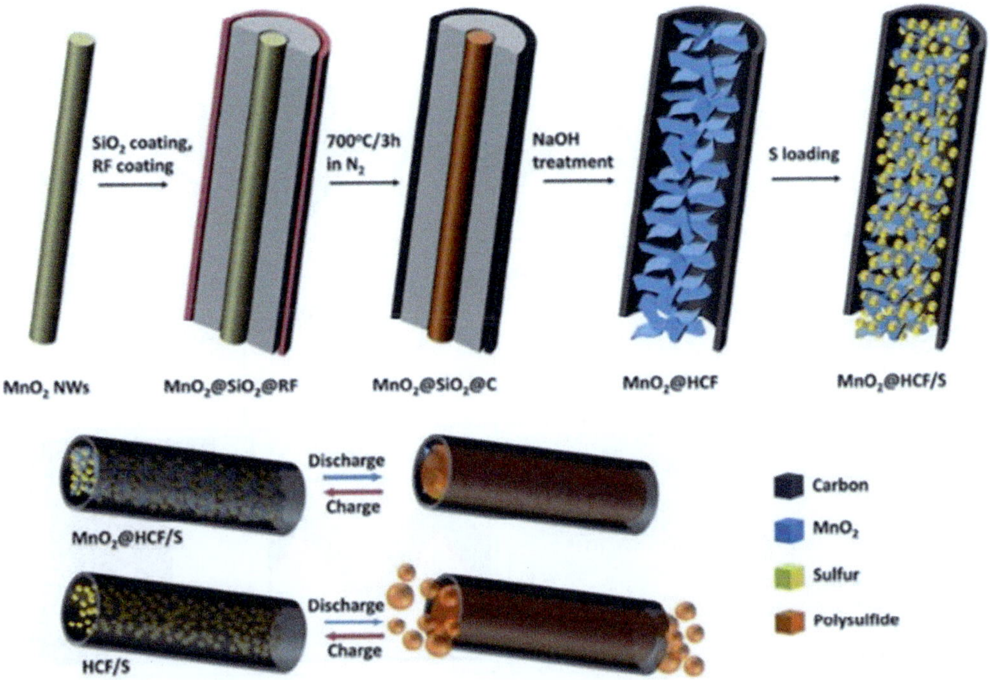

Figure 15.4 Synthesis of the MnO_2@HCF/S composites. Mechanism of advantages of MnO_2@HCF/S composites over HCF/S. Source: *Reprinted (adapted) with permission from Y. Zhang, N. Liu, Nanostructured electrode materials for high-energy rechargeable Li, Na and Zn batteries, Chem. Mater. 29 (22) (2017) 9589–9604. Copyright (2017) American Chemical Society.*

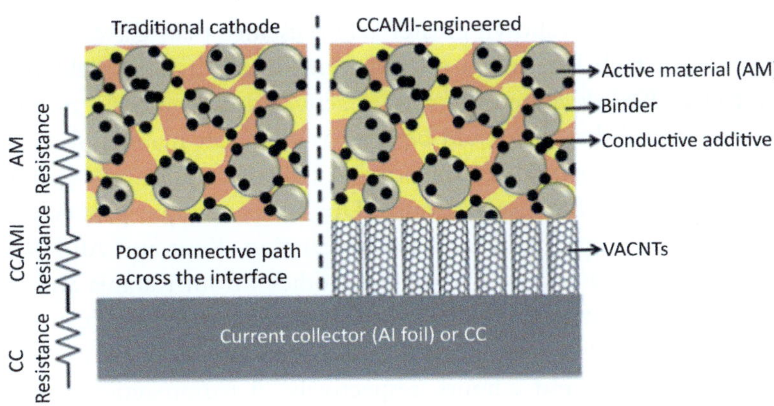

Figure 15.5 Schematic showing various resistances in LIB cathodes. The overall cathode resistance is the sum of the resistances arising from the (1) active material, binder, and conductive additive coatings, (2) CCAMI, and (3) current collector [101]. *LIB*, Li-ion battery; *CCAMI*, Current collector active material interface. Source: *Reprinted (adapted) with permission from L.K. Ventrapragada, J. Zhu, S.E. Creager, A.M. Rao, R. Podila, A versatile carbon nanotube-based scalable approach for improving interfaces in Li-ion battery electrodes, ACS Omega 3 (4) (2018) 4502–4508. Copyright (2018) American Chemical Society.*

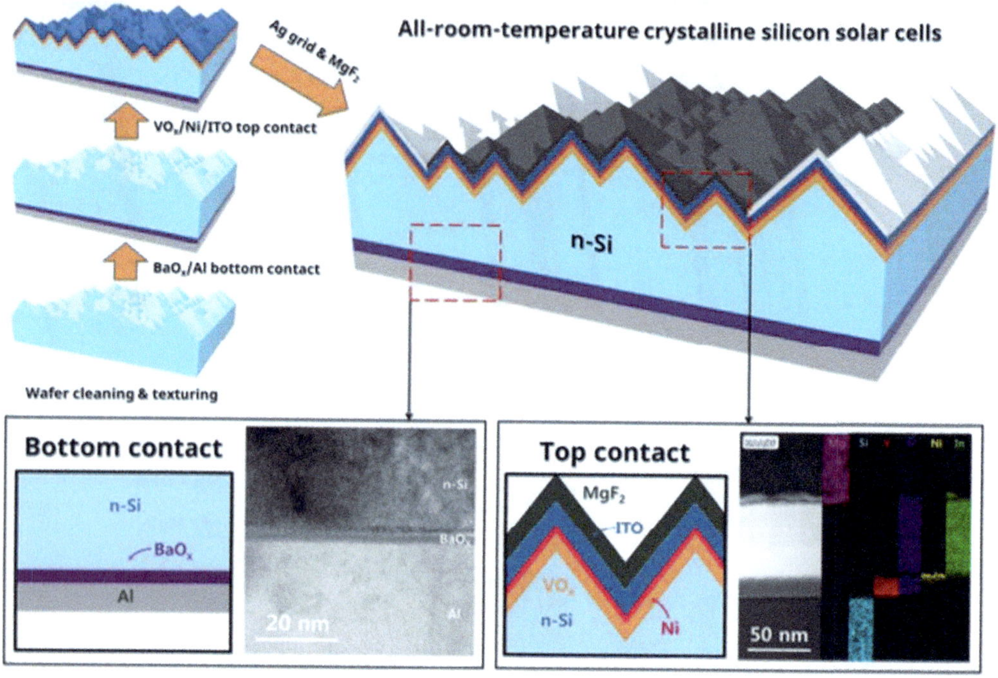

Figure 15.6 Overview of a c-Si solar cell. Fabrication strategy and detailed schematics of heterostructures at the top and bottom contact [102]. Source: *Reprinted (adapted) with permission from S.H. Kim, J.Y. Jung, R. B. Wehrspohn, J. H. Lee, All-room-temperature processed 17.25%-crystalline silicon solar cell, ACS Appl. Energy Mater. 3 (4) (2020) 3180–3185. Copyright (2020) American Chemical Society.*

tallow seed oil. KF/CaO has shown greater acid resistance than Ca/Mg catalyst toward acidic tallow seed, which led to less oil requirement and a high yield of more than 96%. Konwar et al. [98] described the role of activated carbon (AC) in biodiesel production. He mentioned that AC could be used as a support material for various active catalysts and due to its structural resemblance to that of graphite and graphene, it can host both acidic and basic functional groups. Due to this unique property, AC can be used as a heterogeneous catalyst in biodiesel production. Kaur and Ali [99] prepared a Li^+ ion impregnated CaO catalyst to produce biodiesel from Karanja and Jatropha oil. They identified an optimum reaction time for complete transesterification of Karanja and Jatropha oils with 12:1 methanol/oil and a reaction temperature of 65°C with >99% conversion to be 1 and 2 hours, respectively. A nanomagnetic solid base catalyst KF/CaO-Fe_3O_4 was studied by Hu and Han [100], which showed good reusability with >90% catalyst recovery. For a reaction temperature of 65°C and methanol/oil molar ratio of 12:1, the yield of biodiesel exceeded 95% (Fig. 15.7).

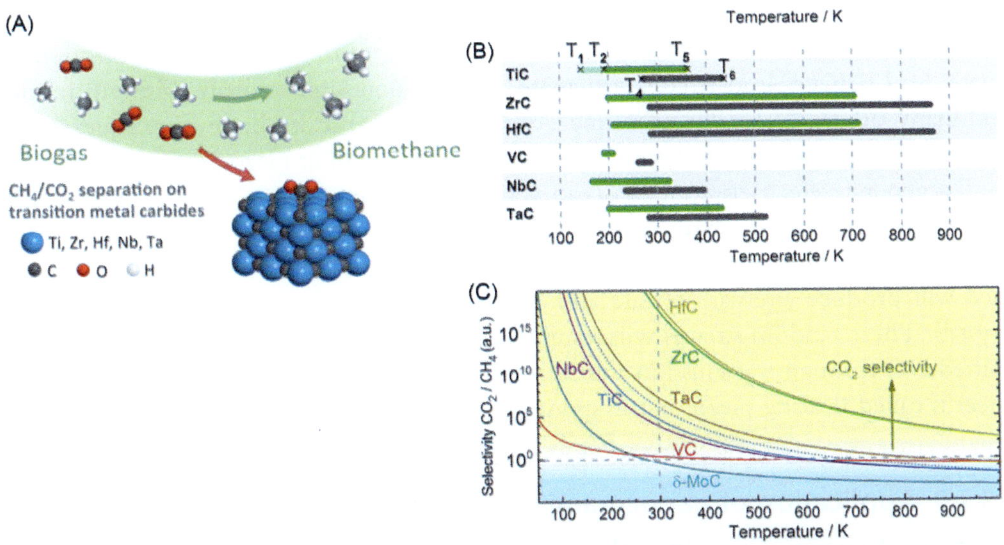

Figure 15.7 (A) CH_4/CO_2 separation on transition metal carbides (B) temperature ranges for different TMC (0 0 1) surfaces, (C) where CO_2 adsorption is predicted high with nearly no CH_4 capture. Solid green (*dark gray*) correspond to partial pressures of 0.01 bar (1.0 bar) evaluated according to free model. Light green give [103]. *TMC*, Titanium matrix composites. Source: *Reprinted (adapted) with permission from C.R. Kunkel, F. Vines, F. Illas, Biogas upgrading by transition metal carbides, ACS Appl. Energy Mater. 1 (1) (2018) 43–47. Copyright (2016) American Chemical Society.*

15.7.8 Nanoceramics used in harnessing tidal/ocean wave energy

Renewable energy is indeed green energy as it does not pose a threat to the environment. But one of the demerits it has is that it often requires a lot of land and space to make its effect visible. For instance, a typical nuclear power plant of 1000 MW facility needs about 1 square mile, whereas to produce the same amount of power a solar power plant would require 75 times more land, and a wind farm would require even more to about 360 times. So, if not land, then the other alternative would naturally be water. Earth's surface, with 70% covered in water, comes with a potential of 10,000–15,000 TWh per annum for wave energy generation. This energy can be utilized for water desalinization, hydrogen production, ocean mining, and ice production. Qu et al. [104], in their review, have mentioned the various opportunities for next-generation water supply systems. The rotation of the earth leads to the rise and fall of tides, and this process is continuous without a break along the sea-coast, which leads to the idea of a continuous source of renewable energy [105]. So, tidal energy bears a promising future in terms of its contribution to global energy generation.

To harness the energy of the waves, special types of crystals are used that produce electricity when subjected to mechanical stress. Such crystals are called piezoelectric crystals, and the electricity it produces is called piezoelectricity. Several crystals show piezoelectric effects, such as quartz, barium titanate, tourmaline, gallium orthophosphate, lead titanate, lead zirconate titanate, sodium potassium niobate, bismuth ferrite, and many others. Naturally occurring crystal berlinite and dry bone shows piezoelectric effects. Polymers like polyvinylidene fluoride also exhibit piezoelectricity.

To generate electricity out of ocean waves, these nanocrystals are embedded inside a rubber sheet. Rubber sheets are used due to their flexibility and inertness to seawater. These sheets move parallel to the wavefront, and its random deflections in the surface will produce alternate tensile and compressive stresses on the piezoelectric Nan crystals. These random stresses will result in electricity production, which is then transmitted DC through transmission cables connected to the sheet. Such a floating rubber sheet is called floating piezoelectric sponge surface.

15.7.9 Application of nanoceramics in geothermal energy generation

Geothermal energy is the thermal energy extracted from the earth's crust from 5 to 10 km below the surface. At such depths, the temperature is exceptionally high. Conventionally thermal fluids such as water are used to extract the heat. But for water to extract heat, the temperature must be high enough, and thus, deep bore length is required, and therefore, cost increases. Certain nanofluids have been identified to have superior thermal conductivity [106–108]. The higher thermal conductivity of the nanofluids accounts for the dispersion of solid nanostructures, which have higher thermal conductivity compared to liquids [109]. Therefore, nanofluids can be used for cooling purposes to cool the pipes exposed to the high temperature below. It can also cool various other accessories like electronics and sensors in drilling machines that are working under high friction and high-temperature environment. Based on the temperature of its source, geothermal energy is classified as low (<100°C), medium (100°C–150°C), and high temperature (>150°C) as they extracted by using ground heat exchangers [110].

The continuous availability of geothermal energy compared to solar and wind energy makes it a more reliable energy source. Also, geothermal energy has an advantage in electricity generation compared to fossil fuels plant, solar PV modules, and hydropower, owing to its low carbon dioxide emission. Fig. 15.8 compares the amount of CO_2 emission from various energy resources for electricity generation shows that the cost of production of electricity is relatively low.

Diglio et al. [112] investigated the efficacy of using nanofluid as a heat carrier instead of conventional fluids such as ethylene glycol–water on a borehole heat exchanger. In their study, various nanoparticles such as copper, graphite, alumina,

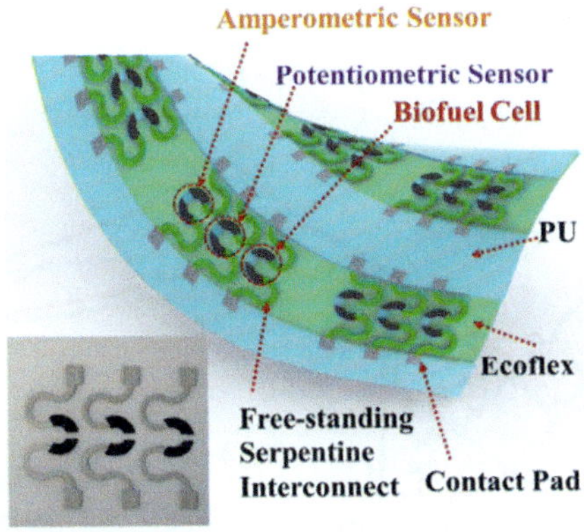

Figure 15.8 Inset shows an image of a printed CNT-based array device [111]. *CNT*, Carbon nanotube. Source: *Reprinted (adapted) with permission from A.J. Bandodkar, I. Jeerapan, J. You, R. Nunez-Flores, J. Wang, Highly stretchable fully printed CNT based electrochemical sensors and biofuel cells: combining intrinsic and design induced stretchability, Nano Lett. 16 (1) (2015) 721–727. Copyright (2015) American Chemical Society.*

aluminum, etc., were used in low concentrations (0.1–1.0 vol.%). It was observed that the application of nanofluids reduces borehole thermal resistance. The results with different nanofluids are shown in It is observed that copper gives an excellent result. Daneshipour et al. [113] carried out their study on two nanofluids, including Al_2O_3–water and CuO–water as borehole heat exchanger circuit fluids.

Geothermal energy has various applications. More than 72 countries have reported direct use of geothermal energy. Iceland is considered one of the cleanest countries worldwide due to geothermal energy applications. Hot water through pipes is distributed throughout the community. About (93%) of its homes are heated with geothermal energy and are saving over $100 million annually over oil imports. One of the advantages of geothermal power plants is that they can be operated 24 hours a day. The world's potential for geothermal power generation is estimated to be 85 GW over the next 30 years; nevertheless, currently, geothermal power is accessible only in limited areas of the world.

Despite the various advantages in extracting energy through geothermal means, the major challenges encountered lie in the requirement of deep borings. Recent researches on nanomaterials have revealed that certain nanoparticles can improve the heat retaining capacity and thus making it possible to extract heat near the earth's surface even at a lower temperature. According to Ganguly [114], this observation opens a range of future research prospects of using shallow borings to utilize geothermal energy (Figs. 15.9–15.11).

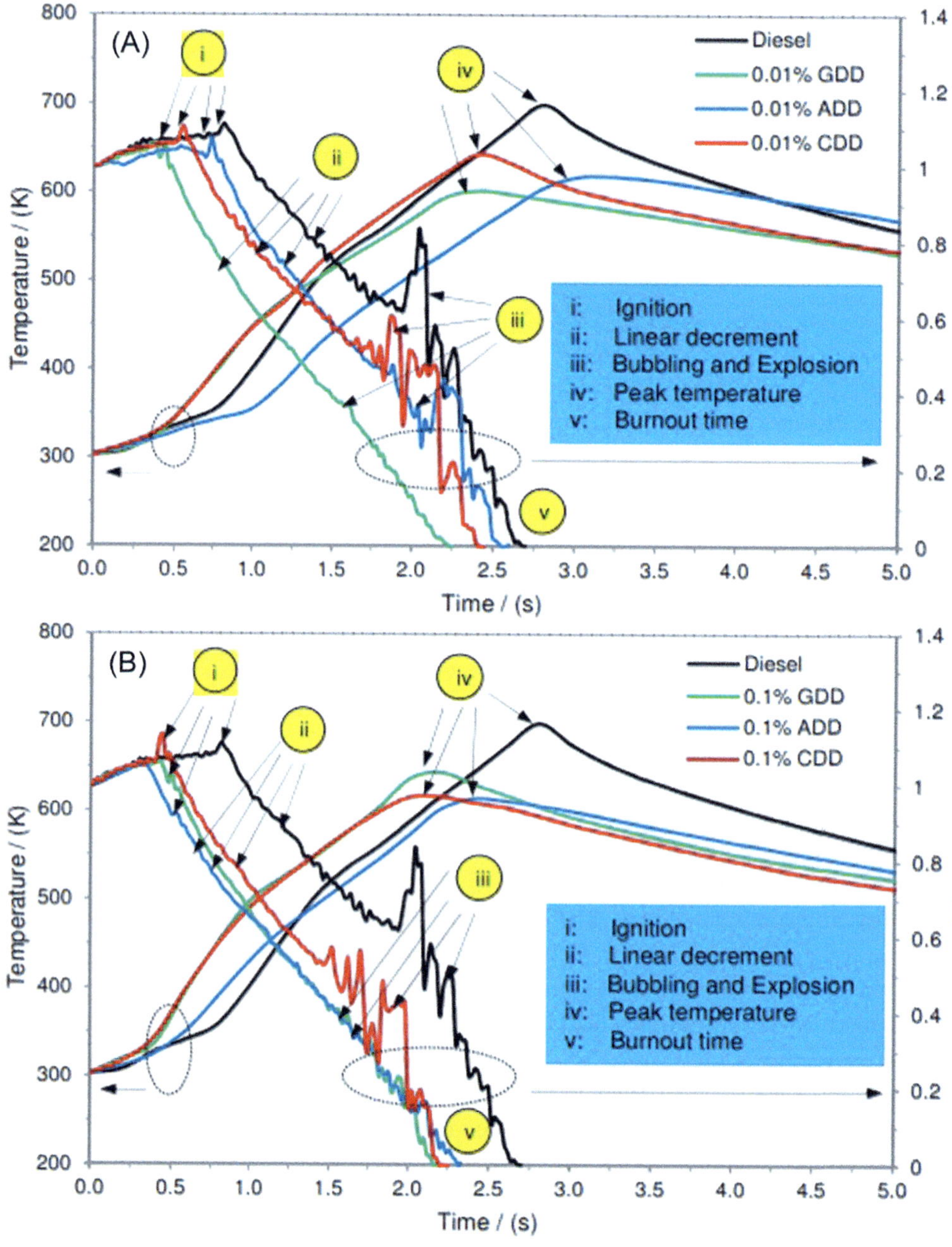

Figure 15.9 Normalized droplet area and temperature histories for the baseline fuel and the dosed diesel fuels at (A) 0.01% dosing and (B) 0.1% dosing [115]. Source: *Reprinted (adapted) with permission from J.B. Ooi, H.M. Islami, V. Swamy, X. Wang, A.K. Swain, J.R. Rajanren, Graphite oxide nanoparticles as diesel fuel additive for cleaner emissions and lower fuel consumption, Energy Fuels 30 (2) (2016) 1341–53. Copyright (2016) American Chemical Society.*

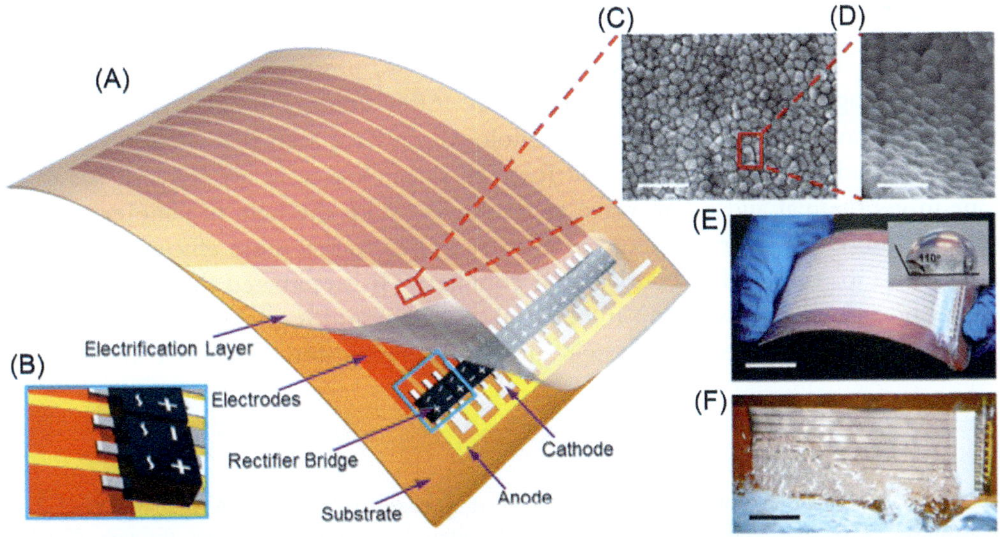

Figure 15.10 TF-TEG. (A) Schematic diagram of an integrated TF-TEG. (B) Enlarged sketch of the arrayed bridge rectifiers. (C) SEM image of PTFE nanoparticles on the electrification layer (scale bar = 1 μm). (D) Enlarged view of nanoparticles at a tilted angle of 60° (scale bar = 500 nm). (E) Picture of a bendable TF-TEG. Inset: Water contact angle on the nanostructured surface (scale bar = 25 mm). (F) Picture of TF-TEG that is interacting with the water wave (scale bar = 25 mm) [116]. *TF-TEG*, Thin film thermoelectric generators; *PTFE*, Polytetrafluoroethylene. Source: *Reprinted (adapted) with permission from X. J. Zhao, G. Zhu, Y.J. Fan, H.Y. Li, Z.L. Wang, Triboelectric charging at the nanostructured solid/liquid interface for area-scalable wave energy conversion and its use in corrosion protection, ACS Nano 9 (7) (2015) 7671–7677. Copyright (2015) American Chemical Society.*

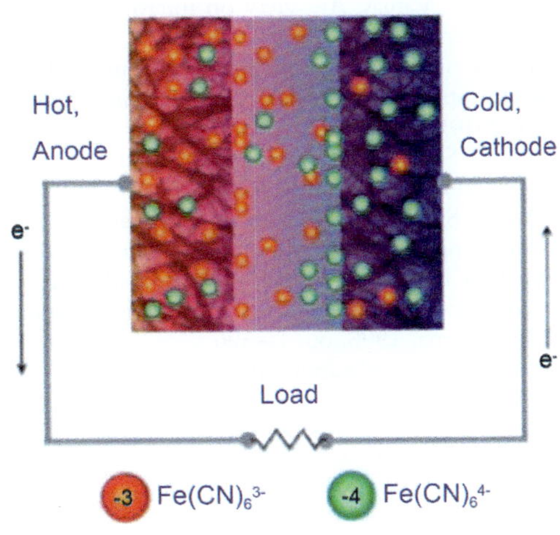

Figure 15.11 Schematic of a thermocell with nanostructured electrodes showing concentration gradients of the ferri-ferrocyanide redox ions during power generation [117]. Source: *Reprinted (adapted) with permission from R. Hu, B.A. Cola, N. Haram, J.N. Barisci, Harvesting waste thermal energy using a carbon nanotube based thermo electrochemical cell, Nano Lett. 10 (3) (2010) 838–846. Copyright (2010) American Chemical Society.*

15.8 Conclusion

This review aims to discuss the progress, advantages, prospects, and issues related to nanotechnology in renewable energy sectors. After going through various research works that have been done on nanotechnology, we have seen that there has been an extensive headway made in solar and fuel cell technologies. A lot of work has to be done in geothermal and ocean wave technologies. With the help of nanotechnology, it will be possible to extract energy from these sources much more efficiently. Biofuels like bioethanol and biodiesel are gaining popularity. It is estimated that by 2040 there will be no vehicles that will run mineral oil. As the world is gradually shifting toward a greener society, nanotechnology will surely be stepping stones for this conversion.

Acknowledgment

This work was supported by Science and Engineering Research Board (SERB) grants funded by Department of Science and Technology (DST) Central, Government of India through Teachers Associateship for Research Excellence (TAR/2018/000195) (S Roy). Author (S. Roy) would also like to acknowledge 'Scheme for Transformational and Advanced Research in Sciences (STARS)' (MoE-STARS/STARS-2/2023-0175) by Ministry of Education for promoting translational India-centric research in sciences implemented and managed by Indian Institute of Science (IISc), Bangalore, for their support.

References

[1] A. Demirbas, Global renewable energy projections, Energy Sources Part. B 4 (2009) 212–224.
[2] K. Satyanarayana, A. Mariano, J. Vargas, Are view on micro algae, a versatile source for sustainable energy and materials, Int. J. Energy Res. 35 (2011) 291–311.
[3] H. Lund, Renewable energy strategies for sustainable development, Energy 32 (2007) 912–919.
[4] D. Taub, Effects of rising atmospheric concentrations of carbon dioxide on plants. Nature education knowledge, Nature 3 (10) (2010) 21.
[5] M. Wilson, K. Kannangara, G. Smith, M. Simmons, B. Raguse, Nanotechnology Basic Science and Emerging Technologies, Chapman & Hall/CRC, 2002, pp. 56–87.
[6] M. Parashar, V.K. Shukla, R. Singh, Metal oxides nanoparticles via sol-gel method: a review on synthesis, characterization and applications, J. Mater. Sci.: Mater. Electron. 31 (2020) 3729–3749.
[7] D. Hairston, "Nano-primed," Chemical Engineering., (1999) 39–41.
[8] A.S. Edelstein, R.C. Cammarata, Nanomaterials: Synthesis, Properties and Applications, IOP Publishing Ltd, Bristol/Philadelphia, 1996, pp. 1–596.
[9] M.T. Swihart, Vapor-phase synthesis of nanoparticles, Curr. Opin. Colloid Interface Sci. 8 (2003) 127–133.
[10] S. Roy, G.G. Botte, Perovskite solar cell for photocatalytic water splitting with a TiO_2/Co-doped hematite electron transport bilayer, RSC Adv. 8 (10) (2018) 5388–5394.
[11] W. Suh, K. Suslick, G. Stucky, Y. Suh, Nanotechnology, nanotoxicology, and neuroscience, Prog. Neurobiol. 87 (2009) 133–170.
[12] G. Gardner, Alternative energy and nanotechnology. PCOST Project, 2008, 1–11.
[13] C. Andre, R. Aljhani, T. Gharbi, Y.C. Guillaume, Incorporation of carbon nanotubes in a silica HPLC column to enhance the chromatographic separation of peptides: theoretical and practical aspects, J. Sep. Sci. 34 (2011) 1221–1227.

[14] S. Sandron, B. Paull, P.N. Nesterenko, Investigation on selectivity of mercaptopropylsilica and silver nanoparticles-functionalized mercaptopropylsilica, Curr. Chromatogr. 2 (2015) 122–135.
[15] C. André, G. Lenancker, Y.C. Guillaume, Non-covalent functionalisation of monolithic silica for the development of carbon nanotube HPLC stationary phases, Talanta 99 (2012) 580–585.
[16] A. Dey, S. Datta, Sk. A. Moyez, A. Kamila, K. Mukherjee, S. Roy, Blending of dielectric perovskite with electron transport materials: a case study towards improving bio-molecular devices for energy harvest, ECS J. Solid State Sci. Technol. 10 (1) (2021) 013003.
[17] P. Zegers, Fuel cell commercialization: the key to a hydrogen economy, J. Power Sources 154 (2) (2006) 497–502.
[18] L. Schlapbach, A. Zuttel, Nature 414 (2001) 353–358.
[19] Weaver, Fuel cell technology review, World Fuel Cells, Elsevier Science Ltd, Oxford, UK, 2002, pp. 81–96.
[20] J.H.E.A. Hirschchenhofer, Fuel Cell Handbook, 2, Academic Press, 1998, pp. 1–27.
[21] F. Ning, X. He, Y. Shen, H. Jin, Q. Li, D. Li, et al., Flexible and lightweight fuel cell with high specific power density, ACS Nano 11 (6) (2017) 5982–5991.
[22] K.E. Martin, J.P. Kopasz, K.W. McMurphy, Status of fuel cells and the challenges facing fuel cell technology today, Fuel Cell Chemistry and Operation, 1040, American Chemical Society, 2010, pp. 1–13.
[23] D.A. Stevens, J.R. Dahn, Electrochemical characterization of the active surface in carbon-supported platinum electrocatalysts for PEM fuel cells, J. Electrochem. Soc. 150 (2003) A770.
[24] S. Garbarino, A. Pereira, C. Hamel, E. Irissou, M. Chaker, D. Guay, Effect of Size on the electrochemical stability of Pt nanoparticles deposited on gold substrate, J. Phys. Chem. C. 114 (7) (2010) 2980–2988.
[25] M. Chen, Y. Xing, Polymer-mediated synthesis of highly dispersed Pt nanoparticles on carbon black, Langmuir 21 (2005) 9334.
[26] Y. Lu, Y. Jiang, W. Chen, PtPd porous nanorods with enhanced electrocatalytic activity and durability for oxygen reduction reaction, Nano Energy (2013) 836–844.
[27] V.T.T. Ho, N.G. Nguyen, C.J. Pan, J.H. Cheng, J. Rick, W.N. Su, et al., Advanced nanoelectrocatalyst for methanol oxidation and oxygen reduction reaction, fabricated as one-dimensional Pt nanowires on nanostructured robust $Ti_{0.7}Ru_{0.3}O_2$ support, Nano Energy, 1 (5) (2012) 687–695.
[28] P.M. Urban, A. Funke, J.T. Muller, M. Himmen, A. Docter, Catalytic processes in solid polymer electrolyte fuel cell systems, Appl. Catal. A, Gen. 221 (1–2) (2001) 459–470.
[29] M. Bron, J. Radnik, M. Fieber-Erdmann, P. Bogdanoff, S. Fiechter, EXAFS, XPS and electrochemical studies on oxygen reduction catalysts obtained by heat treatment of iron phenanthroline complexes supported on high surface area carbon black, J. Electroanal. Chem. 535 (2002) 113.
[30] M.M. Hussain, D. Song, Z.S. Liu, Z. Xie, Modeling an ordered nanostructured cathode catalyst layer for proton exchange membrane fuel cells, J. Power Sources 196 (10) (2011) 4533–4544.
[31] J. Kim, S.W. Lee, C. Carlton, Y. Shao-Horn, Pt-covered multiwall carbon nanotubes for oxygen reduction in fuel cell applications, J. Phys. Chem. Lett. 2 (11) (2011) 1332–1336.
[32] S. Mukerjee, S.J. Lee, E.A. Ticianelli, J. McBreen, B.N. Grgur, N.M. Markovic, Investigation of enhanced CO tolerance in proton exchange membrane fuel cells by supported PtMo alloy catalyst, Electrochem. Solid-State Lett. 2 (1) (1999) 12.
[33] S.M. Lyth, Y. Nabae, N.M. Islam, S. Kuroki, M. Kakimoto, S. Miyata, Electrochemical oxygen reduction activity of carbon nitride supported on carbon black, J. Electrochem. Soc. 158 (2) (2011) B194–B201.
[34] Y. Zheng, Y. Jiao, J. Chen, J. Liu, J. Liang, A. Du, et al., Nanoporous graphitic-C_3N_4@carbon metal-free electrocatalysts for highly efficient oxygen reduction, J. Am. Chem. Soc. 133 (50) (2011) 20116–20119.
[35] B. Halford, Nanotube catalysts, Chem. Eng. N. Arch. 87 (6) (2009) 7.
[36] K. Gong, F. Du, Z. Xia, M. Durstock, L. Dai, Nitrogen-doped carbon nanotube arrays with high electrocatalytic activity for oxygen reduction, Science 323 (5915) (2009) 760–764.
[37] B. Smitha, S. Sridhar, A.A. Khan, Solid polymer electrolyte membranes for fuel cell applications, J. Membr. Sci. 259 (1–2) (2005) 10–26.

[38] M. Watanabe, H. Uchida, M. Emori, Polymer electrolyte membranes incorporated with nanometer-size particles of Pt and/or metal-oxides: experimental analysis of the self-humidification and suppression of gas-crossover in fuel cells, J. Phys. Chem. B 102 (17) (1998) 3129–3137.
[39] M. Wakihara, Recent developments in lithium ion batteries, Mater. Sci. Eng.: R: Rep. 33 (4) (2001) 109–134.
[40] J.M. Tarascon, M. Armand, Issues and challenges facing rechargeable lithium batteries, Nature 414 (6861) (2001) 359–367.
[41] S. Brutti, J. Hassoun, B. Scrosati, C.Y. Lin, H. Wu, H.W. Hsieh, A high power Sn-C/C-$LiFePO_4$ lithium ion battery, J. Power Sources 217 (2012) 72–76.
[42] D.S. Kim, C.K. Lee, H. Kim, Preparation of nano-sized $LiCoO_2$ powder by the combination of sonication and modified Pechini process, Solid State Sci. 12 (1) (2010) 45–49.
[43] J. Yamaki, M. Makidera, T. Kawamura, M. Egashira, S. Okada, Voltage prediction of nano-sized $LiNiO_2$ cathode for use in Li-ion cells, J. Power Sources 153 (2) (2006) 245–250.
[44] W. Chang, S.J. Kim, I.T. Park, B.W. Cho, K.Y. Chung, H.C. Shin, Low temperature performance of $LiFePO_4$ cathode material for Li-ion batteries, J. Alloy. Compd. 563 (2013) 249–253.
[45] X. Zhang, H. Zheng, V. Battaglia, R.L. Axelbaum, Electrochemical performance of spinel $LiMn_2O_4$ cathode materials made by flame-assisted spray technology, J. Power Sources 196 (7) (2011) 3640–3645.
[46] Y. Zhang, N. Liu, Nanostructured electrode materials for high-energy rechargeable Li, Na and Zn batteries, Chem. Mater. 29 (22) (2017) 9589–9604.
[47] F.F.C. Bazito, R.M. Torresi, Cathodes for lithium ion batteries: the benefits of using nanostructured materials, J. Braz. Chem. Soc. 17 (4) (2006) 627–642.
[48] C.L. Chen, K.F. Chiu, H.J. Leu, C.C. Chen, Iron hexacyanoferrate based compound modified $LiMn_2O_4$ cathodes for lithium ion batteries, J. Electrochem. Soc. 160 (5) (2013) A3126–A3130.
[49] X. Zhao, C.M. Hayner, H.H. Kung, Self-assembled lithium manganese oxide nanoparticles on carbon nanotube or graphene as high-performance cathode material for lithium-ion batteries, J. Mater. Chem. 21 (43) (2011) 17297–17303.
[50] M. Kotobuki, Improved performance of hydrothermally synthesized $LiMnPO_4$ by ball-milling as a positive electrode for Li ion battery, ISRN Electrochem. (2013) 1–5.
[51] D.B. Le, S. Passerini, J. Guo, J. Ressler, B.B. Owens, W.H. Smyrl, High surface area V_2O_5 aerogel intercalation electrodes, J. Electrochem. Soc. 143 (1996) 2099–2104.
[52] H.J. Muhr, F. Krumeich, U.P. Schoholzer, F. Bieri, M. Niederberger, L.J. Gauckler, Vanadium oxide nanotubes—a new flexible vanadate nanophase, Adv. Mater. 12 (2000) 231–234.
[53] C. Casas, Las De, W.Z. Li, A review of application of carbon nanotubes for lithium ion battery anode material, J. Power Sources 208 (2012) 74.
[54] N.A. Kaskhedikar, J. Maier, Lithium storage in carbon nanostructures, Adv. Mater. 21 (2009) 2664.
[55] X.X. Wang, J.N. Wang, H. Chang, Y.F. Zhang, Preparation of short carbon nanotubes and application as an electrode material in Li-ion batteries, Adv. Funct. Mater. 17 (2007) 3613–3618.
[56] W. Wang, M. Tian, Y. Wei, S.H. Lee, Y.C. Lee, R. Yang, Binder-free three-dimensional silicon/carbon nanowire networks for high performance lithium-ion battery anodes, Nano Energy 2 (2013) 943–950.
[57] L.F. Cui, L. Hu, J.W. Choi, Y. Cui, Carbon-silicon core-shell nanowires as high capacity electrode for lithium ion batteries, ACS Nano. 4 (2010) 3671.
[58] K.S. Park, K.M. Min, S.D. Seo, G.H. Lee, H.W. Shim, D.W. Kim, Self-supported multi-walled carbon nanotube-embedded silicon nanoparticle films for anodes of Li-ion batteries, Mater. Res. Bull. 48 (2013) 1732.
[59] J. Li, J. Guo, X. Zhang, R. Peng, Tin oxide nano-flower-anchored graphene composites as high-performance anode materials for lithium-ion batteries, Ionics 22 (2016) 2307–2313.
[60] A.L.M. Reddy, M.M. Shaijumon, S.R. Gowda, P.M. Ajayan, Coaxial MnO_2/carbon nanotube array electrodes for high-performance lithium batteries, Nano Lett. 9 (3) (2009) 1002–1006.
[61] N. Du, H. Zhang, B. Chen, J.B. Wu, X.Y. Ma, Z.H. Liu, et al., Porous Co_3O_4 nanotubes derived from $Co_4(CO)_{12}$ clusters on carbon nanotube templates: a highly efficient material for Li battery applications, Adv. Mater. 19 (24) (2007) 4505–4509.

[62] Z. Zhang, X. Zhang, X. You, M. Zhang, M.D. Walle, J. Wang, et al., 3D well-interconnected NiO-graphene-carbon nanohybrids as high-performance anode materials for Li-ion batteries, J. Nanopart. Res. 18 (2016) 247.
[63] S. Roy, G.S. Han, H. Shin, J.W. Lee, J. Mun, H. Shin, et al., Low temperature synthesis of rutile TiO_2 nanocrystals and their photovoltaic and photocatalytic properties, J. Nanosci. Nanotechnol. 15 (6) (2015) 4516–4521. Available from: https://doi.org/10.1166/jnn.2015.9778.
[64] A. Bhaskar, M. Deepa, T.N. Rao, MoO_2/multiwalled carbon nanotubes (MWCNT) hybrid for use as a Li-ion battery anode, ACS Appl. Mater. Interfaces 5 (2013) 2555.
[65] Y. Wu, Y. Wei, J.P. Wang, K.L. Jiang, S.S. Fan, Conformal Fe_3O_4 sheath on aligned carbon nanotube scaffolds as high-performance anodes for lithium ion batteries, Nano Lett. 13 (2013) 818.
[66] X.W. Zhou, G.M. Wu, G.H. Gao, C.J. Cui, H.Y. Yang, J. Shen, et al., The synthesis, characterization and electrochemical properties of multi-wall carbon nanotube-induced vanadium oxide nanosheet composite as a novel cathode material for lithium ion batteries, Electrochim. Acta 74 (2012) 32–38.
[67] J.H. Lee, B.S. Kong, S.B. Yang, H.T. Jung, Fabrication of single-walled carbon nanotube/tin nanoparticle composites by electrochemical reduction combined with vacuum filtration and hybrid co-filtration for high-performance lithium battery electrodes, J. Power Sources 194 (2009) 520.
[68] Y. Yu, L. Gu, C.L. Wang, A. Dhanabalan, P.A. Aken, J. van, Maierm, Encapsulation of Sn@carbon nanoparticles in bamboo-like hollow carbon nanofibers as an anode material in lithium-based batteries, Angew. Chem. Int. Ed. 48 (2009) 6485.
[69] X.F. Li, Y. Zhong, M. Cai, M.P. Balogh, D.N. Wang, Y. Zhang, et al., Tin-alloy heterostructures encapsulated in amorphous carbon nanotubes as hybrid anodes in rechargeable lithium ion batteries, Electrochim. Acta 89 (2013) 387.
[70] Y. Wang, J.Y. Lee, One-step, confines growth of bimetallic tin-antimony nanorods in carbon nanotubes grown in situ for reversible Li^+ ion storage, Angew. Chem. Int. Ed. 45 (2006) 7039.
[71] Z.H. Bi, M.P. Paranthaman, P.A. Menchhofer, R.R. Dehoff, C.A. Bridges, M.F. Chi, et al., Self-organized amorphous TiO_2 nanotube arrays on porous Ti foam for rechargeable lithium and sodium ion batteries, J. Power Sources 222 (2013) 461.
[72] S. Datta, A. Dey, N.R. Singha, S. Roy, Enhanced performance of dye-sensitized solar cell with thermally stable natural dye-assisted TiO_2/MnO_2 bilayer-assembled photoanode, Mater. Renew. Sustain. Energy 9 (4) (2020). Available from: https://doi.org/10.1007/s40243-020-00185-3.
[73] M.A. Green, The path to 25% silicon solar cell efficiency: history of silicon cell evolution, Prog. Photovoltaics: Res. Appl. 17 (3) (2009) 183–189.
[74] Y. Ye, L. Dai, Graphene-based Schottky junction solar cells, J. Mater. Chem. 22 (2012) 24224–24229.
[75] J. Wei, Y. Jia, Q. Shu, Z. Gu, K. Wang, D. Zhuang, Double-walled carbon nanotube solar cells, Nano Lett. 7 (8) (2007) 2317–2321.
[76] X. Li, H. Zhu, K. Wang, A. Cao, J. Wei, C. Li, Graphene-on-silicon Schottky junction solar cells, Adv. Mater. 22 (25) (2010) 2743–2748.
[77] Y. Shi, K.K. Kim, A. Reina, M. Hofmann, L.J. Li, J. Kong, Work function engineering of graphene electrode via chemical doping, ACS Nano 4 (5) (2010) 2689–2694.
[78] S. Spiering, D. Hariskose, M. Powalla, N. Naghavi, D. Lincot, Cd-free Cu(In,Ga)Se_2 thin-film solar modules with In_2S_3 buffer layer by ALCVD, Thin Solid Films 431–432 (2003) 359–363.
[79] S. Yu, L. Shu, Y. Qian, Y. Xie, J. Yang, L. Yang, Hydrothermal preparation and characterization of nanocrystalline powder of β-indium sulfide, Mater. Res. Bull. 33 (5) (1998) 717–721.
[80] B. Sahu, Preparation, structure, composition, optical and photoelectrochemical properties of vacuum annealed In-Se thin films, Thin Solid Films 261 (1–2) (1995) 98–106.
[81] M. Grätzel, Photoelectrochemical cells, Nature 414 (2001) 338–344.
[82] E.L. Narsaiah, M.J. Reddy, U.V.S. Rao, Study of a new polymer electrolyte (PVP + KYF4) for solid-state electrochemical cells, J. Power Sources 55 (1995) 255.
[83] S.A. Haque, E. Palomares, H.M. Upadhyaya, L. Otley, R.J. Potter, A.B. Holmes, et al., Flexible dye sensitised nanocrystalline semiconductor solar cells, Chem. Commun. 3008 (2003).

[84] R. Singh, N.A. Jadhav, S. Majumder, B. Bhattacharya, P.K. Singh, Novel biopolymer gel electrolyte for dye-sensitized solar cell application, Carbohyd. Polym. 91 (2013) 682–685.
[85] G.M. Whitesides, J.K. Kriebel, B.T. Mayers, Self-assembly and nanostructured materials, in: W.T. S. Huck (Ed.), Nanoscale Assembly: Chemical Techniques, Springer, Berlin, 2005, pp. 217–239.
[86] G.M. Whitesides, J.K. Kriebel, B.T. Mayers, Self-assembly and nanostructured materials, in: W.T. S. Huck (Ed.), Nanoscale Assembly Chemical Techniques, Springer, USA, 2005, p. 217.
[87] G. Kartal, O. Kahvecioglu, S. Timur, Investigating the morphology and corrosion behavior of electrochemically borided steel, Surf. Coat. Technol. 200 (11) (2006) 3590–3593.
[88] A. Greco, K. Mistry, V. Sista, O. Eryilmaz, A. Erdemir, Friction and wear behaviour of boron based surface treatment and nano-particle lubricant additives for wind turbine gearbox applications, Wear 271 (2011) 1754–1760.
[89] S. Fontes, Biofuels: policies, standards and technologies, World Energy Counc. (2010) 1–152.
[90] T. Mahmood, S. Hussain, Nanobiotechnology for the production of biofuels from spent tea, Afr. J. Biotechnol. 9 (2010) 858–868.
[91] W. Goh, V. Makam, J. Hu, L. Kang, M. Zheng, S. Yoong, Iron oxide filled magnetic carbon nanotube-enzyme conjugates for recycling of amyloglucosidase: toward useful applications in biofuel production process, Langmuir 28 (2012) 16864–16873.
[92] F. Maa, M. Hanna, Biodiesel production: a review, Bioresour Technol. 70 (1999) 1–15.
[93] G. Chen, C. Guo, H. Qiao, M. Ye, X. Qiu, C. Yue, Well-dispersed sulfated zirconia nanoparticles as high-efficiency catalysts for the synthesis of bis(indolyl)methanes and biodiesel, Catal. Commun. 41 (2013) 70–74.
[94] V. Sajith, C. Sobhan, G. Peterson, Experimental investigations on the effects of cerium oxide nanoparticle fuel additives on biodiesel. (ArticleID581407), Adv. Mech. Eng. 2 (2010) 1–6.
[95] F. Qiu, Y. Li, D. Yang, X. Li, P. Sun, Heterogeneous solid base nanocatalyst: preparation, characterization and application in biodiesel production, Bioresour. Technol. 102 (2011) 4150–4156.
[96] D. Zhang, D. Wei, W. Ding, X. Zhang, Carbon-based nanostructured catalyst for biodiesel production by catalytic distillation, Catal. Commun. 43 (2014) 121–125.
[97] L. Wen, Y. Wang, D. Lu, S. Hu, H. Han, Preparation of KF/CaO nanocatalyst and its application in biodiesel production from Chinese tallow seed oil, Fuel 89 (2010) 2267–2271.
[98] L. Konwar, J. Boro, D. Deka, Review on latest developments in biodiesel production using carbon-based catalysts, Renew. Sustain. Energy Rev. 29 (2014) 546–564.
[99] M. Kaur, A. Ali, Lithium ion impregnated calcium oxide as nanocatalyst for the biodiesel production from Karanja and Jatropha oils, Renew. Energy 36 (2011) 2866–2871.
[100] S. Hu, Y. Guan, Y. Wang, H. Han, Nano-magnetic catalyst KF/CaO–Fe_3O_4 for biodiesel production, Appl. Energy 88 (2011) 2685–2690.
[101] L.K. Ventrapragada, J. Zhu, S.E. Creager, A.M. Rao, R. Podila, A versatile carbon nanotube-based scalable approach for improving interfaces in Li-ion battery electrodes, ACS Omega 3 (4) (2018) 4502–4508.
[102] S.H. Kim, J.Y. Jung, R.B. Wehrspohn, J.H. Lee, All-room-temperature processed 17.25%-crystalline silicon solar cell, ACS Appl. Energy Mater. 3 (4) (2020) 3180–3185.
[103] C.R. Kunkel, F. Vines, F. Illas, Biogas upgrading by transition metal carbides, ACS Appl. Energy Mater. 1 (1) (2018) 43–47.
[104] X. Qu, J. Brame, Q. Li, P. Alvarez, Nanotechnology for a safe and sustainable water supply: enabling integrated water treatment and reuse, Acc. Chem. Res. 46 (2013) 834–843.
[105] S. Gillett, Nanotechnology: clean energy and resources for the future. White Paper. Foresight Institute, 2002, 1–92.
[106] M.H. Ahmadi, M.A. Ahmadi, M.A. Nazari, O. Mahian, R. Ghasempour, A proposed model to predict thermal conductivity ratio of Al_2O_3/EG nanofluid by applying least squares support vector machine (LSSVM) and genetic algorithm as a connectionist approach, J. Therm. Anal. Calorim. 135 (2018) 271–281.
[107] C. RamReddy, P.V.S.N. Murthy, A.J. Chamkha, A.M. Rashad, Soret effect on mixed convection flow in a nanofluid under convective boundary condition, Int. J. Heat. Mass. Transf. 64 (2013) 384–392.

[108] P.R. Mashaei, M. Shahryari, S. Madani, Numerical hydrothermal analysis of water-Al_2O_3 nanofluid forced convection in a narrow annulus filled by porous medium considering variable properties, J. Therm. Anal. Calorim. 126 (2016) 891−904.
[109] A. Sivakumar, N. Alagumurthi, T. Senthilvelan, Experimental and numerical investigation of forced convective heat transfer coefficient in nanofluids of Al_2O_3/water and CuO/EG in a serpentine shaped microchannel heat sink, Int. J. Heat. Technol. 33 (2015) 155−160.
[110] S. Kalagirou, Seawater desalination using renewable energy sources, Prog. Energy Combust. Sci. 31 (2005) 242−281.
[111] A.J. Bandodkar, I. Jeerapan, J. You, R. Nunez-Flores, J. Wang, Highly stretchable fully printed CNT based electrochemical sensors and biofuel cells: combining intrinsic and design induced stretchability, Nano Lett. 16 (1) (2015) 721−727.
[112] G. Diglio, C. Roselli, M. Sasso, C.U. Jawali, Borehole heat exchanger with nanofluids as heat carrier, Geothermics 72 (2018) 112−123.
[113] M. Daneshipour, R. Rafee, Nanofluids as the circuit fluids of the geothermal borehole heat exchangers, Int. Commun. Heat. Mass. Transf. 81 (2017) 34−41.
[114] S. Ganguly, K. Kargupta, D. Banerjee, Nanotechnology and nanomaterials for new and sustainable energy engineering, In: Proc. Int. Conf. Nanomaterials: Appl. Prop. 1 (4) (2012) 1−5.
[115] J.B. Ooi, H.M. Islami, V. Swamy, X. Wang, A.K. Swain, J.R. Rajanren, Graphite oxide nanoparticles as diesel fuel additive for cleaner emissions and lower fuel consumption, Energy Fuels 30 (2) (2016) 1341−1353.
[116] X.J. Zhao, G. Zhu, Y.J. Fan, H.Y. Li, Z.L. Wang, Triboelectric charging at the nanostructured solid/liquid interface for area-scalable wave energy conversion and its use in corrosion protection, ACS Nano 9 (7) (2015) 7671−7677.
[117] R. Hu, B.A. Cola, N. Haram, J.N. Barisci, Harvesting waste thermal energy using a carbon nanotube based thermo electrochemical cell, Nano Lett. 10 (3) (2010) 838−846.

CHAPTER 16

Utilization of nanoceramics and its limitations in construction materials—a global perspective

S. Ajith[1] and V. Arumugaprabu[2]
[1]Department of Mechanical Engineering, Stella Mary's College of Engineering, Aruthenganvilai, Tamil Nadu, India
[2]Department of Mechanical Engineering, Kalasalingam Academy of Research and Education, Krishnankoil, Tamil Nadu, India

16.1 Introduction

Nanoceramics are known to be a type of nanoparticle composed of ceramics that has a size of 100 nm or less but has unique physical and chemical properties. Ceramics is a Greek word that means "Pottery." The classification of ceramics is shown in Fig. 16.1. Nanoceramics are typically categorized as nonmetallic, heat resistant, and inorganic solids that can be made either by metallic or nonmetallic compounds. Ceramics are rigid, brittle, and can be broken easily, however, nanoceramics have diverse properties which include ferroelectric, electrooptical, magnetoresistive, superconductive, ferromagnetic, pyroelectric, dielectric, piezoelectric, wear resistance, high thermal shock resistance, fracture toughness, corrosion resistance, chemical inertness, and low density. Ceramics are categorized into two types such as traditional and advanced ceramics. Nanoceramics were discovered over five decades ago, but still their role in industrial applications is at the experimentation stage. The usage of nanoceramics from lab to industries has developed nowadays due to their higher-end properties [1]. Several industrial applications of nanoceramics are in construction materials, transportation,

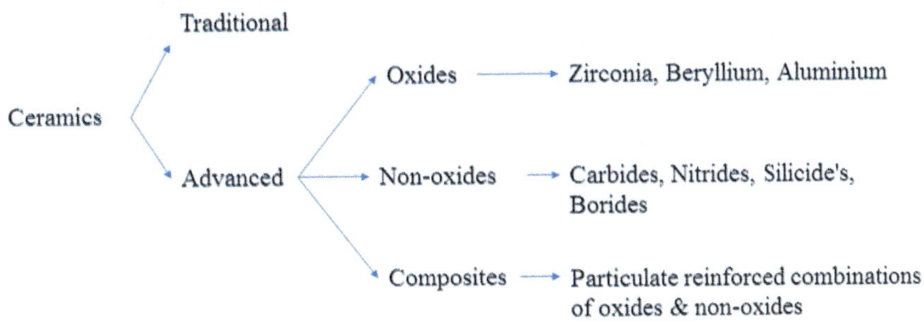

Figure 16.1 Classification of ceramics.

Industrial Applications of Nanoceramics
DOI: https://doi.org/10.1016/B978-0-323-88654-3.00021-4

energy supply, communication systems, and storage. Around 30% of the ceramic materials produced from the industries are dumped as waste which is a significant concern for the environment. Recycling or reusing ceramic waste remains challenging [2]. The purpose of nanoceramics in a construction site is not only to enhance the strength of the construction materials but also to conserve energy in various forms. Energy conservation is the major factor that has to be focused on as 41% of energy in the United States is consumed by residential buildings and commercial buildings through air conditioning, heat, and light. This plays a major role in building materials with high performance, durability, ease in maintenance, and energy-efficiency. Therefore, nanotechnology has become significant in the field of construction engineering [3].

Nanoceramics have piqued researchers' interest as prospective materials since their discovery in 1980, owing to their improved and unique features compared to traditional bulk ceramic materials, which are brittle, hard, and break on impact. Due to their tiny grain size, crystallinity, and abundant grain boundaries, nanoceramics have unique processing, mechanical, and surface properties such as machinability, toughness, strength, superplasticity, and bioactivity [4]. Nanoceramics are used in a variety of current and developing technologies, because of their strength, biocompatibility, hydrophilicity, and wear resistance. Nanoceramic materials have been used in biological, health, and environmental applications. As the constituent particles of nanoceramics are smaller, they provide more surface area to almost anything they come into contact with. This is advantageous in any application that requires a high reaction rate since more of it can react fast with the intended reactant [5].

16.2 Nanoceramics in bitumen

Bitumen is the key material for highway projects as it has to bear traffic loads as well as changes in climatic conditions. To decrease shoving and rutting and to lead better adhesion between aggregate and asphalt in the presence of moisture to decrease stripping, a supreme bitumen binder should have a high relative stiffness at extreme service temperatures (particularly during the summer) [6]. Nanoclay and carbon microfiber modification were used to investigate water sensitivity and tensile strength. Nanomaterials can significantly increase the mechanical strength of nanoclay (1.5%) and alteration of the asphalt mixture offers increased tensile strength and moisture susceptibility [7]. Alternative modifier materials, on the other hand, have gained attention as the cost of bitumen and the overall cost of modified bitumen with a standard addition have continued to rise. The cost of modified bitumen with commercial additives is often between 60% and 150% more than the cost of original bitumen. As a result, recycling materials as bitumen modifiers can be a cost-effective alternative to traditional modifiers. Furthermore, repurposed materials are preferable to typical bitumen modifiers which have detrimental enironmental consequences [8,9].

16.3 Nanoceramics in tiles and walls

Nanoceramics in tiles and floors are used both indoor and outdoor with various shapes and sizes. There are various types of floor tiles such as decorated floor tiles, quarry tiles, and floor tiles. The characteristics of these tiles vary in terms of weather stains, abrasion, and high resistance. For instance, glass-ceramics such as aluminum silicate and alumina silicon nitride are used in elevated temperatures, whereas porcelain and terra cotta are used at around room temperature. The nanoparticles of ceramic materials such as quartz, limestone, and clay showed better mechanical results. The resulting ceramic units or products have sharp and well-developed crystals. The morphology, or look of the fired units from the outside, is clearly defined [10].

16.4 Nanoceramics in concrete

Concrete is a key material for any construction and it is composed of fine aggregates, coarse aggregates, cement, and water. The mechanical and physical properties of concrete can be improved by adding nanoceramics, such as nano-ZrO_2 (nanozirconia) as well as nano-Al_2O_3 (nanoalumina powder). Properties of the concrete such as split tensile strength, compressive strength, water absorption test, and workability have been analyzed in which both the materials showed better results when compared with normal concrete. The workability is determined using the slump cone test which indicates that the slump is reduced by 1.1% when nano-ZrO_2 is added. The same result is obtained with nano-Al_2O_3. The split tensile strength and the compression strength increased by 0.7% when nano-ZrO_2 is added. The water absorption is decreased by 0.12% at 28 days and 0.1% at 60 and 90 days when nano-ZrO_2 is added [11]. Some of the nanoceramics are used in different types of cement to achieve better properties of concrete. Nanomaterials are used in different types of Portland cement such as blast furnace, clinker, Pozzolan, masonry, composite, white, fly ash, gypsum, and ordinary Portland cement. The usage of nanomaterials in Portland cement is high when compared to non-Portland cement. However, some of the non-Portland cements in which nanomaterials are being used are calcium aluminate, Pozzolan lime cement, supersulfated, slag lime, and geopolymer cement [12]. The major properties of nanoceramics when mixed with cement are shown in Fig. 16.2.

Addition of nanocellulose to nanoceramics increases the modulus of elasticity, bonding strength, and decreases moisture absorption. Nanozinc oxide increases the aging resistance and compressive strength, nanotitania resists corrosion, abrasion, and flame, nanoalumina increases the compressive strength and elastic modulus, and resists thermal shock, nanocalcium carbonate improves hardness and compressive strength, carbon nanotubes increase compressive and flexural strength, and cause improved durability and blast resistance [13]. When manufactured with polymers the

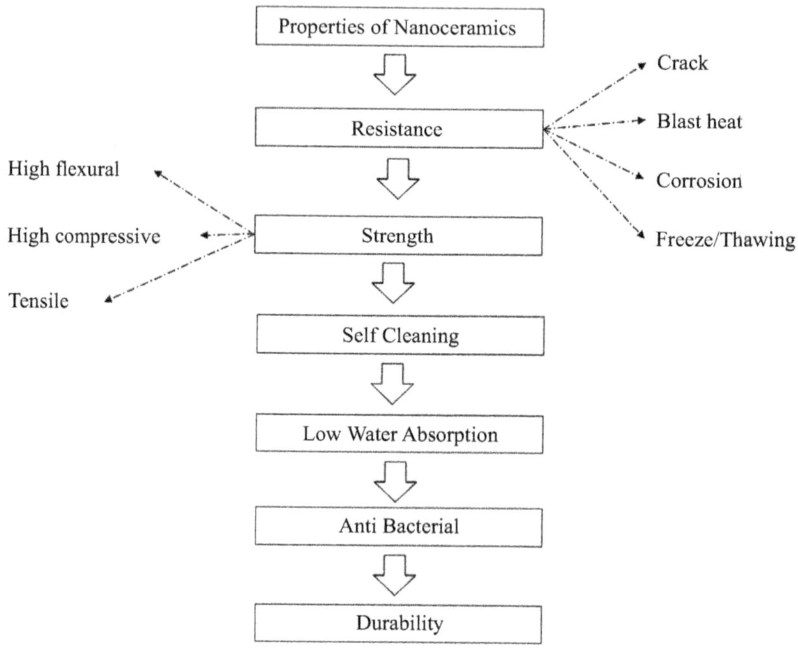

Figure 16.2 Properties of nanoceramics.

nanocomposites show increased tensile and flexural strength which further leads to resistance to wear and tear in buildings. Nanoceramic coating is identified as one of the suitable materials for lowering the cooling loads in buildings. Nanocrystalline materials in the form of coating films tend to minimize corrosion. Nanocrystalline cellulose can be utilized as a component of building materials to provide reinforcement and to extend the life of paints and varnishes. Nanosilica, in addition to having a low-cost, increased tensile strength and compressive strength, ability to prevent silicosis, high surface area, and a lower percentage of CO_2, when combined with recycled concrete aggregates assists in minimizing solid waste pollution [14,15].

16.5 Ceramic wastes in concrete

The key ingredient for manufacturing ceramic products is clay which is rich in silicate (SiO_2). Clay minerals when heated between 600°C and 900°C will exhibit high pozzolanic property. Ceramic waste obtained during the manufacturing of ceramic roof tiles is fired up to a temperature of 1000°C which is used during cement manufacturing. Also, waste from the ceramic floor, brick can be used as a pozzolanic material. Ceramic waste can also be used in cement mortar to minimize disposal problems. Ceramic waste is used as a raw ingredient in cement clinkers which possesses better performance in terms of durability and chemical, physical, and morphological properties when compared with existing cement. It is also reported that 20% replacement of ceramic materials in cement improved mechanical properties and 40% replacement of ceramic materials showed better results in chemical resistance in conventional concrete [16–18]. There will be no change in the characteristics of M-30 grade concrete when 20% of ceramic waste is mixed. From the total raw materials used, it is estimated that 15%–30% of the ceramic waste is produced which is used for excavation pit refilling in the construction site. Ceramic powders can replace less than 40% of the ordinary Portland cement to enhance the property of the microstructure. Ceramic materials can also be used as coarse aggregate in conventional concrete and fine aggregate in geopolymer concrete which has equal strength [19,20].

16.6 Asphalt in concrete

Asphalt concrete is a type of composite material that is used to cover airport runways, driveways, and roads. Asphalt is a valuable building material because of its capacity to sustain high traffic loads as a surface cover for the road foundation. It does, however, have some drawbacks. At the moment, meeting the design requirements for severe temperatures is tough. The asphalt's resilience to high and low temperatures is poor, resulting in melting or cracking. To adjust for the high temperatures, nanoparticles can be added to adjust the material qualities of ductility and elasticity of asphalt, with the most common modifiers being polyethylene, styrene–butadiene–styrene and styrene–butadiene–rubber [21–23].

16.7 Nanoceramics in bricks

For many years, bricks have been utilized as a building material. Currently, clay makes up little more than 50% of earth bricks, with sand and other granular materials making up the remaining portion. These minerals are joined together at high temperatures to form a high-compressive-strength material, making earth brick an excellent residential building material [24,25]. Existing earth bricks, on the other hand, do not have high compressive strength. To determine the impact of nanoclays on earth bricks, compressive strength tests were determined and indicated that the utilization of different nanoparticles in bricks is based on the chemical makeup of the brick. Nanoclays are layered mineral silicates of nanoparticles [26–28].

16.8 Health effects of nanoceramics

Researchers have also identified that nanoparticles can create extensive damage to human organs when they are highly exposed in the manufacturing process. There are many ways in which nanoparticles can enter the human body such as ingestion, inhalation, skin absorption, injection, and transportation. Some of the toxicity effects of nanoceramics in various human organs are pulmonary toxicity, cardiovascular toxicity, nephrotoxicity, dermal toxicity, spermatic toxicity, neurotoxicity, and genotoxicity. The severity of these effects can be minimized by using proper personal protective equipment, such as a suitable coverall, breathing apparatus, goggles, face shield, hand gloves, and footwear, during manufacturing [29].

16.9 Conclusion

Nanoparticles have been highlighted as the next frontier of progress in several sectors, with applications that produce previously unimaginable results. There are different types of nanoceramics used in various industrial applications but the utilization of nanoceramics in construction materials has not been fully utilized. Increasing the performance of construction materials and lowering energy consumption, nanoceramics can help to reduce the use of natural resources. The use of nanoceramics in construction provides a more cost-effective, faster, and safer method of producing building materials. Nanoceramics allow concrete to be precisely engineered for specific uses, lowering costs while improving performance. Focusing the use of nanoceramics for cement and concrete will thus have a long-term and significant impact on the future of construction materials.

References

[1] H.T. Ting, K.A. Abou-El-Hossein, H.B. Chua, Review of micromachining of ceramics by etching, Trans. Nonferrous Met. Soc. China 9 (2009) s1−s16. Available from: https://doi.org/10.1016/S1003-6326(10)60237-3, https://www.sciencedirect.com/science/article/abs/pii/S1003632610602373.

[2] A.A. Hussein, R.P. Jaya, N.A. Hassan, H. Yaacob, G.F. Huseien, M.H.W. Ibrahim, Performance of nanoceramic powder on the chemical and physical properties of bitumen, Constr. Build. Mater. 156 (496) (2017) 505. Available from: https://doi.org/10.1016/j.conbuildmat.2017.09.014.

[3] M.A. Ariffin, M.W. Hussin, M. Samadi, N.H.A.S. Lim, J. Mirza, D. Awalluddin, et al., Effect of ceramic aggregate on high strength multi blended ash geopolymer mortar, J. Teknologi 77 (16) (2015) 33−36. Available from: https://doi.org/10.11113/jt.v77.6391.

[4] Q. Liu, D. Ding, C. Ning, Anodic fabrication of Ti-Nb-Zr-O nanotube arrays, J. Nanomater. 2014 (2014) 1−7. Available from: https://doi.org/10.1155/2014/240346.

[5] O. Van der Biest, Nanoceramics: issues and opportunities, Int. J. Appl. Ceram. Technol. 10 (4) (2013) 565−576. Available from: https://doi.org/10.1111/ijac.12074.

[6] L.M. Costa, H.M.R.D. Silva, J.R. Oliveira, S.R. Fernandes, Incorporation of waste plastic in asphalt binders to improve their performance in the pavement, Int. J. Pavement Res. Technol. 6 (4) (2013) 457−464. Available from: https://doi.org/10.6135/ijprt.org.tw/2013.6(4)0.457.

[7] S.W. Goh, M. Akin, Z. You, X. Shi, Effect of deicing solutions on the tensile strength of micro-or nano-modified asphalt mixture, Constr. Build. Mater. 25 (1) (2011) 195−200. Available from: https://doi.org/10.1016/j.conbuildmat.2010.06.038.

[8] M. Pérez-Martínez, F. Moreno-Navarro, J. Martín-Marín, C. Ríos-Losada, M.C. Rubio-Gámez, Analysis of cleaner technologies based on waxes and surfactant additives in road construction, J. Clean. Prod. 65 (2014) 374–379. Available from: https://doi.org/10.1016/j.jclepro.2013.09.012.
[9] A. Modarres, P. Ayar, Coal waste application in recycled asphalt mixtures with bitumen emulsion, J. Clean. Prod. 83 (2014) 263–272. Available from: https://doi.org/10.1016/j.jclepro.2014.07.082.
[10] H.H.M. Darweesh, Nanoceramics: materials, properties, methods and applications—Part II, J. Nanosci. 1 (1) (2018) 40–66.
[11] Z.J. Kadhim, A.R. Al-Adhadh, The effect of nano ceramic powders on mechanical and physical properties of normal concrete, IOP Conf. Ser.: Mater. Sci. Eng. 739 (1) (2020) 012020. Available from: https://doi.org/10.1088/1757-899X/739/1/012020.
[12] B.B. Das, A. Mitra, Nanomaterials for construction engineering—a review, Int. J. Mater. Mech. Manuf. 2 (1) (2014) 41–46.
[13] L. Raki, J. Beaudoin, R. Alizadeh, J. Makar, T. Sato, Cement and concrete nanoscience and nanotechnology, Materials 3 (2) (2010) 918–942. Available from: https://doi.org/10.3390/ma3020918.
[14] S. Maheswaran, B. Bhuvaneshwari, G.S. Palani, R. Nagesh, S. Kalaiselvam, An overview on the influence of nano silica in concrete and a research initiative, Res. J. Recent Sci. 2 (ISC-2012) (2013) 17–24.
[15] A.K. Rana, S.B. Rana, A. Kumari, V. Kiran, Significance of nanotechnology in construction engineering, Int. J. Recent. Trends Eng. 1 (4) (2009) 46–48.
[16] F. Puertas, I. García-Díaz, A. Barba, M.F. Gazulla, M. Palacios, M.P. Gómez, et al., Ceramic wastes as alternative raw materials for Portland cement clinker production, Cem. Concr. Compos. 30 (9) (2008) 798–805. Available from: https://doi.org/10.1016/j.cemconcomp.2008.06.003.
[17] R.D. Toledo Filho, J.P. Gonçalves, B.B. Americano, E.M.R. Fairbairn, Potential for use of crushed waste calcined-clay brick as a supplementary cementitious material in Brazil, Cem. Concr. Res. 37 (9) (2007) 1357–1365. Available from: https://doi.org/10.1016/j.cemconres.2007.06.005.
[18] E. Vejmelková, M. Keppert, P. Rovnaníková, M. Ondráček, Z. Keršner, R. Černý, Properties of high-performance concrete containing fine-ground ceramics as supplementary cementitious material, Cem. Concr. Compos. 34 (1) (2012) 55–61. Available from: https://doi.org/10.1016/j.cemconcomp.2011.09.018.
[19] M. Samadi, M.W. Hussin, H.S. Lee, A.R.M. Sam, M.A. Ismail, N.H.A.S. Lim, et al., Properties of mortar containing ceramic powder waste as cement replacement, J. Teknologi 77 (12) (2015) 93–97. Available from: https://doi.org/10.11113/jt.v77.6315.
[20] M. Limbachiya, M.S. Meddah, Y. Ouchagour, Use of recycled concrete aggregate in fly-ash concrete, Constr. Build. Mater. 27 (1) (2012) 439–449. Available from: https://doi.org/10.1016/j.conbuildmat.2011.07.023.
[21] C. Zhu, H. Zhang, G. Xu, C. Shi, Aging rheological characteristics of SBR modified asphalt with multi-dimensional nanomaterials, Constr. Build. Mater. 151 (2017) 388–393. Available from: https://doi.org/10.1016/j.conbuildmat.2017.06.121.
[22] M. Mubaraki, S.I.A. Ali, A. Ismail, N.I.M. Yusoff, Rheological evaluation of asphalt cements modified with ASA polymer and Al_2O_3 nanoparticles, Procedia Eng. 143 (2016) 1276–1284. Available from: https://doi.org/10.1016/j.proeng.2016.06.135.
[23] M.T. Rahman, M.R. Hainin, W.A.W.A. Bakar, Use of waste cooking oil, tire rubber powder and palm oil fuel ash in partial replacement of bitumen, Constr. Build. Mater. 150 (2017) 95–104. Available from: https://doi.org/10.1016/j.conbuildmat.2017.05.216.
[24] A. Mohajerani, A. Ukwatta, T. Jeffrey-Bailey, M. Swaney, M. Ahmed, G. Rodwell, et al., A proposal for recycling the world's unused stockpiles of treated wastewater sludge (biosolids) in fired-clay bricks, Buildings 9 (1) (2019) 1–22. Available from: https://doi.org/10.3390/buildings9010014.
[25] A. Mohajerani, A.A. Kadir, L. Larobina, A practical proposal for solving the world's cigarette butt problem: recycling in fired clay bricks, Waste Manag. 52 (2016) 228–244. Available from: https://doi.org/10.1016/j.wasman.2016.03.012.

[26] H. Niroumand, M.F.M. Zain, S.N. Alhosseini, The influence of nano-clays on compressive strength of earth bricks as sustainable materials, Procedia—Social Behav. Sci. 89 (2013) 862–865. Available from: https://doi.org/10.1016/j.sbspro.2013.08.945.
[27] A. Mohajerani, L. Burnett, J.V. Smith, H. Kurmus, J. Milas, A. Arulrajah, et al., Nanoparticles in construction materials and other applications, and implications of nanoparticle use, Materials 12 (19) (2019) 1–25. Available from: https://doi.org/10.3390/ma12193052.
[28] M. Stefanidou, A. Karozou, Testing the effectiveness of protective coatings on traditional bricks, Constr. Build. Mater. 111 (2016) 482–487. Available from: https://doi.org/10.1016/j.conbuildmat.2016.02.114.
[29] G. Pattan, G. Kaul, Health hazards associated with nanomaterials, Toxicol. Ind. Health 30 (6) (2014) 499–519. Available from: https://doi.org/10.1177/0748233712459900.

CHAPTER 17

Fundamentals of nanoceramics and their composites

V. Selvam[1], R. Sarika[2], D. Silambarasan[3], S. Sofiya Lawrence Mary[4] and K.S. Prakash[4]
[1]Department of Chemistry, The Madurai Diraviyam Thayumanavar Hindu College (Affiliated to Manonmaniam Sundaranar University, Tirunelveli), Tirunelveli, Tamil Nadu, India
[2]Department of Chemistry, Fatima College (Affiliated to Madurai Kamaraj University, Madurai), Madurai, Tamil Nadu, India
[3]Department of Physics, The Madurai Diravsiyam Thayumanavar Hindu College (Affiliated to Manonmaniam Sundaranar University, Tirunelveli), Tirunelveli, Tamil Nadu, India
[4]Department of Chemistry, Bharathidasan Government College for Women (Autonomous) (Affiliated to Pondicherry University, Pondicherry), Puducherry, India

17.1 Introduction

17.1.1 Fundamentals and brief history

The evolution of science and technology always demands new materials with novel/special properties to satisfy human needs. In this context, over the past few decades, by possessing dynamical properties, ceramic materials have gained attention for countless applications, such as sensors, electronic devices, fuel cells, manufacturing, and transportation [1]. Hence, there is immense interest among researchers to document the fundamentals, properties, methods of preparation of various nanoceramics, and their applications with regard to the aircraft industry. In this chapter, we present the fundamentals of nanoceramics and their important applications in the aircraft industry.

To understand the concepts of nanoceramics, one should know the basics of nanoscience/nanomaterials. Therefore, before detailing ceramics/nanoceramics, a brief note on nanomaterials is included in this chapter. As already known, nanomaterials are usually defined as materials having a dimension ranging from 1 to 100 nm in at-least one direction. In this dimension, the properties of the material are quite different from its bulk nature. Generally, the functionality of the nanomaterials depends on its size and shape. The electronic properties of nanosized material behave anomalously from a bulk material. This is because the surface to volume ratio, that is, the surface/properties of a large number of atoms can be exposed at this nanoscale range. The field of nanoscience provides a wide range of options for producing nanomaterials with diverse shapes and sizes. These nanomaterials have significant applications in different sectors, including agriculture, energy, environment, electronics, medicine, and other areas [2].

Industrial Applications of Nanoceramics
DOI: https://doi.org/10.1016/B978-0-323-88654-3.00022-6

Ceramics is one type of the materials classified structurally in addition to metals and polymers. The word ceramic comes from a Greek word with the meaning "of pottery" or "for pottery" and may be used to describe a material or a product. A ceramic is generally defined as "any various hard, brittle, heat-resistant and corrosion-resistant solid material made up of either metal or nonmetal compounds that have been shaped to different structure and hardened by applying temperature" [3]. Examples include porcelain, brick, and earthenware.

Ceramics can be prepared in both crystalline and noncrystalline forms. The preparation of crystalline ceramics involves in situ or forming techniques. The forming methods include shaping by hand, casting, injection molding, dry pressing, and other variations. Noncrystalline ceramics, being glass in nature, can be formed by a melting process. Any of the processes such as melting, casting, or blowing into a mold will result in partly crystalline or noncrystalline ceramics [3].

Ceramics are broadly classified into two types: (1) naturally occurring ceramics; and (2) artificially made ceramics. The former are known as natural or traditional ceramics and the latter are known as advanced or engineered ceramics. The advanced ceramics can be further classified into three categories: (1) oxides: alumina, beryllia, ceria, and zirconia; (2) nonoxides: carbide, boride, nitride, and silicide; (3) composite materials: particulate-reinforced, fiber-reinforced, combinations of oxides, and nonoxides [3].

Though metals have remarkable properties, being cheap, tough, strong, and reactive, their high operating temperatures have limited practical applications. At the other end, polymers cannot be operated above 300°C even though they are light and easily approachable for applications. Being hard and brittle, ceramics can endure high operating temperatures. In addition, the chemical inertness/stability can protect the ceramic from other environmental issues. Also, ceramics fabricated by various different materials, can offer a wide range of properties. With all these advantages, ceramic materials have turned out to be suitable for a wide range of practical applications.

Humans have fused clay and silica by applying immense heating to form ceramic materials for at least 26,000 years. Initially, sculpted figures were found along with the later found pottery. After the invention of the wheel, smooth, porous, water-absorbing potteries were produced using the wheel-forming technique. Various techniques/efforts, such as glazing, and coating pottery with silicon, ash, or other materials that can blend upon melting onto the surface, have been used to make the ceramic less porous.

Nanoceramics are ceramic materials with a dimension in the nanometer range (1−100 nm) in at least one direction. The nanosized entities may have various structures: particle, tube, rod, fiber, or thin film. Ceramic materials are naturally brittle and hard but break upon impact. The unique properties, such as fine grain size, numerous grain boundaries, controllable crystallinity, and high surface area, are the beneficial factors of nanoceramics finding application in emerging technological fields, including health, environment, coatings, electronics, electrolytes, weapon manufacturing, etc. [4].

These materials are desirable for aerospace applications (commercial, defense, and vehicles for space explorations), because these materials are lightweight, have high temperature resistance, have good electrical insulation, have a high energy of ablation, suffer from less corrosion, can withstand vibration, and show chemical stability. The advantages of each property are summarized below.

Lighter weight: enabling faster speed, reduced fuel consumption, larger payloads, and longer times in space exploration vehicles

High temperature resistance: aircraft engines can run hotter, thus reducing CO_2 and NOx emissions (it is critical for domes and radomes used in weapon systems)

Electrical insulation: to avoid electromagnetic interference with the instrumentation aboard and the communications between pilot and ground control.

High energy of ablation: to control or protect objects flying in outer space.

Resistance to corrosion and chemical stability: to protect parts of the aircraft from contact with corrosive and hazardous materials, such as fuel.

Vibration- and noise-less: comfort the passengers, hard to detect (in military planes).

17.2 What is the need of nanoceramics?

Ceramic materials have various anomalous and interesting properties, including high stiffness, high mechanical strength, and stability at high temperatures, making them useful for cutting tools, biomedical implants, electronic, defense, space, and automotive applications. It should be noted that the monolithic ceramics also possess very interesting properties, but unfortunately these are highly brittle in nature and unreliable. To overcome this issue, researchers are developing the nanoceramics composites targeting the positive aspects in order to develop highly applicable materials in the above fields. Moreover, the issue becomes bigger with glass as the amorphous structure does not provide any obstacle to crack propagation, and leads to very low fracture toughness values. Hence, nanoceramics are a good alternative to solve these problems. Hence, nanoceramic composites began to attract significant research interest in the 1990s. Niihara first reported significant improvements in mechanical properties such as the fracture toughness and strength of materials embedding nanosized particles (20–300 nm) in grain boundaries of a matrix with larger grains [5]. Due to their superior properties over simple structures, nanoceramic composites are extensively used in different applications. Some of the typical advantages of such composites include improved mechanical properties (e.g., strength, hardness, and fatigue resistance), decreased permeability (to gases, water, and hydrocarbons), greater thermal stability, greater chemical resistance, higher electrical conductivity, and smoother surface appearance.

According to Niihara, there are two categories of nanoceramic composites: (1) nanocomposites having micron-sized matrix grains; and (2) nanocomposites with nanosized matrix grains. He observed that in nanoceramic composite, about 200% better retention of strength at high temperatures and better creep properties were possible compared to pure-phase ceramics. The introduction of nanoceramic particles, even at a small volume fraction loading, into a ceramic matrix causes the development of residual stresses around the particles. These residual stresses result in significant strengthening of the matrix. The toughening of two-phase nanoceramic composites depends on the volume fractions of the dispersed particles and their types. It has been observed that the intergranular fractures seen in some monolithic matrices are suppressed in nanoceramic composites. Crack extension along the grain boundaries is prevented by the strong bonding of the matrix with the dispersion interfaces. A transgranular fracture is thus a common mode of fracture in nanoceramic composites. Tensile stresses in the intergranular particles induce extension of the cracks to the particles, resulting in toughening by particle bridging. Intragranular particles can cause crack deflection and crack impediment, which results in significant toughening. The presence of extensive grain boundaries in nanoceramic composites makes the microstructure relatively stronger [6].

17.3 Synthesis of nanoceramics

Most commonly the preparation of nanoceramics requires chemically synthesized material in powder form. Additional materials are used in order to attain the desired form during its preparation. The preparation of nanoceramics should yield the product of unagglomerated nanosized uniformly distributed powder without any grain growth. This task of the fabrication of nanoceramics is the fundamental criteria. The following sections elaborate upon the methods of synthesis of nanoceramics being widely used. Recent and important methods of synthesis are discussed in this part.

17.3.1 Sol–gel method

Sol–gel is a simple, effective, low-cost, well-established technique that can produce a wide range of materials with novel properties. The "sol" is the name representing solid particles suspended in liquid phase made from colloidal solution, and "gel" corresponds to solid macromolecules immersed in a solvent. Fabrication of nanoceramics by the sol–gel method involves the conversion of a precursor (alkoxide/inorganic metal salt) solution into an inorganic solid through inorganic polymerization reactions induced by water.

Basic steps involved in the process are:

1. Partial hydrolysis: to form reactive monomers.
2. Polycondensation of monomers: to form colloid-like oligomers (sol).
3. Hydrolysis: to promote polymerization and cross-linking leading to 3-D matrix (gel).

The above processes can be represented by the following equations

Partial hydrolysis: $m(OR)_n + mH_2O \rightarrow m(OR)_{n\text{-}m}\ (OH)_m + ROH$

Condensation: $2m(OR)_{n-m}\ (OH)_m \rightarrow (OR)_{n-m}M\text{-}O\text{-}(OR)_{n\text{-}m} + HOH$

$(OR)_{n-1}M\text{-}(OR) + (OH)\text{-}M(OR)_{n-1} \rightarrow (OR)_{n-1}M\text{-}O\text{-}M(OR)_{n-1} + ROH$

3D cross-linkage between $(OR)_{n-m}$ M-O-$(OR)_{n-m}$ molecules [7].

During the polymerization and cross-linkage, the viscosity of the sol increases steadily until the sol to gel transition, where the viscosity suddenly increases and gelation occurs. Further by applying heat, the products are produced. In general, the sol–gel method yields highly pure and well-distributed desired material at ambient conditions. The utmost care should be taken in this process, since the properties highly depend on the precursor solution. In addition, by mixing sols of various compounds, controlled stoichiometry multiple compounds can be produced. These advantages have allowed the sol–gel method to have been widely used to fabricate the nanoceramics materials [8,9]. Some important examples of nanoceramics synthesized by the sol–gel process are shown in Table 17.1.

Table 17.1 Nanoceramics synthesized by sol–gel process.

Sl. no.	Nanoceramics	Research group	References
1	$SrFe_{12}O_{19}$	Garcia-Cerda L.A. et al.	[7]
2	Fe/Ba(Sr)	Surig C. et al.	[10]
3	$SrFe_{12}O_{19}$	Choy J.H. et al.	[11]
4	$BaFe_{11.9}Mn_{0.1}O_{19}$	Kim C.S. et al.	[12]
5	$BaFe_{12}O_{19}$	Jianxun Q. et al.	[13]
6	$[Ba, Sr]Me^{2+}TiFeO_{19}$ with $Me^{2+} = Zn^{2+}$, CO^{2+}	Surig C. et al.	[14]
7	$BaFe_{12}O_{19}$	Zhong W. et al.	[15]
8	$SrFe_{12}O_{19}$	Yongfei W. et al.	[15]
9	$SrFe_{12}O_{19}$	Yang X. et al.	[16]
10	$SrCr_xFe_{12-x}O_{19}$ ($x = 0.0, 0.5, 1.0, 1.5$)	Samira M. et al.	[17]
11	$BaTiO_3$	More S.P. et al.	[18]
12	$BaTiO_3$	Yuan Ting Wu et al.	[19]
13	Pure and Fe-doped $BaTiO_3$	Mohammed Tihtih et al.	[20]
14	$Sr_{0.5}Ba_{0.5}Ti_{1-x}Fe_xO_3$ ($0.00 \leq x \leq 0.25$ in step of 0.05	Dhananjay N. Bhoyar et al.	[21]
15	$Bi_{0.9}Nd_{0.1}FeO_3$	Das S. et al.	[22]
16	Mn doped $BiFeO_3$	Abushad M. et al.	[23]
17	Ni doped $BiFeO_3$	Nadeem M. et al.	[24]
18	$BiFeO_3$	Xu J.H. et al.	[25]
19	$La_2O_2SO_4$	Shaterian Maryam et al.	[26]
20	$Ba_{0.5}Sr_{0.5}Co_{0.2}Fe_{0.8}O_3$	Yousef M. et al.	[27]
21	Mg-doped $LaCrO_3$	Ghouse M. et al.	[28]
22	$La_{0.6}Ba_{0.4}Co_{0.2}Fe_{0.8}O_3$	Al-Yousef Y. et al.	[29]
23	$Ba_{0.5}Sr_{0.5}Co_xFe_{1-x}O_{3-\delta}$ ($x = 0.2–0.8$)	Lim Y.H. et al.	[30]
24	$La_{0.6}Sr_{0.4}Co_{0.2}Fe_{0.8}O_3$	Ghouse M. et al.	[29]
25	$La_{0.7}Ca_{0.3}CrO_3$	Ghouse M. et al.	[27]

17.3.2 Sintering process

Nanocrystalline ceramic materials were expected to have great mechanical behavior, which may improve significantly when the grain size is in nanometer scale. In order to make nanosized ceramic powders, numerous technologies have been the focus of research. Obtaining nanocrystalline ceramic materials with average grain size around <100 nm was an immense challenge because of the difficulties of controlling the grain growth during the sintering process. Therefore, certain parameters like temperature and time have to be optimized in the sintering process to control the grain growth. The following section will give an overview of the task and results of conventional sintering and advanced sintering processes, respectively.

17.3.2.1 Conventional sintering process

The solid nanocrystalline ceramic materials were attained from nanopowders by pressing and conventional sintering process using pressure assisted methods, such as hot pressing, sinter forgoing, hot isostatic pressing, etc. [31]. For the consolidation of ceramic powders, high-temperature and long-time sintering was required. This high-temperature sintering and long-time sintering results in extreme grain coarsening, decomposition of ceramics, and deterioration of the mechanical properties of ceramics [32]. Compared with conventional sintering, hot pressing of ceramic materials allows low densification at high temperatures [33]. Hence, the merits of hot pressing are enhancement of densification kinetics and lowering of grain growth. Whereas the demerits are limited geometry of the end product and the need for cost-effective equipment. In the conventional hot press, the powder sample container is heated by radiation. The heat transfer in the conventional hot press occurs by conduction from the external surface of the container to the powders. Due to the aforementioned heat transfer mechanism, the pace of heating experiences a notable deceleration throughout the final hours of the procedure. Additionally, the whole volume of the space is heated and the compact indirectly receives heat from the hot environment.

17.3.2.2 Advanced sintering process

Consolidation of nanocrystalline powder materials with a density near to the theoretical value and smaller grain size in parallel is a big task. In consolidation of nanocrystalline powders, the densification process and microstructure coarsening are to be considered. Out of various versatile methods, fast heating, high pressure, and addition of numerous agents which accelerate shrinkage and prevent grain growth are the most well-known. Comparing unconventional process and densification techniques, advanced sintering was found to be an adaptable method in ceramic process by inhibiting the grain growth [34]. This potential advanced sintering process includes the use of grain growth inhibitors in heterogeneous solution, shock densification, high-frequency induction heating, and magnetic pulse compaction [35].

17.3.2.3 Microwave sintering

For the past three decades, microwave sintering of ceramic materials has shown huge advantages like rapidness, heating the materials selectively, inhibiting grain size, and reducing processing time, in addition to the required energy [36]. This microwave sintering was applied in various areas including process control, drying of ceramic sanitary wares, calcination, decomposition of gaseous species by microwave plasma, and powder synthesis. This technique was limited to certain materials, such as semiconductors, ceramics, inorganic, and polymeric materials. Most of the bulk metals in microwave sintering cannot be heated due to the limited involvement of microwave radiation (cause of plasma formation) [37].

This microwave sintering process has a series of benefits, like excellent control in microstructure formation, no restriction in geometry of materials, requires low temperature and a shorter process time, achieves great mechanical properties, and has a low manufacturing cost with effective energy savings. In this sintering process, materials couple with microwaves, which is the form of electromagnetic energy with the frequency in the range of 300 MHz to 300 GHz that is transformed into heat energy. The mechanism behind this process is conduction, radiation, and convection (heat transforms between the objects). In the conventional method, the surface is heated initially and then this heat energy is transferred inside of the material. Whereas in this microwave heating, heat is produced inside the material and then heat is transferred to the whole of the volume space available [38]. This mechanism of heating, increases the diffusion process, decreases energy consumption, enhances heat rate, reduces processing time, maintains low temperature, has potential physical and mechanical properties, and has low environmental risks [39]. The schematic of a microwave furnace is shown in Fig. 17.1.

Along with a thermal pod, thermocouple, and controller, a microwave furnace also includes insulation box and susceptors, respectively. In order to achieve microwave transparency, the insulation box incorporates a small chamber constructed from low-density rigid insulation board. This design choice facilitates both low density and minimal dielectric loss, hence enabling the passage of microwaves. Inside the box, the microwave passes through the material and the contents are heated. At room temperature, most of the ceramic material does not absorb microwaves—2.45 GHz [40]. For initial heating of ceramics, susceptors were suitable. Susceptors are made in such a way that they absorb microwaves at room temperature and have enough dielectric loss where the ceramic materials directly interact with the field. Table 17.2 details a few ceramic materials with their respective methods.

17.3.2.4 Spark plasma sintering

This method was investigated by Inoue K in 1960 and patented (US Patent NO. 3 241 956) for metal powders. Due to low efficiency and high cost, this spark plasma

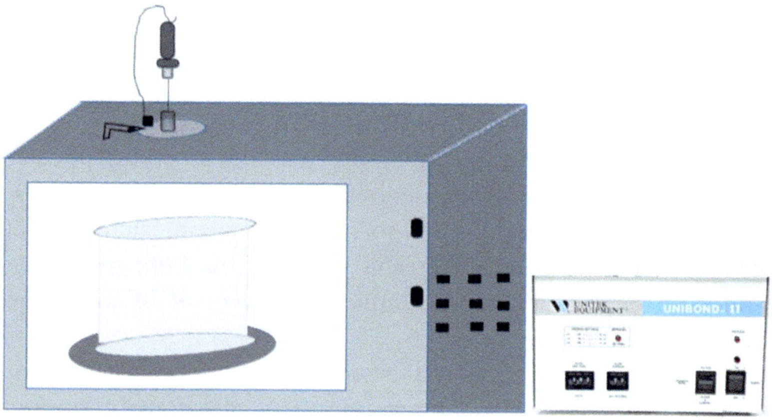

Figure 17.1 Schematic of microwave furnace with thermal pod, thermocouple and controller.

Table 17.2 Ceramic materials and their sintering methods.

S. no.	Material	Method	References
1	0.1–1 μm size of WC particles with Co	Microwave sintering and conventional sintering	[41]
2	W, C, and Co yield WC-6Co with 0.6 μm average grain size	Microwave sintering	[42–44]
3	WC-Co	Microwave postsintering	[45,46]

sintering was not extended in a wider way compared to the above said sintering process. The schematic of spark plasma sintering is given in Fig. 17.2. This typical unit comprises a pressure device, water-cooled chamber with electrodes, pulsed DC generator, and position and temperature regulating systems. Like the hot pressing process, this spark plasma sintering process also loads the precursor powder in a die and uniaxial pressure is applied; instead of external heating, a pulse DC is passed through a conducting electrical die and also into the sample. Here, the die acts as a heating source and the sample is heated both outside and inside, hence the samples are exposed to the pulsed electric field during the process. In this sintering process, the powder sample placed in the graphite die is pressed in the vacuum condition and heated by a pulse current which induces spark plasma for the very quick process. The generated spark plasma increases the surface activation and purification of the in situ sample particles. Therefore, relatively very low temperature, that is, heat and mass transfer between the particles with the help of electric current, makes the process very attractive for densification and preservation to increase the nanocrystalline character of the ceramic powder samples [47].

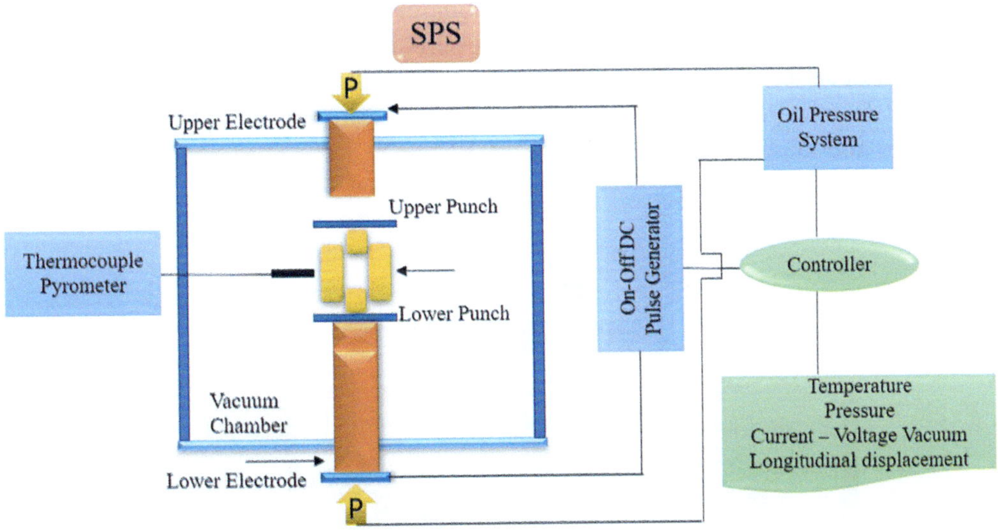

Figure 17.2 Schematic of typical spark plasma sintering process.

There are large research investigations into maintaining the fine grain size and phase homogeneity with the help of this spark plasma sintering process [48]. The enhanced kinetics of the various thermally activated processes, such as densification [49], reactive sintering [50], liquid and solid-state crystal growth [51], raised queries as to whether the atomistic mechanisms are active by conventional heating change due to the nature of spark plasma sintering process. Therefore, this process has links between the mechanical, electrical properties, and densification nature of the materials. Hence, this spark plasma sintering is the most versatile process for the consolidation of nanocrystalline ceramics and ceramic composites at low temperatures [52].

17.3.2.5 High-frequency induction heat sintering

This sintering process is the most promising and effective technique with high temperature, along with pressure, for the successful consolidation of rapid ceramic and metallic powders with near theoretical densification or density. Similar to the hot press, heat is generated from a source of high-frequency electricity that drives a large AC through a coil, known as the work coil, and it is carried out in a graphite die. The current passage is very intense rapid changing the magnetic field. The work piece to be heated is placed in this intense alternating magnetic field and the alternating magnetic field induces the passage of current in the conductive work piece.

The typical high-frequency induction heat sintering unit consists of an uniaxial pressure device, a graphite die with a height of 40 mm and an inner and outer diameter of around 20 and 45 mm, respectively, a water cooler chamber, an induced current of frequency about 50 KHz, and pressure–position–temperature regulating systems.

The schematic is shown in Fig. 17.3. This technique is similar to the hot press, but the intense magnetic field is applied through a pressure die or sometimes through the samples instead of an external heat source. The die acts as a heating source and the samples are heated from both inside and outside. Certain parameters should be followed for this process and these are shown in Table 17.3.

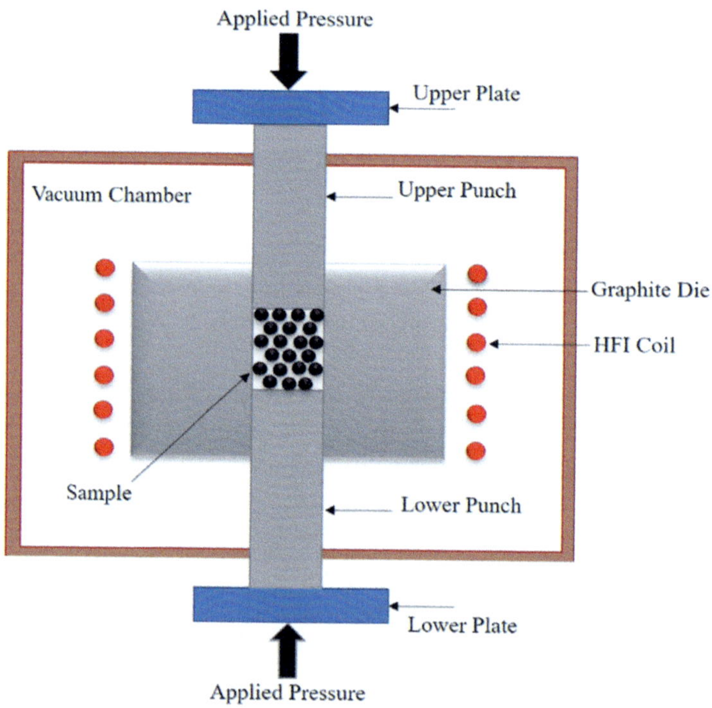

Figure 17.3 Schematic diagram of high-frequency induction heated sintering device.

Table 17.3 Characteristic parameters of high-frequency induction heated sintering process.

S. no.	Parameters	Applied values
1	Vacuum level	1×10^{-3} Torr
2	Heating rate	100–1200ºC/min
3	Total power capacity	15 kW
4	Resistance heating frequency	50 kHz
5	Applied pressure	10–300 MPa
6	Cooling rate	500ºC/min
7	Output of total power	0%–100%
8	Duration time	~10 min

Table 17.4 Ceramic materials and its effect of sintering temperature range on microstructures using high-frequency induction heated sintering process.

S. no.	Material	Sintering temperature range (°C)	Achieved theoretical density values (%)	References
1	Al_2O_3−8YSZ	1300−1500	99	[53]
2	Alumina−3YSZ	1100−1400	99	[54]
3	Mg/HAp	550−580	99.7	[55]
4	Ti−6Al−4V	1000−1100	98.8	[40]

The four major stages of this high-frequency induction heat sintering are vacuum, applied pressure, induction heating, and cooling. These four perfect stages of the high-frequency induction heated sintering process increase accurate performance, production, and reduce distortion. It is worthy to mention that few research performances based on this process involves rapidness in sintering nanostructured hard metal, in a short time [40]. Table 17.4 details some ceramic material composites produced using the high-frequency induction heated sintering process and their respective temperature range.

17.4 Two-photon lithography

This technique uses a laser for the fabrication of a three-dimensional nanoscale structure. In this process, two incident photons from light are absorbed by a precursor material named the photoinitiator (PI) and decompose into radicals (R*) in the excited state (PI*) which then combine with monomers (M), to become monomer radicals (RMn*). These monomer radicals are polymerized to form the desired three-dimensional structure. The process is entirely dependent on the incident laser beam parameters and the properties of the precursor material. The whole process is as follows [56]:

$$PI \xrightarrow{h\nu + h\nu} PI^* \rightarrow R^* + R^*R^* + M \rightarrow RM^* \xrightarrow{M} RMM^*\ldots\ldots\ldots$$
$$\rightarrow RM_n{}^*RM_n{}^* + RM^* \rightarrow RM_{n+m}R \qquad (17.1)$$

Various nanoceramics have been synthesized via this technique [57].

17.4.1 Carbon nanotube-reinforced nanoceramic composites

Recently, the spark plasma sintering (SPS) method has been used to fabricate MWCNT ceramic nanocomposites. Rul et al. fabricated CNT−alumina nanocomposites with molecular level mixing, using an in situ SPS [58]. These nanocomposites possess enhanced hardness, due to the load transfer properties of the CNTs, and increased fracture toughness, due to the bridging behavior of the CNTs during crack propagation.

Balazsi et al. observed that the addition of CNTs improved the mechanical properties of Si_3N_4 significantly [59]. They fabricated MWCNT–Si_3N_4 nanocomposites by the hot isostatic pressing method and observed that the microstructure of the sintered MWCNT composite consisted mainly of -Si_3N_4 and CNTs. With an increase in the sintering time and pressure, MWCNTs from the composite structure disappeared gradually. After sintering at 2 MPa, a well-connected Si_3N_4-bearing microstructure with CNTs developed, with a significant improvement in the mechanical properties compared to other carbon-filled composites. The surface of the carbon fibers remain covered by crystallites or cavities developed due to chemical reactions.

Lim et al. developed alumina–CNT nanocomposites with superior tribological properties by tape casting, followed by lamination and hot pressing [60]. With the addition of CNTs up to 12 wt.%, the wear loss of the composites decreased significantly but the friction coefficient remained the same. Agglomeration of CNTs, a frequent problem with the hot-pressing technique, was significantly reduced in this process and a more uniform distribution of the CNTs can be obtained. Effective dispersion of the CNTs resulted in improved densification of the composites giving them superior tribological and mechanical properties. All the features associated with the toughening of typical fiber-reinforced composites such as crack deflection, crack bridging, and fiber pullouts are also present in CNT-reinforced ceramic matrix composites. In these composites, debonding occurs at the atomic scale with significant residual sliding behavior. For MWCNTs, inner wall sliding takes place easily, thus reducing the toughening. In partially defective nanotubes, a load is transferred effectively from the outer to the inner walls resulting in enhanced strength and toughening. Therefore, the internal nanotube structure can be engineered to develop an adequate toughness for these composites.

17.4.2 Si_3N_4–SiC nanocomposites

Nanosized SiC particles dispersed in Si_3N_4 matrices are termed as Si_3N_4–SiC nanocomposites. Silicon nitride (Si_3N_4) ceramics have been widely used in the industry as structural components due to their excellent mechanical, thermal, and chemical properties. Si_3N_4–SiC nanocomposites are mostly prepared by hot pressing [61]. Si_3N_4 is mixed with Si and C powders and during the course of the carbothermal reduction, the precursor forms metastable phases. Above 1500°C, these phases are crystallized into Si_3N_4 and SiC, along with free carbon. SiC is formed in situ by the carbothermal reduction of Si_3N_4 according to Eq. (17.1) [62].

$$Si_3N_4 + 3C \rightarrow 3SiC + 2N_{2(g)}$$

A maximum increase in strength of ~85% (1500 MPa) and toughness of ~30% was observed for Si_3N_4–SiC (25 vol.%) nanocomposites compared to pure Si_3N_4.

17.4.3 Al_2O_3–SiC nanocomposites

Alumina (Al_2O_3) ceramics exhibit excellent properties, such as hardness, good chemical stability, and high wear resistance. Moreover alumina is abundantly available in nature. Thus due to these two attributes this material is being used in various engineering applications, such as manufacturing of cutting tools, insulators, and biomedical implants [63,64]. However, alumina has a highly ionic nature of Al–O bond, and owing to this the bond fracture toughness is low which restricts its application. The development of composite materials has been found to be one of the solutions to overcome the inherent brittleness of alumina ceramics. Experiments have shown that the reinforcement of nanosized silicon carbide (SiC) particles into an alumina matrix leads to excellent improvement of various properties of the latter [65]. The addition of a relatively small amount of SiC nanoparticles (5 vol.%) into an alumina matrix has been found to improve hardness [66], fracture strength [67], both at room temperature and at elevated temperature [68], as well as creep resistance. It has been shown that reinforcement of SiC nanoparticles into an alumina matrix can improve the functional properties. The thermal and electrical conductivity of Al_2O_3–SiC nanocomposites used both nano- and microsized SiC particles for the synthesis of Al_2O_3–SiC nanocomposites by hot pressing in the atmosphere of argon. It was observed that the mechanical properties, such as fracture strength and fracture toughness, of the composite were increased with the addition of SiC particles. Again with increasing temperature, the thermal conductivity decreased, whereas the DC electrical conductivity was increased with the addition of SiC.

17.4.4 Nanoceramic composite coatings

The coating process is the deposition of a layer on the surface and its thickness normally is in the submicrometer to nanometer range. Properties like thermal, mechanical, chemical stability, and durability are enhanced by the modification of morphology via nanocoatings. Titanium nitride (TiN), titanium carbide (TiC), chromium nitride (CrN), diamond-like carbon (DLC), tungsten carbide/carbon (WC/C), and alumina (Al_2O_3) are the ceramic materials most used in the coating process for engineering applications. Hard coatings such as TiN and Ti-C-N-DLC have significant advantages in aerospace systems. Corrosive fluids like engine oil and lubricants are used in aircraft engines and these may cause the corrosion of engine parts. However, these coatings resist corrosive attacks as well as having strength, toughness, and low-friction properties. Due to the high volume fraction of internal interfaces, nanostructured coatings show excellent properties compared with the usual microstructured coatings. Moreover, the thermal stability, the toughness, and the environmental compatibility of the composite materials were also enhanced with these coatings, as observed by Hauert and Patscheider [69]. In the nanoceramic composite coating, two or more ceramic phases remain combined as multiple layers or as a homogeneous isotropic multiphase mixture. Subramanian and Stafford earlier reported that a multilayer coating provided several advantages, including improvement of the toughness through crack deflection at many interfaces, improvement in oxidation resistance, and better tribological properties [70]. It is worth mentioning that recent research groups have developed wear and corrosion resistance, high-temperature inertness, and antioxidation properties. The fabrication of nanocomposites was improved by electrodeposition on metal surfaces [71,72]. A very hard TiN-α-Si_3N_4-based nanocomposites coating was produced by Verpek et al. using PACVD (plasma-assisted chemical vapor deposition) via the combination of $TiCl_4$, $SiCl_4$-SiH_4, and H_2 at about 600°C [73]. The nanocomposite coating contained nanocrystalline TiN (4–7 nm) in a matrix of amorphous Si_3N_4.

17.5 Conclusion

Nanoceramics are a type of nanoparticle composed of ceramics that were first discovered in the early 1980s. These materials are made from highly pure, well-chosen particles from silica carbide, and alumina, and are a far cry from the traditional tiles, pots, and tableware made of clay. Nanosized ceramic particles (less than 100 nm in diameter) are classified as inorganic, heat-resistant, nonmetallic solids. The most notable ceramic materials are systems of metal oxides, carbides, borides, and nitrides. In this chapter, we summarized the fundamentals of nanoceramics and the importance of their composites. More specifically, various methods of synthesis of nanoceramics have been clearly documented. It is to be noted that in the last two decades there has been immense interest in the research of nanoceramics and their derivatives, and various positive outcomes have been published in the literature. These advanced materials have a wide variety of applications in the fields of electronics, medicine, and the nuclear industry also. We highlighted the standard and related with regard to the applications of nanoceramics within the field of aerospace. In this chapter, we outlined the synthetic procedures for making nanoceramics using various methods. The applications of nanoceramics in aircraft industry are plenty. For example, nanoceramics are being used in engines, brake systems, insulating tiles, nose cones, high-temperature coatings, wind shields, panels and mirrors for the cabin interior, and also in LED lighting.

Acknowledgments

We sincerely thank the College management board, Principal, department of Chemistry and Physics for providing the Research facilities to conduct this work. SS and KSP grateful to Bharadhidasan Government College for Women (Autonomous), Puducherry UT, India for the infrastructure.

References

[1] R.K. Brow, M.L. Schmitt, A survey of energy and environmental applications of glass, J. Eur. Ceram. Soc. 29 (7) (2009) 1193–1201.

[2] J. Rödel, A.B. Kounga, M. Weissenberger-Eibl, D. Koch, A. Bierwisch, W. Rossner, et al., Development of a roadmap for advanced ceramics: 2010–2025, J. Eur. Ceram. Soc. 29 (9) (2009) 1549–1560.

[3] M. Kotobuki, S. Song, C. Chen, L. Lu, Ceramic Electrolytes for All-Solid-State Li Batteries, 2018.

[4] A. Mukhopadhyay, B. Basu, Consolidation–microstructure–property relationships in bulk nanoceramics and ceramic nanocomposites: a review, Int. Mater. Rev. 52 (5) (2007) 257–288.

[5] K. Niihara, New design concept of structural ceramics ceramic nanocomposites, J. Ceram. Soc. Jpn. 99 (1154) (1991) 974–982.

[6] V. Tomar, Analyses of the role of grain boundaries in mesoscale dynamic fracture resistance of $SiC-Si_3N_4$ intergranular nanocomposites, Eng. Fract. Mech. 75 (15) (2008) 4501–4512.

[7] L.A. Garcıa-Cerda, O.S. Rodrıguez-Fernández, P.J. Reséndiz-Hernández, Study of $SrFe_{12}O_{19}$ synthesized by the sol–gel method, J. Alloy. Compd. 369 (1–2) (2004) 182–184.

[8] M.M. Vijatović, J.D. Bobić, B.D. Stojanović, History and challenges of barium titanate: Part II, Sci. Sinter. 40 (3) (2008) 235–244.

[9] A.K. Mishra (Ed.), Sol-Gel Based Nanoceramic Materials: Preparation, Properties and Applications, Springer International Publishing, 2017, pp. 1–20.

[10] C. Sürig, K.A. Hempel, D. Bonnenberg, Formation and microwave absorption of barium and strontium ferrite prepared by sol-gel technique, Appl. Phys. Lett. 63 (20) (1993) 2836–2838.

[11] C. Jin-Ho, H. Yang-Su, S. Seung-Wan, Preparation and magnetic properties of ultrafine $SrFe_{12}O_{19}$ particles derived from a metal citrate complex, Mater. Lett. 19 (5–6) (1994) 257–262.

[12] J. Qiu, M. Gu, Magnetic nanocomposite thin films of $BaFe_{12}O_{19}$ and TiO_2 prepared by sol–gel method, Appl. Surf. Sci. 252 (4) (2005) 888–892.

[13] C. Surig, K.A. Hempel, D. Bonnenberg, Hexaferrite particles prepared by sol-gel technique, IEEE Trans. Mag. 30 (6) (1994) 4092–4094.

[14] W. Zhong, W. Ding, N. Zhang, J. Hong, Q. Yan, Y. Du, Key step in synthesis of ultrafine $BaFe_{12}O_{19}$ by sol-gel technique, J. Mag. Mag. Mater. 168 (1–2) (1997) 196–202.

[15] W. Yongfei, L. Qiaoling, Z. Cunrui, J. Hongxia, Preparation and magnetic properties of different morphology nano-$SrFe_{12}O_{19}$ particles prepared by sol–gel method, J. Alloy. Compd. 467 (1–2) (2009) 284–287.

[16] X. Yang, Q. Li, J. Zhao, B. Li, Y. Wang, Preparation and magnetic properties of controllable-morphologies nano-$SrFe_{12}O_{19}$ particles prepared by sol–gel self-propagation synthesis, J. Alloy. Compd. 475 (1–2) (2009) 312–315.

[17] S. Mandizadeh, F. Soofivand, S. Bagheri, M. Salavati-Niasari, $SrCrxFe_{12}$-xO_{19} nanoceramics as an effective catalyst for desulfurization of liquid fuels: green sol-gel synthesis, characterization, magnetic and optical properties, PLoS One 12 (5) (2017) e0162891.

[18] S.P. More, M.V. Khedkar, S.A. Jadhav, B. Sandeep, A.V. Somvanshi, K.M. Humbe, et al., Wet chemical synthesis and investigations of structural and dielectric properties of BaTiO3 nanoparticles, J. Phys.: Conf. Ser. 1644 (2020) 012007-1-8

[19] Y.T. Wu, X.F. Wang, C.L. Yu, E.Y. Li, Preparation and characterization of barium titanate (BaTiO3) nano-powders by pechini sol-gel method, Mater. Manuf. Process. 27 (12) (2012) 1329–1333.

[20] T. Mohammed, L. Karoum, A. Yahya, S. Salaheddine, J.E.F.M. Ibrahim, Sol-gel synthesis and structural characterization of Fe doped barium titanate nanoceramics, J. Silicate Based Composite Mater. 71 (2019) 190–193.

[21] D.N. Bhoyar, S.B. Somvanshi, P.B. Kharat, A.A. Pandit, K.M. Jadhav, Structural, infrared, magnetic and ferroelectric properties of Sr0 · 5Ba0 · 5Ti1-xFexO3 nanoceramics: modifications via trivalent Fe ion doping, Phys. B: Condens. Matter 581 (2020) 411944.

[22] S. Das, K.P. Bera, T.K. Nath, Synthesis of Bi0. 9Nd0. 1FeO3 multiferroic nanoceramic by sol-gel method and its structural, surface morphology and dielectric analysis, in: AIP Conference Proceedings (1832, 1, p. 050072). AIP Publishing LLC, 2017.

[23] M. Abushad, W. Khan, S. Naseem, S. Husain, M. Nadeem, A. Ansari, Influence of Mn doping on microstructure, optical, dielectric and magnetic properties of $BiFeO_3$ nanoceramics synthesized via sol–gel method, Ceram. Int. 45 (6) (2019) 7437–7445.
[24] M. Nadeem, W. Khan, S. Khan, M. Shoeb, S. Husain, M. Mobin, Significant enhancement in photocatalytic performance of Ni doped $BiFeO_3$ nanoparticles, Mater. Res. Exp. 5 (6) (2018) 065506.
[25] J.H. Xu, H. Ke, D.C. Jia, W. Wang, Y. Zhou, Low-temperature synthesis of $BiFeO_3$ nanopowders via a sol–gel method, J. Alloy. Compd. 472 (1–2) (2009) 473–477.
[26] M. Shaterian, M.A. Rezvani, V. Shahsavandi, K. Qasemi, Synthesis, Characterization and Electrochemical Properties of Lanthanum Oxysulfate Nanoceramic, 2017.
[27] Y.M. Al-Yousef, M. Ghouse, Synthesis of Ba0. 5Sr0. 5Co0. 2Fe0. 8O3 (BSCF) nanoceramic cathode powders by sol-gel process for solid oxide fuel cell (SOFC) application, World J. Nano Sci. Eng. 1 (4) (2011) 99.
[28] M. Ghouse, A. Al-Musa, Y. Al-Yousef, M.F. Al-Otaibi, Synthesis of Mg doped LaCrO3 nano powders by sol-gel process for solid oxide fuel cell (SOFC) application, J. N. Mater. Electrochem. Syst. 13 (2) (2010) 99–106.
[29] M. Ghouse, Y. Al-Yousef, A. Al-Musa, M.F. Al-Otaibi, Preparation of La0. 6Sr0. 4Co0. 2Fe0. 8O3 nanoceramic cathode powders for solid oxide fuel cell (SOFC) application, Int. J. Hydrog. Energy 35 (17) (2010) 9411–9419.
[30] Y.H. Lim, J. Lee, J.S. Yoon, C.E. Kim, H.J. Hwang, Electrochemical performance of Ba0. 5Sr0. 5CoxFe1 − xO3 − δ (x = 0.2–0.8) cathode on a ScSZ electrolyte for intermediate temperature SOFCs, J. Power Sources 171 (1) (2007) 79–85.
[31] J.R. Groza, Nanosintering, Nanostruct. Mater. 12 (5–8) (1999) 987–992.
[32] Y.W. Gu, K.A. Khor, P. Cheang, Bone-like apatite layer formation on hydroxyapatite prepared by spark plasma sintering (SPS), Biomaterials 25 (18) (2004) 4127–4134.
[33] R. Halouani, D. Bernache-Assolant, E. Champion, A. Ababou, Microstructure and related mechanical properties of hot pressed hydroxyapatite ceramics, J. Mater. Sci.: Mater. Med. 5 (8) (1994) 563–568.
[34] S. Vijayan, H. Varma, Microwave sintering of nanosized hydroxyapatite powder compacts, Mater. Lett. 56 (5) (2002) 827–831.
[35] A.J. Allen, S. Krueger, G. Skandan, G.G. Long, H. Hahn, H.M. Kerch, et al., Microstructural evolution during the sintering of nanostructured ceramic oxides, J. Am. Ceram. Soc. 79 (5) (1996) 1201–1212.
[36] J.D. Katz, Microwave sintering of ceramics, Annu. Rev. Mater. Sci. 22 (1) (1992) 153–170.
[37] D. Agrawal, Microwave sintering, brazing and melting of metallic materials, in: Sohn International Symposium; Advanced Processing of Metals and Materials Volume 4: New, Improved and Existing Technologies: Non-Ferrous Materials Extraction and Processing (Vol. 4, pp. 183–192), 2006, August.
[38] P. Yadoji, R. Peelamedu, D. Agrawal, R. Roy, Microwave sintering of Ni–Zn ferrites: comparison with conventional sintering, Mater. Sci. Eng.: B 98 (3) (2003) 269–278.
[39] C. Leonelli, P. Veronesi, L. Denti, A. Gatto, L. Iuliano, Microwave assisted sintering of green metal parts, J. Mater. Process. Technol. 205 (1–3) (2008) 489–496.
[40] K.A. Khalil, Advanced sintering of nano-ceramic materials, Ceram. Mater.-Progress Mod. Ceram. (2012) 65–82.
[41] E. Breval, J.P. Cheng, D.K. Agrawal, P. Gigl, M. Dennis, R. Roy, et al., Comparison between microwave and conventional sintering of WC/Co composites, Mater. Sci. Eng.: A 391 (1–2) (2005) 285–295.
[42] T. Gerdes, M. Whxert-Porada, Microwave sintering of metal-ceramic and ceramic-ceramic composites, MRS Online Proc. Library (OPL) 347 (1994).
[43] K. Ro, K. Dreyer, T. Gerdes, M. Willert-Porada, Microwave sintering of hardmetals, Int. J. Refractory Met. Hard Mater. 16 (4–6) (1998) 409–416.
[44] H. Kolaska, M. Willert-Porada, K. Rodiger, T. Gerdes, G.M.B.H. Widia, Composite and process for the production thereof, U.S. Patent 6,124,040, 2000.
[45] J. Ramkumar, S. Aravindan, S.K. Malhotra, R. Krishnamurthy, Enhancing the metallurgical properties of WC insert (K-20) cutting tool through microwave treatment, Mater. Lett. 53 (3) (2002) 200–204.

[46] S. Aravindan, J. Ramkumar, S.K. Malhotra, R. Krishnamurthy, Enhancement of cutting performance of cemented carbide cutting tools by microwave treatment, in: D.E. Clark, D.C. Folz, C.E. Folgar, M.M. Mahmoud (Eds.), Microwave Solutions for Ceramic Engineers, The American Ceramic Society;, Ohio, 2005, pp. 255–262.

[47] R. Chaim, M. Kalina, J.Z. Shen, Transparent yttrium aluminum garnet (YAG) ceramics by spark plasma sintering, J. Eur. Ceram. Soc. 27 (11) (2007) 3331–3337.

[48] P. Angerer, L.G. Yu, K.A. Khor, G. Krumpel, Spark-plasma-sintering (SPS) of nanostructured and submicron titanium oxide powders, Mater. Sci. Eng.: A 381 (1–2) (2004) 16–19.

[49] Z.A. Munir, U. Anselmi-Tamburini, M. Ohyanagi, The effect of electric field and pressure on the synthesis and consolidation of materials: a review of the spark plasma sintering method, J. Mater. Sci. 41 (3) (2006) 763–777.

[50] R. Licheri, R. Orrù, C. Musa, G. Cao, Combination of SHS and SPS Techniques for fabrication of fully dense ZrB2-ZrC-SiC composites, Mater. Lett. 62 (3) (2008) 432–435.

[51] M. Omori, Sintering, consolidation, reaction and crystal growth by the spark plasma system (SPS), Mater. Sci. Eng.: A 287 (2) (2000) 183–188.

[52] K.A. Khor, X.J. Chen, S.H. Chan, L.G. Yu, Microstructure-property modifications in plasma sprayed 20 wt.% yttria stabilized zirconia electrolyte by spark plasma sintering (SPS) technique, Mater. Sci. Eng.: A 366 (1) (2004) 120–126.

[53] A.K. Khalil, W.K. Sug, Synthesis and densification of ultra-fine Al2o3-(Zro2 + 3%Mol Y2o3) bioceramics by high-frequency induction heat sintering, Mater. Sci. Forum 534–536 (2007) 1033–1036.

[54] S.W. Kim, K.A.R. Khalil, High-frequency induction heat sintering of mechanically alloyed alumina–yttria-stabilized zirconia nano-bioceramics, J. Am. Ceram. Soc. 89 (4) (2006) 1280–1285.

[55] K.A. Khalil, A.A. Almajid, Effect of high-frequency induction heat sintering conditions on the microstructure and mechanical properties of nanostructured magnesium/hydroxyapatite nanocomposites, Mater. Des. 1980–2015 (36) (2012) 58–68.

[56] X. Zhou, Y. Hou, J. Lin, A review on the processing accuracy of two-photon polymerization, AIP Adv. 5 (3) (2015) 030701.

[57] J. Bauer, C. Crook, A.G. Izard, Z.C. Eckel, N. Ruvalcaba, T.A. Schaedler, et al., Additive manufacturing of ductile, ultrastrong polymer-derived nanoceramics, Matter 1 (6) (2019) 1547–1556.

[58] S. Rul, F. Lefevre-Schlick, E. Capria, C. Laurent, A. Peigney, Percolation of single walled carbon nanotubes in ceramic matrix composites, Acta Mater. 52 (2004) 1061–1067.

[59] C.S. Balazsi, Z. Konya, F. Weber, L.P. Biro, P. Arato, Preparation and characterization of carbon nanotube reinforced silicon nitride composite, Mater. Sci. Eng. C. 23 (2003) 1133–1137.

[60] D.S. Lim, D.H. You, H.J. Choi, S.H. Lim, H. Jang, Effect of CNT distribution on tribological behaviour of alumina-CNT composites, Wear 259 (2005) 539–544.

[61] Y.S. Oh, C.S. Kim, D.S. Lim, D.S. Cheong, Fracture strengths and microstructures of Si_3N_4/SiC nanocomposites fabricated by in-situ process, Scr. Mater. 44 (2001) 2079–2081.

[62] M. Friess, J. Bill, J. Golczewski, A. Zimmermann, et al., Crystallization of polymer derived silicon carbonitride at 1873K under nitrogen overpressure, J. Am. Ceram. Soc. 85 (2002) 2587–2589.

[63] R.H. Castro, On the thermodynamic stability of nanocrystalline ceramics, Mater. Lett. 96 (2013) 45–56.

[64] P. Palmero, F. Kern, F. Sommer, M. Lombardi, et al., Issues in nanocomposite ceramic engineering: focus on processing and properties of alumina-based composites, J. Appl. Biomater. Funct. Mater. 12 (2014) 113–128.

[65] B. Baron, C. Kumar, G.L. Gonidec, S. Hampshire, Comparison of different alumina powders for the aqueous processing and pressureless sintering of Al2O3eSiC nanocomposites, J. Eur. Ceram. Soc. 22 (2002) 1543–1552.

[66] Y.L. Dong, F.M. Xu, X.L. Shi, C. Zhang, Z.J. Zhang, J.M. Yang, et al., Fabrication and mechanical properties of nano/micro sized Al2O3/SiC composites, Mater. Sci. Eng. A 504 (2009) 49–54.

[67] Y.K. Jeong, A. Nakahira, P. Morgan, K. Niihara, Effect of milling conditions on the strength of alumina-silicon carbide nanocomposites, J. Am. Ceram. Soc. 80 (1997) 1307–1309.

[68] Y. Xu, A. Zangvil, A. Kerber, SiC nanoparticle-reinforced Al_2O_3 matrix composites: role of intra- and intergranular particles, J. Eur. Ceram. Soc. 17 (1997) 921–928.
[69] R. Hauert, J. Patscheider, From alloying to nanocomposites improved performance of hard coatings, Adv. Eng. Mater. 2 (2000) 247–259.
[70] C. Subramanian, K.N. Stafford, Review of multicomponent and multilayer coatings for tribological applications, Wear 165 (1993) 85–95.
[71] W.X. Cheng, J.P. Tu, H.Y. Gan, Z.D. Xu, Electroless preparation and tribological properties of Ni-P-carbon nanotube composite coatings under lubricated condition, Surf. Coat. Technol. 160 (2002) 68–73.
[72] L. Shi, C.F. Sun, F. Zhou, W.M. Liu, Electrodeposited nickel-cobalt composite coating containing nano-sized Si3N4, Mater. Sci. Eng. A 397 (2005) 190–194.
[73] S. Verpek, S. Reiprich, L. Shizhi, Super hard nanocrystalline composite materials, the TiN/Si3N4 system, Appl. Phys. Lett. 66 (1995) 2640–2643.

CHAPTER 18

Nanoceramics in the electronics and electrical industry

J. John Donald Raj[1], S. Kishore Babu[1], Shilpa Rajkumar[1], Deepankara Vrushabhadas Shastri[1] and Kantha D. Arunachalam[2]

[1]Department of Physics and Nanotechnology, SRM Institute of Science and Technology, Kattankulathur, Tamil Nadu, India
[2]Faculty of Sciences, Marwadi University, Rajkot, Gujarat, India

18.1 Introduction

Ceramic materials with any of their dimensions in the nanometer range, that is, 1–100 nm, are said to be nanoceramics. They cover a wide range of ceramics with at least one nanoscale dimension, including nanosized and nanostructured particles, nanofibers, nanotubes, 2D sheets, nanorods, thin films, nanostructured surfaces, and bulk materials created for a wide variety of industries and applications. Nanotechnology, which includes nanoparticles, nanocrystals, nanostructure, and all ceramic materials at the nanoscale, particularly nanoceramics, is one of the most exciting fields, encompassing materials sciences, cement, concrete, and others. Ceramics are commonly characterized as inorganic, heat-resistant, nonmetallic solids comprising both metallic and nonmetallic chemicals. They are used to make nanoceramics [1,2]. Because of their unique nanoscale features, nanoceramics have received a lot of attention in research into building materials such as bricks, ceramic wall and floor tiles, and advanced bioceramics. Nanoceramics were first found in the early 1980s and are made via a method known as sol–gel, which involves the mixing of nanoparticles in a suitable solution and then gelling them to form nanoparticles. Sintering, which is the production of a solid through pressure and heat, was used in later approaches. Advancements in nanoceramics research have been made feasible because of these technologies. Nanoceramic materials are classified into different types based on its features such as dielectric, pyroelectric, ferroelectric, piezoelectric, magnetoresistive, ferromagnetic, superconductive, and electrooptical. Nanoceramics have traditionally been utilized in small-scale electronics, but higher strength materials, such as those used in aircraft engines and aerospace technology, have become more common in recent years. Ceramics have recently emerged as a strong contender for high-temperature, high-chemical-resistance applications and they show high thermal conductivity, high resistivity, and oxidation resistance. Researchers have been focusing on developing new materials for use in cooling and

Industrial Applications of Nanoceramics
DOI: https://doi.org/10.1016/B978-0-323-88654-3.00023-8

energy conversion systems for the past few years. Due to the research on the structure and magnetoelectric properties of nanomaterials, such as their interrelation, nanosized ceramic materials are the most important for further studies. Spinel structure ceramics are being studied because of their outstanding magnetic and electrical properties; these are the most important study materials in terms of fundamental electronic components. Ceramics used in the spinel structure are quite stable and are used in a variety of technological applications, such as magnetic recording, sensors, magnetic resonance imaging, transformers, and so on [3]. Due to their high resistivity, low magnetic, and dielectric losses, nanosized spinel ceramics are significant dielectric materials in high-frequency applications. At low frequencies (102−105 Hz), these materials have a high dielectric constant and low conductivity. As a result, they're used in microwave applications as well as a variety of other devices such as high-frequency transformer cores and switches. Materials like silicon carbide function as a semiconductor, while chromium oxide is an electrical conductor and aluminum oxide is an insulator. As a result of their electrical properties, ceramic materials have been used in a range of electronic applications. Several electrical properties are affected by particle size and composition. Dielectric constant and electrical resistance, for example, are increased mainly due to small particle size [4]. Some oxides when combined with ceramics, such as lithium aluminosilicate, have a conductivity that is higher than the constituent oxides. Pristine $CoCr_2O_4$ nanoceramics are a type-II multiferroic material that exhibits magnetoelectric coupling and has recently sparked a lot of curiosity [5]. In electronic memory devices, these nanoceramics are more efficient at storing charges. $NiFe_2O_4$ (NFO) has recently been found as a potential material for negative electrodes in Li-ion batteries and supercapacitors [6]. The importance of particle/grain size and shape in modifying the electrical, magnetic, and electrochemical properties of materials for device applications is now well recognized.

The focus of this book chapter will be on the nanoceramic materials that have attained remarkable application in the modern electronics and electrical industry.

18.2 Properties of nanoceramics

Because of their size and molecular structure, nanoceramics offer unique characteristics. These characteristics are frequently expressed in the form of numerous magnetic and electric physics phenomena, such as dielectric, ferroelectric, piezoelectric, pyroelectric, ferromagnetic, magnetoresistive, superconductive, and electrooptical properties [7]. Nanoceramics are extremely robust, with significant resistance to compression and bending. Their tensile strength is comparable to that of steel, and most ceramics have the ability to maintain their tensile strength even at higher temperatures. Their brittleness, on the other hand, is the most significant technical impediment to their practical deployment, particularly in load-bearing applications. Nanoceramics are extremely light, robust, ductile, and long-lasting. The fractal nanotrust, a nanostructure architecture invented by Julia Greer, is the cause for all of these properties being present in a single material. Due to the low price, higher bandwidth, and good actuation capabilities, lead-based ceramics such as $Pb(Zr_xTi_{1-x})O_3$ (abbreviated as PZT) have been widely used for the application of electronic and microelectronic devices in recent decades. However, the toxicity of lead oxide and the high pressure during the sintering process can possibly inflict damage on the environment. Hence, researchers have focused on other nanoceramics such as ZnO, which is a low-cost, environmentally acceptable material that has piqued the curiosity of researchers and technologists for device applications. It has a wide range of applications, including solar cells, photolytic breakdown of organic dyes, antimicrobial action, and cancer treatment. ZnO is also employed in actuators and piezoelectric transducers due to its lack of a center of symmetry [5,8].

18.3 Electrical properties of nanoceramics

Nanostructures comprising ceramic materials have emerged as intriguing replacements to conventional piezoelectric composites in recent years. Ceramic nanocomposites are composites that have crucial nanoscopic characteristics. They have the potential to deliver high-performance features, such as better electrical conductivity, that traditional advanced ceramics have never been able to accomplish [9]. In the case of ZnO, the grain boundaries play a major role in altering the electrical properties. The characteristics of a solid are largely controlled by the surface/interface energy. Interfaces allow for the introduction of nonhomogeneity into the material. The coupling surface/interface model was widely utilized in terms of the effect of surface/interface energy on electrical characteristics. The majority of electrical property experiments have centered on Fourier-transform infrared (FTIR) spectroscopy measurements. The electrical characteristics of ZnO were determined by FTIR in the work of David et al. [10] Interferometry was used to study the piezoelectric characteristics of ferroelectric and antiferroelectric thin films as a function of frequency and DC electric bias by Li et al. [11]. Carbon nanotubes (CNTs) played a major role in improving the electrical properties of nanoceramics. Insulating nanoceramics were converted to metallically conductive composites using single-walled CNTs (SWCNTs). By including CNTs in various matrices such as polymers, metals, and ceramics, people try to generate innovative engineering materials with improved mechanical properties, taking advantage of the nanotube's incredible strength. Zhan et al. [12] used spark-plasma sintering to successfully apply SWCNTs in the reinforcement of ceramic composites, which provided an improvement of 194% in fracture hardness over pure alumina.

18.3.1 Experimental techniques

Nanoceramics have primarily been created via the sol–gel process since their development. Other techniques, like sintering and two-photon lithography, were used after that [13]. Pressure and heat combine to form a solid during the sintering process. Sintering at high temperatures produces a rough substance that degrades the characteristics of ceramics and takes a long time to complete [14]. This process also limits the range of geometries that can be used in the nanoceramic's ultimate form. The sol–gel process for making metal oxide glasses and ceramics involves hydrolyzing chemical precursors to make a sol, which then forms a gel, which produces an amorphous oxide when dried and pyrolyzed. In fact, sol–gel technology is a well-known colloidal chemical technology, which allows the production of a wider range of materials with innovative and predefined characteristics in a fairly simple process at low cost. By this method, the precursor solution is converted into an inorganic solid.

18.3.2 Electrical applications

Certain insulators, semiconductors, and conductors are usually made of ceramics. Aluminum oxide (Al_2O_3), for example, is used to manufacture insulators since it does not conduct electricity. In electronic devices, other ceramics, such as barium titanate ($BaTiO_3$) [7], are employed as semiconductors. $BaTiO_3$ is a lead-free ferroelectric ceramic that is also environmentally beneficial. As a result, it is a suitable fit for a wide range of applications. Electronic components, namely multilayer ceramic capacitors (MLCC) [15], positive temperature coefficient thermistors, and electrosensors, often use barium titanate, which has a perovskite structure [16–18]. At temperatures above the temperature at which the metal becomes superconducting, some copper oxide-based ceramics become superconducting. In electrochemical applications, transition metal-based nitrides, carbides, and borides have potential significance as cathodes. This comes from the material's superior properties such as electronic conductivity and excellent thermal conductivity, combined with its mechanical strength and high melting point, indicating that it can remain stable under various conditions. Materials such as Al_2O_3, $SiC + BeO$ are used in integrated circuit (IC) substrates because of their insulating properties. Low-dielectric-constant electroceramics are utilized as substrates that are used in integrated circuits, while high-dielectric-constant electroceramics are employed in capacitors. Certain semiconducting nanoceramic materials, such as SiC, $LaCrO_3$, SnO_2, $ZnO + Bi_2O_3$, are used in gas detectors, thermistors, and varistors. ZnO and SiO_2 are highly employed for the manufacture of piezolighters, flexible piezodetectors, piezofilters, and surfacewave transducers owing to their piezoelectric property [19]. Because certain materials can store large amounts of electricity in a very small volume, some ceramics, such as strontium titanate ($SrTiO_3$), are used as capacitors in various electronic devices in the form of thin films. Lithium aluminosilicate ceramics have the potential to be used as solid electrolytes in high energy density solid lithium battery systems [20]. Polymer crystalline ceramics with a high dielectric constant are widely used in various applications in electronic products, including power transmission equipment, industrial capacitors, high-energy storage equipment, storage equipment, and so on. Nanoceramics are used in various current and developing technologies. Due to their strength, biocompatibility, hydrophilicity, and wear resistance, nanoceramic materials have been used in biomedicine, healthcare, and environmental applications. Due to their insulating and/or conductive properties, nanoceramics are also used to create next-generation chips. $Bi_{0.487}Na_{0.427}K_{0.06}Ba_{0.026}TiO_{3-x}CeO_2$, $BaCu(B_2O_5)$-doped $SrTiO_3$-doped yttrium lanthanum oxide ceramics and other doped ceramic materials have been used for energy storage. Due to their low cost and good performance, metal electrode MLCC are considered to be one of the most important passive components in the power supply business [21].

18.3.3 Ceramic matrix composites

The inherent brittleness of ceramics limits a wide range of applications. The use of ceramic matrix composites (CMC), especially fiber-reinforced CMC, is one way to solve this problem. Each of the three components of these systems are built with nanoceramic materials. The internal reinforcing fibers are made of amorphous to nanocrystalline ceramics, giving the composite material strength and durability. It is commonly made from polymer-derived ceramics. After that, the fiber is coated with an interface, which is commonly deposited using chemical vapor deposition. This interface enables composite strengthening processes such as fracture deflection and fiber pullout. The matrix or bulk material is the final component of the composite material. Chemical vapor deposition is employed in the production of this nanoceramic material, which serves as the primary component contributing to the strength of composite materials.

18.3.4 Application of ceramic matrix composite's in electronics

Ceramic materials with unique electrical properties can be employed to make portable gas sensors. Ferrites with spinel structures, for example, are used as electronic materials in gas-sensing applications [22]. These are cost-effective, simple, and possess the ability to be modified through structural and compositional parameter control. The use of ceramic materials in H_2 separation systems is another application [23]. These systems make use of dense proton—electron conducting ceramic materials and membranes built of them. A variety of proton—electron conducting materials, as well as related membrane materials, have been examined. Proton-conducting noncomposite materials, tungstate-based, $BaPrO_3$-based, $LaGaO_3$-based, niobate/tantalite, and other metal oxide-based composite ceramic/membrane systems are some of the most important categories. In these systems, self-supporting and supported membranes are used. Surface modification of these materials may be required at times to increase performance. The toxicity of ceramics used in the electronics industry has recently been a matter of concern in this field. Today, many efforts have been made to adopt nanomaterials in various industries. CNTs, fullerenes, quantum dots, dendrimers, metal oxides, silica, alumina, and metal groups, such as gold, silver, nickel, and iron nanoparticles, are all being studied [24].

18.3.5 Nanoceramics' future potential

Because of their unique mechanical and physiochemical properties, such as excellent wear resistance in harsh environments, high ballistic resistance, high thermal shock resistance, superior strength, corrosion/oxidation resistance, low density, electrical, optical, and/or magnetic properties, and chemical inertness, nanoceramic materials are considered promising materials for the future. It is worth noting that silicon nitride (Si_3N_4) ceramic has yet to fulfill its promise in the biological area, despite its invention half a century ago. Improvements in research on existing nanoceramic materials, as well as new ones, are expected to increase in terms of their qualities and appropriateness for a variety of applications that are not possible with other materials.

18.4 Conclusion

In this chapter, the electronic properties and the applications of nanoceramic materials in the electrical industries were summarized. The abilities of certain nanoceramic materials are being studied and certain unique properties have yet to be discovered.

References

[1] K.A. Khalil, Advanced sintering of nano-ceramic materials, Ceram. Mater.: Progress Mod. Ceram. (2012) 65–82.
[2] H.H.M. Darweesh, Nanoceramics: materials, properties, methods and applications—part II, J. Nanosci. 1 (2018) 40–66.
[3] R. Sridhar, D. Ravinder, J. Laxman Naik, K. Vijaya Kumar, N. Maramu, S. Katlakunta, Investigation of structural, magnetic and electrical properties of chromium substituted nickel ceramic nanopowders. IntechOpen, 2021 10.5772/intechopen.94941.
[4] R. Shilpa, R. Mudachathi, Increasing the free spectral range of Fabry-Perot cavities using dissimilar 1D photonic crystals, Mater. Lett. 299 (2021) 130089. Available from: https://doi.org/10.1016/j.matlet.2021.130089.
[5] P. Choudhary, P. Saxena, A. Yadav, V.N. Rai, A. Mishra, Dielectric and ferroelectric properties of $CoCr_2O_4$ nanoceramics, J. Adv. Dielectr. 9 (3) (2019). Available from: https://doi.org/10.1142/S2010135X19500152, http://www.worldscientific.com/worldscinet/jad.
[6] S. Kishore Babu, M. Jayachandran, T. Maiyalagan, T. Vijayakumar, B. Gunasekaran, Metal-organic framework (MOF-5) incorporated on $NiCo_2O_4$ as electrode material for supercapacitor application, Mater. Lett. 302 (2021) 130338. Available from: https://doi.org/10.1016/j.matlet.2021.130338.
[7] P. Sharma, P. Kumar, R.S. Kundu, J.K. Juneja, N. Ahlawat, R. Punia, Structural and dielectric properties of substituted barium titanate ceramics for capacitor applications, Ceram. Int. 41 (10) (2015) 13425–13432. Available from: https://doi.org/10.1016/j.ceramint.2015.07.131.

[8] E.- H. Mobarak, S.C. Paul, M.S. Islam, S. Akter, K. Sohag, Study on the frequency and temperature dependent electrical properties of MnO and PbO doped ZnO nanoceramics, J. Adv. Chem. Eng. 8 (1) (2018).
[9] A. Regoutz, R.G. Egdell, D.J. Morgan, R.G. Palgrave, H. Téllez, S.J. Skinner, et al., Electronic and surface properties of Ga-doped in 2O3 ceramics, Appl. Surf. Sci. 349 (2015) 970–982. Available from: https://doi.org/10.1016/j.apsusc.2015.04.106, http://www.journals.elsevier.com/applied-surface-science/.
[10] C. David, F. Paumier, B.P. Tinkham, D. Eyidi, M. Marteau, P. Guérin, et al., Modification of electrical properties induced by annealing of ZnO:B thin films deposited by chemical vapour deposition: kinetic investigation of evolution, Mater. Sci. Eng.: B 178 (9) (2013) 698–702. Available from: https://doi.org/10.1016/j.mseb.2012.12.012.
[11] J.F. Li, D.D. Viehland, T. Tani, C.D.E. Lakeman, D.A. Payne, Piezoelectric properties of sol-gel-derived ferroelectric and antiferroelectric thin layers, J. Appl. Phys. 75 (1) (1994) 442–448. Available from: https://doi.org/10.1063/1.355872.
[12] G.D. Zhan, J.D. Kuntz, J.E. Garay, A.K. Mukherjee, Electrical properties of nanoceramics reinforced with ropes of single-walled carbon nanotubes, Appl. Phys. Lett. 83 (6) (2003) 1228–1230. Available from: https://doi.org/10.1063/1.1600511.
[13] Z. Gao, B. Xu, M. Ma, A. Feng, Y. Zhang, X. Liu, et al., Electrostatic self-assembly synthesis of $ZnFe_2O_4$ quantum dots ($ZnFe_2O_4$@C) and electromagnetic microwave absorption, Compos. B: Eng. 179 (2019). Available from: https://doi.org/10.1016/j.compositesb.2019.107417, https://www.journals.elsevier.com/composites-part-b-engineering.
[14] A. Kiani, M. Rahmani, S. Manickam, B. Tan, Nanoceramics: synthesis, characterization, and applications, J. Nanomater. 2014 (2014) 1–2. Available from: https://doi.org/10.1155/2014/528348.
[15] H. Kishi, Y. Mizuno, H. Chazono, Base-metal electrode-multilayer ceramic capacitors: past, present and future perspectives, Jpn. J. Appl. Phys. 42 (1) (2003) 1–15. Available from: https://doi.org/10.1143/JJAP.42.1.
[16] C. Jia, Z. Shao, H. Fan, R. Feng, F. Wang, W. Wang, et al., Barium titanate as a filler for improving the dielectric property of cyanoethyl cellulose/antimony tin oxide nanocomposite films, Compos. A: Appl. Sci. Manuf. 86 (2016) 1–8. Available from: https://doi.org/10.1016/j.compositesa.2016.03.016.
[17] T. Zheng, J. Wu, Quenched bismuth ferrite-barium titanate lead-free piezoelectric ceramics, J. Alloys Compd. 676 (2016) 505–512. Available from: https://doi.org/10.1016/j.jallcom.2016.03.205.
[18] H. Gong, X. Wang, S. Zhang, L. Li, Synergistic effect of rare-earth elements on the dielectric properties and reliability of $BaTiO_3$-based ceramics for multilayer ceramic capacitors, Mater. Res. Bull. 73 (2016) 233–239. Available from: https://doi.org/10.1016/j.materresbull.2015.07.010, http://www.sciencedirect.com/science/journal/00255408.
[19] X.Q. Fang, J.X. Liu, Electrical properties of macro-, micro-, and nanoceramic and nanocomposite coatings and materials, Handbook of Nanoceramic and Nanocomposite Coatings and Materials, Elsevier Inc, China, 2015, pp. 333–344. Available from: https://doi.org/10.1016/B978-0-12–799947-0.00015–8, http://www.sciencedirect.com/science/book/9780127999470.
[20] R. Laishram, O.P. Thakur, D.K. Bhattacharya, Dielectric and piezoelectric properties of La doped lead zinc niobate–lead zirconium titanate ceramics prepared from mechano-chemically activated powders, Mater. Sci. Eng.: B. 172 (2) (2010) 172–176. Available from: https://doi.org/10.1016/j.mseb.2010.05.007.
[21] T.Y. Tseng, H.S. Nalwa, Handbook of Nanoceramics and Their Based Nanodevices Synthesis and Processing, 2009.
[22] A. Šutka, K.A. Gross, Spinel ferrite oxide semiconductor gas sensors, Sens. Actuators B: Chem. 222 (2016) 95–105. Available from: https://doi.org/10.1016/j.snb.2015.08.027.
[23] Z. Tao, L. Yan, J. Qiao, B. Wang, L. Zhang, J. Zhang, A review of advanced proton-conducting materials for hydrogen separation, Prog. Mater. Sci. 74 (2015) 1–50. Available from: https://doi.org/10.1016/j.pmatsci.2015.04.002.
[24] V. Srivastava, D. Gusain, Y.C. Sharma, Critical review on the toxicity of some widely used engineered nanoparticles, Ind. Eng. Chem. Res. 54 (24) (2015) 6209–6233. Available from: https://doi.org/10.1021/acs.iecr.5b01610, http://pubs.acs.org/journal/iecred.

CHAPTER 19

Emerging trends of electrochemical applications of nanoceramics

Khairunnisa Amreen and Sanket Goel
MEMS, Microfluidics and Nanoelectronics Lab, Department of Electrical and Electronics Engineering, Birla Institute of Technology and Science, Hyderabad, Telangana, India

19.1 Introduction

Advancement in nanotechnology has induced the evolution of a novel category of materials with extraordinary control over their structure, morphology, composition, properties, and defects. Nanoceramics (NCs) are the output of nanotechnology. They include various types of ceramics with the nanoscale range dimension, including rods, fibers, tubes, particles, films, and sheets [1]. NCs were first discovered in the 1980s, however, in the last two decades, a significant interest within industry and academia with regard to their fabrication and application has been observed. NCs are basically nanomaterials prepared from very fine particles, preferably less than 100 nm in size. These are inorganic, nonmetallic solid particles that are specifically heat resistant. Based on the size requirement of NCs, the most important ceramic materials are usually the metal oxides, borides, carbides, and nitrides [2]. In addition, the morphological characteristic of the NCs can be fine powders, crystalline, thick or thin films, fibers, etc. These are strategically fabricated from materials like silicon, alumina, and carbide on a large scale. These NCs are enormously different from conventional ceramics not just in terms of material characterization but also in the synthesis process, properties, and applications. The bulk ceramics used in artistic objects, pottery, tiles, etc. are made from materials like minerals and clay. NCs, on the other hand, require various nonmetallic fine particles. NCs bear some remarkable chemical, mechanical, physical, magnetic, and electrical properties like their dielectric nature, ferromagnetism, magnetoresistance, etc. In addition, they show exceptional features like superplasticity, robustness, toughness, strength, machinability, and bioactivity, owing to their morphology [3]. A wide range of applications of NCs in diverse fields like electronics and electrical, mechanical, industrial, magnetic applications, sensors, energy devices, healthcare, tissue engineering, bone implants, and coatings has emerged lately [3].

Industrial Applications of Nanoceramics
DOI: https://doi.org/10.1016/B978-0-323-88654-3.00024-X

19.2 Properties of nanoceramics

The properties of bulk materials change drastically when they are downsized to a nano-scale diameter. NCs have characteristic properties that are quite different from their bulk counterparts. These properties are subjective to their composition. Nevertheless, the size plays the key role in the properties. The applications of NCs in turn are dependent on these properties. Some of the noteworthy properties of NCs are: (1) Mechanical properties: The mechanical strength and rigidity of the NCs are prominently improved with their size. They exhibit substantial tolerance toward compression and bending. In fact, a few of them resembles strength as comparable as steel. Their strength is maintained at temperature as high as 1400°C. (2) Electrical properties: NCs are made up of materials with varied electrical properties. For instance, NCs from aluminum oxide are insulators in nature, silicon carbides are semiconductors, on the other hand, and chromium oxide are electrical conductors. Similarly, NCs with a mixed composition have advanced conductivity. Therefore, there electrical properties are greatly influenced by the materials they are made of. These properties are also reliant on the particle size. The electrical resistance and dielectric constant change with particle size. (3) Magnetic properties: NCs are generally highly resistant toward demagnetization. Due to their nanometer size, the surface–interface interaction, band structure and arrangement of atoms is significantly different from the bulk atoms, hence the magnetic behavior is enhanced significantly. NCs made from iron, barium, nickel, chromium, and manganese oxides are magnetic in nature. (4) Optical properties: Just like all the other characteristics, the optical properties of the NCs are size based. The absorption, emission of light, photons, and radiation depend upon the structural arrangements of atoms which in the case of nanosized particles are mostly on the surface. Therefore, NCs of the same metal oxides in different physical states can show different absorption behavior. For example, for TiO_2NCs, the nanoparticles are more efficient absorbers of UV light than their powder form. NCs forming oxides of metals like yttrium, thorium, neodymium, europium show luminescence properties. (5) Chemical properties: NCs are basically inert in nature; particularly those with fewer defects and a perfect structure. Therefore, these can be employed for various applications. Owing to their inert behavior, they tend to be good anticorrosive materials, hence they are useful as coatings for corrosion resistance. (6) Miscellaneous: NCs have tremendous superplasticity and ductility, i.e., they can be deformed without damaging their structure. In addition, as compared to the bulk ceramics, NCs have impressively reduced brittleness, and henceforth are applied for load-supporting applications. Unlike bulk ceramics, these have enhanced hardness and sintering [4].

19.3 Fabrication of nanoceramics

Fabrication of NCs encompasses considerable artificial inventiveness. Though one can design a stable method, there is repeatedly a component of error. Diverse NCs have been produced in the last few decades by the conventional ceramic approach that includes mixing and crushing powders of the starting materials like oxides, carbonates, etc., followed by heating them at higher temperatures and mechanical grinding. The existing route avoids physical force to attain improved control of morphology. In this context, various techniques have been explored including both physical and chemical methods. The synthesis of NCs through chemical and physical routes has seen a noteworthy advancement in recent decade. Fig. 19.1 gives the schematic overview of some of the general preparation methods adapted on a large scale Fig. 19.1.

19.3.1 Physical method

In general, three physical methods of fabricating NCs are widely adapted:

1. Flame decomposition: This is a most commonly preferred method for preparing nano-diameter particles. It is a type of chemical vapor condensation method. A liquefied precursor material is used as a starting reagent. A combination of a fuel−oxidant like methane−air, propane−oxygen, is used to vaporize and oxidize a liquid precursor via combustion. Using a carrier gas, thermophoretic deposition of the decomposed precursor over a cold substrate takes place. An elevated temperature of 1200K−3000K is utilized which aids chemical gas phase transitions. Different types of precursors are employed in this. For instance, $TiCl_4$ is used as a chemical precursor to prepare TiO_2 particles; $SiCl_4$ for SiO_2. In addition to these, silanes, alkoxides, hydrides, etc. are also used. Since flame decomposition is an efficient approach in terms of yield and economy, it is used at industry scale. However, the biggest drawback lies with the controllability factor of the particle size and morphology [5].
2. Spray pyrolysis: This method gives high purity NCs. It is also referred as phase transition decomposition, plasma evaporation, or aerosol thermolysis. Usually a sol or suspension precursor reagent is deployed in this technique. Herein, the aerosol droplets

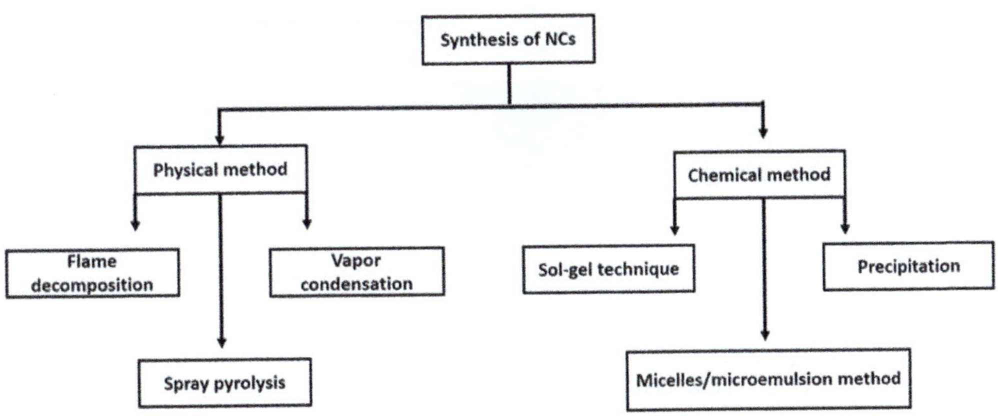

Figure 19.1 Schematic overview of the common nanoceramics synthetic routes.

of the starting reagent are formed via nebulization. These droplets further undergo an evaporation process followed by condensation in a cooling environment. Then, drying and thermolysis of the precipitated particles take place at a raised temperature which gives microporous particles. These particles are subjected to sintering (heating and condensation without liquefaction), producing nanoparticles. Varied methods of nebulization like ultrasonic and electrostatic are utilized. Based on the type of atomizer used, the size of the aerosols, velocity, and rate of formation varies. The parameters like pH, solution concentration, viscosity, precursor reagent, heating temperature, resistance time, etc. decide the fate of the aerosol and in turn the particle size. An aqueous medium is generally preferred owing to cost-effectivity, being low hazard, and, most importantly, the greater availability of water soluble precursors. Reagents like nitrate salts and metal chlorides are used commonly. Sparingly soluble, insoluble precursors are avoided to arrest impurity generation. Homogenized, impurity free nanosize NCs can be obtained with this. However, the use of excess solvents is not cost effective especially when large production is done [6]. Fig. 19.2 is a typical schematic representation of flame spray pyrolytic method of fabrication.

3. Vapor condensation: This approach involves the formation of NCs from a supersaturated vapor of metals. It comprises two major steps. First, a metal-based

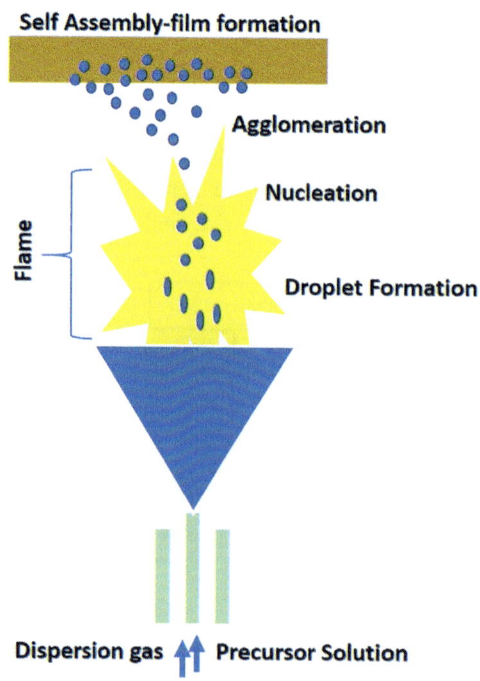

Figure 19.2 Schematic representation of flame spray pyrolysis fabrication approach.

nanopowder is made to condense in an inert ambient convection gas post obtaining the supersaturated metal vapor in a closed chamber. Second, the oxidation of this metal powder occurs via oxygen–gas interaction to produce metal oxide. High-temperature-dependent annealing is needed for the completion of this step. The setup includes a metal vapor source inside a full-vacuum chamber. The vapor source is usually an inert gas like helium or argon, depending upon the material being prepared. These inert gases are made to react with oxygen for generating metal oxides. Similar way, ammonia is used for metal nitrides and alkane or alkene for metal carbides. NCs are formed when supersaturation occurs above the vapor source. A collecting platform, kept over this source, cooled by liquefied nitrogen is used as the base surface for deposition of these nanoparticles formed. Later, these particles are scraped off from the surface. Vaporization techniques like sputtering, thermal evaporation, and laser ablation are used to form supersaturated vapor. A variety of NCs based upon vapor source can be obtained by this method resulting in high-purity NCs. This approach is also useful to form thin films and coatings of NCs. Nevertheless, this method has limitations in terms of cost-effectiveness and low yields compared to the other methods [7]. Fig. 19.3 is a representation of the vapor condensation method.

19.3.2 Chemical method

Chemical routes for fabrication of NCs are usually preferred due to their ease of adaption. The common chemical methods of synthesis adapted for large-scale production of NCs are as follows:

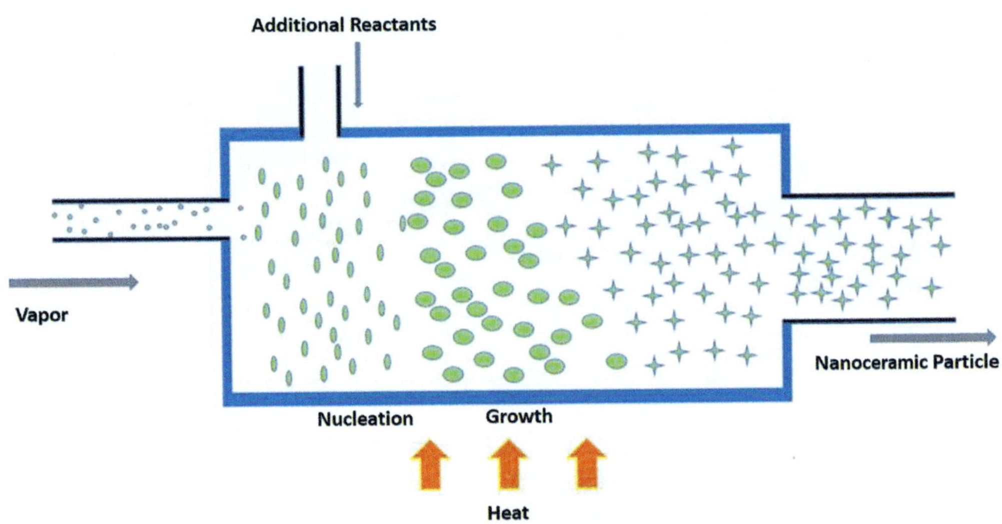

Figure 19.3 Diagrammatic sketch of vapor condensation method.

1. Sol–gel method: This is the pioneer approach used for fabricating nanosized NCs especially of metal oxides. It is a type of hydrolysis reaction wherein, metal alkoxide precursors are hydrolyzed in an alcoholic phase. This converts the alkoxides to hydroxides. Dehydrating the hydroxides results in condensation forming a net framework. These network-like structures link together to undergo a gelation process giving a dense gel. Basically, it is a polymer with a 3D structure with interlinked pores. From the obtained gel, solvent is removed, followed by drying to give super fine powder which upon being subjected to heat gives the respective metal oxide powder. To speed up the process, as alkoxides are prone to low hydrolysis rate, acid or a base catalyst can be added. Upon evaporating the gel, a xerogel is formed, leading to shrinking of the network structure. To avoid this, high pressure and temperature is applied that protect the porous structure from shrinkage. This product is called as an aerogel and is more porous than xerogels. This method gives homogeneous and robust NCs, however scaling up the process is challenging [8]. Fig. 19.4 is a typical representation of the sol–gel approach.
2. Reverse micelles/microemulsion methods: This approach has recently emerged as one of the most promising methods for fabricating nanocrystalline particles comprising ceramics. Reverse micelles are basically nanometer-sized aqueous droplets that are stabilized by surfactants dissolved in organic phases. This results in spherical masses called reverse micelles. The polar heads of the surfactants, in the aqueous

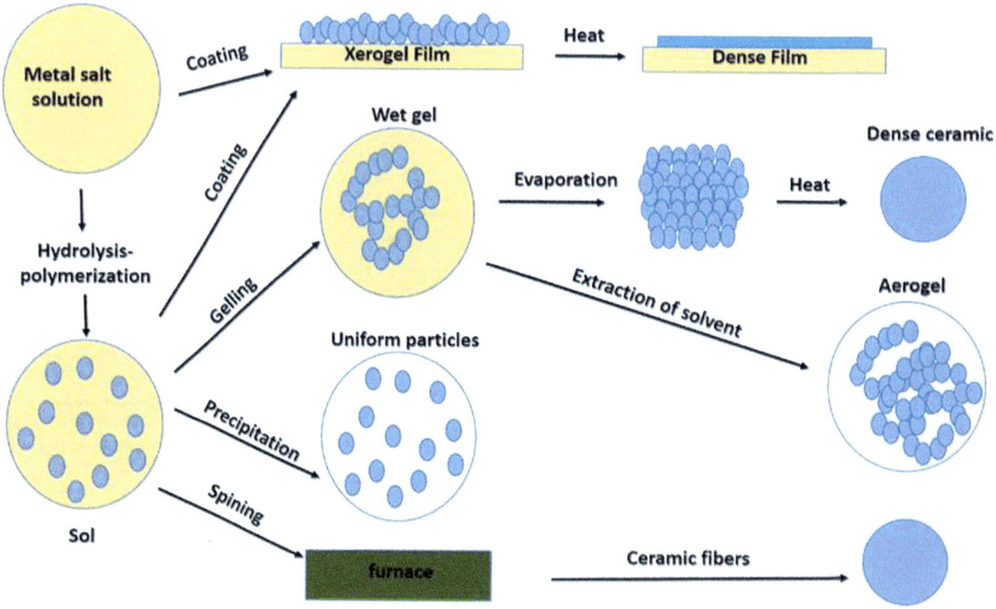

Figure 19.4 Schematic representation of sol–gel approach of nanoceramics synthesis.

phase, organize themselves around small water regions causing the dispersion of aqueous phase in organic phase. These reverse micelles act as nanoreactors for the synthesis of NCs with controlled morphology. The size of nanoparticles prepared here depend upon the size of the reverse micelles. Several methods like hydrolysis of precursors, precipitation of metal salt solution, removal of solvent, calcination, etc. can be carried out inside these micelles to prepare NCs. The parameters, like precursor concentration and aqueous phase weight percentage, in the microemulsion also determine the size, distribution, and morphology of the formed produced. The biggest limitation of this method is that is uses a large quantity of reactants and has a low yield [9]. Fig. 19.5A is a typical structure of a reverse micellar, Fig. 19.5B is the water in oil microemulsion diagram, and Fig. 19.5C is the oil in water microemulsion diagram [10].

3. Precipitation: This approach is the most simplest and a very traditional method of NCs preparation. Herein, precursor metal salt is taken and dissolved nitrate, oxychlorides, and chlorides, for instance, can be used to prepare Al_2O_3 and $AlCl_3$. Likewise, for ZrO_2, $ZrCl_4$ is used. Upon addition of strong basic solutions like sodium hydroxide or ammonium hydroxide, the metal hydroxides precipitates in the aqueous metal salt solution. The other dissociated ions can be washed off and hydroxide precipitate is obtained. This precipitate is subjected to multiple times of washing for impurity removal followed by calcination at high temperature. This method is also employed to prepare ceramic precursors which further upon pyrolysis give NCs. Although this is the simplest method and can also be applied at a large scale, this method has a drawback in terms of controlling size. The precipitation reaction is a spontaneous reaction and cannot be controlled, hence often it ends up preparing larger size microparticles than nanosized particles [11–13].

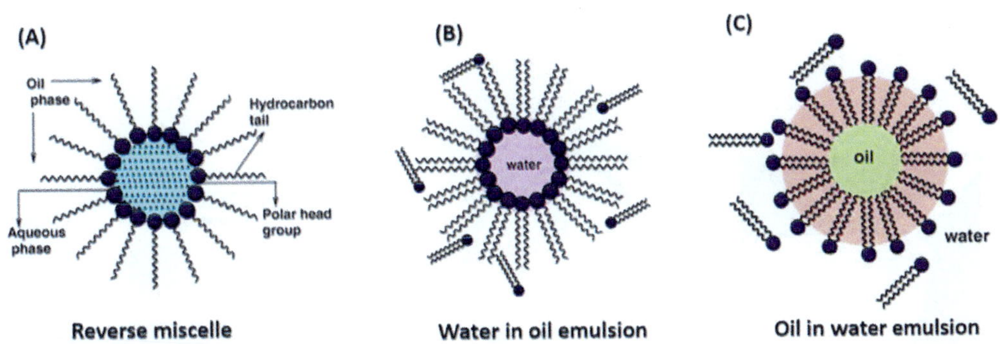

Figure 19.5 (A) Structure of typical reverse micelle. (B) Structure of water in oil microemulsion and (C) oil in water microemulsion. Source: *Reprinted with copyright permission from M.A. Malik, M.Y. Wani, M.A. Hashim, Microemulsion method: a novel route to synthesize organic and inorganic nanomaterials. 1st Nano Update, Arab. J. Chem. 5 (4) (2012) 397–417. https://doi.org/10.1016/j.arabjc.2010.09.027.*

19.3.3 Other approaches

There are quite a few approaches wherein chemical and physical methods of preparation are combined. For example, the synthesis of preceramic polymers [14]. Herein, a chemical reaction is carried out to form a polymer, which, in turn, is then subjected to pyrolysis given the desired NCs. This process is an excellent way to overcome the limitation of low yields due to physical processes. NCs of silicon carbide and silicon nitride has been prepared successfully through this. Polymers with elements like Si, N, C, and Al have been prepared and subjected to pyrolysis to attain metal carbides and metal nitrides. These strategically prepared polymers are called preceramic polymers and are made by simple chemical reactions. Silicon carbides and nitrides are the most explored ceramic materials through this route. For this, polycarbosilanes and polysilazanes, prepolymers are exposed to higher temperatures, 1000ºC–1200ºC. Similarly, another method called mechanochemical [15] has also emerged lately. In this method, the precursor salts or starting reagents are mechanically milled under a specific temperature and pressure which not only breaks them down to nanosize but also allows the chemical reactions to take pace while milling. NCs of Al and Zr have been prepared successfully via this method. There are three basic ways of mechanochemical milling, namely (1) grindstone based using a mortar and pestle, (2) planetary ball milling, and (3) high-speed vibration mixer milling. Fig. 19.6 gives the schematic representation of these processes.

19.4 Types of nanoceramics

Due to the advancement of the strategic synthesis of NCs, several types of NCs driven by their applications are being fabricated. Furthermore, NC nanocomposites are also being explored. A few of the remarkably designed NCs are discussed here.

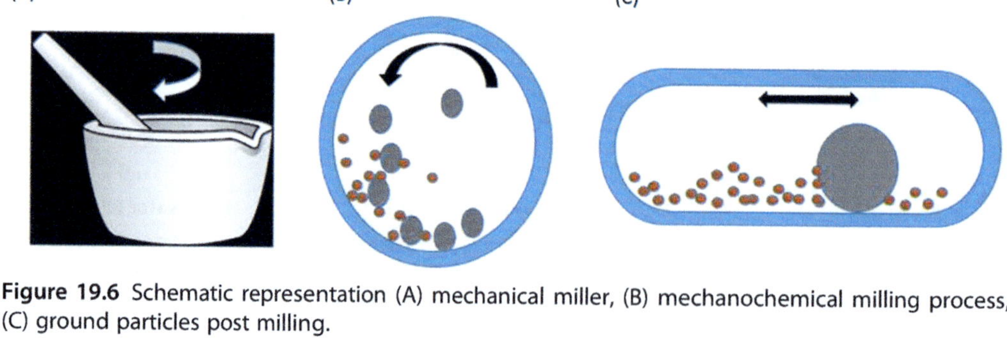

Figure 19.6 Schematic representation (A) mechanical miller, (B) mechanochemical milling process, (C) ground particles post milling.

19.4.1 Chiral nanoceramics

Chirality is a geometric feature where a molecule or an object is nonsuperimposable over its mirror image. So far, significant chiral NCs with novel properties have been of interest. NCs with metals, nitrides, and carbides comprising chiral geometries ranging from atomic level to molecular level have been reported. NCs with nanomaeter, mesoprous and micrometer dimensions are also explored. Nevertheless, essential findings and technological advances of chiral NCs are yet to be explored. The structural diversity and chirality of NCs are dependent upon their enantioselectivity, biocompatibility, catalytic activity, and chiroptical activity, which require monitoring of the chemical composition of these. Commonly studied chiral NCs are chiral ceramic oxides, chiral zeolites, and chiralpolyoxometalates. Hydrothermal, vapor deposition, solvothermal, synthesis with hard and soft templates, chiral biomineralization, etc. are some of the synthetic strategies used. Ceramic oxides like silicon dioxide (SiO_2), zinc oxide, copper oxide (CuO), tungsten (WO_3), and molybdenum oxide (MoO_3) are a few of the chiral NCs fabricated. In addition to these, polyoxometalates, a packed 3D framework of an anion with more than three transition metals linked by a common oxygen atom (oxyanion), have also been processed. Furthermore, zeolites, microporous tetrahedral structures with silicon, aluminum, and oxygen-based NCs, have been studied. These porous structures are useful to trap large number of cations like sodium, potassium, and calcium. In the future, computational chemistry will enable the design of chiral NCs based upon their applications [16].

19.4.2 Composite nanoceramics

The composites of NCs have been discovered to be exceptional engineering materials lately. Niihara et al. were the first to discover that by preparing the composites of NCs, their mechanical properties like their strength and toughness improved considerably. They broadly classified these composites into two categories: (1) NCs with microsize matrix grains, and (2) NCs with nanosize matrix grains [17]. Mukhopadhyay et al. gave a detailed review of the structural preparation, classification, and properties of these NCs composites. They classified these into five major types. They are as follows: (1) Intergranular: wherein nanosize diameter reinforcements get incorporated at the boundaries and junctions of the matrix grains. Hence, they become fracture resistant. (2) Intragranular: wherein nanosize diameter reinforcements get incorporated inside the matrix grains improving its mechanical strength. (3) Inter—intra granular: wherein the nanosize diameter reinforcements get incorporated both inside and at the boundaries of the matrix grains. As a result, greatly enhanced toughness, robustness and stability at high temperature is obtained. (4) Nano/nanocomposites: wherein both the grain matrix and the substance being incorporated are of nanosize diameter. This results in providing superelasticity and machinability as the number of matrix grain boundaries increases. (5) Nano/microcomposite: wherein, the microsized particle is incorporated into the nanosize grain matrix. This augments mechanical strength significantly. Fig. 19.7 is a schematic of the various types of NCs nanocomposites that are fabricated.

These tend to afford resistance from pressure, heat, stress, fracture, and fatigue, and therefore, they can be applied in various industrial applications without wear and tear. However, there are certain limitations associated with the preparation of these, e.g., the sintering technique has to be sophisticated, and agglomeration of particulates has to be controlled. To address these issues, methods like sol–gel, inert gas condensation, and high-energy ball milling are being used. A technique called activated sintering is also being utilized, wherein an electric field is imposed with applied pressure while sintering is performed. This is called plasma-activated or field-assisted sintering. Mukhopadhyay et al. further discussed various methods of preparation, the challenges associated with them, and their future prospects [18].

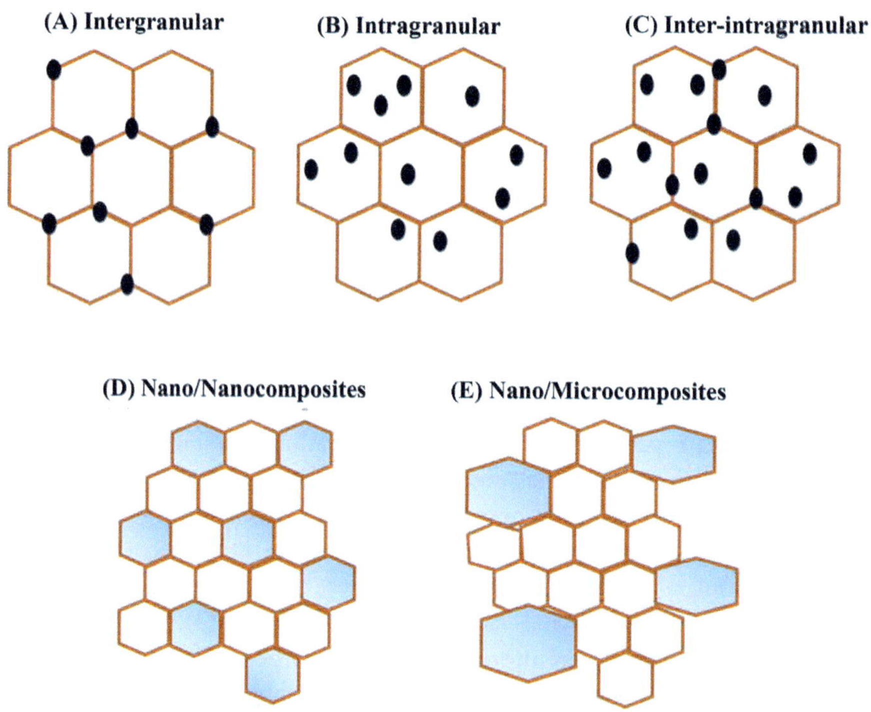

Figure 19.7 Schematic representation of types of NCs nanocomposites: (A) intragranular, (B) intragranular, (C) inter-intragranular, (D) nano/nanocomposites, (E) nano/microcomposites. *NCs*, Nanoceramics.

19.4.3 Bioactive nanoceramics

NCs used in biomedical applications like tissue engineering, bone regeneration, implants etc. are referred to as bioactive NCs. NCs made up of calcium-phosphate like hydroxyapatite, brushite, octacalcium phosphate, biphasic calcium, etc. are bioactive in nature. Their constituents are similar to those of human bones and tissues, and hence these tend to form an apatite layer and bond with the bones, thus increasing regeneration. However, these materials are brittle and cannot withstand stress. To improve this, they are infused with bioinert metals forming bioinert NCs. Oxides of titanium, zirconia, and alumina are some of the materials used in bone reconstruction and tissue generation as they are tough, possess high mechanical strength, and are fracture and corrosion resistant. The properties of these materials are greatly influenced by the synthesis route adapted. Henceforth, for synthesis of these, basic top-down and bottom-up approaches are employed. Despite having improved bioactivity and an implantable nature, these bioactive NCs face challenges in terms of low production yield, higher cost, and lack of reproducibility. Hence, to overcome these challenges, biphasic NCs are being developed. With enhanced features, these are extensively used in the biomedical field for regenerative medicine purposes, tissue engineering, bone implants, reconstructions, nerve repair, skin repair, etc. Some of the examples of these are collagen or gelatin scaffolds, chitin or chitosan scaffolds, alginate scaffolds, etc. The demand for improving the properties of these materials is leading to a continual research for the synthesis of these bioinert NCs through novel routes on a large scale, with high yield and cost-effectiveness [19].

19.4.4 Polymer nanoceramics

In a recent trend, NCs are being used as fillers in polymers to form hybrid material with enhanced properties. For instance, world-renowned automobile company Toyota has progressed the use of Nylon-6- clay hybrid as a covering material in engines and fuel tubes to give more tensile strength. Likewise, NCs dispersed in polymers are being utilized in dental implants. Dental adhesives, restorative dental materials, dental cements, cavity fillers, etc. are some of the commercially available polymeric NCs. Another emerging application of polymer resin NCs composite is in nanofiltration. Specifically designed membranes using polymer NCs are used as gas separators and liquid evaporators, for example, drying of organic solvent in sol–gel synthesis. Herein, the membrane acts as a mediator for pervaporation [1].

19.5 General applications of nanoceramics

Based on the attributes of their properties, especially electrical, mechanical, and magnetic properties, NCs are widely used as semiconductors, insulators, conductors, magnets, etc. in diverse applications in the fields of electronics and electrical, biomedical, aerospace, nuclear, construction, sensors, electrochemical, biosensing, etc. A few of the common industrial applications are discussed here.

19.5.1 Electrical application

As discussed, NCs are useful as conductors, semiconductor, and insulators. Based upon the type of material they are fabricated from, their property can be utilized. For example, NCs prepared from aluminum oxide are nonconducting in nature, and hence are utilized as insulators. On the other hand, NCs prepared from barium, like barium titanate, are semiconductors. Some of the other NCs used as semiconductors are silica and lanthanum, like SiC and $LaCrO_3$. One of the notable properties of NCs is their superconductivity. This refers to the capability of a material to be conductive at extreme low temperatures with zero resistance. Copper oxide-based NCs are superconductive in nature. Similarly, NCs from transition metals, borides, and carbides are useful as electrodes in electrochemical sensing applications. NCs of barium also show a dielectric nature, that is, they can be polarized under the influence of an applied electric field. Hence, they are utilized as capacitors. NCs also display pyroelectric behavior, that is, they can generate voltage upon heating or cooling, for example, lead zirconate titanate is applied in infrared detectors. NCs also portrays ferroelectric behavior, that is, they have electric polarization that can be reversed by the influence of an applied electric field. All the ferroelectric materials are basically pyroelectric in nature. NCs of thorium and yttrium show a luminescent property, hence they are utilized in laser-based devices. Certain NCs are useful in energy storage. For example, strontium titanate is known to store a large amount of electrical energy even when used in a small amount and hence it is used in supercapacitors. Lithium aluminosilicate is another example of NCs for energy storage applications which are used in lithium batteries. Recently, NCs exhibiting piezoelectric effect have been gaining attention in the commercial industry. This denotes the appearance of electric potential over the face of a crystal upon applying pressure mechanically. This is useful in components of televisions, transistors, FM radios, monitors, etc. Because of these manipulative properties, especially electronic and thermal properties, along with their mechanical strength and greater melting points, it is suggested that these materials can be used in robust and tough environments without losing their properties. Therefore, this opens a broader route of applications in different sectors [20].

19.5.2 Mechanical applications

On an industrial scale, NCs are usually utilized in the applications that need fracture resistance, abrasion resistance, robust, strong, and nonbrittle materials. Therefore, NCs like alumina, silicon nitrides and carbides, and nickel alloys are generally used. A few of the strength-dependent applications of these metal-based NCs are for making tools used in metal cutting, grinding, and shaping. Silicon-based NCs are highly resistant at elevated temperatures, and hence are used in machines like turbocharger rotors. A combination of ceramic materials and metal-based ceramics are termed cermets, these are used in aerospace engineering for the manufacture of heat protecting tiles in spacecrafts, space vehicle components, and rocket nose cones used in payloads. Because of the extensive mechanical properties, NCs are useful in construction and architectural materials. For instance, NCs-based floor and wall tiles are emerging as the new favorite building material. NCs are being extensively used as fillers as they can monitor the thermal expansion greatly, enhance the drying process and lessen the shrinkage and plasticity due to drying. Recently, NCs have been used as fluxes. Fluxes are those materials that are capable of reducing the fusion temperature of the materials they are mixed with. Silica-based NCs are generally the most commonly used in construction. Furthermore, NCs tend to have greater tensile property and bending skills which is why their use in monument and statue building has seen an increase. The cost-effectiveness constitutes an advantage in the field of construction [4]. Another significant application of the NCs owing to their hardness and corrosion-resistance is coatings. Materials like titanium carbonitride and nitride are often employed as wear-resistant coating. Furthermore, NCs of tungsten like tungsten carbide are also of great use due to their inert nature and high elasticity. Multilayered, superlattices of more than one type of nitride have been gaining much attention lately. These are useful in the manufacture of pumps, compressors, and bearings. Often, techniques like chemical vapor deposition, thermal, or plasma spraying are utilized for these coatings [21].

19.5.3 Magnetic applications

Metal-based NCs containing magnetic metals like iron and manganese exhibit excellent magnetic behavior. Oxides of iron, manganese, and strontium are some of the examples used as soft and hard magnets. The use of these in water purification and disinfection is also emerging. Ferrites, iron-based NCs, are being applied as economically viable magnets in electric motors. The key role of these types of magnets is to convert electrical energy into mechanical energy. Ferrites are more advantageous than other metal magnets as they lose less power, as ferrites are capable of conducting high-frequency current. A nanocomposite of manganese–zinc ferrites is widely used in magnetic recording media. Some of the commercially available products based on the magnetic behavior of NCs are magnetic tape, computer diskettes, and magnet seal [22].

19.5.4 Biomedical applications

Among the various applications, NCs in biomedical and clinical use are of great significance. The most common use is in bone regeneration, tissue engineering, bone implants, and dentistry. NCs with calcium phosphate, zirconia, silica, and bioactive glass are often used as bone substitutes. It has been found that growth hormones can actually help in bone regeneration and growth, and these bioactive NCs can be involved in physiological processes enhancing the tissue and bone growth. The scaffolds of these NCs, like hydroxyapatite, can cause cell migration, differentiation, and proliferation, hence leading to bone formation from nearby tissues or bone repair through regeneration. Furthermore, in the case of drug delivery application for bone repair, the emerging trend is to design novel formulations of nanosize hydroxyapatite whose properties are close to that of the living bone for biocompatibility. Orthopedic implants for knee, ankle, and hip made of NCs are some of the recent advances of bone tissue engineering. Bone grafting and bone cavity filling with these NCs have been proven to have healing potential. Incorporation of minerals like zinc and magnesium in the calcium phosphate NCs has been proven to promote the activity of osteoblastic cell. Hence, they are useful to improve bone development and aid bone density and calcification. Paste of NCs as injectable bone matrix to be used as fillings in spinal cord is also commercially viable now. Use of zinc phosphate NCs in an insulin protein binding application to attempt delivery via the oral route of insulin has also been investigated. Coating of NCs over synthetic implants made up of cobalt, chromium, titanium, and stainless steel has proven to be a boon in bone repair. Paul et al. gave a detailed review of the applications of NCs specifically in tissue and bone engineering [23]. Another noteworthy application of bioactive NCs is in prosthodontics. This is the branch of dentistry involving the treatment of dental defects, loss of tooth, crown, and enamel. Therefore, NCs like ZrO_2 Al_2O_3 glass ceramic are used as prosthodontic implant and filler material. The major advantages of these over conventional materials are that they have good strength and low thermal and electrical conductivity. In terms of esthetics, their appearance matches with the natural color. Furthermore their anticorrosive, fracture resistance and translucent natures make them ideal material in dentistry. Wang et al. reviewed the advances in nanomaterials in prosthodontics [24]. Fig. 19.8 summarizes the broad applications of NCs in tissue engineering.

19.6 Electrochemical applications of nanoceramics

Despite their extraordinary electrical properties, NCs have been explored little as electrochemical sensors. The plausible reason behind this could be that NCs show a low electron transfer mechanism on their own, hence they have to be incorporated with other conducting materials like carbons, nanomaterials, polymers, composites, etc. to enhance the electron transfer rate. Over the last decade, the use of NCs, especially hydroxyapatite (HAP), incorporated with nanocomposite as various electrochemical applications and biosensors has grown significantly. Herein, we have summarized a few of the recently developed electrochemical applications and biosensor using NCs composites, classified based upon the analytes being sensed.

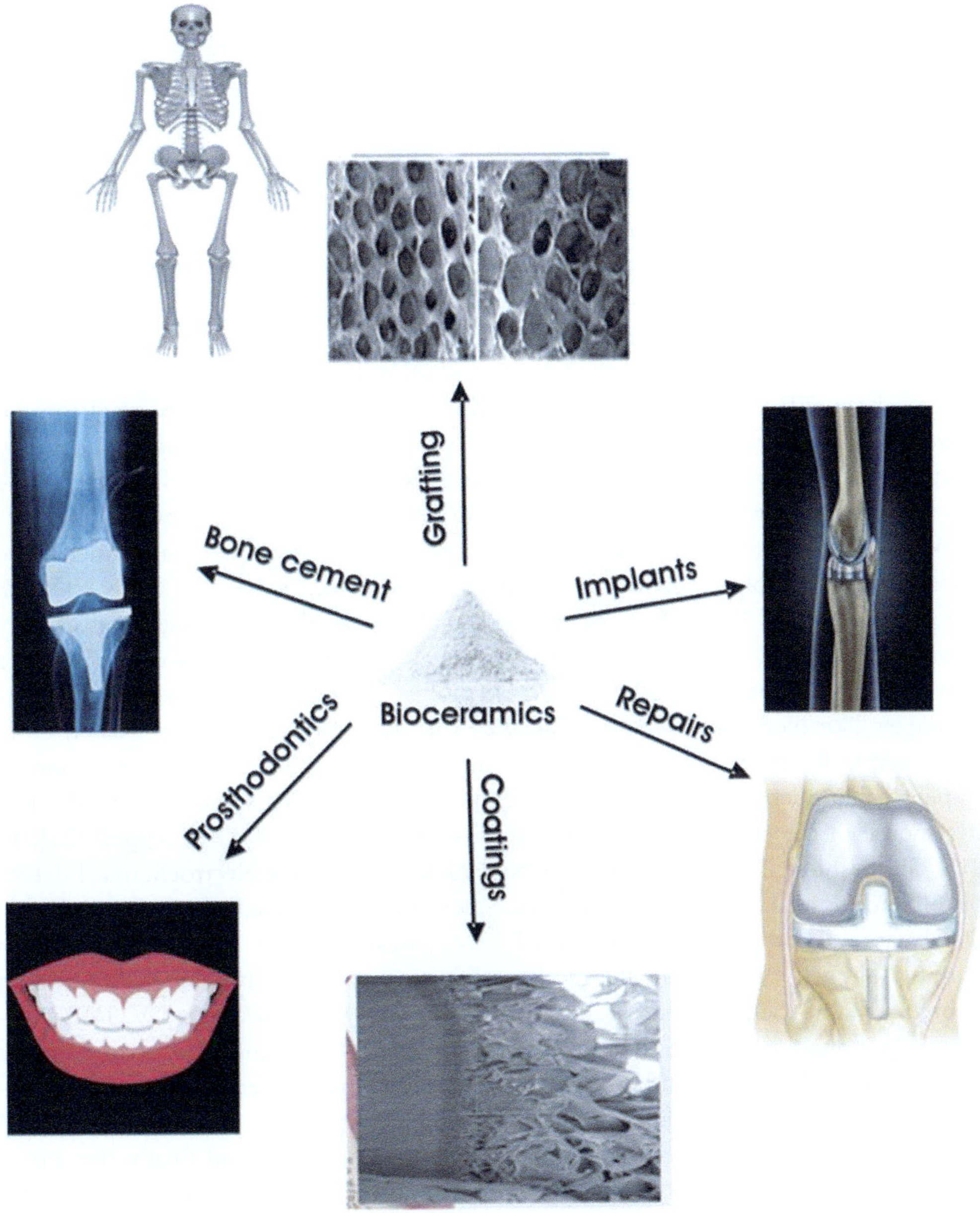

Figure 19.8 Generalized biomedical applications of nanoceramics.

19.6.1 Electroanalytical chemical sensors

Significant literature reports are available that showcase the use of NCs and their composites for the electrochemical sensing of various chemical analytes like environmental pollutants, drugs, organic molecules, heavy metals, polymers, etc. In this section, some of the noteworthy reported works are discussed. For example, in 2009 Mhammedi et al. attempted the detection of para-nitrophenol, an environmental pollutant, in river water. They developed a unique carbon paste electrode incorporated with hydroxyapatite. p-Nitrophenol got adsorbed on the HAP matrix. An open circuit potential along with cyclic voltammetry (CV) and square wave voltammetry (SWV) was used for performing experiments. They could detect a linear concentration range of 2×10^{-7} and 1×10^{-4} mol/L, with the limit of detection (LOD) as low as 8×10^{-9} mol/L [25]. Pang et al. in 2014 prepared a unique nanocomposite of HAP with graphene for electroanalytical detection of luteolin in real peanut hulls samples. Graphene sheets incorporated in Hap were used as modification for a glassy carbon electrode as the base. CV and differential pulse voltammetry (DVP) were used for this study. The various parameters like effect of pH, scam rate, and concentration were examined. The sensor gave remarkable detection in the linear range of 2.0×10^{-8} to 1.0×10^{-5} M, with LOD of 1.0×10^{-8} M. Further, they evaluated real samples with satisfactory recovery values [26]. In the same year, Ajab et al. reported detection of lead (Pb), a heavy metal environmental pollutant that is capable of bioaccumulation in physiological samples like blood, saliva, and urine. For this, they utilized a carbon-based electrode chemically modified with bismuth-incorporated HAP. Square wave anodic stripping voltammetry (SWSV) was the technique utilized. The sensor exhibited a linear response in the concentration range of 10–60 μg/L with LOD of 5 μg/L [27]. In 2015, Ghoreishi et al., explored another NC, nickel titanate for electrochemical detection of salicylic acid, an organic molecule. Since, titanate showed good electrochemical behavior, no other material was used here. Nickel titanate NC was prepared in house by sol–gel, followed by chemical modification of a carbon paste electrode with the prepared and characterized NC. Various techniques were used like CV, DPV, linear sweep voltammetry (LSV), chronoamperometry (CA), and chronocoulometry (CC) for the studies to obtain better sensitivity. The fabricated sensor showed two different linear ranges for low and high concentrations as 3.0–40.0 μM and 40.0–1000.0 μM. An excellent LOD of 68.0 nM was achieved and finally the sensor was applied for detection of salicylic acid in real samples [28]. The same research group also reported a manganese titanate-based electrochemical sensor for the detection of captopril with interference mitigation from aminobenzoic acid. Manganese titanate NC was prepared by sol–gel, followed by chemical modification of a carbon paste electrode with the prepared and characterized NC. The same techniques as stated earlier—CV, DPV, LSV, CA and CC—were used for the studies. The sensor displayed two different linear ranges for low and high concentrations of 1.0×10^{-8} to 1.0×10^{-7} and 1.0×10^{-7} to 1.0×10^{-6}. A remarkable LOD of 1.6 nM was attained and finally the sensor was applied for detection of captopril in pharmaceutical drugs and biological samples [29]. Kanchana et al. synthesized HAP nanoparticles and characterized them using XRD, XPS, Raman, FTIR, SEM, and EDX. Thin films of HAP nanoparticles were dropcasted on a glassy carbon electrode. Further, this electrode was used for folic acid sensing using CV, LSV, and CA. A linear detection range of 1.0×10^{-7} to 3.5×10^{-4} M was determined and LOD was 75 nM. The sensor was selective toward folic acid in the presence of other interfering substances like ascorbic acid and further utilized for real-time sensing in pharmaceutical samples [30]. Kaur et al. developed a new strategy for selective and sensitive multiplexed detection of heavy metals like cadmium, lead, arsenic, and mercury in trace amounts. For this, they fabricated a HAP composite with silver and zeolite-modified glassy carbon electrode. SWS and impedance were the techniques used. The linear concentration ranges obtained were 0.5–1600, 0.6–1600, 0.9–1800, and 0.8–1800 ppb and LOD was 0.1, 0.1, 0.2, and 0.2 ppb for Cd^{2+}, Pb^{2+}, As^{3+}, and Hg^{2+} respectively. Real sample analysis with water bodies with negligible interference was performed, thus proving the fabricated sensor to be highly selective and sensitive toward each analyte even in the presence of others and higher concentrations of other biochemicals. Furthermore, the electrode was highly stable and reproducible [31]. In 2016, Bharath et al. developed a novel sensor for hydrazine. Herein, they used HAP nanorods decorated with gold nanoparticles via a hydrothermal precipitation process. The gold nanoparticles were affixed with HAP nanorods with an overall diameter of 10 nM. The amperometric i–t curve technique was used for the studies. The sensor gave exceptional electrocatalytic behavior toward hydrazine with no interference from common organic and inorganic compounds in a linear range of 0.5–1429 μM with a LOD of 0.017 μM [32]. Similarly, Gao et al. developed a sensor for hydrazine with a reduced graphene oxide and HAP composite. Graphene oxide was reduced by applying a temperature of 85°C and Hap was grown in situ over the reduced graphene oxide. This composite gave notable electrochemical sensing of hydrazine with a chronoamperometric technique in a wide linear range of 2.5 μM to 1.16 mM. The LOD was 0.43 μM with high reproducibility and stability [33]. Another research group, Alam et al. used a similar nanocomposite of reduced graphene oxide and HAP for bisphenol detection. A chemical reduction method was used by them for fabricating the nanocomposite. GCE was used as an underlying carbon. The chemically modified electrode displayed electrocatalytic activity toward bisphenol in a linear range of 0.2–2 mM/L. The LOD obtained was 60 pM/L [34]. Wang et al. developed a unique conducting polymer incorporated by HAP nanocomposite. Poly(3,4-ethylenedioxythiophene) (PEDOT) was used here and doped with nanodiameter HAP. The composite was electrodeposited over a glassy carbon electrode. Amperometry and CV were the techniques applied for this. The prepared sensor was used for the selective detection of nitrite. The linear range obtained was 0.25–1.05 mM with an exceptional LOD of 83 nM. The electrode was highly stable and reproducible and gave a sensitive response toward nitrite in real water samples as well [35]. Mirzajani et al. synthesized a γ-Fe_2O_3 and hydroxyapatite and Cu(II) nanocomposite with magnetic behavior. The prepared nanocomposite was characterized via various techniques like FTIR, SEM, TEM, and EDX. The nanocomposite was used to modify the carbon paste electrode by dropcasting. It was further investigated for the electrochemical detection of metformin (a pharmaceutical drug with broad applications especially in controlling diabetes), in real urine and pharmaceutical drug samples. The linear range obtained was 0.1–80 μM with a LOD of 14 nM. The most interesting part here was that the composite was a core–shell magnetic nanoparticle encapsulating HAP. Fig. 19.9 gives the schematic representation for the synthesis and fabrication procedure of the electrode [36].

Anjum et al. reported HAP NC with doped graphite. The fabricated sensor gave a response to methanol, ethanol, and propanol, without interference from atmospheric air and vapors. A low concentration of 100 ppm was detected successfully. By doping graphite into HAP, the sensitivity toward alcohol was increased six times. Fig. 19.10A shows a schematic of the sensor [37]. Sun et al. performed detection of heavy metals like Pb^{2+}, Cu^{2+}, and Hg^{2+} simultaneously using a HAP-modified indium tin oxide glass electrode (ITO) electrode. Herein, they used DPV as a technique. The electrode exhibited good linear concentration ranges of 1×10^{-7} to 1×10^{-6} and 3×10^{-6} to 1×10^{-5} M for Pb^{2+}, 1×10^{-7} to 1×10^{-6} M and 1×10^{-6} to 1×10^{-5} M for Cu^{2+}, and 1×10^{-7} to 1×10^{-5} M for Hg^{2+}. The LOD obtained were 1×10^{-12}, 1×10^{-12}, and 8×10^{-8} M for Pd^{2+}, Cu^{2+}, and Hg^{2+}, respectively. No interference from other heavy metals was observed.

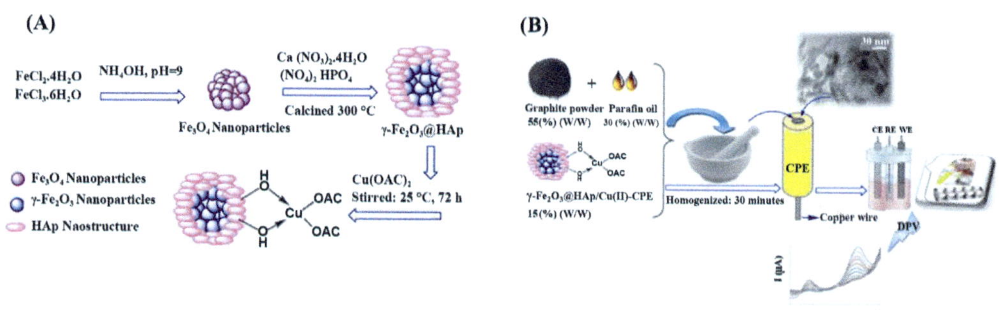

Figure 19.9 (A) Schematic representation of the preparation of composite. (B) Electrode fabrication procedure and electrochemical sensing. Source: *Printed with copyright permit from R. Mirzajani, S. Karimi,. Preparation of Γ-Fe2O3/hydroxyapatite/Cu(II) magnetic nanocomposite and its application for electrochemical detection of metformin in urine and pharmaceutical samples. Sens. Actuators, B: Chem. 270 (2018) 405–416. https://doi.org/10.1016/j.snb.2018.05.032.*

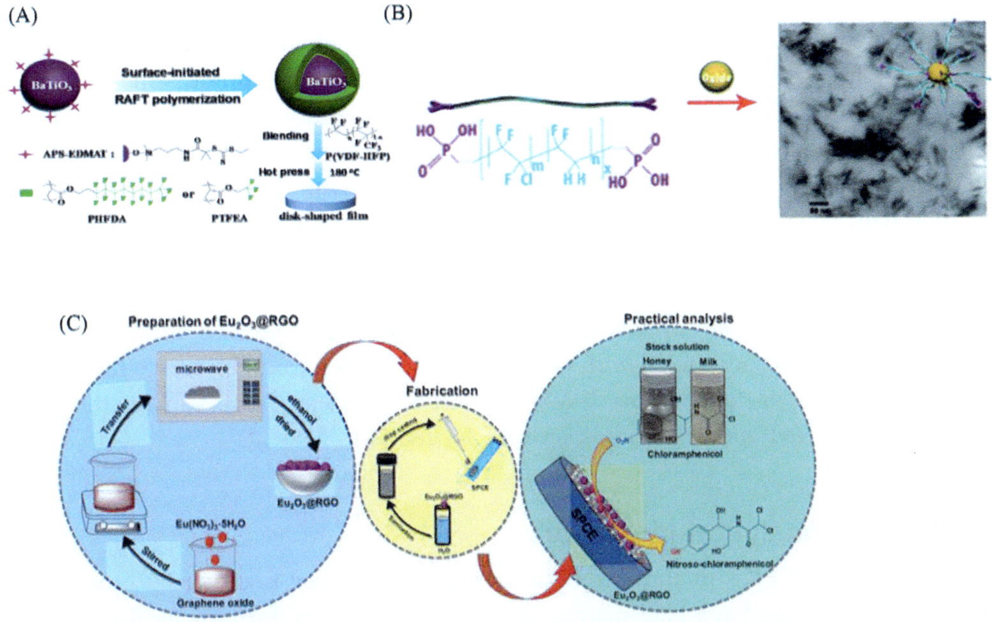

Figure 19.10 (A) Schematic of the HAP doped graphite sensor, (B) Schematic representation of their sensor. (C) Schematic representation of nanocomposite formation and electrode fabrication. *HAP*, Hydroxyapatite. Source: *(A) Reprint with copyright permit from Anjum et Al. [37]; (B) reprinted with copyright permission [38]; (C) reprint from Rajaji et al. with copyright permission [39].*

Fig. 19.10B is the schematic representation of their sensor reprinted with copyright permission [38]. Similarly, Velmurugan et al. established an approach for functionalization of multiwalled carbon nanotubes (MWCNT) via ultrasonic vibrations, and incorporated HAP and chitosan scaffolds for electrochemical detection of nitrofurantoin. A stable composite of chitosan and MWCNT decorated with HAP nanoparticles was prepared by a hydrothermal method. The prepared composite was modified over a glassy carbon electrode surface and used for electroanalytical sensing of nitrofurantoin. CV and amperometry was used for the studies. A linear concentration range of 0.005–982.1 μM with a LOD of 1.3 nM was obtained. The electrode was stable and reproducible [40]. Rajaji et al. developed europium(III) oxide incorporated with reduced graphene oxide nanocomposite and utilized it for the detection of chloramphenicol in food samples. A microwave-assisted approach was used for preparing this. Furthermore, a screen-printed carbon electrode was used as a base electrode and drop-casted with prepared composite. It showed an exceptional linearity in the range of 0.02–800.25 μM toward chloramphenicol with LOD of 1.32 nM. Fig. 19.10C shows the schematic of the nanocomposite preparation and electrode fabrication [39].

Similarly, Kanchana et al. designed manganese-doped HAP for the electrochemical sensing of hydroquinone. A glassy carbon electrode was chosen as the base electrode. The prepared sensor gave a good response toward oxidation of hydroquinone in a linear range of 1.0×10^{-8} to 1.6×10^{-4} M with a lowest detection limit of 11 nM. It was highly stable, reproducible, and gave a noninterfering response even in analysis of real tap and industry water samples [41]. Kashani et al. made a chemically modified carbon paste electrode with nickel titanate NC. CV and DPV were used for electrochemical detection of ortho- and para-hydroxybenzoic acids simultaneously with no interference. The linear concentration ranges used were 10–1000 μM with a detection limit of 0.38 and 0.10 μM for ortho and para-hydroxybenzoic acid, respectively [42]. Recently, Tran et al. developed HAP nanorods mixed with chitosan over a glassy carbon electrode for electrochemical detection of copper ions. SWV was used to detect Cu(II) ions in the linear range of 10–100 μM with a detection limit of 10 μM [43].

19.6.2 Electroanalytical biosensors

Lately, the usage of NCs in and as biosensor has also observed noteworthy progression. NCs have been utilized as a base matrix, or as a trapping entity for biological molecules, living organisms, enzymes, and antibodies to detect other analytes. NCs, owing to their porous nature, have the capability to withhold larger biomolecules. A few of the significant reported biosensors are discussed here. Wang et al. in 2009, developed a hybrid nanocomposite film of HAP and silica. Further, they utilized this as a base matrix over a glassy carbon electrode to immobilize horseradish peroxidase enzyme. The fabricated biosensor was applied for the detection of hydrogen peroxide. The biosensor gave excellent mediator-less reduction of H_2O_2. CV and amperometry were used. A linear concentration range of 1.0–100 μM for H_2O_2 was used. LOD obtained was 0.35 μM [44]. Wang et al. in 2010 developed a HAP nanoarray-based biosensor for the detection of cyanide. The nanoarray was aligned as nanowires consisting of a diameter of 200 nm and length of 1 μM. The HAP nanoarray was directly electrodeposited over a glassy carbon electrode with the help of a membrane. Furthermore, CV was used to study the electrocatalytic activity. The LOD obtained was 0.6 ng/mL [45]. Furthermore, Lu et al. fabricated a novel biosensor for the detection of phenols. Herein, a gold electrode was used as a base electrode. A nanocomposite of HAP and chitosan was prepared and chemically modified as a matrix over a gold electrode. Tyrosinase enzyme was immobilized over the prepared matrix. The invented biosensor was applied for electroanalytical detection of phenolic compounds using CV and impedance studies. Catechol was the phenolic compound chosen here for the investigation. The biosensor gave a larger linear range of 10 nM to 7 μM with LOD of 5 nM [46]. Li et al. developed a glucose biosensor. Herein, a nanocomposite of mesoporous HAP and titanium oxide was prepared by mixing and ultrasonication with MWCNT. A thin film of this nanocomposite was coated onto a glassy carbon electrode that served as an underlying matrix to trap glucose oxidase enzyme. CV and CA were used for the studies. The fabricated biosensor gave electrocatalytic oxidation of glucose in a linear range of 0.01–15.2 mM. The LOD obtained was 2 μM. Zero interference from coexisting biochemicals like dopamine, uric acid, ascorbic acid, and other carbohydrates was observed. The biosensor was extremely stable and used for the detection of glucose in human blood real samples [47]. Kanchana et al. developed a new approach of synthesis of HAP via microwave irradiation using EDTA as an organic modifier. This gave electroactive functionalized HAP nanoparticles which were further coated on the glassy carbon electrode. The prepared sensor was used for enzymeless sensing of biochemical uric acid. CV, amperometry, LSV, and impedance spectroscopy were employed. The sensor showed linear detection of uric acid in the range of 1×10^{-7} to 3×10^{-5} M. The LOD obtained was 142 nM [48]. Yang and his research group designed a bilirubin biosensor based on Fe_3O_4-HAP-polypyrrole which was molecularly imprinted on magnetic nanoparticles. The particles were anchored to a magnetic glassy carbon electrode surface. CV and amperometry were used to detect bilirubin in the linear concentration of 0.1–17 μM and a LOD of 0.007 μM [49]. Bharath et al. designed a biosensor using a combination of 2D graphene oxide and 1D HAP. HAP nanorods were deposited over reduced graphene oxide via a hydrothermal process. The prepared composite was used as a matrix over a glassy carbon electrode for trapping glucose oxidase enzyme. The fabricated biosensor was tested for electrocatalytic oxidation of glucose in a linear range of 0.1–11.5 mM with LOD of 0.03 mM. No interference from coexisting chemicals was seen [50]. Kanchana et al. reported a nanocomposite formation using HAP and iron-doped carbon nanotube particles. These were used for detection of L-dopa in the copresence of uric acid. The hybrid composite was coated on a glassy carbon electrode and used for amperometric detection of L-dopa. The prepared biosensor gave a linear response in the range of 1.0×10^{-7}–1.1×10^{-6} M in the presence of uric acid, ascorbic acid, and other common interfering biochemicals. It exhibited good reproducibility and stability, hence it can be used for real sample detection in pharmaceutical and medicinal plants [51]. Huang et al. demonstrated a very interesting strategy to enhance the electrochemical sensitivity in immunoassays. The signal amplification depended upon the reaction with the phosphate groups that were present on the HAP nanoparticles, which were in turn reacted with redox active molybdophosphate precipitate. An antigen–antibody reaction using a cancer biomarker α-fetoprotein as the antigen was studied. The antibodies against α-fetoprotein were immobilized on the prepared matrix over a glassy carbon electrode. The biosensor showed a linear range of detection from 0.1 pg/mL to 1 ng/mL. The LOD obtained was 50 fg/mL. Further, the biosensor was applied for real serum sample analysis [52]. Fig. 19.11 is the schematic representation of their approach.

Kanchana et al. fabricated Ce-doped HAP nanoparticles for simultaneous determination of norepinephrine, tyrosine, and uric acid. This novel biosensor was prepared

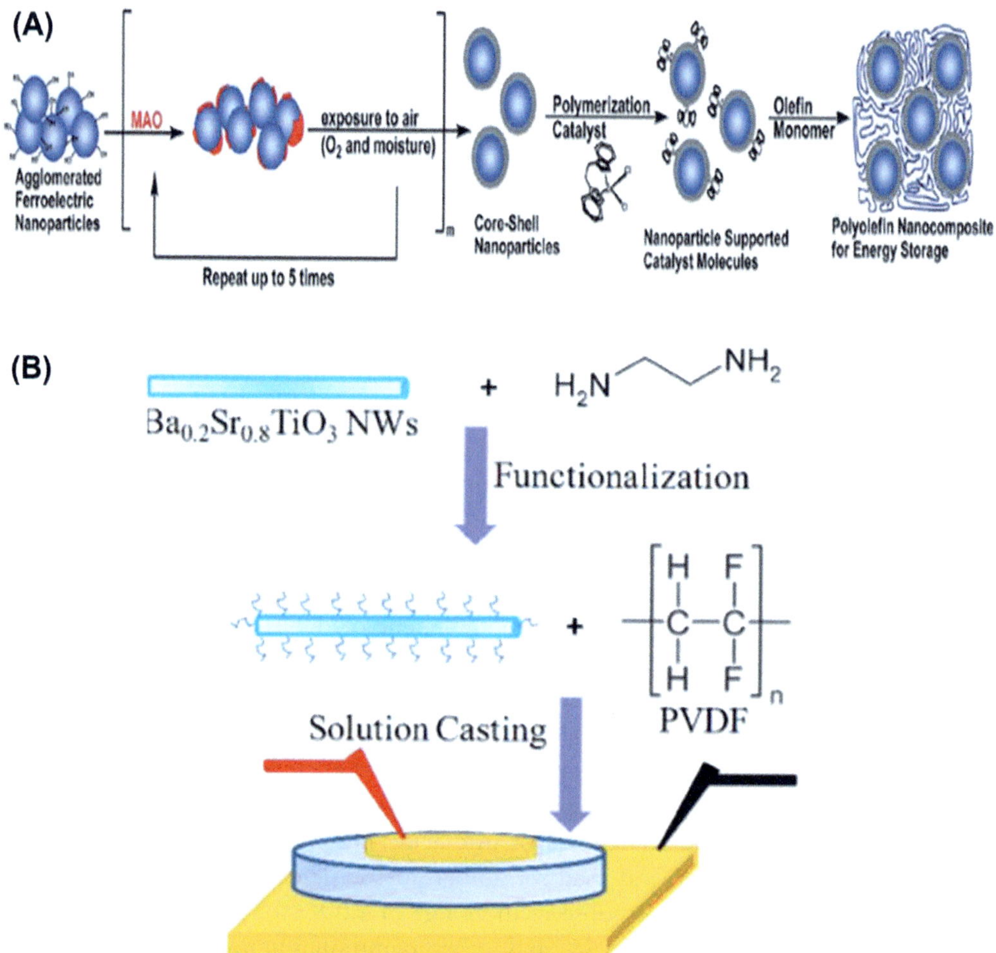

Figure 19.11 Schematic representation of the biosensor fabricated by Huang et al., for cancer biomarker detection using immunoassay approach.

by chemically modifying a glassy carbon electrode with Ce-HAP made by a hydrothermal process. The electrode demonstrated outstanding electrocatalytic oxidation of all the three analytes with well-defined peaks and no cointerference using DPV. The linear concentration range was 0.1–200 μM for norepinephrine and tyrosine whereas it was 0.5–200 μM for uric acid. The LOD obtained were 0.058 μM, 0.072 μM, and 0.39 μM, respectively. Fig. 19.12 shows the schematic [53].

Jia et al. described a strategy for enzymeless detection of glucose using a hybrid material of HAP-nickel foam and polyaniline. HAP was electrodeposited on nickel foam followed by self-assembled layer of polyaniline membrane for enhancing the electron transfer

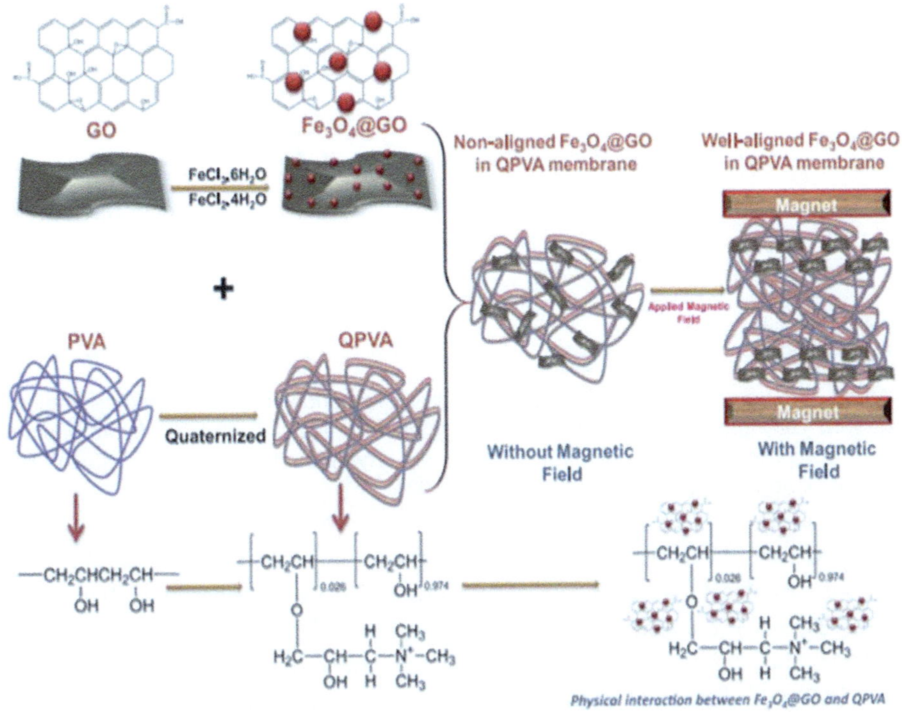

Figure 19.12 Schematic representation of the biosensor and its simultaneous response. Source: *Reprinted copyright permission from P. Kanchana, M. Navaneethan, C. Sekar, Fabrication of Ce doped hydroxyapatite nanoparticles based non-enzymatic electrochemical sensor for the simultaneous determination of norepinephrine, uric acid and tyrosine, Mater. Sci. Eng. B: Solid-State Mater. Adv. Technol., 226 (2017) 132–140. https://doi.org/10.1016/j.mseb.2017.09.015.*

behavior of HAP. The fabricated biosensor gave electrocatalytic oxidation of glucose in the linear range of 1 μmol/dm^3 to 30 mmol/dm^{3}with LOD as 0.32 μmol/dm^3. Antiinterference response from ascorbic acid, sucrose, dopamine and lactose were observed [54]. Erdem et al. made HAP nanoparticles modified graphite electrodes for electroanalytical detection of DNA. This was the first of its kind study were DNA hybridization and sequencing was studied. Pencil graphite electrode was the base used and Hepatitis B virus DNA sequence was used as the bioanalyte [55]. Shahamirifard et al. developed HAP-zinc oxide and Palladium nanoparticle based composite for detection of arbutin and vitamin C simultaneously. Carbon paste electrode was the underlying electrode used. CV and Impedance were used for the studies. The sensor showed good electrocatalytic oxidation of arbutin and vitamin C together with distinct peaks and no interference. The linear response obtained was 0.12–56 μM for arbutin and 0.12–55.36 μM for vitamin C with LOD of 85.7 and 19.4 nM respectively [56]. Similarly, Prongmanee et al. made graphene oxide and HAP composite modified screen-printed electrode for detection of tryptophan.

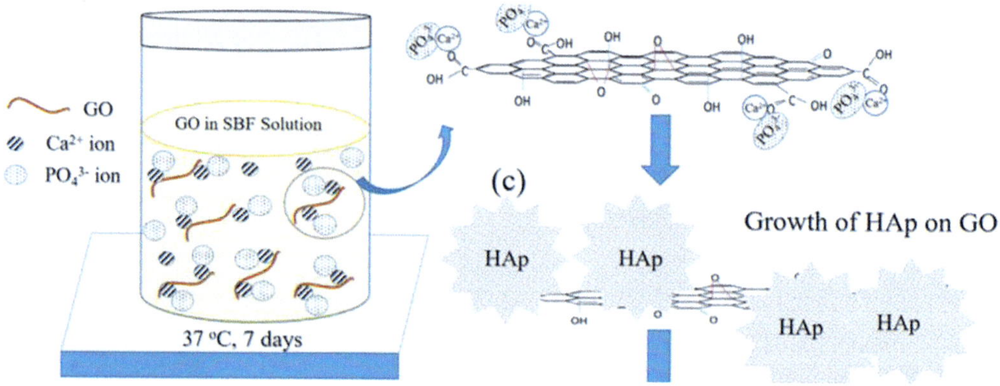

Figure 19.13 Schematic representation of composite preparation from Prongmanee et al. *Source: Reprinted with copyright permission from W. Prongmanee, I. Alam, P. Asanithi, Hydroxyapatite/graphene oxide composite for electrochemical detection of L-Tryptophan, J. Taiwan Inst. Chem. Eng. 102 (2019) 415–423. https://doi.org/10.1016/j.jtice.2019.06.004.*

LSV was used here. HAP was directly grown on GO in simulated body fluid (SBF). The electrode gave tremendous electrocatalytic activity toward tryptophan with high sensitivity and selectivity in a concentration range of 7–1000 μM and LOD as 5.5 μM. The authors have also compared the studies with graphene oxide-screen-printed electrode (without HAP) and discovered that, HAP enhances the sensitivity of the sensor. Fig. 19.13 shows the schematic of the composite preparation for better understanding [57].

Very recently, Iyyappan et al. described the electrochemical sensing of uric acid with synthesized HAP nanoparticles depending upon the role of surface acidity and mesopore volume. HAP were prepared with and without Triton X-100 using three approaches; reflux, hydrothermal and ultrasonication. Two types of particles were obtained, discrete and agglomerated. Electrocatalytic sensing was influenced by surface acidity in discrete HAP whereas, mesopore volume played role in agglomerated HAP. The electrode gave good electrochemical oxidation of uric acid in the concentration linear range of 0.068 – 50.0 μM and the LOD was 0.05 μM. The plausible mechanism of uric acid sensing over electrode is given in the schematic in Fig. 19.14 [58].

Vakili et al. fabricated an important approach of virus detection. Herein, they prepared HAP nanoparticle based DNA biosensor for detection of BK polyomavirus in the urine samples of transplant patients. Sol–gel technique was used for HAP preparation. A glassy carbon electrode, modified with HAP nanoparticle was used as base matrix to immobilize a probe DNA molecule. DNA hybridization between the probe and target sequence was examined via electrochemical response of methylene blue using DPV. Furthermore, DNA from urine samples of the patients with transplant were isolated and studied as a real sample. The biosensor gave a linear response in the concentration range of 50.00×10^{-12} to 1.00×10^{-9} mol/L with LOD of

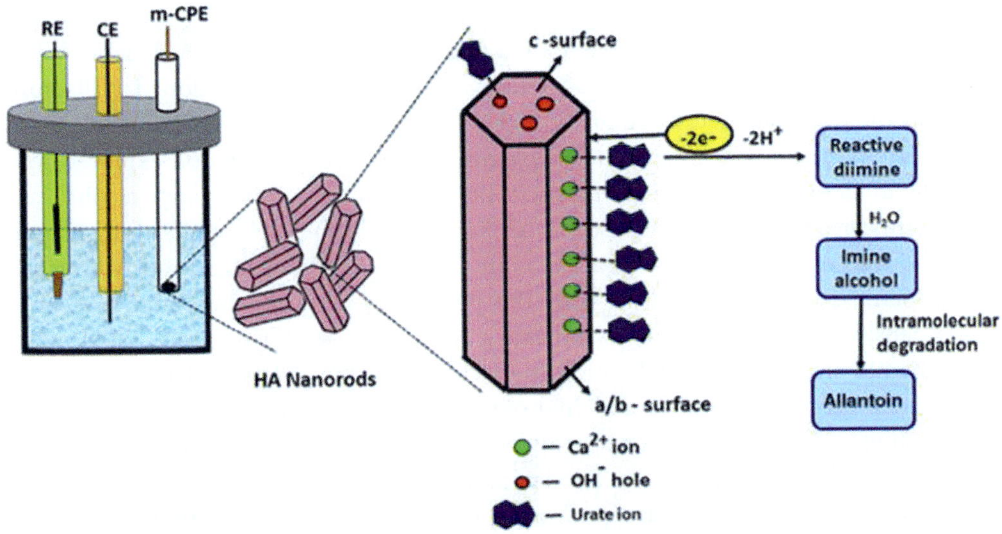

Figure 19.14 Schematic representation of electrochemical experimental setup and electrode modification and uric acid oxidation plausible mechanism. Source: *Reprinted with copyright permission from E. Iyyappan, S.J. Samuel Justin, P. Wilson, A. Palaniappan, Nanoscale hydroxyapatite for electrochemical sensing of uric acid: roles of mesopore volume and surface acidity, ACS Appl. Nano Mater. 3 (8) (2020) 7761–7773. https://doi.org/10.1021/acsanm.0c01322.*

41.08×10^{-12} mol/L [59]. In 2021, Vladislavic et al. developed a novel approach of in situ growth of HAP on glassy carbon electrode and further used for electrochemical detection of cysteine. Base carbon used was glassy carbon electrode, wherein a two-step procedure for electrodepositing the HAP was performed. The prepared electrode was exposed to cysteine using CV and SWV. An exceptional electrochemical oxidation of cysteine in a linear range of 0.5–10 $\mu mol/dm^3$ with LOD as 0.03 $\mu mol/dm^3$ [60]. The development of biosensor using various strategically synthesized and functionalized NCs is continually growing day by day. The future beholds numerous applications and scope with these.

19.6.3 Other electrochemical applications

It is very recent that as well as their chemical and biosensing applications, NCs also have been utilized for gas-sensing and energy-harvesting applications. However, these are not well explored yet and very few reports are available. Some of them are discussed here. Huixia et al. developed composites of HAP and conducting polymers. Polypyrrole and polyaniline, two different polymers with different properties, were used to obtain HAP composites. Ammonia gas was used as testing analyte. It was discovered that pure HAP gave less sensitivity toward ammonia than polymer-infused composite of HAP [61]. Zhu et al. studied CO sensing using Pd-SnO_2 composite

NCs. Even though the mechanism wasn't known, they found that a low content of Pd in the composite gave effective CO sensing [62]. Waghmare et al. fabricated a perovskite bismuth ferrite-based gas sensor with doped and undoped palladium. Nitrogen dioxide was used as an analyte. The prepared sensor gave decent selectivity and sensitivity under 100 ppm. The mechanism for this also has to be explored [63]. Yang et al. demonstrated the preparation of a composite of polyvinyl alcohol and HAP composite membrane and used it for borohydride fuel cell application. Herein, MnO_2/C inks over Ni-foam were used as an air cathode, PtRu black and Au catalysts on Ni-foam as anode were used with the prepared composite membrane. A good power density value was obtained, indicating the potential of a polymer—HAP composite membrane as an electrochemical fuel cell application [64]. Similarly, Zhao et al. developed a carbon nanotubes and hydroxyapatite-based nanocomposite and used it for a biofuel cell application. The setup was a compartment-less glucose-O_2 biofuel cell. Herein, glucose oxidase as anodic and laccase as cathodic biocatalysts were used. The enzymes were entrapped over the matrix of HAP-carbon nanotubes. A remarkable power density of 15.8 $\mu W/cm^2$ with cell voltage of 0.28 V in 10 mM glucose was obtained. This proved to be useful as a novel BFC device [65]. Recently, Ubaidullah et al. made calcium-doped barium zirconate NCs using a metal precursor approach and utilized it for supercapacitor application. Noteworthy dielectric properties were observed in the frequency of 20 kHz to 1 MHz at room temperature. They also found that the calcium ion plays a key role in the dielectric constant, as it was directly proportional to the calcium ion concentration. Owing to the high surface area, outstanding porosity, and great dielectric constant, this material is appropriate for use in energy devices [66]. However, utilizing NCs in energy devices and as gas sensors requires major research gaps to be filled, such as increasing the electron transfer behavior, predicting the gas sensing mechanism, improving power density, etc. Significant work and various approaches and strategies are yet to be explored.

19.7 Conclusion and future aspects

Emerging widespread research into the fabrication of NCs and their nanocomposite with various metals, carbons, polymers, etc. has given us a great deal of knowledge and information about these novel materials. Although ceramics have been used as bulk materials for a long time, reducing their size to nanolevel has greatly improved their chemical, physical, electrical, and mechanical properties. The insights into fabrication strategies are extremely useful for modulating the properties, and therefore can lead to broader applications. Certain limitations pertaining to NCs' fabrication, like contamination during preparation, agglomeration, nonuniform structure and morphology, and size distribution exist, however they can be rectified by optimizing parameters like pH, temperature, and other reaction conditions easily. This chapter summarizes the properties and synthesis methods of NCs and gives a detailed discussion about the electrochemical-based applications. Although NCs have good electronic properties, the electron transfer and diffusion behavior of NCs on their own is a bit slower. Consequently, these NCs are combined with other high conducting materials to form composites. Based on these, the chapter briefly described the types of NCs that are commonly prepared. Innovative processing methods, like activated sintering, have mainly led to the development of these modified NCs. Furthermore, general applications like magnetic, mechanical, biomedical, electrical, luminescent, etc. are also stated here. A detailed discussion about the specific electrochemical applications, like the use of NCs as electroanalytical sensing of chemicals, biosensors, gas sensors, and energy harvesters in the last decade are explored in this chapter. It is found that NCs and their composites have been used for electrochemical sensing of significant analytes like dopamine, uric acid, H_2O_2, glucose, hydrazine, DNA, antigen, drug molecules, alcohol, organic moieties, etc. The research into the electrochemical behavior of NCs for energy and gas-sensing devices is quite young and has a lot more future scope. In the future, various other synthesized materials can be incorporated into NCs to make them useful for commercial applications. Since NCs are mostly biocompatible, fabricating energy devices with them, like fuel cell, capacitors, pacemakers, etc. can be useful for implantation in the physiological system. Furthermore, in the future, the use of NCs in microfluidic and miniaturized devices can also be observed. It is plausible that NCs can be grown in microfluidic devices. Several reports are available that utilize continuous and discontinuous methods of synthesizing NCs in microfluidic channels in a controlled environment to monitor parameters like temperature, pH, reaction mixture, retention time, etc. Since, these devices provide ease of handling and control, it is probable that the NCs obtained can be of uniform morphology and reproducible compared to their counterparts in bulk. Therefore, avoiding the conventional chemical wet synthesis, automated and fully integrated microdevice-based preparation of NCs can be achieved. In addition, the NCs can be used as base electrodes in microdevices. The literature suggests that the electrochemical applications reported have been performed in the bulk volume cells. These reactions can be done in minidevices using microelectrodes, ultramicroelectrodes, nanoelectrodes, etc., modified with various NCs as base electrodes. Also, the microdevices and platforms incorporated with NCs are a great potential tool for point-of-care testing. Hence, NCs are predicted to be extremely useful in fabricating lab on chip devices in the future. Lastly, it must be mentioned that, despite the progress made over the years in preparing and manipulating the properties of NCs and their applications, this remains an emerging field and future research is needed to resolve issues like electrochemical behavior and the miniaturization of the electrode systems to microelectrochemical sensing platforms which could further lead to novel and critical applications.

References

[1] O. Van Der Biest, Nanoceramics: issues and opportunities, Int. J. Appl. Ceram. Technol. 10 (2013) 565–576. Available from: https://doi.org/10.1111/ijac.12074.

[2] A. Kiani, M. Rahmani, M. Sivakumar, B. Tan, Nanoceramics: synthesis, characterization, and applications, J. Nanomater. 2014 (2014) 2–4. Available from: https://doi.org/10.1155/2014/528348.

[3] Z. Singh, Nanoceramics in bone tissue engineering: the future lies ahead, Trends J. Sci. Res. 3 (2018) 120–123. Available from: https://doi.org/10.31586/nanomaterials.0303.03.

[4] H. Hassanien, M. Darweesh, Nanoceramics: materials, properties, Methods Applications-Part II 1 (2018) 40–66.

[5] N. Banapurmath, B. Salimath, J. Yaradoddi, Synthesis techniques for preparation of nanomaterials, 2019. Available from: https://doi.org/10.1007/978-3-319-68255-6.

[6] A. Purwanto, U.S. Maret, Fabrication of functional nanomaterials using flame assisted spray pyrolysis, 2016. Available from: https://doi.org/10.1063/1.4866728.

[7] M.T. Swihart, Vapor-phase synthesis of nanoparticles, Curr. Opin. Colloid Interface Sci. 8 (2003) 127–133. Available from: https://doi.org/10.1016/S1359-0294.

[8] J.D. Mackenzie, E.P. Bescher, Chemical routes in the synthesis of nanomaterials using the sol–gel process, Acc. Chem. Res. 40 (2007) 810–818. Available from: https://doi.org/10.1021/ar7000149.

[9] R.S. Chaurasiya, H.U. Hebbar, Nanoscience in Food and Agriculture 4. Sustainable Agriculture Reviews, 24, Springer, Cham, 2017. Available from: https://doi.org/10.1007/978-3-319-53112-0_5.

[10] M.A. Malik, M.Y. Wani, M.A. Hashim, Microemulsion method: a novel route to synthesize organic and inorganic nanomaterials, 1st Nano Update, Arab. J. Chem. 5 (2012) 397–417. Available from: https://doi.org/10.1016/j.arabjc.2010.09.027.

[11] N. Rajput, Methods of preparation of nanoparticles—a review, Int. J. Adv. Eng. Technol. 7 (2015) 1806–1811. Available from: https://doi.org/10.1016/j.jare.2015.02.007.

[12] A.V. Rane, K. Kanny, V.K. Abitha, S. Thomas, Methods for Synthesis of Nanoparticles and Fabrication of Nanocomposites, Elsevier Ltd, 2018. Available from: https://doi.org/10.1016/b978-0-08–101975-7.00005-1.

[13] C. Zeng, C. Wang, F. Wang, Y. Zhang, L. Zhang, A novel vapor-liquid segmented flow based on solvent partial vaporization in microstructured reactor for continuous synthesis of nickel nanoparticles, Chem. Eng. J. 204–205 (2012) 248–253. Available from: https://doi.org/10.1016/j.cej.2012.07.096.

[14] E. Bernardo, L. Fiocco, G. Parcianello, E. Storti, P. Colombo, Advanced ceramics from preceramic polymers modified at the nano-scale: a review, Materials (Basel) 7 (2014) 1927–1956. Available from: https://doi.org/10.3390/ma7031927.

[15] C. Xu, S. De, A.M. Balu, M. Ojeda, R. Luque, Mechanochemical synthesis of advanced nanomaterials for catalytic applications, Chem. Commun. 51 (2015) 6698–6713. Available from: https://doi.org/10.1039/c4cc09876e.

[16] J. Fan, N.A. Kotov, Chiral nanoceramics, Adv. Mater. 32 (2020) 1–28. Available from: https://doi.org/10.1002/adma.201906738.

[17] New design concept of structural ceramics. Ceramic nanocomposites, in: K. Niihara (Ed.), J. Ceram. Soc. Japan. Int., 99, 1991, pp. 945–952.

[18] A. Mukhopadhyay, B. Basu, Bulk nanoceramic composites for structural applications: a review, Proc. Indian. Natl. Sci. Acad. 72 (2006) 97–111.

[19] N. Kohli, E. García-Gareta, Bioactive nanoceramics, nanomater, Theranostics Tissue Eng. (2020) 233–257. Available from: https://doi.org/10.1016/b978-0-12-817838-6.00009-7.

[20] K.M. Batoo, Nano Ceramics and Their Applications, Springer, Singapore, 2019. Available from: https://doi.org/10.1007/978-981-13-9833-9_7.

[21] E. Ramanathan, S. Balasubramanian, Synthesis and characterization of hexafluorozirconic acid powder and its application in nanoceramic coating, Surf. Coat. Technol. 304 (2016) 228–236. Available from: https://doi.org/10.1016/j.surfcoat.2016.07.009.

[22] A.B. Abou Hammad, A.M. El Nahrawy, B.A. Hemdan, A.L.K. Abia, Correction to: Nanoceramics and novel functionalized silicate-based magnetic nanocomposites as substitutional disinfectants for

water and wastewater purification, Environ. Sci. Pollut. Res. 27 (21) (2020) 26668–26680. Available from: https://doi.org/10.1007/s11356-020-09666-4.

[23] A. Journal, S. Publications, Nanoceramic matrices: biomedical applications Willi Paul and Chandra P. Sharma division of biosurface technology, biomedical technology wing, Sree Chitra Tirunal Institute for Medical Sciences & Technology, Thiruvananthapuram 695012, India, Am. J. Biochem. Biotechnol. 2 (2006) 41–48.

[24] W. Wang, S. Liao, Y. Zhu, M. Liu, Q. Zhao, Y. Fu, Recent applications of nanomaterials in prosthodontics, J. Nanomater. 2015 (2015). Available from: https://doi.org/10.1155/2015/408643.

[25] M.A. El Mhammedi, M. Achak, M. Bakasse, A. Chtaini, Electrochemical determination of para-nitrophenol at apatite-modified carbon paste electrode: application in river water samples, J. Hazard. Mater. 163 (2009) 323–328. Available from: https://doi.org/10.1016/j.jhazmat.2008.06.126.

[26] P. Pang, Y. Liu, Y. Zhang, Y. Gao, Q. Hu, Electrochemical determination of luteolin in peanut hulls using graphene and hydroxyapatite nanocomposite modified electrode, Sens. Actuators, B Chem. 194 (2014) 397–403. Available from: https://doi.org/10.1016/j.snb.2013.12.113.

[27] H. Ajab, A.A.A. Khan, J.O. Dennis, M.A. Abdullah, Bismuth-modified hydroxyapatite carbon electrode for heavy metal detection in biomatrices, Appl. Mech. Mater. 625 (2014) 813–817. Available from: https://doi.org/10.4028/http://www.scientific.net/AMM.625.813.

[28] S.M. Ghoreishi, F.Z. Kashani, A. Khoobi, M. Enhessari, Fabrication of a nickel titanate nanoceramic modified electrode for electrochemical studies and detection of salicylic acid, J. Mol. Liq. 211 (2015) 970–980. Available from: https://doi.org/10.1016/j.molliq.2015.08.035.

[29] S.M. Ghoreishi, E. Karamali, A. Khoobi, M. Enhessari, Preparation of a manganese titanate nanosensor: application in electrochemical studies of captopril in the presence of para-aminobenzoic acid, Anal. Biochem. 487 (2015) 49–58. Available from: https://doi.org/10.1016/j.ab.2015.07.008.

[30] P. Kanchana, C. Sekar, Development of electrochemical folic acid sensor based on hydroxyapatite nanoparticles, Spectrochim. Acta – Part. A Mol. Biomol. Spectrosc. 137 (2015) 58–65. Available from: https://doi.org/10.1016/j.saa.2014.07.087.

[31] B. Kaur, R. Srivastava, B. Satpati, Ultratrace detection of toxic heavy metal ions found in water bodies using hydroxyapatite supported nanocrystalline ZSM-5 modified electrodes, N. J. Chem. 39 (2015) 5137–5149. Available from: https://doi.org/10.1039/c4nj02369b.

[32] G. Bharath, A. Naldoni, K.H. Ramsait, A. Abdel-Wahab, R. Madhu, E. Alsharaeh, et al., Enhanced electrocatalytic activity of gold nanoparticles on hydroxyapatite nanorods for sensitive hydrazine sensors, J. Mater. Chem. A. 4 (2016) 6385–6394. Available from: https://doi.org/10.1039/c6ta01528j.

[33] F. Gao, Q. Wang, N. Gao, Y. Yang, F. Cai, M. Yamane, et al., Hydroxyapatite/chemically reduced graphene oxide composite: Environment-friendly synthesis and high-performance electrochemical sensing for hydrazine, Biosens. Bioelectron. 97 (2017) 238–245. Available from: https://doi.org/10.1016/j.bios.2017.06.005.

[34] M.K. Alam, M.M. Rahman, A. Elzwawy, S.R. Torati, M.S. Islam, M. Todo, et al., Highly sensitive and selective detection of Bis-phenol A based on hydroxyapatite decorated reduced graphene oxide nanocomposites, Electrochim. Acta. 241 (2017) 353–361. Available from: https://doi.org/10.1016/j.electacta.2017.04.135.

[35] G. Wang, R. Han, X. Feng, Y. Li, J. Lin, X. Luo, A glassy carbon electrode modified with poly (3,4-ethylenedioxythiophene) doped with nano-sized hydroxyapatite for amperometric determination of nitrite, Microchim. Acta. 184 (2017) 1721–1727. Available from: https://doi.org/10.1007/s00604-017-2180-9.

[36] R. Mirzajani, S. Karimi, Preparation of Γ-Fe2O3/hydroxyapatite/Cu(II) magnetic nanocomposite and its application for electrochemical detection of metformin in urine and pharmaceutical samples, Sens. Actuators, B Chem. 270 (2018) 405–416. Available from: https://doi.org/10.1016/j.snb.2018.05.032.

[37] S.R. Anjum, V.N. Narwade, K.A. Bogle, R.S. Khairnar, Graphite doped hydroxyapatite nanoceramic: selective alcohol sensor, Nano-Structures Nano-Objects 14 (2018) 98–105. Available from: https://doi.org/10.1016/j.nanoso.2018.01.010.

[38] M. Sun, Z. Li, S. Wu, Y. Gu, Y. Li, Simultaneous detection of Pb^{2+}, Cu^{2+} and Hg^{2+} by differential pulse voltammetry at an indium tin oxide glass electrode modified by hydroxyapatite, Electrochim. Acta. 283 (2018) 1223–1230. Available from: https://doi.org/10.1016/j.electacta.2018.07.019.

[39] U. Rajaji, S. Manavalan, S.M. Chen, M. Govindasamy, T.W. Chen, T. Maiyalagan, Microwave-assisted synthesis of europium(III) oxide decorated reduced graphene oxide nanocomposite for detection of chloramphenicol in food samples, Compos. Part. B Eng. 161 (2019) 29–36. Available from: https://doi.org/10.1016/j.compositesb.2018.10.043.

[40] S. Velmurugan, S. Palanisamy, T.C.-K. Yang, M. Gochoo, S.W. Chen, Ultrasonic assisted functionalization of MWCNT and synergistic electrocatalytic effect of nano-hydroxyapatite incorporated MWCNT-chitosan scaffolds for sensing of nitrofurantoin, Ultrason. Sonochem. 62 (2020) 104863. Available from: https://doi.org/10.1016/j.ultsonch.2019.104863.

[41] P. Kanchana, N. Sudhan, C. Sekar, G. Neri, Manganese doped hydroxyapatite nanoparticles based enzyme-less electrochemical sensor for detecting hydroquinone, J. Nanosci. Nanotechnol. 19 (2018) 2034–2043. Available from: https://doi.org/10.1166/jnn.2019.15760.

[42] F.Z. Kashani, S.M. Ghoreishi, A. Khoobi, M. Enhessari, A carbon paste electrode modified with a nickel titanate nanoceramic for simultaneous voltammetric determination of ortho – and para -hydroxybenzoic acids, Microchim. Acta 186 (2019) 12. Available from: https://doi.org/10.1007/s00604-018-3113-y.

[43] H.V. Tran, C.D. Huynh, T.D. Le, H.S. Hoang, Hydroxyapatite nano-rods/chitosan modified glassy carbon electrode for Cu(II) ions determination, Electron. Mater. Lett. 16 (2020) 396–403. Available from: https://doi.org/10.1007/s13391-020-00222-3.

[44] B. Wang, J.J. Zhang, Z.Y. Pan, X.Q. Tao, H.S. Wang, A novel hydrogen peroxide sensor based on the direct electron transfer of horseradish peroxidase immobilized on silica-hydroxyapatite hybrid film, Biosens. Bioelectron. 24 (2009) 1141–1145. Available from: https://doi.org/10.1016/j.bios.2008.06.053.

[45] S. Wang, Y. Lei, Y. Zhang, J. Tang, G. Shen, R. Yu, Hydroxyapatite nanoarray-based cyanide biosensor, Anal. Biochem. 398 (2010) 191–197. Available from: https://doi.org/10.1016/j.ab.2009.11.029.

[46] L. Lu, L. Zhang, X. Zhang, S. Huan, G. Shen, R. Yu, A novel tyrosinase biosensor based on hydroxyapatite-chitosan nanocomposite for the detection of phenolic compounds, Anal. Chim. Acta. 665 (2010) 146–151. Available from: https://doi.org/10.1016/j.aca.2010.03.033.

[47] J. Li, D. Kuang, Y. Feng, F. Zhang, M. Liu, Glucose biosensor based on glucose oxidase immobilized on a nanofilm composed of mesoporous hydroxyapatite, titanium dioxide, and modified with multi-walled carbon nanotubes, Microchim. Acta. 176 (2012) 73–80. Available from: https://doi.org/10.1007/s00604-011-0693-1.

[48] P. Kanchana, C. Sekar, EDTA assisted synthesis of hydroxyapatite nanoparticles for electrochemical sensing of uric acid, Mater. Sci. Eng. C. 42 (2014) 601–607. Available from: https://doi.org/10.1016/j.msec.2014.05.072.

[49] Z. Yang, X. Shang, C. Zhang, J. Zhu, Photoelectrochemical bilirubin biosensor based on Fe_3O_4/hydroxyapatite/molecularly imprinted polypyrrole nanoparticles, Sens. Actuators, B Chem. 201 (2014) 167–172. Available from: https://doi.org/10.1016/j.snb.2014.05.021.

[50] G. Bharath, R. Madhu, S.M. Chen, V. Veeramani, A. Balamurugan, D. Mangalaraj, et al., Enzymatic electrochemical glucose biosensors by mesoporous 1D hydroxyapatite-on-2D reduced graphene oxide, J. Mater. Chem. B. 3 (2015) 1360–1370. Available from: https://doi.org/10.1039/c4tb01651c.

[51] P. Kanchana, S. Radhakrishnan, M. Navaneethan, M. Arivanandhan, Y. Hayakawa, C. Sekar, Electrochemical sensor based on Fe doped hydroxyapatite-carbon nanotubes composite for L-dopa detection in the presence of uric acid, J. Nanosci. Nanotechnol. 16 (2016) 6185–6192. Available from: https://doi.org/10.1166/jnn.2016.11645.

[52] Y. Huang, C. Tang, J. Liu, J. Cheng, Z. Si, T. Li, et al., Signal amplification strategy for electrochemical immunosensing based on a molybdophosphate induced enhanced redox current on the surface of hydroxyapatite nanoparticles, Microchim. Acta. 184 (2017) 855–861. Available from: https://doi.org/10.1007/s00604-016-2069-z.

[53] P. Kanchana, M. Navaneethan, C. Sekar, Fabrication of Ce doped hydroxyapatite nanoparticles based non-enzymatic electrochemical sensor for the simultaneous determination of norepinephrine, uric acid and tyrosine, Mater. Sci. Eng. B Solid-State Mater. Adv. Technol. 226 (2017) 132−140. Available from: https://doi.org/10.1016/j.mseb.2017.09.015.

[54] L. Jia, X. Wei, L. Lv, X. Zhang, X. Duan, Y. Xu, et al., Electrodeposition of hydroxyapatite on nickel foam and further modification with conductive polyaniline for non-enzymatic glucose sensing, Electrochim. Acta. 280 (2018) 315−322. Available from: https://doi.org/10.1016/j.electacta.2018.05.130.

[55] A. Erdem, G. Congur, Hydroxyapatite nanoparticles modified graphite electrodes for electrochemical DNA detection, Electroanalysis 30 (2018) 67−74. Available from: https://doi.org/10.1002/elan.201700462.

[56] S.A. Shahamirifard, M. Ghaedi, A new electrochemical sensor for simultaneous determination of arbutin and vitamin C based on hydroxyapatite-ZnO-Pd nanoparticles modified carbon paste electrode, Biosens. Bioelectron. 141 (2019) 111474. Available from: https://doi.org/10.1016/j.bios.2019.111474.

[57] W. Prongmanee, I. Alam, P. Asanithi, Hydroxyapatite/Graphene oxide composite for electrochemical detection of L-Tryptophan, J. Taiwan. Inst. Chem. Eng. 102 (2019) 415−423. Available from: https://doi.org/10.1016/j.jtice.2019.06.004.

[58] E. Iyyappan, S.J. Samuel Justin, P. Wilson, A. Palaniappan, Nanoscale hydroxyapatite for electrochemical sensing of uric acid: roles of mesopore volume and surface acidity, ACS Appl. Nano Mater. 3 (2020) 7761−7773. Available from: https://doi.org/10.1021/acsanm.0c01322.

[59] S.N. Vakili, M. Rezayi, M. Chahkandi, Z. Meshkat, M. Fani, A. Moattari, A novel electrochemical DNA biosensor based on hydroxyapatite nanoparticles to detect BK polyomavirus in the urine samples of transplant patients, IEEE Sens. J. 20 (2020) 12088−12095. Available from: https://doi.org/10.1109/JSEN.2020.2982948.

[60] Nives Vladislavić 2021-Cysteine.pdf, n.d.

[61] L. Huixia, L. Yong, L. Lanlan, T. Yanni, Z. Qing, L. Kun, Development of ammonia sensors by using conductive polymer/hydroxyapatite composite materials, Mater. Sci. Eng. C. 59 (2016) 438−444. Available from: https://doi.org/10.1016/j.msec.2015.10.036.

[62] S. Zhu, Y. Liu, G. Wu, L. Fei, S. Zhang, Y. Hu, et al., Mechanism study on extraordinary room-temperature CO sensing capabilities of Pd-SnO_2 composite nanoceramics, Sens. Actuators, B Chem. 285 (2019) 49−55. Available from: https://doi.org/10.1016/j.snb.2019.01.027.

[63] S.D. Waghmare, S.D. Raut, B.G. Ghule, V.V. Jadhav, S.F. Shaikh, A.M. Al-Enizi, et al., Pristine and palladium-doped perovskite bismuth ferrites and their nitrogen dioxide gas sensor studies, J. King Saud. Univ. − Sci. 32 (2020) 3125−3130. Available from: https://doi.org/10.1016/j.jksus.2020.08.024.

[64] C.C. Yang, Y.J. Li, S.J. Chiu, K.T. Lee, W.C. Chien, C.A. Huang, A direct borohydride fuel cell based on poly(vinyl alcohol)/hydroxyapatite composite polymer electrolyte membrane, J. Power Sources. 184 (2008) 95−98. Available from: https://doi.org/10.1016/j.jpowsour.2008.06.042.

[65] H.Y. Zhao, H.M. Zhou, J.X. Zhang, W. Zheng, Y.F. Zheng, Carbon nanotube-hydroxyapatite nanocomposite: a novel platform for glucose/O2 biofuel cell, Biosens. Bioelectron. 25 (2009) 463−468. Available from: https://doi.org/10.1016/j.bios.2009.08.005.

[66] M. Ubaidullah, J. Ahmed, A.M. Al-Enizi, A. Tyagi, S.F. Shaikh, N. Tarannum, et al., Metal organic precursor derived $Ba_{1-x}Ca_xZrO_3$ ($0.05 \leq x \leq 0.20$) nanoceramics for excellent capacitor applications, J. King Saud. Univ. − Sci. 32 (2020) 1937−1943. Available from: https://doi.org/10.1016/j.jksus.2020.02.009.

CHAPTER 20

Photonic applications of nanoceramics

Shilpa Rajkumar[1], Deepankara Vrushabhadas Shastri[1], J. John Donald Raj[1], S. Kishore Babu[1] and Kantha D. Arunachalam[2]
[1]Department of Physics and Nanotechnology, SRM Institute of Science and Technology, Kattankulathur, Tamil Nadu, India
[2]Faculty of Sciences, Marwadi University, Rajkot, Gujarat, India

20.1 Introduction

20.1.1 Nanoceramics

Nanoceramics is an amalgamation of nanotechnology and ceramics. The term "ceramics" is associated with heat-treated clay with additional stabilizing supplements. While the synthesis of ceramics has been mastered over hundreds of years, the advent of nanotechnology created a breakthrough in the ceramics industry. Nanoceramics are essentially multiphase systems fabricated from ultrafine particles less than 100 nm in size [1]. They are categorized based on the unique properties that they offer compared to regular homogeneous materials. They can be classified as nonmetallic, inorganic, and heat-resistant solids. The very nature of nanoceramics with the isotropic grain size that are closely stacked together gives rise to their unique and improved physical properties, such as dielectricity, ferroelectricity, piezoelectricity, pyroelectricity, ferromagnetism, magnetoresistance, and superconductivity.

Nanotechnology provides the added advantage of tuning parameters at the synthesis level. This provides us with the opportunity to design property-specific nanoceramics for the desired application [2]. One can also use nanolevel properties to enhance bulk materials' properties, such as enhanced strength and hardness. Traditional brittle materials can be made more ductile by reducing the size of the grain used in its manufacture. Therefore, nanoceramics' physical and mechanical strength is dependent of the size of the particle used to make them.

Advanced ceramics are used more than ever in a wide range of applications. People are more and more interested in ceramics in the field of healthcare, such as bioceramics [3] and nanoceramic-based drug delivery systems. In the energy field, ceramics are not only fixed as combustible ceramics or waste in the nuclear industry [4], but are also very important in power generation, transportation, and storage of electricity [5,6].

20.1.2 Photonics

Photonics is the subject pertaining to the investigations into light. It is also known by the terms "Photon Science" and "Lightwave Technology," dealing with the science and

Industrial Applications of Nanoceramics
DOI: https://doi.org/10.1016/B978-0-323-88654-3.00025-1

technology related to light, respectively. Photonics provided a breakthrough in various novel technologies for generating, transmitting, amplifying, modulating, detecting, and analyzing light (e.g., with spectroscopy), and particularly using light for various practical purposes. It thus builds heavily on optical technology and the optical physics.

The core of photonics is the technology of generating light (such as using lasers or light-emitting diodes), transmitting, amplifying, modulating, detecting, and analyzing light (such as using spectroscopy), especially the use of light for various practical purposes. Therefore, it is based on optical technology (optics), supplemented by modern developments such as optoelectronics (mainly from semiconductors), laser systems, optical amplifiers, and new materials (such as photonic metamaterials). The scientific basis is primarily physics, especially optical physics and related fields such as laser physics and quantum optics.

20.1.3 Nanoceramics enhancing photonics

It is a well-established fact that glasses and single crystals (SC) have dominated photonics research and industry. Their numerous applications in both passive and active optical components [7,8] have established their versatile nature. Glasses are the most common candidate when it comes to passive components such as optical lenses and filters, while SC fulfill the active component category in their widespread use in lasers and display. Although both materials have acquired their unique niche in the photonics industry, the search for better and novel materials that can complement as well as enhance the functioning of the existing materials in the industry persists.

Nanoceramics are one such class of materials that have offered potential improvement and novel functionality in the photonic industry. This is because ceramic materials are insulators with a large optical bandgap, which are very suitable for optical applications. In addition, the response of ceramics to light propagation is highly dependent on their microstructure. Therefore, its transparency can be adjusted according to the processing conditions. Nanoceramics have found their way into many photonic applications owing to their unique properties arising from their microstructural arrangement. These materials, synthesized from fine particles less than 100 nm in size, provide the opportunity to tune the optical properties of the material. Apart from being fabricated into optical components as such, these materials can be integrated with other optical materials, such as glasses (glass-ceramics), to enhance their optical properties.

Ceramics manufacturing is also less complex than that of SC. As a result, ceramic objects of any shape and size may be produced, which is difficult to do in the fabrication of SC. Ceramic composites may also be doped with greater impurity concentration levels than SC produced via melt solidification, and ceramics with impurity concentration gradients are simple to manufacture. Compared with glass, ceramic has better mechanical resistance, hardness, and toughness. Generally speaking, they also provide higher thermochemical stability than glass. On the contrary, ceramics are opaque, making it difficult for light to travel through them. Yet, in recent years ceramics made of small particles have uniform particle size distribution and very low porosity, and their light transmittance value may be close to that of glass. Transparent ceramics with excellent mechanical and optical properties are used as passive optical components as we shall explore in the upcoming sections.

20.2 Photonic applications

Photonics is regarded as one of the most important technologies of the 21st century. It adds value to electrical goods through optronics and has had rapid market expansion, which is likely to continue in the near future. Yet, photonics has only infiltrated a small portion of the general market, with massive sales of laser diodes (LDs) in CD/DVD players and associated optical data storage technologies. The development of silicon photonics and other LEDs with better power output and efficiency for photonic ICs or mass-production laser types (such as VECSEL) is projected to open up enormous growth potential.

20.2.1 Transparent glass-ceramics

Ceramics are made of grains with grain edges and pores generally between them. From the perspective of light propagation, the beam is reflected and refracted in a heterogeneity of materials, such as grain boundaries and hole walls. Therefore, depending on the scattering of light, ceramic behaves as a transparent, translucent, or opaque medium, which is a function of the ceramic microstructure. The surface, grain boundaries, and pores are the major light scattering sources in a dense and high-purity ceramic. As a result, the ceramic's transparency is determined by their relative density, and therefore by the ceramic processing parameters.

In the synthesis of transparent ceramics [9,10], many sintering techniques are employed.

- Stress-assisted sintering is a sintering method that uses external stresses. External compressive stresses generate extra mass transport driving forces, resulting in higher densification rates and slower grain growth dynamics [11]. The stress is delivered uniaxially through a piston in hot pressing, and the microstructure may indicate preferred orientation. Hot isostatic pressing, in which the powder is put in a membrane that is under isostatic tension using a high-pressure gas, can prevent this.
- Electric fields are used in field-assisted sintering procedures to densify materials in a short amount of time while retaining the ultrafine powder size. Sintering is further aided by modest uniaxial compressive pressures, which are often referred to as spark plasma sintering, despite the fact that the formation of any plasma during the process is debatable.

Therefore, transparent or translucent ceramics are near-fully dense polycrystalline materials with tiny grain sizes, $d < 100$ nm, significantly smaller than the wavelength of visible light, with a density >99.0% of theoretical density. Some examples of transparent ceramics are Al_2O_3, YAG, ALON ($Al_{23}O_{27}N_5$), Mg-spinel ($MgAl_2O_4$), ZrO_2: Y_2O_3, Y_2O_3:ThO_2, and other transparent ceramics such CaF_2 and ZnS.

With the growing interest in nanotechnology and subwavelength photonic structures [12], another type of material that has attracted much attention is glass-ceramics [13], which can be produced by precisely controlling the partial crystallization of glass during the cooling stage after melting, resulting in microcrystalline structures within an amorphous matrix [14–16]. Therefore glass-ceramics are glasses containing nanoscale particles embedded in the glass matrix.

Glass-ceramics are considered intermediate materials between inorganic glass and ceramic [17]. A glass-ceramic material can be highly crystalline or have a large amount of amorphous glass matrix. They have many characteristics in common with glass and ceramics, with low defects, high hardness, high resistance to thermal shock (typical properties of ceramics), and easy fabrication and molding, even with complex shapes (typical property of glass). Glass materials have played an important role in the development of optical waveguide devices and remain of great importance to today's photonic devices and systems. This is because all guided wave applications require the material to have very good transparency, at least in the selected operating wavelength; this means that the material must exhibit low optical scatter and low ion/atom absorption.

Low scattering levels can be obtained by meeting one of two criteria:

1. both crystalline phases and residual glass have closely matched indices of refraction, and microcrystals have low birefringence.
2. the crystal size is significantly smaller than the light wavelength.

The second instance is the most common, however it is not required because other techniques exist to maintain transparency, such as controlling the refractive index (RI) difference between the glass matrix and the crystalline phase, as described in condition (1) above. As a result, large-grain GCs that are transparent have been shown [10,18,19]. The high transparency is explained by the simultaneous change of the glass matrix and crystal composition during the crystallization process, which greatly reduces the difference between the refractive indices of the two phases (this difference can be reduced to 0.015); this characteristic is dominant in terms of dispersion, regardless of crystal size. In general, the transparency of glass-ceramics is greater than Rayleigh's scattering theory predicts.

20.2.2 LASERS

There is increasing interest in the development of new ceramic laser materials, because the material has many significant advantages over SC of the same composition, including increased mechanical strength and resistance to fracture. Additionally, advanced laser ceramics utilizing oxide materials, initially Y_2O_3 and $Y_3Al_5O_{12}$, possess the capability to manufacture sizable samples with a consistent (or specifically tailored) distribution of dopants.

This achievement of high concentrations of active ions is an uncommon feat for single crystals (SC), while maintaining excellent light transmittance. In addition, the more developed laser ceramics based on oxide materials have the ability to produce large-scale samples with a uniform (or specially designed) dopant distribution and achieve high concentration activity. Ions, which is not common in SC and has a high light transmittance. However, in recent years, the development of laser materials based on fluoride crystals has attracted a lot of attention, since compared to oxide materials, their phonon spectrum is short enough, resulting in a much weaker multiphonon fluoride relaxation, which makes it possible to obtain oscillation with high optical efficiency and minimal diffraction. Rare earth ions in fluoride crystals last longer than in oxide crystals, facilitating the construction of efficient laser amplifiers with more cost-effective and compact LD pump sources and longer pump pulses. Due to the high bandgap and small Raman gain coefficient, many overflows and losses can be minimized. The popularity of fluoride nanoceramics is due to their many advantages over comparable SC, which are much more dangerous than oxides. Optical ceramic materials, especially fluorinated optical ceramics, have been known for decades. Since the discovery of these polycrystalline transparent optical materials in the middle of the 20th century, many researchers from many laboratories have been committed to applying these polycrystalline transparent optical materials to various fields. In the 1960s and 1970s, both the United States and the Soviet Union developed processes for producing magnesium, calcium, and barium ceramic fluorides for IR optics (US trademarks IRTRAN and USSR CO trademarks).

20.2.3 Optical filters

Many optoelectronic devices require devices that allow a very small bandwidth of radiation. For high-quality sensing purposes, single-wavelength filters are highly essential. In such cases devices known as optical filters can be used to transmit specific wavelengths. Fabry Perot (FP) cavities have carved out a niche for several predominant optical applications which make them an indispensable component and an exhaustive conceptual resource. FP cavities are essentially two mirrors separated by a distance, L, equivalent to a discrete number of harmonics of the cavity's resonant frequency. The selective sensitivity of the cavity can be improved by adjusting the cavity length to various resonant frequencies. Technologies such as micromachining and microelectromechanical systems provide a degree of freedom to vary FP cavity length and pave the way to applications such as tunable laser [20], tunable optical filter [21–23], wavelength selection in spectrometry [24] and chemical sensors [25]. FP cavity characteristics such as Q-factor, finesse, and free spectral range (FSR) of the cavity modes depend on the reflectivity of the component mirrors as well as the cavity width.

Unlike metallic mirrors, the reflection spectra of distributed Bragg reflectors (DBRs) can be tuned owing to their geometry. A DBR is a stack of high- and low-RI multilayers of subwavelength thickness alternatingly ordered in one dimension.

Such structures are also known as one-dimensional photonic crystal (1D PC). The unique capability of 1D PC to impede the propagation of certain frequencies of light owing to their geometrical structure gives rise to a distinctive photonic bandgap (PBG), which is suitable for designing highly reflective mirrors [26,27]. Layers of alternating high and low RI media can prevent the propagation of light in a specific spectrum. This structure is called a 1D PC. On the contrary, the bandgap of these structures is limited to the RI contrast of the constituent materials causing a limited FSR for the FP cavity utilizing these mirrors.

The most common materials used in the fabrication of 1D PCs are TiO_2/SiO_2, being high- and low-index mediums, respectively. These materials are a common choice as they:

- Provide a high RI contrast.
- Have a broad spectral range from blue to IR [28].
- Are relatively cheaper.
- Can be fabricated using RF sputtering or e-beam evaporation techniques [29,30].

But these materials also require additional postdeposition processing, such as thermal annealing, to provide better adhesion and reflectivity. In addition to deposition time, annealing of up to 60 minutes either in ambient or room temperature is required to provide a stable DBR [30]. Advanced fluoride and sulfide-based ceramics have risen to the occasion to mitigate these limitations. For example, nanoceramic transparent optical materials such as ZnS/CaF_2-based PC filters have the following characteristics:

- They provide sufficient high index contrast to excite large PBG in the visible region (400–600 nm).
- These materials can be fabricated at room temperature with no substrate heating [31,32].
- Materials also show good adhesion to one other without the need of thermal treatment [32].
- Good control over film thickness using vacuum thermal evaporation technique allows us to match theoretical optimized values [31].
- Time of fabrication is considerably reduced without the need for predeposition thermal treatment (substrate heating) or postdeposition thermal treatment (thermal annealing).
- They can be deposited using a relatively gentle deposition technique which is the thermal evaporation method. Here the materials are deposited with comparatively low kinetic energies, which is of importance while dealing with organic compounds.

Following this, the thermal stability quality of these nanoceramic materials has proven to be an advantage in designing thermally stable filters. By using the negative thermooptic properties of CaF_2 to compensate for the positive thermal expansion effect, a defect mode with negligible thermal drift is obtained. This is because the negative thermooptic effect of the fluoride-based nanoceramic ultimately increases the overall RI contrast of the PC as temperature increases, which contributes to the slight enlargement of the absolute PBG.

20.3 Conclusion

Wide optical gaps, monolithic and thick glasses (oxide, fluoride glasses, etc.), semiconductors, and SC have traditionally been the most common optical materials. Ceramics composed of wide bandgap materials, on the other hand, can broaden the range of optical materials available due to their simplicity of manufacturing and superior thermomechanical and chemical resilience.

Incorporation of ceramics into the optical materials field, on the other hand, necessitates the use of ceramics processing technologies in order to produce the proper microstructure for the intended optical application. Today, we can manufacture ceramics with the microstructure control necessary for photonic applications using film deposition techniques, nano- and microassembly, colloidal processing, directed solidification, and other methods.

References

[1] A. Kiani, M. Rahmani, M. Sivakumar, B. Tan, Nanoceramics: synthesis, characterization, and applications, J. Nanomaterials 2014 (2014). Available from: https://doi.org/10.1155/2014/528348, http://www.hindawi.com/journals/jnm/.

[2] P.P. Fedorov, V.V. Osiko, S.V. Kuznetsov, E.A. Garibin, Russian Federation fluoride laser nanoceramics, J. Physics: Conf. Ser. 345 (1) (2012). Available from: https://doi.org/10.1088/1742-6596/345/1/012017. Institute of Physics Publishing, 17426596, http://www.iop.org/EJ/journal/conf.

[3] D. Arcos, I. Izquierdo-Barba, M. Vallet-Regí, Spain promising trends of bioceramics in the biomaterials field, J. Mater. Sci.: Mater. Med. 20 (2) (2009) 447–455. Available from: https://doi.org/10.1007/s10856-008-3616-x. 09574530.

[4] D.V. Shastri, K. Annamalai, N.P. Rajamane, T. Revathi, S. Rajadesingu, K.D. Arunachalam, Structural and physico-mechanical investigations of $Na_2B_4O_7$ geopolymer for γ-radiation attenuating applications, Ceram. Int. 48 (19 (PART B)) (2022) 29359–29365. Available from: https://doi.org/10.1016/j.ceramint.2022.05.381.

[5] Y. Katoh, K.M. Fox, H.-T. Lin, I. Belharouak, S. Widjaja, D. Singh, Ceramic Materials for Energy Applications, 32 (9), Wiley, 2011.

[6] S. Kishore Babu, M. Jayachandran, T. Maiyalagan, T. Vijayakumar, B. Gunasekaran, Metal-organic framework (MOF-5) incorporated on $NiCo_2O_4$ as electrode material for supercapacitor application, Mater. Lett. 302 (2021). Available from: https://doi.org/10.1016/j.matlet.2021.130338, http://www.journals.elsevier.com/materials-letters/.

[7] R. Hui, M. O'Sullivan, Fiber optic measurement techniques, Fiber Optic Measurement Techniques, Elsevier Inc, 2009. Available from: https://doi.org/10.1016/B978-0-12-373865-3.X0001-8, http://www.sciencedirect.com/science/book/9780123738653.

[8] N. Collaert, High mobility materials for CMOS applications, High Mobility Materials for CMOS Applications, Elsevier, 2018, pp. 1–375. Available from: https://doi.org/10.1016/C2016-0-02168-X, http://www.sciencedirect.com/science/book/9780081020616.

[9] R.L. Coble, Sintering alumina: effect of atmospheres, J. Am. Ceram. Soc. 45 (3) (1962) 123–127. Available from: https://doi.org/10.1111/j.1151-2916.1962.tb11099.x.

[10] S. Wen, Y. Wang, B. Lan, W. Zhang, Z. Shi, S. Lv, et al., Pressureless crystallization of glass for transparent nanoceramics, Adv. Sci. 6 (17) (2019). Available from: https://doi.org/10.1002/advs.201901096, http://onlinelibrary.wiley.com/journal/10.1002/(ISSN)2198-3844.

[11] R. Bordia, E. Olevsky, Advances in sintering science and technology, J. Am. Ceram. Soc. 92 (7) (2009) 1383. Available from: https://doi.org/10.1111/j.1551-2916.2009.03241.x.

[12] M. Bardosova, T. Wagner, Nanomaterials and Nanoarchitectures: A Complex Review of Current Hot Topics and their Applications, Springer, 2015.
[13] I. Alekseeva, O. Dymshits, M. Tsenter, A. Zhilin, V. Golubkov, I. Denisov, et al., Russian Federation optical applications of glass-ceramics, J. Non-Cryst. Solids 356 (52-54) (2010) 3042–3058. Available from: https://doi.org/10.1016/j.jnoncrysol.2010.05.103. 00223093.
[14] Holand, Beall, Glass Ceramic Technology, 2012.
[15] V. Marghussian, Nano-Glass Ceramics. Processing, Properties and Applications, William Andrew, 2015.
[16] Lewis, Glasses and Glass-Ceramics, 1989.
[17] A. Quandt, M. Ferrari, G.C. Righini, Advancement of glass-ceramic materials for photonic applications sol-gel based nanoceramic, Materials: Preparation, Properties and Applications, Springer International Publishing, South Africa, 2016, pp. 133–155. Available from: http://doi.org/10.1007/978-3-319-49512-5.
[18] T. Berthier, V.M. Fokin, E.D. Zanotto, New large grain, highly crystalline, transparent glass-ceramics, J. Non-Cryst. Solids 354 (15-16) (2008) 1721–1730. Available from: https://doi.org/10.1016/j.jnoncrysol.2007.08.052.
[19] S.M. Wang, F.H. Kuang, Q.Z. Yan, Q.C. Zhang, C.C. Ge, Crystallization Behavior of a new transparent glass-ceramics, Adv. Mater. Res. 105-106 (1) (2010) 597–599. Available from: https://doi.org/10.4028/http://www.scientific.net/amr.105-106.597.
[20] H. Cai, B. Liu, X.M. Zhang, A.Q. Liu, J. Tamil, T. Bourouina, et al., Opt. Express 16 (21) (2008) 16670–16679. Available from: https://doi.org/10.1364/OE.16.016670.
[21] M. Djavid, M.S. Abrishamian, Photonic crystal channel drop filters with mirror cavities, Opt. Quantum Electron. 39 (14) (2007) 1183–1190. Available from: https://doi.org/10.1007/s11082-007-9168-3.
[22] R. Shilpa, R. Mudachathi, Increasing the free spectral range of Fabry-Perot cavities using dissimilar 1D photonic crystals, Mater. Lett. 299 (2021). Available from: https://doi.org/10.1016/j.matlet.2021.130089, http://www.journals.elsevier.com/materials-letters/.
[23] S. Rajkumar, Abhinaya, Yamuna, R. Mudachathi, Thermal and layer thickness dependence of defect and band edge transmissions in 1D PCs based on ZnS/CaF_2 multilayers, IOP Conf. Ser. Mater. Sci. Eng. 1219 (1) (2022)012015. Available from: https://doi.org/10.1088/1757-899X/1219/1/012015.
[24] G. Lammel, S. Schweizer, P. Renaud, 14th IEEE International Conference on Micro Electro Mechanical Systems, 2001.
[25] M.J. Thorpe, D. Balslev-Clausen, M.S. Kirchner, J. Ye, Opt. Express 16 (4) (2008) 2387–2397. Available from: https://doi.org/10.1364/OE.16.002387.
[26] J.N. Winn, Y. Fink, S. Fan, J.D. Joannopoulos, Omnidirectional reflection from a one-dimensional photonic crystal, Opt. Lett. 23 (20) (1998) 1573–1575. Available from: https://doi.org/10.1364/OL.23.001573, http://ol.osa.org/Issue.cfm.
[27] P. Yeh, M. Hendry, Phys. Today. (1990) 77–78. Available from: https://doi.org/10.1063/1.2810419.
[28] H.Y. Lee, T. Yao, $TiO_2(ZnS)/SiO_2$ one-dimensional photonic crystals and a proposal for vertical micro-cavity resonators, J. Korean Phys. Soc. 44 (2) (2004) 387–392.
[29] L. Martiradonna, L. Carbone, M. De Giorgi, L. Manna, G. Gigli, R. Cingolani, et al., High Q-factor colloidal nanocrystal-based vertical microcavity by hot embossing technology, Appl. Phys. Lett. (2006) 181108. Available from: https://doi.org/10.1063/1.2200748.
[30] I.W. Feng, S. Jin, J. Li, J. Lin, H. Jiang, SiO_2/TiO_2 distributed Bragg reflector near 1.5 μm fabricated by e-beam evaporation, J. Vac. Sci. Technol. A: Vac. Surf. Films 31 (6) (2013). Available from: https://doi.org/10.1116/1.4823705.
[31] T. Kitabayashi, T. Asashita, N. Satoh, T. Kiba, M. Kawamura, Y. Abe, et al., Fabrication and characterization of microcavity organic light-emitting diode with CaF_2/ZnS distributed Bragg reflector, Thin Solid Films 699 (2020) 137912. Available from: https://doi.org/10.1016/j.tsf.2020.137912.
[32] M. Muallem, A. Palatnik, G.D. Nessim, Y.R. Tischler, Room temperature fabrication of dielectric Bragg reflectors composed of a CaF_2/ZnS multilayered coating, ACS Appl. Mater. Interfaces 7 (1) (2015) 474–481. Available from: https://doi.org/10.1021/am506531p, http://pubs.acs.org/journal/aamick.

CHAPTER 21

Application of nanoceramic-based composites in wastewater treatment

Saptarshi Roy and Md. Ahmaruzzaman
Department of Chemistry, National Institute of Technology Silchar, Silchar, Assam, India

21.1 Introduction

One of the most aggravating problems that mankind is facing in the 21st century is the lack of sufficient freshwater resources for consumption and other domestic purposes. An estimated 1.3 billion people are unable to access unpolluted water and over 2.4 billion people lack appropriate sanitation facilities. Every year, an estimate of 5 million people, particularly from the less developed countries of Asia, Africa, and South America become prone to water-related health issues.

Toxic heavy metals which have been in use for a very long time have become one of the major contributors to water pollution, together with hazardous organic dyes and persistent organic pollutants (POPs). Heavy metals are used in the industrial sector as building materials and pigments. These have been reported to be a key environmental concern and also a serious threat to the living organisms, including humans due to their persistent, accumulative, and nonbiodegradable behavior.

The textile industries, tannery, and the plastic industries generate a huge amount of waste every single day and are some of the main culprits of water pollution. Approximately 11.2 million liters of water are consumed weekly by a medium-sized tannery and textile industry [1]. Industrial discharge from pulp mills, printing, and textile industries contain residual dyes that have become a serious environmental concern throughout the world [2–5]. Also, the management of these colored effluents by conventional treatment methods is not convenient. These dyes are mostly toxic amines and contain chromium, resulting in health hazards at an increasingly alarming rate. Indeed, there has been appreciable research into the removal of hazardous dyes from industrial wastewater before its release into aquatic bodies.

Similarly, POPs, a common form of organic pollutants are a menace to the environment. They are mainly carbon-based chemical compounds, such as dichloro-diphenyl-trichloroethane, polychlorinated biphenyls, and various other pesticides. Due to their persistent nature, they are of great concern as they accumulate in the living bodies of animals and humans leading to biomagnification. In addition to their

Industrial Applications of Nanoceramics
DOI: https://doi.org/10.1016/B978-0-323-88654-3.00026-3

toxicity, they can be transported over long distances. The escalated activities of industrialization due to industrial expansion and lack of proper hygienic sanitation facilities have added fuel to the fire and have led to this present state of crisis. This scenario has been worsening and it is expected that in the coming decades, the problem of water scarcity will take a grave turn. In this context, there is the utmost demand for the innovation of new technologies for water remediation that eliminate the persistent pollutants prior to their discharge into the aquatic medium. Several treatment methods, such as ion exchange, degradation, adsorption, membrane filtration, chemical precipitation, and electrochemical techniques, have been employed for the treatment of wastewater. Ideal wastewater treatment methods should be cost-effective and should involve minimum utilization of chemicals, thus simultaneously maintaining the protocols of green chemistry. In this regard, adsorption by solid materials has emerged as a potential candidate for the elimination of several contaminants because of their operational simplicity and high efficiency.

Presently, many adsorbent materials have been exploited for eliminating dyes, heavy metal ions, and POPs from pollutant-laden wastewater without any great degree of success due to low selectivity and ineffective separation after the experiment. Further, in-depth advanced research has to be undertaken for easy recovery of the adsorbents so as to recycle and reuse them and also enhance their removal efficiency. Hence, intensive search has been carried out for adsorbents that are stable, magnetically recyclable, possess high selectivity for pollutant binding, and are also budget and eco-friendly. Ceramic-based materials have emerged as prospective candidates in this regard. Their various attractive characteristics have drawn the fancy of the research community for their utilization in various wastewater treatment applications.

21.2 Ceramics and nanoceramics

Ceramics can be defined as inorganic nonmetallic materials that can be fired at high temperatures for the purpose of hardening and providing shape [6]. The term "ceramic" has been derived from a Greek word that means pottery. The world of ceramics can be widely classified into two categories: traditional ceramics and advanced ceramics. Lately, advanced ceramics can be further divided into three classes: oxide ceramics (beryllium, alumina and zirconia), nonoxide ceramics (silicides, carbides, borides, and nitrides), and composite ceramics (particulate-reinforced combinations of oxides and nonoxide ceramics) [6]. Ceramics are brittle but suitable for withstanding high temperatures. They are blessed with novel characteristics that provide them superiority over metals and polymers. They possess enhanced chemical stability, superior wear resistance, and lower electrical and thermal conductivity, which allow them to act as electrical insulators. The ability to retain their unique property of high temperature strength and brittleness at elevated temperatures makes them a promising class suitable for solving engineering application problems and in other broad areas [7]. The interesting field of ceramics holds huge scope in the production of advanced materials for its large-scale applications.

Nanoceramic refers to ceramic materials comprising at least one dimension at the nanometer scale (1–100 nm). They incorporate ceramic forms ranging from nanostructured and nanosized particles, sheets, fibers, rods, tubes, thin films, and nanostructure surfaces [8]. Since their discovery in 1980, they have received widespread appraisal and attention from researchers for their enhanced potentiality and excellent features in contrast to their conventional bulk counterparts that are rigid and brittle and suffer fractures on slight impact. Nanoceramics exhibit interesting mechanical and surface features such as superplasticity, firmness, enhanced robustness, machinability, bioactivity, and controllable crystallinity [9]. Their high surface area can be attributed to the smaller size of their constituent particles which can be employed in reactions where a high reaction rate is required. Nanoceramics present a broad range of applications in biomedicine, environmental remediation, wastewater treatment, and also in industrial construction of next-generation chips, sensors, corrosion-resistant coatings, electrolytes for fuel cells, etc. They can also be utilized in space exploration, transport, construction, and weapon manufacturing.

21.3 Treatment approaches for pollutant-laden wastewater

There are a multitude of approaches available for the remediation of contaminated wastewater, which can be widely classified into three categories: physical, chemical, and biological. Such treatment techniques include membrane filtration, electrodialysis, precipitation, ion-exchange, photocatalysis, floatation, coagulation/flocculation, and adsorption. Presently, there has been little success in the direction of effective treatment of wastewater by the employment of a single process because of the complex and reluctant behavior of the effluents. In practical applications, several techniques are generally combined for effective treatment purposes. The advantages and disadvantages of the different techniques being embraced for the treatment of pollutant-laden wastewater are listed in Table 21.1 [10,11].

Each of the treatment technologies, besides having advantages, also suffer from inherent limitations. Membrane filtration displays high separation selectivity but is not a good choice because of high residual generation and its expensive nature. Electrodialysis leads to high energy consumption and membrane fouling which increases the operation cost to a great degree and hence is not preferable. Ion exchange technique suffers from the problem of secondary pollution and expensive synthetic resins, besides having the superiority of high treatment efficiency [10]. Photocatalysis requires a long treatment duration which is a major drawback, although this technique is comparatively eco-friendly as it generates less toxic by-products. Floatation is usually not preferred because of high

Table 21.1 Various technologies available for pollutant laden wastewater [10,11].

Treatment technologies	Advantages	Disadvantages
Membrane filtration	High separation selectivity	Expensive and heavy sludge generation
Coagulation/flocculation	Economically feasible	Sludge generation
Precipitation	Low cost and convenient	Large volume of sludge generation
Ion exchange	Enhanced treatment efficiency	High cost of resins, regeneration and disposal problems
Photocatalysis	Less toxic by-products	Prolonged treatment period
Adsorption	Easy operation, flexibility and simplicity of design and insensitivity to harmful pollutants	Regeneration of adsorbent
Floatation	Good treatment efficiency	High maintenance cost
Electrodialysis	Enhanced separation selectivity	Membrane fouling and high energy consumption

Modified from M.J.K. Ahmed, M. Ahmaruzzaman, A review on potential usage of industrial waste materials for binding heavy metal ions from aqueous solutions, J. Water Process Eng. 10 (2016) 39–47. https://doi.org/10.1016/j.jwpe.2016.01.014.

operational and maintenance cost [12]. Coagulation and flocculation is hampered by increased sludge formation, although it is efficient for the removal of turbidity along with contaminants. The technique of precipitation is a cost-effective and convenient method but is undesirable for water that is contaminated with only a trace amount of pollutants and also results in the generation of high volumes of sludge.

Adsorption is an economical, environmentally benign, and efficient technique that has been employed to mitigate the problem of alleviating pollutants from wastewater worldwide.

21.4 Adsorption

The process of adsorption has emerged as one of the most convenient and promising methodologies in wastewater treatment due to its budget-friendly nature and convenient operation with the availability of a vast scope of adsorbents [13]. It is a surface phenomenon concerning the deposition of molecular species at the interface of two phases such as solid–liquid or solid–gas interface. It occurs when the absorbable molecules (adsorbate) come into contact with a solid porous surface (adsorbent) and the molecules are retained on its surface by intermolecular forces of attraction. This process does not result in the generation of hazardous by-products and does not resemble the process of "absorption." It can be classified into two types, that is, physisorption and chemisorption. Adsorption due to weak reversible physical forces (dipole–dipole interactions, H-bonding, van der Waals interactions) is termed as physisorption, whereas that due to strong chemical interactions between the surface of the adsorbent and the adsorbate molecules or ions is referred to as chemisorption. Factors such as temperature, the nature of the adsorbate and the adsorbent, pH, and other experimental factors affect adsorption.

In the first place, activated charcoal was employed in water remediation techniques which have been substituted with many eco-friendly and affordable adsorbents [14]. In the process of adsorption, as pollutants adhere to the adsorbent surface, it should possess high surface area for supplying more space to the contaminant molecules to stick to it. So far, several nanomaterials have been employed as adsorbents, among which ceramic nanocomposites are promising because of their enhanced surface area to volume ratio, free active valencies, porous structure, and their surface energies.

21.5 Abatement of various water pollutants using ceramic nanocomposites

21.5.1 Adsorption of dyes from wastewater

The rapid growth of civilization and industrialization has resulted in the inevitable use of chemicals that have become detrimental for human health as well as for the environment. So, it is of utmost necessity to ensure safe use and disposal methods. Dyes and pigments, which are commonly employed in cosmetics, food, textiles, plastics, leather, rubber, and pharmaceutical industries, are hazardous to all forms of life and also to our environment. Because of the inappropriate coloring techniques in textile industries, the excess pigments blend with water and are dumped without any treatment as effluent into water bodies. This causes serious health problems to all forms of life by penetrating through the pores of skin or by inhalation and ingestion. They are the main culprits of diseases such as dermatitis and kidney disorders, and may lead to cancer on prolonged exposure [15].

Dye molecules are not prone to biodegradation and cannot be easily removed due to their complex chromogen–chromosphere chemical structure. Traditional methods fail to remove them because of their high water solubility [16]. Moreover, they are a source of carcinogens due to the aromatic rings constituting their structure. An estimated amount of 28.9 lakh kg of dye is manufactured in India, and hence the eradication of dyes from wastewater is of significant concern due to their toxic effects on mankind, plants, and animals.

Numerous techniques for the remediation of wastewater contaminated with dyestuff have been adopted such as biodegradation, coagulation-flocculation, ion exchange, Fenton oxidation, reverse osmosis, electrochemical oxidation, ozonation, and adsorption. Degradation of certain dyes results in the formation of products which are as or more toxic than the reactant, and hence not favored. The scientific community is always on a constant search to produce a cheap, convenient, and economic method that simultaneously has high efficiency to remove the pollutants. In this regard, adsorption is a promising and feasible option to eliminate dyes from wastewater due to its high removal efficiency even at lower concentrations. It binds the contaminants on the surface of the adsorbent by virtue of hydrogen bonding, van der Waals forces, or electrostatic interactions [17]. They can be subdivided into ionic and nonionic dyes. Dyes that result in the formation of positively charged ions in aqueous medium are cationic dyes and those that form negatively charged ions are termed anionic dyes. Anionic dyes include acid dyes, reactive dyes, and direct dyes while examples of a few cationic dyes are rhodamine B, malachite green (MG), and methylene blue. Vat dyes and disperse dyes are nonionic dyes [18].

21.5.1.1 Removal of anionic dyes

A novel composite of hydroxyapatite (HAp) and chitosan nanoparticles (CS-MHAp) was successfully fabricated for the elimination of anionic reactive blue 19 dye [19]. The removal efficiency was observed to be higher than the corresponding individual chitosan and HAp nanoparticles. It was observed that the efficiency of adsorption was dependent on the pH displaying higher removal efficiency at lower pH, while the adsorption rate decreased gradually as the pH was increased. Interestingly, the CS-MHAp adsorbent could be easily recovered by an external magnet after the treatment and could be reused a number of times without any successive loss of its adsorption capacity. Similar results were demonstrated by Manatunga et al. in a study where a metal-containing azo dye was adsorbed by four different HAp nanocomposites for its effective application in the removal of dye contaminants from wastewater [20]. HAp/chitosan composite was developed by Hou et al. by embedding HAp into chitosan and was potentially employed for the elimination of Congo red dye from an aqueous medium. The hydrogen bonding between the amino groups of the dye and the hydroxyl ion present on the adsorbent greatly facilitated the adsorption efficiency of the composite [21]. Three anionic azo dyes (methyl orange, methyl yellow, and methyl red) were adsorbed on a novel nanocatalyst, HAp/palladium/iron composite, that displayed enhanced catalytic activity, good magnetic separate ability, recyclability, and stability. Moreover, acidic conditions enhanced the catalytic behavior of the adsorbent, and hence it could be a promising catalyst for the degradation of dyes [22].

In another study by Azzaoui et al. the tendency of an efficient, low-cost HAp microcomposite film for the elimination of methyl orange dye was evaluated and it was observed that this could be a potential alternative for the treatment of industrial wastewater laden with toxic dyes [23]. A competent and affordable adsorbent, HAp/ CeO_2 nanocomposite was employed for the elimination of Erichrome Black T. It was obvious that the adsorption capacity increased with the fall of pH, as the deprotonation of phosphate ions at higher pH leads to repulsion. An increase in the concentration of the dye and the surface area enhanced the removal capacity of the adsorbent [24]. Synthesis and utilization of HAp/chitosan and HAp/carboxymethyl cellulose nanocomposites for the elimination of lead ions and acid yellow dye waste was achieved by Manatunga et al. [25]. It was found that the maximum adsorption efficiency recorded for the acid yellow 220 dye was 200 and 303 mg/g for CMC@HAp and CTS@HAp, respectively. Reactive red 141 dye was adsorbed on HAp and HAp/ titanium calcined pellets with a maximum adsorption efficiency of 20.62 and 28.41 mg/g for HAp and TiHAp, respectively [26]. The addition of titanium modified the structure of pristine HAp making it more desirable for adsorption at elevated pH. Enhanced adsorption capability can be greatly attributed to the hydrogen bonding between the nitrogen of the dye and HAp.

21.5.1.2 Removal of cationic dyes

In an investigation by Sarma et al. the adsorption efficiencies of two cationic dyes, Brilliant green and Crystal violet, on kaolinite clay material were analyzed. It was revealed that the efficiencies for both the dyes increased on acid treatment of the clay material [27]. The electrostatic force of attraction was mainly accountable for its adsorption. Crystal violet displayed higher removal capacity than brilliant green and the adsorbent displayed a reusability nature up to a third batch cycle. The adsorption of methylene blue dye on magnetic hydroxyapatite-immobilized oxidized multiwalled carbon nanotubes (mHAp-oMWCNT) was demonstrated by Wang et al. and it was found that the removal of the dye depended on the ion exchange property of the hydroxyapatite and the functional groups −OH and −COOH of the carbon nanotubes [28]. Langmuir isotherm presented a maximum adsorption of 328.4 mg/g for methylene blue. A novel cost-effective nanocomposite was successfully fabricated by Ragab et al. from hydroxyapatite and chitosan and it was employed for the elimination of brilliant green dye which presented optimum adsorption at pH 7 [29]. Factors such as dosage, dye concentration and contact time had an effect on the capacity of adsorption. Rhodamine B was adsorbed on HAp/alginate nanocomposite beads in a study explored by Oladipo and Gazi which revealed highest adsorption of 480 mg/g for the dye [30]. In a continuous study, the flow rate of dye and the depth of adsorbent affected the adsorption. Hosseinzadeh et al. reported an efficient removal method of MG dye from an aqueous medium using a novel nanocomposite adsorbent prepared by a simple free radical graft copolymerization of acrylamide monomer on to starch backbones in the presence of graphene oxide nanosheets and nanohydroxyapatite (n-HAp) [31].

21.5.2 Adsorption of inorganic pollutants

Inorganic pollutants constitute a multitude of heavy metal ions [Pb(II), Ni(II), Cr(II), Cd(II), As(III)]; anions (F^-, NO^{3-}, S^{2-}); and radioactive ions (Cs^+, I^-) that are accountable for causing subsequent toxicities in human beings and other living organisms, as listed in Table 21.2. Industrial discharge constitutes an excessive concentration of heavy metal ions while domestic and agricultural practices may release anions into the environment [34]. Accidental mishaps may sometimes result in the release of the radioactive wastes into water bodies from nuclear reactors, leading to severe damage to the ecosystem [35]. Addressing this issue, it has become mandatory and an immediate task to recycle water by alleviating these toxic contaminants up to a certain level that is permissible by WHO standards. Ceramic nanofibers have attracted the interest of the global research community and are successfully utilized in the potential elimination of heavy metal ions from contaminated wastewater.

Table 21.2 Permissible discharge limits of industrial effluents and health hazards of toxic heavy metals [10,32,33].

Metals	Permissible limits for industrial effluents discharge (mg/L) WHO std.	Health effects
Arsenic	–	Carcinogenic, causes liver tumors and gastrointestinal effects
Mercury	–	Excess use may result in headache, pain in the abdominal and diarrhea, paralysis, gum inflammation and loss of appetite
Cadmium	0.1	Highly toxic and causes lung fibrosis and dyspnea
Chromium	–	Suspected human carcinogen and causes lung tumors
Copper	0.05–1.5	Long term exposure causes stomach ache, irritation of eyes, mouth and headache
Nickel	–	Causes chronic bronchitis, reduced lung function and lung cancer
Zinc	5.0–15.0	Causes "metal fume fever" and restlessness
Lead	0.1	Suspected carcinogen, anemia, high blood pressure, muscle and joint pains and kidney problem

Modified from M.J.K. Ahmed, M. Ahmaruzzaman, A review on potential usage of industrial waste materials for binding heavy metal ions from aqueous solutions, J. Water Process Eng. 10 (2016) 39–47. https://doi.org/10.1016/j.jwpe.2016.01.014.

Mahapatra and his team were able to successfully synthesize electrospun iron oxide-alumina nanocomposite fibers and employed them as adsorbents for the elimination of heavy metal ions from the aqueous system. A maximum removal percentage of 90% for Hg^{2+} ions was noted with the percentage decreasing in the order $Hg^{2+} > Ni^{2+} > Pb^{2+} > Cu^{2+}$ [36]. The adsorption behavior of the different ions was also studied as a function of initial concentration, contact time, and pH. It was found from the Langmuir equation that the capacities of absorption were 4.98 mg/g for Cu^{2+}, 32.36 mg/g for Ni^{2+}, 23.75 mg/g for Pb^{2+}, and 63.69 mg/g for Hg^{2+} ions. To enhance the adsorption capability of the alumina nanofibers, their surface was immobilized with thiol and octyl functional groups that have a strong affinity toward the pollutants. Grafting the external surface improved the accessibility of the adsorption sites to the adsorbates, thus making it more efficient even for the adsorption of contaminants at very trace concentrations. 3-Mercaptopropyltrimethoxysilane and *n*-octyltriethoxysilane acted as grafting agents. The hybrid nanofibers were observed to be more potent and selective in contrast to the parent alumina nanofibers and displayed high adsorption efficiency for Pb^{2+} and Cd^{2+} ions [37]. The Pb^{2+} removal efficiency of the thiol-functionalized alumina nanofibers increased up to 74% from the initial 40% of pure alumina nanofibers. In another study by Yang et al. a layered structure composed of titanate nanofibers, $Na_2Ti_3O_7$, and $Na_{1.5}H_{0.5}Ti_3O_7$ was fabricated by the reaction between titania and caustic soda for the successful elimination of hazardous radioactive wastes and heavy metal ions, such as Sr^{2+}, Ra^{2+}, and Pb^{2+}, from contaminated wastewater. Their high adsorption rate could be attributed to the ion exchange mechanism that allows them to selectively adsorb the radioactive bivalent ions from the aqueous solution. Layered titanate nanofibers were revealed to be better adsorbents than conventional zeolites or clays. Moreover, the hazardous ions can be disposed safely after adsorption without any ensuing secondary treatment [36]. Iron oxides and ferrites have been used widely for their application in water treatment and remediation. Ren et al. produced a novel low-cost adsorbent comprising α-Fe_2O_3 nanofibers for potential application in the elimination of Cr(VI) from aqueous systems. The adsorbent was synthesized by a simple hydrothermal technique followed by calcination. It was demonstrated that the nanofibers displayed excellent and faster adsorption capacities for the removal of Cr(VI) with a maximum adsorption efficiency of 16.17 mg/g. Moreover, the adsorption activity of these fibers was undoubtedly found to be better than the commercial powders with no observable decrease in the adsorption capacity after repeated recycling and reuse [38].

21.5.3 Adsorption of phenolic compounds

Phenol and phenolic compounds constitute one of the most hazardous forms of organic chemical pollutants and are generally present in aqueous effluents of oil refineries, petrochemicals and steel plants, coal conversion processes, plastics, and drug industries in high concentrations. Various phenolic compounds are employed as raw materials and solvents in industrial activities and are therefore a very obvious contaminant in the industrial effluent contaminating the natural water resources. They are reputed to be carcinogenic and harmful for living organisms even in low concentrations. Short-term effects include respiratory problems, headaches, and irritation of the eyes. Long-term exposure may be detrimental towards all forms of life and leads to serious problems of weakness, muscle pain, anorexia, weight loss, fatigue, heart disease, and ultimately respiratory cancer [39].

A multitude of treatment technologies are utilized for the removal of phenol from wastewater, among which granular activated carbon filtration, ozonation, chlorination, advanced oxidation processes (H_2O_2/UV, O_3/UV, Fenton's reagent, photocatalytic processes), solvent extraction, and membrane separation are notable [40,41]. Different varieties of membranes have also been utilized for the effective removal of these contaminants [42]: liquid membranes [43—45], anion exchange membranes [46], nanofiltration/reverse osmosis membranes [47,48], and pervaporation membranes [49,50].

With the intention of enhancing the removal accomplishments of various membrane processes, a combination of various other treatment technologies with these membrane filtration processes has been increasingly developed in the recent years. These hybrid techniques contain either polymeric or ceramic membranes.

The development of separation methodologies using a new porous approach has been able to eliminate this toxic species from aqueous medium. Active and in-depth research has been focused on the utilization of the abundant natural materials like clays and natural apatites for application in phenol elimination from wastewaters [51,52], but limited surveys were performed on the elimination of phenolic compounds using natural adsorbents [53,54]. The feasibility of doping surface modifiers such as Al^{3+} ions to hydroxyapatite can improve the adsorption performance for organic contaminants. The capacity of alumina to adsorb organic and inorganic compounds depends on the pH of the solution and its surface modification. This process is much more economical than anion exchange resins. The association of alumina with apatite to form porous nanocomposites may be cost-effective and environment-friendly. This approach relies on the preparation of nanocomposites associating hydroxyapatite with alumina, leading to homogeneous mesoporous materials. For this purpose, a novel porous, low-cost adsorbent, Al_2O_3—hydroxyapatite composite were synthesized by Bouiahya et al. using natural phosphate as calcium and phosphorus precursors in the presence of Al^{3+} ions [55]. The supplementary active sites on the surface of the adsorbent may enhance the phenol adsorptive capacity over pure hydroxyapatite. The retention capacity of these nanocomposites is dependent on the complex interaction between intermolecular and molecule—solid interactions. In another study by Lin et al. hydroxyapatite nanopowders were synthesized and employed as a potential adsorbent for phenol adsorption from aqueous medium [56]. The nanopowders were prepared by a simple chemical precipitation technique. Factors affecting the adsorption of phenol included contact time, initial phenol concentration, pH, adsorbent dosage, temperature of the solution, and adsorbent calcining temperature. The highest phenol adsorption efficiency was recorded to be 10.33 mg/g for 400 mg/L initial phenol concentrations at pH 6.4 and 60°C. Narwade et al. synthesized a graphite–hydroxyapatite (GraHAp) nanocomposite by a wet precipitation method and studied its potential for the adsorption of phenol from an aqueous medium in batches [57]. The doping of graphite in hydroxyapatite supported interaction between the aromatic rings and graphite sheets enhanced the adsorption process. Adsorption of phenol from aqueous medium was demonstrated by Narwade et al. using very interesting and novel bio-based solvent-cast films fabricated from cellulose nanofibrils and hydroxyapatite (HAp) by the method of wet chemical precipitation [58]. Maximum phenol adsorption of 64 mg/g was observed at pH 2. The hydrophilic behavior of the films with high water diffusing, transport properties, and low swelling could be advantageous in the development of an efficient adsorbent and filter membranes for water purification in the future.

21.5.4 Adsorption of natural organic matter

Besides phenolic compounds, another major threat is the existence of natural organic matter (NOM) in consumable water which causes a wide range of complications in the water treatment process. The formation of suspected carcinogenic disinfection by-products as a result of reactions between NOM and oxidants/disinfectants such as chlorine is a serious concern that cannot be abandoned. A multitude of treatment technologies have been employed to eliminate NOM by coagulation, adsorption, ion exchange, oxidation, filtration, and membrane separation. The lack of durability and stability in extreme conditions of pH and temperature, premature degradation due to chemical and biological agents, and fouling are considered to be the major pitfalls of polymeric membranes compared with ceramic membranes. In this regard, extensive research has been conducted on ceramic membranes in recent years for their potential utility in a wider range of filtration techniques. This is mainly due to their superior characteristics of mechanical strength, their ability to resist harsh chemicals, extreme temperatures, withstand attacks from microorganisms, and their ease of cleaning. Ceramic membranes fabricated from inorganic oxides of alumina, silica, zirconia, titania, and, more recently, stannic oxide are notable [59–61].

21.6 Future prospects

Nanoceramics as adsorbents have received significant interest from the researchers across the world due to their attractive features, such as the bioactivity, eco-friendly and cost-effective nature, excellent performance of adsorption, low water solubility, and high stability under harsh conditions. To enhance the adsorption and catalytic efficiencies, many functional groups have been grafted for modification by codoping and impregnation on the surface of the ceramic materials. These modified functionalized materials have the novelty of easy regeneration, fast recovery rate, and strong binding ability for the competitive adsorption of contaminants. Hence, their potential should be employed in the presence of organic, inorganic, and other hazardous contaminants for wide-scale commercial implementation. Also, further studies should be conducted toward the employment of ceramic-based materials in catalytic degradation of pollutants. Moreover, the unique approach of one-pot synthesis of the materials should be encouraged and a comprehensive cost-friendly treatment of industrial effluents should be cultivated. To promote the applicability of ceramic nanomaterials-based water filtration in the forthcoming days, certain factors should be contemplated: (1) the development of advanced nanocomposites for better adsorption; (2) fabrication of cost-friendly ceramic-based filtration units for easy affordability to the poor people; (3) effects of pH, temperature, etc. need to be considered for proper treatment; and (4) complete analysis of the nanomaterials to prevent any secondary water pollution and thus any significant health problems. The ultimate objective of the entire research directed toward water remediation is to serve the people with their basic right of clean consumable water. Reformation of the current available techniques can mark a new perspective in the field of water remediation in the coming decades.

21.7 Conclusion

This chapter basically drafts the potential of ceramic nanocomposites as dynamic materials to settle the ongoing global issue of wastewater treatment, which is one of the greatest problems that the world is facing today. The utilization of ceramic-based materials for the treatment of water contaminated with pollutants can be a promising alternative to the conventional techniques, which suffer the problem of high sludge generation as by-products. Ceramics are potent materials of interest because of their novel electronic structure, light absorption efficiency, charge transport behavior, and porous structure. Therefore, ceramic nanomaterials have been continuously exploited by scientists as excellent adsorbents and photocatalysts to alleviate the hazardous contaminants from wastewater. The low cost of fabrication and enhanced surface area afford them better utility for their use as adsorbents. Moreover, these nanomaterials are designed so as to enable their recycling and reuse, thus eliminating the possibility of any negative impacts to the environment. To further enhance the activity of the ceramic nanomaterials, several modifications have been implemented such as synthesis of composites and doping. Factors such as pH, particle size, crystallinity, initial concentration of the pollutant, and ionic strength of the competitive ions affect the adsorption process. Altogether, ceramic-based nanomaterials can be considered as highly efficient and advantageous in the arena of environmental remediation. Advanced studies and in-depth research concerning the fabrication of recyclable, magnetic-enabled, and high adsorptive ceramic-based materials will not only pave the way for the effective treatment of pollutant-laden wastewater treatment at a commercial scale, but also serve the people with clean and fresh drinking water in the days to come.

References

[1] A.A. Oladipo, M. Gazi, Two-stage batch sorber design and optimization of biosorption conditions by Taguchi methodology for the removal of acid red 25 onto magnetic biomass, Korean, J. Chem. Eng. 32 (2015) 1864−1878.

[2] A.A. Oladipo, M. Gazi, S. Saber-Samandari, Adsorption of anthraquinone dye onto eco-friendly semi-IPN biocomposite hydrogel: equilibrium isotherms, kinetic studies and optimization, J. Taiwan. Inst. Chem. Eng. 45 (2014) 653−664.

[3] S. Chakraborty, S. De, J.K. Basu, S. DasGupta, Treatment of a textile effluent: application of a combination method involving adsorption and nanofiltration, Desalination 174 (2005) 73−85.

[4] G. Crini, Non-conventional low-cost adsorbents for dye removal: a review, Bioresour. Technol. 97 (2006) 1061−1085.

[5] R. Sanghi, B. Bhattacharya, Review on decolorisation of aqueous dye solutions by low cost adsorbents, Color Technol. 118 (2002) 256−269.

[6] H.T. Ting, K.A. Abou-El-Hossein, H.B. Chua, Review of micromachining of ceramics by etching, Trans. Nonferrous Met. Soc. China. 19 (2009) s1−s16.

[7] I.P. Tuersley, A. Jawaid, I.R. Pashby, Review: various methods of machining advanced ceramic materials, J. Mater. Process. Technol. 42 (1994) 377−390.

[8] O.V.D. Biest, Nanoceramics: issues and opportunities, Int. J. Appl. Ceram. Technol. 10 (2013) 565−576.

[9] A. Kiani, M. Rahmani, S. Manickam, B. Tan, Nanoceramics: synthesis, characterization and applications, J. Nanomat. 2014 (2014) 1−2.

[10] M.J.K. Ahmed, M. Ahmaruzzaman, A review on potential usage of industrial waste materials for binding heavy metal ions from aqueous solutions, J. Water Process Eng. 10 (2016) 39−47.

[11] A.K. Mishra, Sol-gel Based, Nanoceramic materials: preparation, properties and applications | | application of hydroxyapatite-based nanoceramics in wastewater treatment: synthesis, Characterization, and Optimization (Chapter 8), Springer International Publishing, Switzerland, 2017, pp. 231−251. ISBN: 978-3-319-49510-1, 978-3-319-49512-5 (eBook).

[12] I.A. Abbas, A.M. Al-Amer, T. Laoui, M.J. Al-Marri, M.S. Nasser, M. Khraisheh, et al., Heavy metal removal from aqueous solution by advanced carbon nanotubes: critical review of adsorption applications, Sep. Purif. Technol. 157 (2016) 141−161.

[13] M. Khajeh, S. Laurent, K. Dastafkan, Nanoadsorbents: classification, preparation, and applications (with emphasis on aqueous media), Chem. Rev. 113 (2013) 7728−7768.

[14] J.S. Hu, L.S. Zhong, W.G. Song, L.J. Wan, Synthesis of hierarchically structured metal oxides and their application in heavy metal ion removal, Adv. Mater. 20 (2008) 2977−2982.

[15] F. Akarslan, H. Demiralay, Effects of textile materials harmful to human health, Acta Phys. Pol. A. 128 (2015) 407−408.

[16] M. Yusuf, Synthetic dyes: a threat to the environment and water ecosystem. In Textiles and clothing. (Chapter 2), Scrivener Publishing LLC, 2019, pp. 11−26. ISBN: 9781119526599(Online), 9781119526315 (Print).

[17] A.B. dos Santos, F.J. Cervantes, J.B. van, Review paper on current technologies for decolourisation of textile wastewaters: perspectives for anaerobic biotechnology, Bioresour. Technol. 98 (2007) 2369−2385.

[18] P.S. Vankar, Chemistry of natural dyes, Resonance 5 (2000) 73−80.

[19] V.C. Nguyen, Q.H. Pho, Preparation of chitosan coated magnetic hydroxyapatite nanoparticles and application for adsorption of reactive blue 19 and Ni^{2+} ions, Sci. World J. 2014 (2014) 1–9.

[20] D.C. Manatunga, R.M. de Silva, K.M. Nalin de Silva, N. de Silva, E.V.A. Premalal, Metal and polymer-mediated synthesis of porous crystalline hydroxyapatite nanocomposites for environmental remediation, R. Soc. Open Sci. 5 (2018) 171557.

[21] H. Hou, R. Zhou, P. Wu, W. Lan, Removal of Congo red dye from aqueous solution with hydroxyapatite/chitosan composite, Chem. Eng. J. 211–212 (2012) 336–342.

[22] A. Safavi, S. Momeni, Highly efficient degradation of azo dyes by palladium/hydroxyapatite/Fe_3O_4 nanocatalyst, J. Hazard. Mater. 201–202 (2012) 125–131.

[23] K. Azzaoui, E. Mejdoubi, A. Lamhamdi, M. Lakrat, O. Hamed, S. Jodeh, et al., Preparation of hydroxyapatite biobased micro composite film for selective removal of toxic dyes from wastewater, Desalin. Water Treat. 149 (2019) 194–208.

[24] S. Chaudhary, P. Sharma, R. Renu, et al., Hydroxyapatite doped CeO_2 nanoparticles: impact on biocompatibility and dye adsorption properties, RSC Adv. 6 (2016) 62797–62809.

[25] D.C. Manatunga, R.M. De Silva, K.M. Nalin De Silva, R. Ratnaweera, Natural polysaccharides leading to super adsorbent hydroxyapatite nanoparticles for the removal of heavy metals and dyes from aqueous solutions, RSC Adv. 6 (2016) 105618–105630.

[26] F. Asjadi, E. Salahi, I. Mobasherpour, Removal of reactive red 141 dye from aqueous solution by titanium hydroxyapatite pellets, J. Dispers. Sci. Technol. 37 (2015) 14–22.

[27] G. Sarma, S. Sen Gupta, K. Bhattacharyya, Removal of hazardous basic dyes from aqueous solution by adsorption onto kaolinite and acid-treated kaolinite: kinetics, isotherm and mechanistic study, SN Appl. Sci. 1 (2019). Available from: https://doi.org/10.1007/s42452-019-0216-y.

[28] Y. Wang, L. Hu, G. Zhang, T. Yan, L. Yan, Q. Wei, et al., Removal of Pb(II) and methylene blue from aqueous solution by magnetic hydroxyapatite-immobilized oxidized multi-walled carbon nanotubes, J. Colloid Interface Sci. 494 (2017) 380–388.

[29] A. Ragab, I. Ahmed, D. Bader, The removal of brilliant green dye from aqueous solution using nano hydroxyapatite/chitosan composite as a sorbent, Molecules. 24 (2019) 847.

[30] A.A. Oladipo, M. Gazi, Uptake of Ni^{2+} and rhodamine B by nanohydroxyapatite/alginate composite beads: batch and continuous flow systems, Toxicol. Environ. Chem. 98 (2015) 189–203.

[31] H. Hosseinzadeh, S. Ramin, Fabrication of starch-graft-poly (acrylamide)/graphene oxide/hydroxyapatite nanocomposite hydrogel adsorbent for removal of malachite green dye from aqueous solution, Int. J. Biol. Macromol. 106 (2018) 101–115.

[32] General Standards for Discharge of Environment Pollutants. Effluent. Gazette Notification of MoEF, G.S.R. 422(E) Gazette No. 174 (1993) 545–560.

[33] R.A. Corbitt, Standard Hand Book of Environmental Engineering, 2nd ed., McGraw – Hill, New York, 1999.

[34] K. Krishna, R.M. Kurakalva, Risk assessment of heavy metals and their source distribution in waters of a contaminated industrial site, Environ. Sci. Pollut. Res. 21 (2013) 3653–3669.

[35] D. Yang, Z. Zheng, H. Liu, H. Zhu, X. Ke, Y. Xu, et al., Layered titanate nanofibers as efficient adsorbents for removal of toxic radioactive and heavy metal ions from water, J. Phys. Chem. C. 112 (2008) 16275–16280.

[36] A. Mahapatra, B.G. Mishra, G. Hota, Electrospun $Fe_2O_3-Al_2O_3$ nanocomposite fibers as efficient adsorbent for removal of heavy metal ions from aqueous solution, J. Hazard. Mater. 258 – 259 (2013) 116–123.

[37] D. Yang, B. Paul, W. Xu, Y. Yuan, E. Liu, X. Ke, et al., Alumina nanofibers grafted with functional groups: a new design in efficient sorbents for removal of toxic contaminants from water, Water Res. 44 (2010) 741–750.

[38] T. Ren, P. He, W. Niu, Y. Wu, L. Ai, X. Gou, Synthesis of α-Fe_2O_3 nanofibers for applications in removal and recovery of Cr(VI) from wastewater, Environ. Sci. Pollut. Res. 20 (2012) 155–162.

[39] S. Mishra, J. Bhattacharya, Batch studies on phenol removal using leaf activated carbon, MJChem 9 (2007) 51–59.

[40] S. Cheng, S.-J. Tsai, Y.-F. Lee, Photocatalytic decomposition of phenol over titanium oxide of various structures, Catal. Today 26 (1995) 87–96.

[41] A. Barrios-Martinez, E. Barbot, B. Marrot, P. Moulin, N. Roche, Degradation of synthetic phenol-containing wastewaters by MBR, J. Membr. Sci. 281 (2006) 288–296.
[42] A. Bodalo, E. Gomez, A.M. Hidalgo, M. Gomez, M.D. Murcia, I. Lopez, Nanofiltration membranes to reduce phenol concentration in wastewater, Desalination 245 (2009) 680–686.
[43] Y.H. Wan, X. Wang, X.J. Zhang, Treatment of high concentration phenolic wastewater by liquid membrane with N503 as mobile carrier, J. Membr. Sci. 135 (1997) 263–270.
[44] P.F.M.M. Correia, J.M.R. de Carvalho, Recovery of phenol from phenolic resin plant effluents by emulsion liquid membranes, J. Membr. Sci. 225 (2003) 41–49.
[45] M.T.A. Reis, O.M.F. de Freitas, M.R.C. Ismael, J.M.R. Carvalho, Recovery of phenol from aqueous solutions using liquid membranes with Cyanex 923, J. Membr. Sci. 305 (2007) 313–324.
[46] T. Kojima, K. Nishijima, M. Matsukata, Removal and recovery of phenol from FCC effluent, J. Membr. Sci. 102 (1995) 43–47.
[47] C.F. Schutte, The rejection of specific organic compounds by reverse osmosis membranes, Desalination 158 (2003) 285–294.
[48] J.M. Arsuaga, M.J. Lopez-Munoz, A. Sotto, G. Rosario, Retention of phenols and carboxylic acids by nanofiltration/reverse osmosis membranes: sieving and membrane-solute interaction effects, Desalination 200 (2006) 731–733.
[49] C.C. Pereira, A.C. Habert, R. Nobrega, C.P. Borges, New insights in the removal of diluted volatile compounds from dilute aqueous solution by pervaporation process, J. Membr. Sci. 138 (1998) 227–235.
[50] P. Wu, R.W. Field, R. England, B.J. Brisdon, A fundamental study of organofunctionalised PDMS membranes for the pervorative recovery of phenolic compounds from aqueous streams, J. Membr. Sci. 190 (2001) 147–157.
[51] I. Horvat, S. Radeka, T. Plavša, I. Lukić, Bentonite fining during fermentation reduces the dosage required and exhibits significant side-effects on phenols, free and bound aromas, and sensory quality of white wine, Food Chem. 285 (2019) 305–315.
[52] M. Ahmaruzzaman, D.K. Sharma, Adsorption of phenols from wastewater, J. Colloid Interface Sci. 287 (2005) 14–24.
[53] H. Bouyarmane, S. El Asri, A. Rami, C. Roux, M.A. Mahly, A. Saoiabi, et al., Pyridine and phenol removal using natural and synthetic apatites as low-cost sorbents: influence of porosity and surface interactions, J. Hazard. Mater. 181 (2010) 736–741.
[54] A. Bahdod, S. El Asri, A. Saoiabi, T. Coradin, A. Laghzizil, Adsorption of phenol from an aqueous solution by selected apatite adsorbents: kinetic process and impact of the surface properties, Water Res. 43 (2009) 313–318.
[55] K. Bouiahya, I. Es-saidi, C. El Bekkali, A. Laghzizil, D. Robert, J. Nunzi, et al., Synthesis and properties of alumina-hydroxyapatite composites from natural phosphate for phenol removal from water, Colloids Interface Sci. Commun. 31 (2019) 100188.
[56] K. Lin, J. Pan, Y. Chen, R. Cheng, X. Xu, Study of the adsorption of phenol from aqueous solution on apatite nanopowders, J. Hazard. Mater. 161 (2008) 231–240.
[57] V. Narwade, R. Khairnar, V. Kokol, Phenol adsorption on graphite-hydroxyapatite nanocomposites: kinetic and isotherm study, JMSSE 5 (2017) 544–548.
[58] V. Narwade, R. Khairnar, V. Kokol, In-situ synthesised hydroxyapatite-loaded films based on cellulose nanofibrils for phenol removal from wastewater, Cellulose 24 (2017) 4911–4925.
[59] P. Manno, P. Moulin, J.C. Rouch, M. Clifton, P. Aptel, Mass transfer improvement in helically wound hollow fibre ultrafiltration modules: yeast suspensions, Sep. Purif. Technol. 14 (1998) 175–182.
[60] L.R.B. Santos, C.V. Santilli, A. Larbot, M. Persin, S.H. Pulcinelli, Influence of membrane–solution interface on the selectivity of SnO_2 ultrafiltration membranes, Sep. Purif. Technol. 22–23 (2001) 17–22.
[61] P. Narong, A.E. James, Efficiency of ultrafiltration in the separation of whey suspensions using a tubular zirconia membrane, Desalination 219 (2008) 348–357.

CHAPTER 22

Industrial applications of nanoceramics: from lab to real-time utilization—the environmental, legal, health, and safety issues of nanoceramics

Sameera Khathoon A.[1], Darul Raiyaan G.I.[2], Mohamed Khalith S.B.[3] and Kantha D. Arunachalam[4]
[1]Post Graduate and Research Department of Zoology, J. B. A. S. College for Women, Chennai, Tamil Nadu, India
[2]Department of Biotechnology, College of Science and Humanities, SRM Institute of Science and Technology, Kattankulathur, Tamil Nadu, India
[3]Department of Chemistry, Sadakathullah Appa College, Tirunelveli, Tamil Nadu, India
[4]Faculty of Sciences, Marwadi University, Rajkot, Gujarat, India

22.1 Introduction

Ceramic material fabricated from the ultrafine particles is termed nanoceramics and encompasses various forms of ceramic materials that are nanosized and nanostructured particles, nanofibers, nanotubes, nanoribbons, nanosheets, nanorods, nanofluids, and thin films. These are produced for many applications in many industrial sectors based on their properties [1]. Nanoceramics are widely used because of their unique properties compared to conventional bulk ceramic materials. Because of the small grain size, nanoceramics exhibit other characteristics like having a superplastic nature [2]. A lot of interest has emerged in recent years focusing on the nanosized ceramic particles and or materials. Nanoceramics are very important and trusted materials because of their special mechanical and physiochemical properties, like excellent wear resistance in a hard unfriendly environment, high ballistic resistance, high thermal shock resistance, superior strength, corrosion/oxidation resistance, low density, good electrical property, optical and magnetic properties, fracture toughness, chemical inertness, nontoxicity, excellent bioactivity, chemical resistance, mechanical resistance, refractory properties, and hardness within normal and high temperatures [3].

Nanoceramics have a wide range of applications in health, environment, and other areas. Despite the many advantages of nanoceramics, there is increasing concern with regard to the adverse effects on human health and environment. The composition of

Industrial Applications of Nanoceramics
DOI: https://doi.org/10.1016/B978-0-323-88654-3.00027-5

nanoceramics can be oxide and/or nonoxide nanoceramic materials, silicate materials, and or hard metals, which are exclusively used in architectural and industrial works because of their mechanical and physical properties. Some nanoceramics products that are widely used in the research field and industrial sectors are silica materials, titanium nanoceramics, alumina products, zirconia, tungsten carbide, oxides of zinc, silicon nitride, magnesia products, boron nitride, ferric oxide materials, ceriananoceramics, and silicon carbide [4]. Due to the presence of hydroxyl radicals, nanoceramics are chemically active and have a high adsorption capacity but these characteristics provide a potential risk to the environment and health of human beings [5].

Several methods were used to apply ceramic coatings, including glazing, spraying, laser-based techniques, inkjet printing, deposition procedures, and injection of nano-powders, which might release coarse and fine nanoparticles and cause deposits on the surface area, disrupting the environment [6]. Nanoceramics are usually chemically active materials as they have a high adsorption capacity but during the synthesis and application of these nanoceramics they may become potential hazards to the environment [7].

22.2 Environmental issues

22.2.1 Water pollution

When nanoceramics are used in wastewater treatment, the particulate matter of nano-ceramics may contaminated the water to a greater extent and finally reach into the human body leading to various waterborne diseases [6]. The chemical pollutants may be present in natural water sources like rivers and lakes, where the high specific surface area of NPs favors the circulation or movement of such chemical substances, which finally causes damage to the organisms in the environment [4].

22.2.2 Air pollution

Environmental impacts caused by the nanoceramics industries include when these nanoceramics' ultrafine particles enter into the environment through air, resulting in air pollution [8].

Some literature has proven the adverse effect of dust from nanoceramic industries, such as airborne NPs (ANPs) and engineered NPs (ENPs), through a series of key biological mechanisms. In the occupational work environment they cause cardiopulmonary effects on human health [6].

22.2.3 Noise pollution

During the nanoceramics production process, an enormous amount of noise is released out into the environment from the industries, thus it results in noise pollution [9].

22.2.4 Odor

When ENPs from the ceramic industries are dumped in waste, then an unusual odor may be released that ultimately disrupts the environment and the surrounding organisms [10].

22.2.5 Economic loss

Compared with their coarse-grained counterparts, nanoceramics possess different physical and structural properties. According to a report, cubic to tetragonal phase transitions occurred under reversible pressure at room temperature when $MgAl_2O_4$ nanoceramics were examined under high pressure to determine their structural ability and the effects of particle size on their strength and structural characteristics. $MgAl_2O_4$ exhibits a cracking behavior above 60-nm-sized grain whereas below 60 nm it shows a metal-like plastic deformation behavior. This report indicates that this structural stability of nanoceramics fails to provide strengthening material which results in the loss of economic value and affects the production of those commercial products [11].

The nanocomposites (NCs) of nanoceramic TiC/a-Si_3N_4 can cause fractures depending on their particle size and temperature in the Si_3N_4 amorphous layer having an intergranular plastic flow. This cracking behavior affects the commercial value of nanoceramics used in pottery and other industrial applications. The primary basis for these nanoceramic materials is particle size; hence, care should be taken while manufacturing them [12].

For the treatment of water, nanostructured nanoceramics and NCs are synthesized by the sol–gel method as they are widely helpful in environmental remediation and water treatment, compared with chemical methods. Different mesoporous nanocrystalline materials are used. Research shows that this approach is nontoxic to humans and environment, safe, and eco-friendly and it is a cost-effective method for bacterial disinfection in wastewater treatment [13].

A composite nanoceramic (Prime & Bond NT Ceram X), an ormocer-based NC (Admira), a resin nanofilled NC, and a microhybrid resin NC (Tetric Ceramic) were used in a clinical study to evaluate the clinical issues or performance of these restoring materials when placed on Class I and Class II cavities for 3 years. The use of nanoceramic such as the ormocer, resin nanofilled, and nanoceramic composites for restoration provides acceptable clinical performance as it requires complete 3 years to restore cavities of both class. Among them, nanoceramic composite showed 97% positive correlation results. But if these restoration process carried out for less than 3 years, may show lesser restoring capacity then the above-mentioned result [14].

22.2.6 Radionuclide toxicity and radiation-induced damage due to nanoceramics

Alumina nanoceramics possess various industrial applications as they have good abrasive and dielectric characteristics. Gamma-alumina is a nanoporous material that can be used in membrane separation as an adsorbent compound and as a supporting metal in reactors/catalytic reactors [6]. Al_2O_3 nanoceramic material is coated on steel samples for corrosion resistance and protection from radiation damage, which results in nickel, manganese, and chromium leaching out. At 550°C temperature and radiation exposed to 4000 h, alumina nanoceramics specifically retain their structural integrity when fuel cladding imposes strain without damaging the sample by cracking. It becomes challenging to balance the stress and damage from radiation exposure [15]. Oxide nanoceramics have attracted various researchers' attention and are now considered as an essential candidate material for various applications in nuclear systems development from fuels to radioactive waste [16].

22.3 Health issues of nanoceramics

The exposure to nanosized particles from the ceramic industries leads to various effects on human health, including impaired lung functions, lung cancer, bronchitis, chronic obstructive pulmonary disease, adverse acute respiratory asthma wheezing, breathlessness and dry cough, and cardiovascular effects like infections in the cardiac system, ischemia, atherosclerosis, cardiac arrhythmia, and hypertension [4].

22.3.1 Toxicity to microorganisms

Vibrio fischeri is used in Microtox—an in vitro test technique—for identification of toxic effects on ecological samples. It can identify the toxicity of environmental samples or other chemicals and the effect of various NPs. The BOD5 values using microtox in the biodegradable mixture are reduced due to the addition of Ni/γ-Al_2O_3 (NiNC) 43.8% and γ-Al_2O_3 (NC) 31.6%. This indicates that Ni/γ-Al_2O_3 (NiNC) is more toxic than γ-Al_2O_3 NC concerning EC50s values such as 16.1% and 29.9%, respectively [7].

22.3.2 Effects on larvae in water bodies

Amphibians play a major role in the food web and they are affected directly or indirectly as they live near or and on the water reservoirs and possess extremely sensitive skin that readily adsorbs chemical substances from the environment. They act as good indicators of the environmental health as do their larval forms or eggs because they are more sensitive to contamination than the adults. As they graze, the NPs enter into their eggs through incorporation from the sediments. Thus it has been suggested by various authors to assess the toxicity of such incorporated NPs so that we may obtain further information about the toxic effects in order to prevent the hazardous nanoceramics causing harm in the future [7].

NPs of alumina (NPs-Al_2O_3) are exclusively utilized in various applications as they possess many advantageous characteristics and properties. These NPs can become potentially dangerous because their environmental effects are still unknown. In a recent study Svartz evaluated the toxic characteristics of a 25 mg/L nickel gamma-alumina NC (Ni/γ-Al_2O_3) (NC) and the NPs involved during the synthesis of nanoceramics—γ-Al_2O_3 support (SPC) and nickel oxide/gamma-alumina NiO/γ-Al_2O_3 precursor (PC)—on embryo–larval development of *Rhinella arenarum*. When the embryos of *R. arenarum* were exposed to alumina nanoceramics they exhibited improper development, flexures and changes in behavior, and at the cellular level the changes includes alterations of pharyngeal cells, the epithelia of intestine were altered by dissociating, thus killing the cells by a process called apoptosis, and many lysosomes were found by light microscopy and transmission electron microscopy. The presence of Ni as oxide in the epithelium of the intestine is revealed by the scanning electron microscope images. The PC increases toxicity due to the presence of Ni as oxide as it plays a vital function in breathing and feeding mechanisms. It also inhibits glutathione *S*-transferase activity. Oxidative stress damage (lipid peroxidation) was observed in tadpoles treated with irreversible genotoxic damage. When these NPs are produced exponentially, it increases the risk of pollution levels, which may lead amphibians as the targeted animals to this toxicity [5].

22.3.2.1 Lethal effects

When the larvae and eggs of *R. arenarum* are exposed with acute exposure the value of PC nanoceramic material in lethal concentration LC50 for 96 h is twice that of NC and it is also more toxic than the alumina support nanoceramic material (SPC). During chronic exposure, the toxicity range falls with similar sensitivity for both PC and NC. But very low sensitivity is shown by SPC. These are some lethal effects obtained during the exposure of nanoceramics to the living species [5].

22.3.2.2 Sublethal effects

From the first hour and even at low concentration the sublethal effects are shown by the exposed embryos, whereas SPC and NC showed results when the organism was exposed to the highest concentration. The occurrence of cellular dissociation is the significant sublethal effect during the segmentation and gastrulation period. Severe alterations in morphology occurred after 168 h of exposure including late development, small cephalic condition or microcephalic head, edema in the cardiac surface, edema in the abdominal surface, and axial flexures. There were alterations in behavior noted such as erratic swimming, restlessness, and involuntary contractions but these effects were mainly dependent on the concentration. The toxic effect of PC was up to twice and 10 times more than NC and SPC. Almost all the analyzed tissue of larvae showed lipid vacuoles in a higher range, and hyaline vacuoles were also found in large numbers when they were exposed to 10 mg/L PC. The intestine epithelium is highly disorganized and obliterates the lumen cavity in treated larva compared to control. The exposed larval oropharyngeal epithelium was observed with the following alterations: protrusions occurring at the surface of the cell, cellular disorganization, apoptotic cells in oropharynx dissociates and detaches, irregular effects occurring on the cells of cartilage, toxicity in the neural system, alteration in morphology and changes in histology [5].

The exposed liver of the larva showed essential changes, such as highly unorganized liver tissues and a high number of hypertrophic cells in the sinusoid lumen compared to the control species of larvae conserved in its hepatic structure. There were many apoptotic cells found on the lumen of the exposed larval blood vessels system [5].

22.3.2.3 Analysis observation

The AMPHITOX bioassay test results showed a significant increase in the toxicity of eggs and larvae of *R. arenarum* over time. The Ni NC toxic range is always greater than the gamma-alumina nanoceramics, which were 1.58 and 0.83 mg/L and 14.5 and 10.5 mg/L for samples. In the final results, the amphibian larvae showed collapsed cavities, edema, flexures, and behavioral alterations like hyperkinesia and reduction in movements. Through the above results, we can provide proof that wildlife is vulnerable to xenobiotics and specific standardized ecotoxic tests must be produced to conserve natural species and to provide a sustainable environment [7].

22.3.3 Effects on human health

Various types of nanoceramics are used as cosmetics substances. The ormocer technology and nanotechnology have combined to modify an organic NC nanoceramic material that contains methacrylate-modified, silicon dioxide–containing nanofiller, and resin matrix that is replaced by a matrix full of highly dispersed methacrylate-modified polysiloxane particles, thus these particles causes skin damage to the sensitive person when used in beauty cosmetics [14].

Ceramic bionanohybrid materials are used in the medical field as drug delivery or carrier agents for gene and cancer therapy purposes. Layered double hydroxides are a type of bioceramic hybrid delivery carrier material that possesses unique characteristics in the support of drug delivery into the diseased person. When a high dose of a drug is introduced to maintain the cellular uptake it may result in drug resistance and adverse toxic effects to the infected person [17].

Various types of nanodevices used in medicines, for example, NPs, nanogels, nanocapsules, and nanoscaffolds, have been developed for different applications, such as magnetic resonance imaging, tissue repair, bone repair, but they also harm human health. NPs with a diameter less than 200 nm can cross the blood–brain barrier and cause brain diseases or cross the lung–blood barrier to cause pulmonary disorders [18].

22.3.4 Cytotoxicity

Nanoceramics are widely used in medical application as a coating material for various implants. Titanium substrate is coated with a unique TiO_2 nanoceramic to treat peri-implantitis by complete removal of bacterial deposits and other organic residues, including bacterial extracellular protein, polysaccharides, lipopolysaccharides, and nucleic acids, which are associated with the peri-implant inflammation, bone resorption, cytotoxicity, biocompatibility, etc. Ultraviolet light (UV) is required as it activates the pure crystalline TiO_2 nanoceramic coating material to produce various reactive oxygen species (ROS) which then induce photocatalytic decontamination of bacterial residues from the surface of the implant. The photocatalytic activity of titanium substrate is initiated by UV light but this UV light possesses a harmful risk to the cells and tissues of the implant surface area [19].

The in-vitro studies on human cells and rat cells are used by most researchers in their work to focus on the impact of Al_2O_3 nanoceramic NP toxicity. Ecotoxicity studies have shown that these alumina nanocomposite NPs causes toxic effect on autotrophic and heterotrophic organisms. The toxicity mechanism of this type of alumina NPs may induce oxidative stress in algae, and it even affects the microinvertebrates and various fishes of the aquatic system. Metal oxide nanoceramics materials have raised much concern because of their nonbiodegradable characteristics which may allow their rapidly be distributed through the environment. The toxic range of oxide nanoceramic NPs is mainly dependent on their chemical composition, size, shape of the particle, and surface charges. Periplasmic accumulation and toxicity can occur due to the interaction of bacteria with the metal oxide NPs [7].

Bioceramics cause autoimmune diseases by minimizing the immune response against foreign agents that enter the body [20].

Nanometer grains of nanoceramics implanted into the body may produce health risks after implantation, for example, calcium phosphate degrades with the involvement of various physical, chemical, and biological factors including erosion and mechanical abrasion. Calcium phosphate nanoceramics may breakup and form grains followed by free particle release, and osteoclasts and macrophages resorbtion by releasing acids. The degraded nanoceramic materials which then enters into the cells and blood causes potential risk, hence it is essential to identify the adverse effect on human health caused by free particles of nanoceramic before clinical use [4].

Silica-based nanoceramics contain amorphous particles and proton-donating silanol groups which cause hemolysis in mammalian red blood cells and cell membrane protein denaturation through electrostatic interaction with silicate. When a human inhales silica-based nanoceramics, it results in the pulmonary disease called silicosis as the particles elicit harmful toxicological effects. Silica-based nanoceramics undergo penetration into the skin and cell nuclei and cause cytotoxic effects in mice, and also affect the lymph nodes and liver-induced cellular damage to the skin, and inhibit cell growth. A mutagenicity test done using *Salmonella typhimurium* (Ames test) found that there was an increase in mutagenicity with decreasing particle size as the particle size is mainly involved in the biological effects. The continuous accumulation of smaller nanoceramic particles in the nuclei of the cell induces genotoxicity and also causes pregnancy complications in mice [21].

Silica-based bioactive glass-ceramics are widely used in the medical field but they possess a risk of tissue damage when introduced for restoration processes, so these materials are structured to the nanoscale to form nanoceramics to overcome such limitations. Though these nanoceramics have been used in successful restoration, ultimately it ends with a potential risk assessment. When silica and calcium-based nanoceramics are introduced for restoration, damage is caused to the erythrocytes as the nanoceramics attach to the walls of red blood corpuscles (RBC), generate ROS, and produces antioxidant behavior against nanoceramics, leading to a hemolytic effect on human beings [22].

In in vitro testing of nanoceramics, when the concentration of nanoceramics was lower there were no cytotoxic effects shown in living organisms but the higher concentration of nanoceramics led to the reduction of cell viability in biological organisms, hence the adverse effect is dependent on the concentration of the nanoscaled ceramic materials. There are chitin-based NCs, such as PUBNCs (polyurethane bio NCs), normally used as nanoceramic materials for packaging food substances, which are mainly made using nanoclay, especially bentonite, as it is rich montmorillonite, that result in the increase of toxicity to the cells through damaging the DNA of the liver cell line and through an increase of concentration beyond 88 ng/mL after 48 hours of exposure [23].

A study carried out to detect the effect of TiO_2 nanoceramic concentration on water buffalo (*Bubalus bubalis*), showed a decrease in cell viability and membrane integrity after 6 hours of incubation with NPs. The functionality of spermatozoa and DNA integrity was also affected as TiO_2 nanoceramic accumulates in the head and plasma of buffalo spermatozoa causing a high amount of DNA fragmentations in the sperm of such organisms [24].

22.3.5 Toxicity on transgenic *Drosophila*

In recent studies, the toxicity of Graphene zinc oxide NC (GZNC) was evaluated on transgenic *Drosophila melanogaster* (*hsp70-lacZ*) Bg^9 in its third instar of larvae by exposing it to nearly 0.199 and 3.996 μg/μL for 24 and 48 h, which resulted in the tissue damage and DNA damage. But when the range of GZNC exposure was lower than the above mentioned range it did not show any potential toxic effect on the organisms [25].

22.3.6 Challenges in purifying blood from virus

There is a risk of association of blood and blood products with contaminated viruses like HIV, Hepatitis, etc. A potential application of nanoceramics, such as NPs, nanofibers, nanopores, or membranes, is the removal of viruses from infected blood. Some viruses freely move in the blood and infect the tissue and other cells so small-sized nanoporous membranes are used to remove the viruses via filtration. However, there are disadvantages such as when nanopores are produced, there is the risk of crack formation, which can be then rectified by nanofibers. But these nanoporous membranes are for single-time use and need to be prepared again for the second time, and so on. Nanoporous membranes also tend to clog due to protein aggregation so it is a challenging task to rectify and purify the blood from viruses [26].

22.3.7 Hazards of nanoceramic-coated vessels

The materials used as food preparation vessels and storage often undergo hard or stressful conditions during their lifetime, for example, scraping and abrasive pads are used for cleaning. Many of the food preparation vessels are made of nanoceramic NPs or ceramic-coated materials, where these potential NPs migrate toward food substances and cause ill effects. In recent studies, researchers evaluated the NP migration under three different consumer conditions, such as washing using scrub pads, scouring with the help of steel wool, and scratching using tungsten carbide burr attachments. Three different nanoceramic particles were identified: aluminum (Al), silicon (Si), and titanium (Ti). Generally, the titanium concentration and silicon concentration were higher in comparison with the nonused pan after various usage methods by the consumer. Tungsten carbide burr attachment helps detect TiO_2 and SiO_2 NPs by its most aggressive use showing visible damage to surface metals and ceramic parts through glazing exposure. TiO_2 and SiO_2 particle number concentrations were 10^8 and 10^7 particles dm^{-2}, with median diameters of 250 nm and 460 nm, respectively, suggesting that particle migration mainly depends on matrix degradation with the abrasion process mechanically. The concentration of aluminum (Al) migration was higher after use compared to unused pans. Thus the migration of these NPs from the nanoceramic-coated food preparation materials may lead to the entry of these nanoceramic NPs into the food and ultimately into the human intestine, causing colon cancer and digestive disruptions [27].

22.4 Legal issues

Over the past years NPs have been considered to be Occupational Hazardous substances that arise from the nanoceramic industries [6]. The European Union ceramic industry produces 23% of nanoceramic materials worldwide [10]. The growth and development of nanotechology industries, with an estimated population of 6 million workers in 2020 has raised concern on the adverse health effect and environmental damage with regard to exposure of NPs such as ENPs and ANPs [28]. These industries produce nanoceramic NPs legally for commercial purposes but there is no proper maintenance of occupational ENPs from such industries [29].

Government and some private sectors are investing in the growth of nanotechnology, thus increasing the commercial value in the market, but most of the hybrid nanoceramics production methods are cost-effective so these investments are restricted and are becoming less. Nanocoating nanoceramic materials are used in glass panels for building in Western Europe where *U* values are given to differentiate the application of thin multilayer nanocoatings. These *U* values differ among common houses and passive houses, for example, one side nanoceramic coated glass panels have a *U* value of 6 W/m^2. For long-time use double side nanoceramic coated glass panels are preferred with *U* values of 3 W/m^2. Now, highly efficient glass panels in the newly constructed buildings may have a *U* value of 1.1 W/m^2. Panels with 0.6 W/m^2 K measured *U* value used for passive houses are already on the market. This type of *U*-valued glass panel around the houses discriminates the society with economic status and cost-effectiveness. Only rich people can afford these nanocoated nanoceramic glass panels [30].

The complete implementation of nanomaterials and nanoceramics in biological and biomedical applications must be investigated deeply by giving priority to humans as well as to the other organisms of the surrounding environment. Hence nanoceramics and their associated toxicity are poorly understood and studied by various researchers and they are yet to improve the biocompatibility of multifunctional composites and ceramics [23].

22.5 Safety issues

Nanoceramics are usually strong and provide resistance against compression and bending, maintaining strength even at high temperatures, but their brittleness is the biggest technical barrier that prevents a practical application, for example, a load-bearing application. When constructed from nanoparticles, ceramics can be superplastically deformed at a modest temperature before being heat-treated at a higher temperature for high-temperature strengthening. Densification is one of the major disadvantageous processes that occur during the application of pressure when nanoceramics are produced; an outside layer of the particle densifies into a hard impervious shell which constrains the inside sample from shrinking, thus resulting in cracking with strain incompatibility [31].

NPs obtained from ceramic industries do not pose any harm to human beings unless they are inhaled into the body [6].

Hyaluronic acid NPs are widely used for clinical purposes, as they are a CaP bioceramic material. The dose range of these particles at lower than 26 mg/kg can be considered as safe and possess less cytotoxicity but when it increases beyond 26 mg/kg then it becomes a toxic material to human health [18].

The US Environmental Protection Agency (EPA) is working on the ecotoxicology and legislation of nanoceramics together with the Food and Drug Administration. Previously the EPA encouraged nanomaterials producers and manufacturers to provide information in a program of nanoscale materials, but later the EPA worked with other US agencies to collect the information about the manufacturing, usage, and disposal of such nanomaterials to allow the safer use of nanomaterials [32].

Under Toxic Substances Control Act norms, it is essential to test the materials before they enter the market because the use of chemicals before packing may possess a risk to human health and the environment [23].

EPA norms have made it compulsory to give detailed information about the production of nanomaterials, that is, the manufacturing method, the volume of production, and how does it is exposed. EPA is mainly focused on carbon nanotubes, such as single-walled and multiwalled carbon nanotubes, and when focusing on nanoscale materials like clays and silica, it details some test rules for commercial use [23].

Some agencies and organization working for the safety of nanoscale nanomaterials in the United States include the Consumer Product Safety Commission as an essential commission working on the safety of nanomaterials, Occupational Safety and Health Administration, and National Institute for Occupational Safety and Health. In Europe, there are some organizations that control the nanosafety measures, including registering, evaluating, authorizing, and restricting the chemical use and also classifying, labeling, and packing. In Japan, the control organizations with regard to nanosafety measures include Ministry of Economy, Trade and Industry (METI), National Institute of Advanced Industrial Science and Technology. The National Nanotechnology Center (NANOTEC) in Thailand provides "Nano Q" as a specific nanomark for selected Thai nanoproducts to identify nanoparticle types, sizes, and properties. Australia is also involved in the controlled measure of nanosafety via an assessment scheme called the National Industrial Chemicals Notification and Assessment Scheme. While checking nanomaterial biosafety, there are some international agencies and other organizations that do contribute toward safety measures, for example, the Working Party on Manufactured Nanomaterials under the auspices of for the Economic Co-operation and Development, the WHO, and also the American Society for Testing Materials where the International Agency for Research on Cancer operates [23].

22.6 Conclusion

From the above it is evident that nanoceramics have a negative side though they have various applications. Hence there is a need to study the adverse effects of these materials in all aspects of application. Continuous monitoring of nanoceramic materials and the manufacturing industries is critical to determine toxicity on the biota and environment. Guidelines and protocols should be implemented on the usage and manufacturing of these ceramic materials. Further research is needed to overcome the problems caused by this material.

References

[1] S.A. Pauline, Nanomaterials and their biomedical applications, Springer Series in Biomaterials Science and Engineering 16, Springer (2021), pp: 131–133.

[2] A.K. Mishra, Sol-gel based nanoceramic materials: preparation, Properties and Applications (eBook), Springer International Publishing, 2017.

[3] E. Vunain, S.B. Mishra, A.K. Mishra, B.B. Mamba, Nanoceramics: Fundamentals and Advanced Perspectives, Springer International Publishing AG, 2017.

[4] J. Zhao, V. Castranova, Toxicology of nanomaterials used in nanomedicine, J. Toxicol. Environ. Health Part. B 14 (8) (2011) 593–632.

[5] G. Svartz, C. Aronzon, S. Ṕerez Cat́an, S. Soloneski S, C. Ṕerez Coll, Oxidative stress and genotoxicity in *Rhinella arenarum* (Anura: Bufonidae) tadpoles after acute exposure to Ni-Al nanoceramics, Environ. Toxicol. Pharmacol. 80 (2020).

[6] M.J. Bessa, F. Brandao, M. Viana, J.F. Gomes, E. Monfort, F.R. Casseeh, et al., Nanoparticle exposure and hazard in the ceramic industry: an overview of potential sources, toxicity and health effects, Environ. Res. 184 (2020).

[7] G. Svartz, M. Papa, M. Gosatti, M. Jordan, A. Soldati, P. Samter, et al., Monitoring the ecotoxicity of γ-Al_2O_3 and Ni/γ-Al_2O_3 nanomaterials by means of a battery of bioassays, Ecotoxicol. Environ. Saf. (2017) 0147–6513.

[8] R.A. Baan, Carcinogenic hazards from inhaled carbon black, titanium dioxide, and talc not containing asbestos or asbestiform fibers: recent evaluations by an IARC Monographs Working Group, Inhal. Toxicol. 19 (2007) 213–228.

[9] M. Bozsin, Environmental challenges for the ceramic industry, Ceramic Engineering and Science, Springer, 1974, pp. 93–101.

[10] European Commission, Reference Document on Best Available Techniques in the Ceramic Manufacturing Industry, 2007.

[11] J. Xu, T. Huang, H. Dong, D. Tan, Z. Feng, T. Wei, et al., Structural and mechanical properties of magnesium aluminate nanoceramics under high pressure, Appl. Phys. Lett. 117 (2020) 041901.

[12] S.L. Suib, A review of nanoceramic materials for use in ceramic matrix composites, in: A.K. Mishra (Ed.), Sol-gel Based Nanoceramic Materials: Preparation, Properties and Applications, © Springer International Publishing AG. 2017.

[13] A.B. Abou Hammad, A.M. El Nahwary, B.A. Hemdan, A.L. King Abia, Nanoceramics and novel functionalized silicate-based magnetic nanocomposites as substitutional disinfectants for water and wastewater purification, Environ. Sci. Pollut. Res. Int, 2020.

[14] S.H. Mahmoud, A.E. Embaby, A.M. Abdallah, Clinical performance of ormocer, nanofilled, and nanoceramic resin composites in class I and class II restorations: a three-year evaluation, Oper. Dent. 39 (1) (2014) 32–42.

[15] F.G. Ferré, A. Mairov, D. Iadicicco, M. Vanazzi, S. Bassini, M. Utili, et al., Fonzo, Corrosion and radiation resistant nanoceramic coatings for lead fast reactors, Corros. Sci. (2017).

[16] W.J. Weber, A. Navrotsky, S. Stefanovsky, E.R. Vance, E. Vernaz, Materials science of high-level nuclear waste immobilization, MRS Bull. 34 (2009).

[17] J.H. Choy, J.M. Oh, S.J. Choi, Nanoceramics-biomolecular conjugates for gene and drug delivery, Adv. Sci. Technol. 45 (2006).

[18] Y. Hong, H. Fan, B. Li, B. Guo, X. Zhang, Fabrication, biological effects, and medical applications of calcium phosphate nanoceramics, Mater. Sci. Eng.: R Rep. 70 (2010).

[19] H. Wu, L. Xei, M. He, R. Zhang, Y. Tian, S. Liu, et al., A wear-resistant TiO_2 nanoceramic coating on titanium implants for visible-light photocatalytic removal of organic residues, Acta Biomater. 97 (2019) 597−598.
[20] A. Obata, T. Kasuga, Chapter 2: Silica-based nanoceramics, Nanobioceramics for Healthcare Applications (2016).
[21] S.J. Kalita, A. Bhardwaja, H.A. Bhatta, Nanocrystalline calcium phosphate ceramics in biomedical engineering, Mater. Sci. Eng. C. 27 (2007) 441−449.
[22] I. Tsamesidis, K. Kazeli, E. Lyperaki, G.K. Pouroutzidou, I.M. Oikonomou, P. Komninou, et al., Effect of sintering temperature of bioactive glass nanoceramics on the hemolytic activity and oxidative stress biomarkers in erythrocytes, biomedical engineering society − cellular and molecular, Bioengineering (2020).
[23] S. Sanand, A. Hussain, G. Kaul, Chapter 22: Biosafety of nanomaterials used in nanoceramics and nanocomposites, Handbook of Nanoceramic and Nanocomposite Coatings and Materials (2015) 476−480.
[24] K. Pawar, G. Kaul, Toxicity of titanium oxide nanoparticles cause functionality and DNA damage in buffalo (*Bubalus bubalis*) sperm in vitro, Toxicol. Ind. Health (2012).
[25] Y.H. Siddique, K. Wasi, K. Saba, J. Smita, N. Falaq, R., R.S. Braj, et al., Toxic potential of synthesized graphene zinc oxide nanocomposite in the third instar larvae of transgenic *Drosophila melanogaster* (hsp70-lacZ) Bg9, Toxic. Nanomater. (2014). Volume 2014 |Article ID 382124.
[26] Y. Zhao, S. Sugiyama, T. Miller, X. Miao, Nanoceramics for blood-borne virus removal, Expert. Rev. Med. Devices 5 (3) (2008) 395−398.
[27] S.A. Ntim, S. Norris, K. Scott, T.A. Thomas, G.O. Noonan, Consumer use effects on nanoparticle release from commercially available ceramic cookware, Food Control. 87 (2017) 31−39.
[28] M.C. Roco, The Long View of Nanotechnology Development: the National Nanotechnology Initiative at 10 Years, Springer, 2011.
[29] P.I. Dolez, M. Debia, Overview of workplace exposure to nanomaterials, Nanoengineering (2015) 428.
[30] O. Van der Biest, Nanoceramics: issues and opportunities, Int. J. Appl. Ceram. Technol. (2013) (2013) 1−12.
[31] K.T. Smith, Azonano Tech, 2019. http://www.azananon.com.
[32] M.N. Duvall, A.M. Wyatt, Regulation of Nanotechnology and Nanomaterials at EPA and Around the World: Recent Developments and Context, Beveridge & Diamond, Washington, DC (USA), 2011, pp. 1−6.

CHAPTER 23

Nanomaterial as an emerging green catalyst in environmental remediation

Md. Ahmaruzzaman, Gaurav Yadav and Th. Babita Devi
Department of Chemistry, National Institute of Technology Silchar, Silchar, Assam, India

23.1 Introduction

An increase in urbanization, industrialization, and the fast-growing population has caused severe issues related to the environment. Environmental pollution is one of the crucial issues that we are facing today. Various new techniques are being explored for the environmental remediation of contamination in water, air, and soil [1]. Heavy metals, fertilizers, toxic gases, dyes, pesticides, herbicides, particulate matter, sewage, industry effluents, and many other compounds cause environmental contamination [2,3]. Remediation of contamination is quite challenging due to the mixing of the different compounds, their low reactivity, and their high volatility. This chapter discusses the use of nanoparticles (NPs) in the development of environment remediation techniques [4].

There is urgent need for technology that is clean, nontoxic, biocompatible, and environment-friendly. Nanotechnology provides promising solutions to such environmental issues due to the various properties of NPs, like their morphology (shape and size), microstructure (hollow, sphere, tube, porous, belt), composition, and other chemical properties (catalytic, magnetic, optical, electrochemical).

Nowadays, carbon-based nanostructured substances like carbon nanotubes (CNTs), and graphene have gained attention due to their mechanical, electrical, and chemical properties. Carbon-based NPs and graphene efficiently adsorb organic solvents and oils. Carbon aerogels which are obtained from resorcinol-formaldehyde resin and graphene show high-performance adsorption, catalytic applications, and capacitive deionization. CNTs are used in water treatment. Photocatalytic degradation of various pollutants is also an effective method for environment remediation, such as air purification [5], water cleansing [6] and soil remediation [7]. Photocatalysis helps in the degradation of toxic compounds like pesticides, dyes, harmful gases, volatile organic compound, and bacteria [7–12]. Many materials like zinc oxide, titanium oxide, metal sulfide, and iron oxide are produced with different techniques including hydrothermal methods, sol–gel method, and chemical deposition method [13–16]. Green nanotechnology has gained

Industrial Applications of Nanoceramics
DOI: https://doi.org/10.1016/B978-0-323-88654-3.00028-7

intense attention due to its environment benefits. This technology helps in the reduction of pollutants without causing harmful effects on the environment or human health. Nanotechnology helps to obtain the desired materials with low cost and toxicity. Some key challenges that must be consider during the synthesis of new NPs are cost-effectiveness, nontoxicity, target capture, recyclability, biodegradability, and potential to recover them after use for environment remediation.

This chapter aims to facilitate the advancement of recent nanotechnology used for environment remediation of various type of pollutants. By the help of various possible ways, we can remove the pollutants from the environment, for example, by adsorption, photocatalysis, chemical reactions, and filtration [1,17–22]. The various different types of materials are used for the approaches given in Fig. 23.1

23.2 Green synthesis of nanoparticles

The synthesis of nanomaterials by various types of physical and chemical methods is not so good and causes harmful effect on humans and the environment. In recent

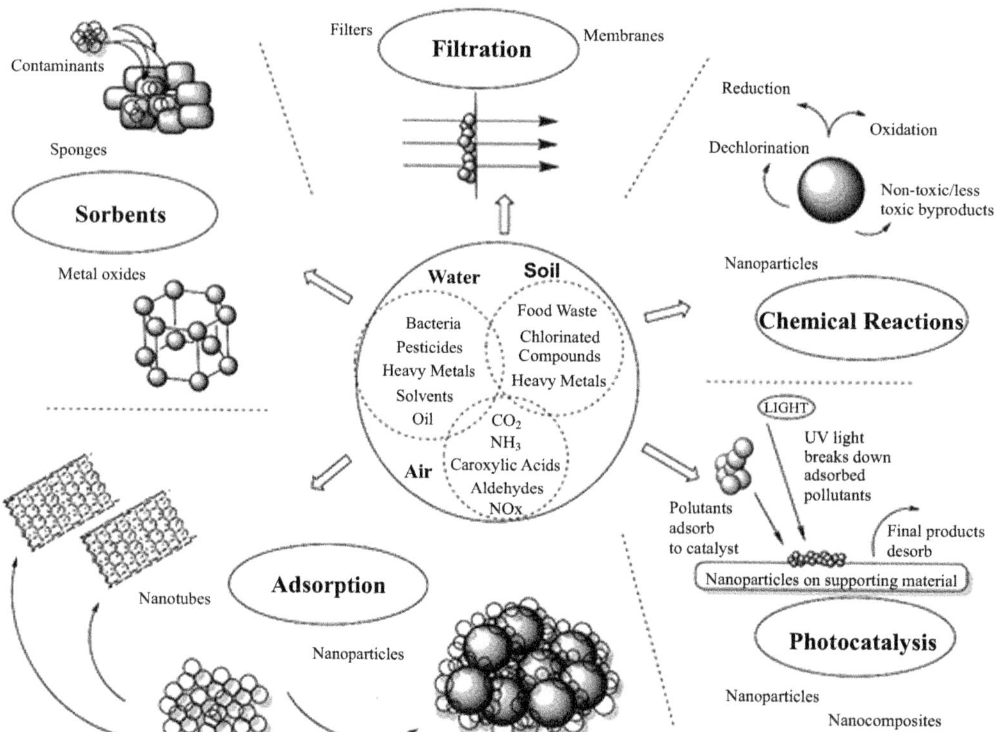

Figure 23.1 Environment remediation approach.

years, green synthesis has gained intense attention for the preparation of various types of materials. Green synthesis of NPs produced for cleanup, control, and remediation will help environment. The main aim of green synthesis is to avoid the manufacture of harmful and unwanted by-products through the development of eco-friendly synthesis. The green synthesis of nanomaterials involves the use of fungi, yeast, bacteria, algae, plants, and plant extracts. The green synthesis process is simpler, economical, more efficient, and can easily perform large operations. Different nanomaterials like palladium, gold, silver, iron, and ZnO are synthesized with the help of green methods (Table 23.1). The generated nanomaterials possess various properties like catalytic activity, antimicrobial action, and antioxidant behavior.

23.3 Carbon-based nanoparticles

The use of activated carbon for environment remediation has benefits over other conventional materials. This is due to the large surface area and other physical and chemical properties. The benefits of carbon-based NPs for environment remediation are due to (1) excellent capacity of adsorption of pollutants; (2) great surface area; and (3) fast kinetics. Activated carbon yields various structural arrangements, like SWCNT, MWCNT, fullerenes C_{60}, fullerenes C_{540}, and graphene [19]. Single-walled CNTs (SWCNT) and multiwalled CNTs (MWCNT) have been studied for various environment remediation techniques. A new adsorption technique uses CNT clusters. This helps in the removal of bacteria through the adsorption process in water. Adsorption is one of the best methods to remove contamination from wastewater. Carbon black

Table 23.1 Various nanomaterials synthesized by using green process.

Serial number	Species	Size (nm)	Nanoparticle obtained	References
1	*Trichosporon montevideense*	12–53	Gold	[23]
2	*Eucalyptus leaf*	95 ± 5	Iron	[24]
3	*Reetha* and *Shikakai*	30	Silver	[25]
4	*Rosa canina*	>50	Zinc oxide	[26]
5	*Vibrio natriegens*	136 ± 31	Selenium	[27]
6	*Phaseolus lunatus*	18	SnO_2	[28]
7	*Coccinia grandis*	40–50	Copper	[29]
8	*Alchornea laxiflora*	2.12 ± 0.21	Bimetallic Pt/Cu	[30]
9	*Saccharimyces cerevisae* broth	4–5	Gold, silver	[31]
10	*Verticillium*	21–25	Silver	[32]
11	*Rhizopus nigricans*	35–40	Silver	[33]
12	*Desulfovibrio desulfuricans*	20–50	Gold	[34]

was modified via oxidization with 65% HNO_3 by refluxing 10 g black carbon with 65% of HNO_3 in 120 minutes at 110oC. This helps to increase the adsorption of Cd^{2+} and Cu^{2+} due to the increase in the amount of functional groups. Adsorption of Cd^{2+} and Cu^{2+} increases with the increase in pH. Metal other than Cd^{2+} and Cu^{2+} are also efficiently adsorbed by modified carbon black above the pH level of 5.5. Both Cu^{2+} and Cd^{2+} are pH dependent so the adsorption of these ions is less at low pH because of competition between H^+ and Cd^{2+} or Cu^{2+} for adsorption sites. As the pH level increases the concentration of H^+ decreases and Cd^{2+} and Cu^{2+} ion adsorption increases. Zhou et al. observed the adsorption of Cd(II), Cu(II), Pd(II), and Zn(II) on nanoscale carbon black and hydroxyapatite [35]. The smaller particle of carbon nanoscale has a greater tendency for adsorption than larger activated carbon and carbon black because of the internal surface's micropores.

Carbon NPs can also help in environment remediation through a photocatalytic technique. In the presence of UV radiation, the energy of photons will have be equal or greater than the bandgap of nanotubes, thus promoting the conduction band electrons and valence band holes. The holes are the cause of the manufacture of hydroxyl radicals that help in the oxidation of chlorine containing organic compounds. Fig. 23.2 shows the systematic representation of environmental remediation by a photocatalytic process.

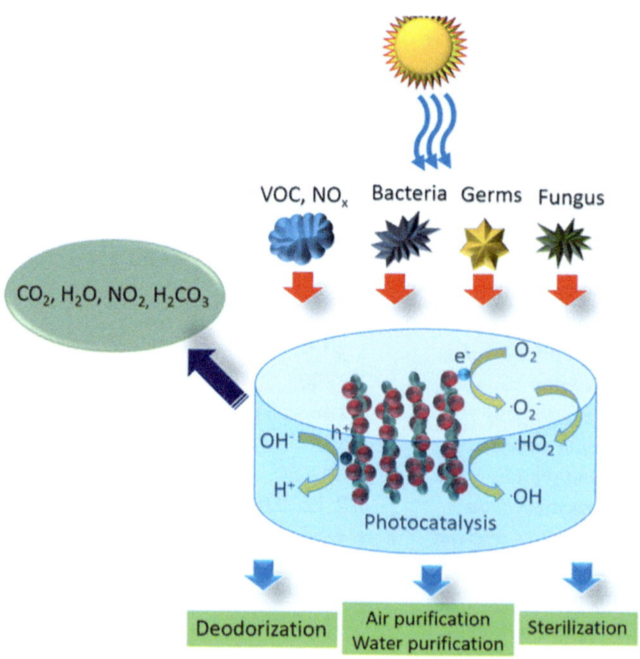

Figure 23.2 Systematic representation of environment remediation by photocatalytic approach.

23.4 Carbon nanotubes

CNTs are microcrystals of graphite which have a cylindrical hollow surface. There are mainly two forms: (1) MWCNT, and (2) SWCNT. MWCNTs are constructed with concentric cylinders where the spacing between adjacent layers is 3.4 Å. CNTs have attracted several researchers due to the great efficiency of adsorption and various potential applications in environment remediation. They are used widely for the removal of pollutants like organic compounds, heavy metals, pharmaceuticals, endocrine-disrupting chemicals [36], arsenic compounds, and other personal care products.

Another factor that affects the adsorption of CNTs is oxygen content. Different chemicals like H_2O_2, $KMnO_4$, KOH, HNO_3, NaOCl, NaOH, and H_2SO_4 are used to oxidize CNTs. The increase in the efficiency of adsorption of Ni(II), Pb(II), Cu(II), and Cd (II) was observed when CNTs oxidized with HNO_3 [37]. The adsorption of dyes like methylene blue (MB) and methyl violet is also possible with the help of CNTs. This is due to the increase in the pore surface volume and surface area of CNTs. Due to this, CNTs show exceptionally high adsorption efficiency for various organic contaminants, like ethyl benzene [38], 1,2-dichlorobenzene [39], and benzene. Peng et al. found that in less than 40 minutes there is maximum adsorption (9.98 mg/g) of 1,2-dichlorobenzene by CNT [40] when the initial concentration is 10 mg/L. Ethyl benzene adsorption depends upon the porosity and chemical nature of CNTs. SWCNTs show more adsorption of ethyl benzene than MWCNTs due to electrostatic attraction between the ethyl benzene and MWCNTs [40].

Li et al. [41] observed the number of cationic dyes adsorbed increased with the pH because of the electrostatic attraction between the positive charge dyes and CNTs' surface. Many CNTs adsorbates' properties play influential roles in the adsorption process, like electric dipole moment, molecular weight, and critical temperature [42].

SWCNTs and MWCNTs also play an important role in the adsorption of Zn^{2+} ions. With the increase in temperature, the quantity of adsorption of Zn^{2+} increases [43]. CNTs can be reapplied multiple times for the treatment of water. The treatment of water is the major complication but with the help of CNT we can remove the metal ion carrier. MWCNT with TiO_2 composite is used as a photocatalyst for the treatment of water. A small quantity of CNT in TiO_2 can increase the effect of TiO_2. The applications of CNTs nanocomposites for environment remediation are listed in Table 23.2.

23.5 Graphene materials

Pristine graphene and its modifications are used for environment remediation. Li et al. observed the adsorption of fluoride with the help of pristine graphene in aqueous solution [56]. Graphene oxide is widely used for the adsorption of various

Table 23.2 Applications of carbon nanotubes (CNTs) nanocomposites for environment remediation.

Material in nanocomposite	Target materials	Benefits	References
CNT with Ag and Cu nanoparticles	Elimination of *Staphylococcus aureus* and *Escherichia coli* from waste water	The increment is due to large surface area	[44,45]
Nanocarbon colloids	Elimination of Cu^{2+}, Cd^{2+}, Zn^{2+}, Pb^{2+}, Ni^{2+}, Hg^{2+}, Cr^{6+} from contaminated water	With the formation of filtration nanocomposites, and the coagulation metal ions detain other contaminated material	[46]
CNT with iron oxide	Elimination of Ni^{2+}, Sr^{2+}, and Co^{2+} from water	Increase in the adsorption efficiency and surface area	[47]
CNT with TiO_2 and P-25	Elimination of phenol and its derivatives, dyes, and metals from water	With the help of band-gap modification, visible light catalysis and deceleration of electron-hole recombination	[48–55]

contaminants in water, such as H_2S, NH_3, SO_x, pesticides, heavy metals, and volatile organic compounds. Oil cleansing from water is one of the biggest challenges. Currently, many procedures like adsorption, skimming, are biodegradation are used [57]. The rGO/CNT aerogels possess super high adsorption capability and fast rate of absorption of oils and organic solvents. Porous graphene oxide has high absorption efficiency for heavy metals [58]. The adsorption process on GO of metals ion is dependent upon pH. To increase the efficiency of GO, many other materials are added into GO to form composite material like chitosan–gelatin so they possess high adsorption efficiency [59,60]. It can be easily recycled many times due to its stability. CNT is also used to enhance the capacity of GO to eliminate metal ions from water. In a recent study it was found that bulky graphene material helps in the adsorption of dyes with high efficiency and great recycling ability [61–64]. CDS–graphene and ZnO–graphene composites show photocatalytic efficiency towards contaminants of water. With the addition of graphene in ZnO, the elimination rate of Cr^{6+} increased 40% in the presence of UV radiation [65]. Applications of graphene materials are displayed in Table 23.3.

Table 23.3 Application of graphene materials.

Materials	Application	References
Graphene oxide	Gas/water—NH_3, H_2, SO_3, pesticides, heavy metals, pharmaceuticals	[66–70]
TiO_2 graphene	Benzene removal	[71,72]
Pristine graphene	Fluoride removal in water	[56]
ZnO-Graphene and CdS-graphene	Heavy metals in water	[73]

23.6 Polymer-based nanoparticles

Polymers are widely used for the elimination and detection of contaminants present in water, for example, manganese, arsenic, iron, nitrate, heavy metals arsenic, and organic pollutants (aromatic and aliphatic hydrocarbon, volatile organic compounds, pharmaceuticals). Polymer hosts like emulsifiers, surfactants, surface-functionalized materials, and stabilizing materials are used to increase the stability and overcome the limitations of other NPs and provide other beneficial properties like thermal stability, mechanical strength, recyclability, and durability to the materials.

The polymer substrates are generally fluoropolymers and saturated carbon chains such as PDMS (polydimethylsiloxane), PE (polyethylene), PVP (polyvinylpyrrilodone), PANI (polyaniline), P3HT (poly 3-hexylthiophene), PP (polypropylene), and Nafion. PNA/TiO_2 was prepared by Ameen et al. who used an in situ mechanism to increase the photocatalytic activity to remove MB in the presence of visible light. APU (amphiphilic polyurethane) is used to degrade polynuclear aromatic hydrocarbons like phenanthrene from contaminated soils. PAMAM (dendrimers) and polyamidoamine have been used for sewage water remediation of metal ion contamination. The dendritic nanopolymers have functional groups like hydroxamates, carboxylates, and primary amines, which are able to remove a variety of solutes present in water like Au^+, Ag^+, Cu^{2+}, Fe^{2+}, Ni^{2+}, Fe^{3+}, U^{6+}, and Zn^{2+} [74]. Several metals and metals ion like Fe, Zn, Cu, Fe/Ni, Fe/Pd, and Pd/Zn are used for the degradation of various contaminants like brominated methane, chlorinated ethene, chlorinated methane, trihalomethane, chlorinated benzene, dyes, and pesticides; they can be loaded on a polymer matrix to increase its efficiency. Water pollution is the major problem that we face. Some heavy metals like mercury present in water is harmful for living beings. Sumesh et al. [75] observed that water-soluble silver NPs with a core diameter of 20 ± 5 and 9 ± 5 nm secured by MSA (mercaptosuccinic acid) carried on alumina could be used to remove mercurate ion from waste water. The rate of removal of mercury is the maximum at pH 5–6. At higher pH, there is a formation of mercury hydroxo complexes which are stable and do not interact with the surface of the nanocomposite. Lisha et al. [76] observed the gold NP established on alumina matrix is used to eliminate inorganic mercury from wastewater. Tong et al. investigated iron NPs immobilized in PEG/Nylon 66 to eliminate nitrobenzene. Many polymer-supported NPs have been investigated and some are listed in Table 23.4.

23.7 Metal and metal oxide-based nanoparticles

Metal and metal oxides are the most widely studied NPs for environmental remediation. The shape and size of metal and metal oxide NPs are the key factors affecting their performance. Metal and metal oxide NPs are synthetically prepared by two methods: (1) physical procedures, including severe plastic deformation, ultrasound shot peening, high-energy ball milling, and inert gas condensation; and (2) chemical procedures, including chemical vapor condensation, liquid phase reduction, reverse micelle, liquid flame spray, and gas phase reduction. Of these, thermal decomposition, coprecipitation, and hydrothermal synthesis are mostly used with high yield.

Table 23.4 Polymer-supported nanoparticles (NPs) for removal of various pollutants.

Materials	Target pollutant	References
Polyacrylic acid/polyvinyl acid	Trichloroethylene	[77]
PAMAM dendrimers	Removal of heavy metals	[74]
Cu on chitosan matrix	Cr^{6+}	[78]
Amphiphilic polyurethane NPs	Aromatic hydrocarbons	[79]
Hydrated ferric oxide on cation exchange	Cd(II), Pb(II), Zn(II)	[80]
Fe_2O_3 on polymeric anion exchangers	Cd(II), Cu(II), Pb(II)	[81]
Fe_3O_4 on alginate matrix	Methyl orange, methylene blue	[82]
Polyamine modified cellulose	Volatile organic pollutants	[83]
Magnetite montmorillonite	Cr^{6+}	[84]
Fe^0 with polyvinyl pyrrolidone	Bromate ion	[85]
TiO_2 with PANI	Phenol	[86]
TiO_2 on P3HT	Methylene orange	[87]
Fe^0 with polystyrene-divinylbenzene	Nitrate	[88]
Chlorinated methane, chlorinated benzene, trihalomethanes, chlorinated ethene,		
Pesticides—DDT, Lindane, TNT, Pentachlorophenol		
Organic dyes—chrysoidin ($C_{12}H_{13}ClN_4$), orange II ($C_{16}H_{11}N_2NaO_4S$), trapaeolin O ($C_{12}H_9N_2NaO_5S$)		
Heavy metals—chromium, lead, nickel, mercury, cadmium		
Inorganic anions—nitrates, perchlorate		

23.8 Silver nanoparticles

Silver NPs are widely used for their antibacterial, antiviral, and antifungal properties. Ag groups assist in the inactivation of respiratory enzymes and inhibit the production of ROS (reactive oxygen species) [89]. Silver ions can also stop DNA replication and affect the permeability and structure of cell membranes [90]. Particles of size smaller than 10 nm are generally more toxic to the bacteria like *Pseudomonas aeruginosa* and *Escherichia coli* [91,92]. Silver nanomaterials also inhibit the binding of viruses. Silver NPs incorporated with polymer matrix like poly(L-lactic acid)-*co*-poly(ε-caprolactone) and polymethoxybenzyl have also antibacterial property for *Aspergillus niger*, *Salmonella entrica*, *Staphylococcus aureus*, and *E. coli* [93–96]. Fig. 23.3 shows the various factors affecting the function of Ag and their mechanisms of action for antibacterial, antiviral, and antifungal properties.

23.9 Nano zerovalent iron

Elemental iron with zero oxidation state (Fe^0) is termed zerovalent iron (ZVI). Heavy metals and chlorinated hydrocarbons are adsorbed or degraded with the help

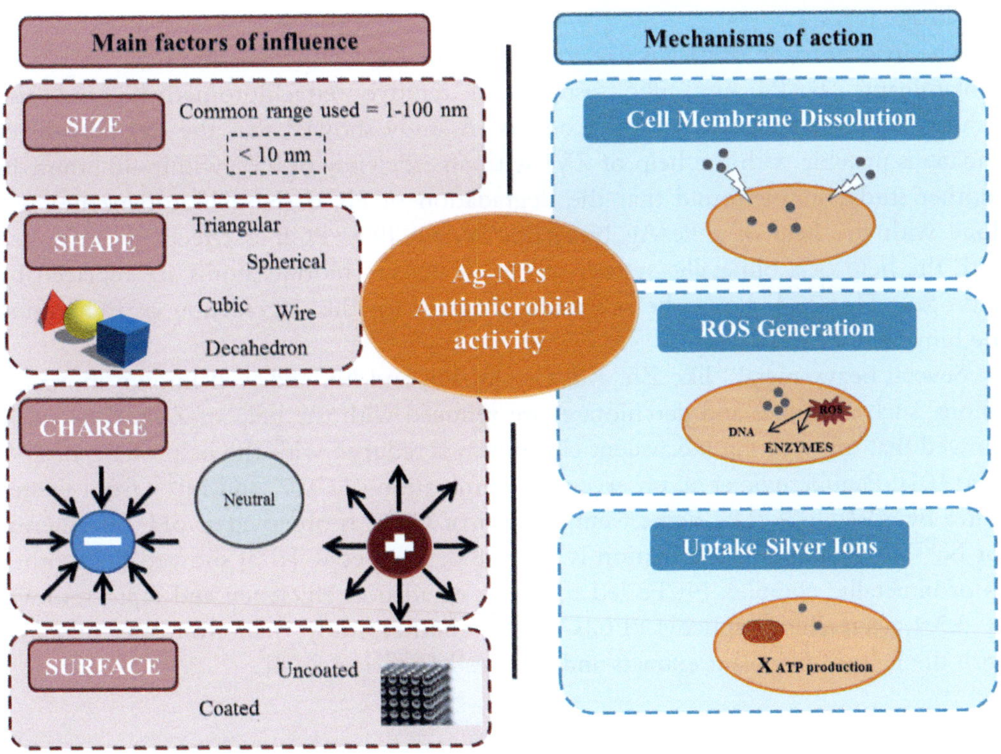

Figure 23.3 Factors affecting the function of Ag and their mechanism of action.

of zerovalent iron. Nano zerovalent iron (nZVI) effectively helps in the degradation of different organic contaminants like organochlorines, organic dyes, chlorinated solvents, and polychlorinated biphenyls by electron donating. Firstly, iron donates the electron and the hydrocarbon accepts the electron followed by reductive dechlorination [97,98].

$$Fe^0 \rightarrow Fe^{2+} + 2e^-$$

$$RCl + H^+ + 2e^- \rightarrow RH + Cl^-$$

Dechlorination of tetrachloroethene is done with the help of ZVI. More than 99% of C_2Cl_4 is converted into ethane in 24 hours.

$$C_2Cl_4 + 4Fe^0 + 4\,H^+ \rightarrow C_2H_4 + 4Fe^{2+} + 4Cl^-$$

Other compounds including TCE (trichloroethylene), PCE (tetrachloroethylene), t-DCE (trans-dichloroethene), and tetrachloromethane were observed. All these compounds are effectively degraded with the help of ZVI and more than 95% degradation efficiency is observed when the initial concentration of compounds are 10 mg/L and the dosage is 5 g/L.

The incorporation of metals like Pt, Pd, Ni, or Ag can increase the removal rate of contaminants. Fe/Pd bimetallic system can remove tetrachloromethane and also reduce the formation of toxic by-products. A study showed that the degradation of lindane is possible with the help of ZVI with an efficiency of 95% within 40 hours. In another study, it was found that the degradation of HCB (hexachlorobenzene) was done with the help of a Fe/Ag bimetallic system [99]. PCE is effectively degraded with the help of a bimetallic system of Fe/Ni where chlorine atoms are replaced by hydrogen. Fig. 23.4 shows the schematic diagram for the degradation of PCE using the bimetallic Fe/Ni system.

Several heavy metals, like Zn, Ag, Cr, As, Pb, and Ni, and well-known inorganic anions, such as nitrate and perchlorate, are reduced with the help of ZVI. One study showed that 65–110 mg hexavalent chromium is reduced with the help of 1 g of ZVI [100,101]. Paguberovic et al. observed the elimination of Cu^{2+} and Ni^{2+} from wastewater by ZVI [102]. The highest elimination of Cu^{2+} is observed at pH 7.0 whereas for Ni^{2+} the maximum elimination is at pH 8.0. Ding et al. [103] showed the synthesis of bimetallic complex Ni/Fe led to high degradation efficiency and rapid removal of 3,3,4,4-tetrachlorobiphenyl (PCB77). In another study, nanoiron was prepared with the help of five plant extracts and removed Cr^{6+} ion [104].

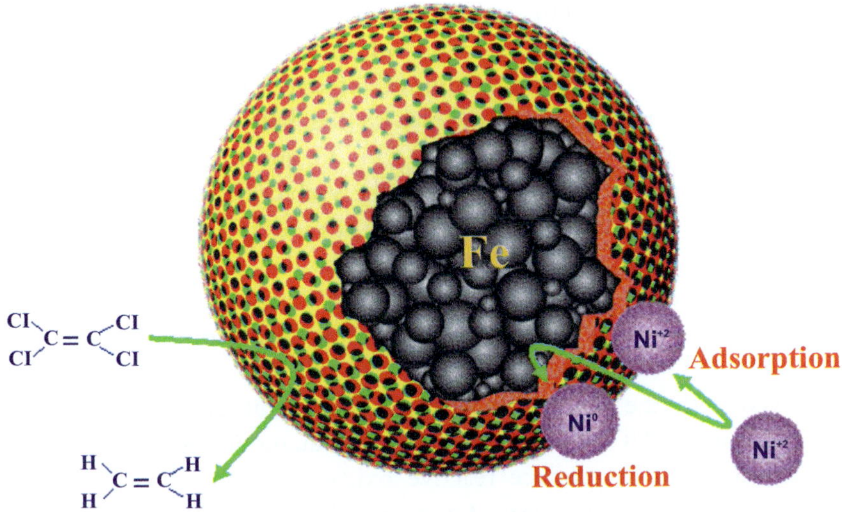

Figure 23.4 Degradation of tetrachloroethylene using bimetallic system of Fe/Ni.

23.10 TiO_2 photocatalyst

TiO_2 nanomaterials are used for reductive and oxidative transformation of inorganic and organic contaminants in water. TiO_2 NPs have been successfully used for water purification, self-cleaning surfaces, and water remediation. SO_2 is continuously released into the environment through power plants, factories, automobiles, etc. The addition of SO_2 to water helps in formation of acid rain which causes corrosion. TiO_2 is used to convert sulfur dioxide into elemental sulfur.

$$SO_2 + 2H_2S \rightarrow 2H_2O + 3S_{solid}$$

TiO_2 is used as a photocatalytic for the removal of water contamination. Photocatalytic activity can be increased with the help of a polymer matrix and metal ions. TiO_2 is used for photolytic degradation of various organic compounds like organochlorine pesticides and triazines, and also helps in the adsorption of heavy metal ions such as Zn, Cu, Cd, Ni, and Pb from water [105].

TiO_2 photocatalyst has a bandgap of 3.2 eV. The efficiency of the photocatalyst greatly depends on the adsorption of the contaminant. The adsorbative effectiveness of the surface is due to the film's properties like porosity, roughness, and surface area. The surface area of TiO_2 can be increased during the sol−gel process by assimilating a surfactant [106,107]. When an impurity is doped with a photocatalyst, the bandgap decreases, for example, nitrogen and fluorine-doped titanium oxide have a bandgap of

2.75. For doping, carbon, sulfur and other various rare earth metals and transition metals are used, including Cr, Fe, Co, Mn, Gd, Pr, Er, Sm, La, and Ce. Doped TiO_2 NPs and 1D TiO_2 are used for the removal of dyes and other halogenated compounds. TiO_2 nanomaterials are activated with the help of UV radiation. A recent study showed that doping of nonmetals like S, F, C, and N increases the efficiency of a photocatalyst for the degradation of dyes in the visible region. Skubal et al. observed the removal of aqueous cadmium with the help of modified bidentate chelating ligand on the surface of titanium dioxide NPs. TiO_2 is widely used for the elimination of various organisms like fungi, viruses, bacteria, and algae. With the help of N-doped TiO_2, we can remove 95% of AO7 within 1 hour and complete degradation was achieved in 2 hours [108] in the presence of UV radiation. The degradation of *E. coli* is also achieved with the help of N-doped titanium dioxide within 2 hours [108]. Various applications of TiO_2 for degradation of pollutants are listed in Table 23.5.

TiO_2 NPs are used to produce reactive oxidants, such as hydroxyl radicals, and serve as disinfectant for bacteria, yeast, fungi, and algae [115,116]. Ag-doped TiO_2 and nanofiber TiO_2 are used to degrade 2-chlorophenol in the presence of UV radiation. The rate of degradation of 2-chlorophenol with the help of Ag-doped TiO_2 is higher compared to TiO_2 nanofiber. Binary mixed oxide $TiO_2{-}SiO_2$ is used for the photodegradation of MB dyes. These types of mixed oxides allow the remediation of a wide range of contaminants.

23.11 ZnO nanoparticle

Nowadays, ZnO nanostructures are used for the removal of heavy metals [117] such as Cu^{2+} ions. The ZnO obtained by green synthesis (*Corymbia citriodora* leaves) [118] is used for the degradation of MB in the presence of UV radiation. Parthiban et al. [119] synthesized ZnO with the help of *Pyrus pyrifolia* leaves that was used to degrade more

Table 23.5 Applications of TiO_2 for removal of pollutants.

Types of nanoparticles	Contaminants	References
Ag-doped TiO_2	2,4,6-trichlorophenol,	[109,110]
One-dimensional TiO_2	Water borne pathogenic viruses	[111,112]
Ag doped TiO_2 nanofiber	Lead, arsenic	[113]
Ag doped TiO_2 nanofiber	Methylene blue	[113]
Crystalline TiO_2	*Escherichia coli* and *Staphylococcus aureus* in water Methylene blue, creatinine	[114]

than 80% of MB within 210 minutes. A nanocomposite of ZnO/PbS also plays an effective role in environmental remediation. ZnO NPs exhibit a strong antimicrobial effect for a wide variety of microorganisms [120]. Ag-doped ZnO is used for the degradation of acetaminophen and atenolol in the presence of visible light. Approximately 90% of acetaminophen and 70% of atenolol is degraded with the help of ZnO [121].

23.12 Other metal and metal oxides nanoparticles

Like the metals and metals oxides mentioned above, there are more metals that aid in environmental remediation whose fabrication is undertaken via green methods. Cerium oxide is used efficiently for the removal of Cr(VI) through adsorption on the surface [122]. Martin et al. observed the degradation of phosphate from wastewater with the help of ferric oxide [123]. A bimetallic complex of Cu/Al shows better degradation efficiency of CH_2Cl_2 and CCl_4 than ZVI [124]. CuO NPs are used to remove fluoride from the aqueous solution [125].

23.13 Conclusion

NPs provide great technology for the removal of contamination. Metal-based NPs are used for environmental remediation. Apart from metal-based NPs, nanomaterials like CNTs, graphene-based materials, and other polymer-supported NPs play an influential role in removing pollutants. Most metals and metal-based oxides are fabricated by green synthesis, which reduces metals from the environment without having a severe negative effect on living beings. Photocatalytic degradation of pollutants provides a cleaner approach for environmental remediation. TiO_2-based nanomaterials show high selectivity. The CNTs/metal oxides are promising materials for large-scale ecological remediation.

A number of studies have been done to observe the use of nanotechnology but at a small scale. So, a variety of study should be undertaken at large scales in order to understand the real environmental problems.

References

[1] T. Masciangoli, W. Zhang, Environmental technologies, Environ. Sci. Technol. 37 (2003) 102–108.

[2] A. Vaseashta, M. Vaclavikova, S. Vaseashta, G. Gallios, P. Roy, O. Pummakarnchana, Nanostructures in environmental pollution detection, monitoring, and remediation, Sci. Technol. Adv. Mater. 8 (2007) 47–59. Available from: https://doi.org/10.1016/j.stam.2006.11.003.

[3] F.I. Khan, A.K. Ghoshal, Removal of volatile organic compounds from polluted air, J. Loss Prev. Process Ind. 13 (2000) 527–545. Available from: https://doi.org/10.1016/S0950-4230(00)00007-3.

[4] P.G. Tratnyek, R.L. Johnson, Nanotechnologies for environmental cleanup, Nano Today 1 (2006) 44–48. Available from: https://doi.org/10.1016/S1748-0132(06)70048-2.

[5] P. Pichat, Appl. Catal., B 99 (2010) 428–434.

[6] V.J.P. Vilar, A.I.E. Gomes, V.M. Ramos, M.I. Maldonado, R.A.R. Boaventur, Photochem. Photobiol. Sci. 8 (2009) 691–698.
[7] M.M. Higarashi, W.F. Jardim, Catal. Today 76 (2002) 201–207.
[8] L. Zoua, Y. Luo, M. Hooper, E. Hu, Chem. Eng. Process. 45 (2006) 959–964.
[9] J.S. Dalton, P.A. Janes, N.G. Jones, J.A. Nicholson, K.R. Hallam, G.C. Allen, Environ. Pollut. 120 (2002) 415–422.
[10] I.K. Konstantinou, T.A. Albanis, Appl. Catal., B 49 (2004) 1–14.
[11] P. Hajkova, P. Spatenka, J. Horsky, I. Horska, A. Kolouch, Plasma Process. Polym. 4 (2007) S397–S401.
[12] W.F. Jardim, S.G. Moraes, M.M.K. Takiyama, Water Res. 31 (1997) 1728–1732.
[13] P. Xu, G.M. Zeng, D.L. Huang, C.L. Feng, S. Hu, M.H. Zhao, et al., Sci. Total. Environ. 424 (2012) 1–10.
[14] J.C. Comenares, R. Luque, J.M. Campelo, F. Colmenares, Z. Karpiński, A.A. Romero, Materials 2 (2009) 2228–2258.
[15] N. Bao, L. Shen, T. Takata, K. Domen, Chem. Mater. 20 (2008) 110–117.
[16] J.-S. Hu, L.-L. Ren, Y.-G. Guo, H.-P. Liang, A.-M. Cao, L.-J. Wan, et al., Angew. Chem., Int. Ed. 44 (2005) 1269–1273.
[17] H. Tong, S. Ouyang, Y. Bi, N. Umezawa, M. Oshikiri, J. Ye, Nano-photocatalytic materials: possibilities and challenges, Adv. Mater. 24 (2012) 229–251. Available from: https://doi.org/10.1002/adma.201102752.
[18] X. Zhao, L. Lv, B. Pan, W. Zhang, S. Zhang, Q. Zhang, Polymer-supported nanocomposites for environmental application: a review, Chem. Eng. J. 170 (2011) 381–394. Available from: https://doi.org/10.1016/j.cej.2011.02.071.
[19] M. Mauter, M. Elimelech, Environmental applications of carbon-based nanomaterials, Environ. Sci. Technol. 42 (2008) 5843–5859. Available from: https://doi.org/10.1021/es8006904.
[20] Y.C. Sharma, V. Srivastava, V.K. Singh, S.N. Kaul, C.H. Weng, Nano-adsorbents for the removal of metallic pollutants from water and wastewater, Environ. Technol. 30 (2009) 583–609. Available from: https://doi.org/10.1080/09593330902838080.
[21] X. Gui, J. Wei, K. Wang, A. Cao, H. Zhu, Y. Jia, et al., Carbon nanotube sponges, Adv. Mater. 22 (2010) 617–621. Available from: https://doi.org/10.1002/adma.200902986.
[22] L.Y. Ng, A.W. Mohammad, C.P. Leo, N. Hilal, Polymeric membranes incorporated with metal/metaloxide nanoparticles: a comprehensive review, Desalination 308 (2013) 15–33. Available from: https://doi.org/10.1016/j.desal.2010.11.033.
[23] W. Shen, Y. Qu, X. Pei, X. Zhang, Q. Ma, Z. Zhang, et al., Green synthesis of gold, nanoparticles by a newly isolated strain *Trichosporon montevideense* for catalytic hydrogenation of nitroaromatics, Biotechnol. Lett. 38 (9) (2016) 1503–1508.
[24] Y. Liu, X. Jin, Z. Chen, The formation of iron nanoparticles by *Eucalyptus leaf* extract and used to remove Cr (VI), Sci. Total. Environ. 627 (2018) 470–479.
[25] U.K. Sur, B. Ankamwar, S. Karmakar, A. Halder, P. Das, Green synthesis of silver nanoparticles using the plant extract of Shikakai and Reetha, Mater. Today Proc. 5 (1) (2018) 2321–2329.
[26] S. Jafarirad, M. Mehrabi, B. Divband, M. Kosari-Nasab, Biofabrication of zinc oxide nanoparticles using fruit extract of *Rosa canina* and their toxic potential against bacteria: a mechanistic approach, Mater. Sci. Eng. C. 59 (2016) 296–302.
[27] H. Fernández-Llamosas, L. Castro, M.L. Blázquez, E. Díaz, M. Carmona, Speeding up bioproduction of selenium nanoparticles by using *Vibrio natriegens* as microbial factory, Sci. Rep. 7 (1) (2017) 16046.
[28] S. Begum, M. Ahmaruzzaman, Green synthesis of SnO2 nanoparticles loaded on activated carbon and its application as photocalayst in the degradation of alizarin red S dye, Mater. Today Proc. 5 (1) (2018) 2314–2320.
[29] T.B. Devi, M. Ahmaruzzaman, Facile preparation of copper nanaoparticles using *Coccinia grandis* fruit extract and its application towards the reduction of toxic nitro compound, Mater. Today Proc. 5 (1) (2018) 2098–2104.

[30] A.A. Olajire, A. Kareem, A. Olaleke, Green synthesis of bimetallic Pt@Cu nanostructures for catalytic oxidative desulfurization of model oil, J. Nanostruct. Chem. 7 (2) (2017) 159–170.
[31] A. Mourato, M. Gadanho, A.R. Lino, R. Tenreiro, Biosynthesis of crystalline silver and gold nanoparticles by extremophilic yeasts, Bioinorg. Chem. Appl. 1 (2011) 1. Available from: https://doi.org/10.1155/2011/546074.
[32] P. Mukherjee, A. Ahmad, D. Mandal, et al., Fungus-mediated synthesis of silver nanoparticles and their immobilization in the mycelial matrix: a novel biological approach to nanoparticle synthesis, Nano Lett. 1 (2001) 515–519. Available from: https://doi.org/10.1021/nl0155274.
[33] B.K. Ravindra, A.H. Rajasab, A comparative study on biosynthesis of silver nanoparticles using four different fungal species, Int. J. Pharm. Pharm Sci. 6 (1) (2014) 372–376.
[34] K. Deplanche, L.E. Macaskie, Biorecovery of gold by *Escherichia coli* and *Desulfovibrio desulfuricans*, Biotechnol. Bioeng. 99 (2008) 1055–1064.
[35] D.-M. Zhou, Y.-J. Wang, H.-W. Wang, S.-Q. Wang, J.-M. Cheng, J. Hazard. Mater. 174 (2010) 34–39.
[36] C. Jung, J. Heo, J. Han, N. Her, S.-J. Lee, J. Oh, et al., Hexavalent chromium removal by various adsorbents: powdered activated carbon, chitosan, and single/multi-walled carbon nanotubes, Sep. Purif. Technol. 106 (2013) 63–71.
[37] M.M. Khin, A.S. Nair, V.J. Babu, R. Murugan, S. Ramakrishna, A review on nanomaterials for environmental remediation, Energy Environ. Sci. 5 (2012) 8075–8109. Available from: https://doi.org/10.1039/c2ee21818f.
[38] Y.L. Zhao, J.F. Stoddart, Acc. Chem. Res. 42 (2009) 1161–1171.
[39] Y.-H. Li, J. Ding, Z.K. Luan, Z.C. Di, Y.F. Zhu, C.L. Xu, et al., Carbon 41 (2003) 2787–2792.
[40] X. Peng, Y. Li, Z. Luan, Z. Di, H. Wang, B. Tian, et al., Chem. Phys. Lett. 376 (2003) 154–158.
[41] Y.-H. Li, Z. Luan, Z. Di, Y. Zhu, C. Xu, D. Wu, et al., Competitive adsorption of Pb, Cu and Cd ions from aqueous solutions by multiwalled carbon nanotubes, Carbon. 41 (2003) 2787–2792. Available from: https://doi.org/10.1016/S0008-6223(03)00392-0.
[42] G.P. Lithoxoos, A. Labropoulos, L.D. Peristeras, N. Kanellopoulos, J. Samios, I.G. Economou, Adsorption of N, CH, CO and CO gases in single walled carbon nanotubes: a combined experimental and Monte Carlo molecular simulation study, J. Supercrit. Fluids. 55 (2010) 510–523. Available from: https://doi.org/10.1016/j.supflu.2010.09.017.
[43] B. Bina, H. Pourzamani, A. Rashidi, M.M. Amin, J. Environ. Public Health (2012) 1. Available from: https://doi.org/10.1155/2012/ 817187, in press.
[44] L. Brunet, D.Y. Lyon, K. Zodrow, J.-C. Rouch, B. Caussat, P. Serp, et al., Environ. Eng. Sci. 25 (2008) 565–576.
[45] R.A. Khaydarov, R.R. Khaydarov, O. Gapurova, Water Res. 44 (2010) 1927–1933.
[46] H. Hyung, J.D. Fortner, J.B. Hugues, J.-H. Kim, Environ. Sci. Technol. 41 (2007) 179–184.
[47] Y. Yu, J.C. Yu, J.-G. Yu, Y.-C. Kwok, Y.-K. Che, J.-C. Zhao, et al., Appl. Catal., A 289 (2005) 186–196.
[48] T. Liu, H.Q. Tang, X.M. Cai, J. Zhao, D.J. Li, R. Li, Nucl. Instrum. Methods Phys. Res., Sect. B 264 (2007) 282–286.
[49] T. Liu, T. Huiqin, Z. Jie, L. Dejun, L. Ruying, S. Xueliang, Front. Mater. Sci. China 1 (2007) 147–150.
[50] S. Mu, Y. Long, S.-Z. Kang, J. Mu, Catal. Commun. 11 (2010) 741–744.
[51] S. Battiston, M. Minella, R. Gerbasi, F. Visentin, P. Guerriero, A. Leto, Carbon 48 (9) (2010) 2470–2477.
[52] X.L. Tan, M. Fang, X.K. Wang, J. Nanosci. Nanotechnol. 8 (2008) 5624–5631.
[53] H. Wang, X. Quan, H. Yu, S. Chen, Carbon 46 (2008) 1126–1132.
[54] C.-H. Wu, J. Hazard, Mater 144 (2007) 93–100.
[55] Z. Zhu, Y. Zhou, H. Yu, T. Nomura, B. Fugetsu, Chem. Lett. 35 (2006) 890–901.
[56] Y. Li, P. Zhang, Q. Du, X. Peng, T. Liu, Z. Wang, et al., Adsorption of fluoride from aqueous solution by graphene, J. Colloid Interface Sci. 363 (2011) 348–354. Available from: https://doi.org/10.1016/j.jcis.2011.07.032.
[57] M.Z. Iqbal, A.A. Abdala, Environ. Sci. Pollut. Res. 20 (2013) 3271.
[58] X. Mi, G.B. Huang, W.S. Xie, W. Wang, Y. Liu, J.P. Gao, Carbon 50 (2012) 4856.

[59] N.N. Zhang, H.X. Qiu, Y.M. Si, W. Wang, J.P. Gao, Carbon 49 (2011) 827.
[60] Y.Q. Chen, L.B. Chen, H. Bai, L. Li, J. Mater. Chem. A 1 (2013) 1992.
[61] T. Wu, M.X. Chen, L. Zhang, X.Y. Xu, Y. Liu, J. Yan, et al., J. Mater. Chem. A 1 (2013) 7612.
[62] F. Liu, S. Chung, G. Oh, T.S. Seo, ACS Appl. Mater. Interfaces 4 (2012) 922.
[63] J.P. Zhao, W.C. Ren, H.M. Cheng, J. Mater. Chem. 22 (2012) 20197.
[64] Y.X. Xu, Q.O. Wu, Y.Q. Sun, H. Bai, G.Q. Shi, ACS Nano 4 (2010) 7358.
[65] X. Liu, L. Pan, T. Lv, T. Lu, G. Zhu, Z. Sun, et al., Microwave-assisted synthesis of ZnO-graphene composite for photocatalytic reduction of Cr(vi) *Catal*, Sci. Technol. 1 (2011) 1189–1193. Available from: https://doi.org/10.1039/c1cy00109d.
[66] Q. Bao, D. Zhang, P. Qi, Synthesis and characterization of silver nanoparticle and graphene oxide nanosheet composites as a bactericidal agent for water disinfection, J. Colloid Interface Sci. 360 (2011) 463–470. Available from: https://doi.org/10.1016/j.jcis.2011.05.009.
[67] M. Seredych, T.J. Bandosz, Removal of ammonia by graphite oxide via its intercalation and reactive adsorption, Carbon 45 (2007) 2130–2132. Available from: https://doi.org/10.1016/j.carbon.2007.06.007.
[68] B. Song, C. Zhang, G. Zeng, J. Gong, Y. Chang, Y. Jiang, Antibacterial properties and mechanism of graphene oxide-silver nanocomposites as bactericidal agents for water disinfection, Arch. Biochem. Biophys 604 (2016) 167–176. Available from: https://doi.org/10.1016/j.abb.2016.04.018 [PubMed] [CrossRef] [ ]Google Scholar.
[69] P. Gao, K. Ng, D.D. Sun, Sulfonated graphene oxide-ZnO-Ag photocatalyst for fast photo degradation and disinfection under visible light, J. Hazard. Mater. 262 (2013) 826–835. Available from: https://doi.org/10.1016/j.jhazmat.2013.09.055 [PubMed] [CrossRef] [ ]Google Scholar.
[70] C.H. Deng, J.L. Gong, P. Zhang, G.M. Zeng, B. Song, H.Y. Liu, Preparation of melamine sponge decorated with silver nanoparticles-modified graphene for water disinfection, J. Colloid Interface Sci. 488 (2017) 26–38. Available from: https://doi.org/10.1016/j.jcis.2016.10.078 [PubMed] [CrossRef] [ ]Google Scholar.
[71] Y. Zhang, Z.-R. Tang, X. Fu, Y.-J. Xu, TiO graphene nanocomposites for gas-phase photocatalytic degradation of volatile aromatic pollutant: is TiO graphene truly different from other TiO carbon composite materials? ACS Nano 4 (2010) 7303–7314. Available from: https://doi.org/10.1021/nn1024219 [PubMed] [CrossRef] [ ]Google Scholar.
[72] M. Yang, N. Zhang, Y. Xu, Synthesis of fullerene-, carbon nanotube-, and graphene-TiO nanocomposite photocatalysts for selective oxidation: a comparative study, ACS Appl. Mater. Interfaces. 5 (2013) 1156–1164. Available from: https://doi.org/10.1021/am3029798.
[73] N. Zhang, M.-Q. Yang, Z.-R. Tang, Y.-J. Xu, CdS-graphene nanocomposites as visible light photocatalyst for redox reactions in water: a green route for selective transformation and environmental remediation, J. Catal. 303 (2013) 60–69. Available from: https://doi.org/10.1016/j.jcat.2013.02.026.
[74] M.S. Diallo, S. Christie, P. Swaminathan, J.H. Johnson, W.A. Goddard, Dendrimer enhanced ultra filtration. 1. Recovery of Cu(II) from aqueous solutions using PAMAM dendrimers with ethylene diamine core and terminal NH groups, Environ. Sci. Technol. 39 (2005) 1366–1377. Available from: https://doi.org/10.1021/es048961r.
[75] E. Sumesh, M.S. Bootharaju, Anshup, T. Pradeep, et al., J. Hazard. Mater. 189 (2011) 450–457.
[76] K.P. Lisha, Anshup, T. Pradeep, Gold. Bull. 42 (2009) 144–152.
[77] H. Uchida, S. Hatoh, M. Watanabe, Electrochim. Acta 43 (1998) 2111–2116.
[78] Q. Wang, H.J. Qian, Y.P. Yang, Z. Zhang, C. Naman, X.H. Xu, J. Contam. Hydrol. 114 (2010) 35–42.
[79] W. Tungittiplakorn, L.W. Lion, C. Cohen, J.Y. Kim, Engineered polymeric nanoparticles for soil remediation, Environ. Sci. Technol. 38 (2004) 1605–1610. Available from: https://doi.org/10.1021/es0348997.
[80] L. Cumbal, A.K. SenGupta, Environ. Sci. Technol. 39 (2005) 6508–6515.
[81] P. Yuan, M. Fan, D. Yang, H. He, D. Liu, A. Yuan, et al., J. Hazard. Mater. 166 (2009) 821–829.
[82] B.C. Pan, B.J. Pan, W.M. Zhang, Q.J. Zhang, Q.R. Zhang, P.J. Jiang, et al., Chinese Pat., CN, 2007, 10191355.3, 2007.

[83] F.D. Guerra, M.L. Campbell, D.C. Whitehead, F. Alexis, Capture of aldehyde VOCs using a series of amine-functionalized cellulose nanocrystals, ChemistrySelect 3 (2018) 5495–5501. Available from: https://doi.org/10.1002/slct.201703149.
[84] L.F. Wu, S.M.C. Ritchie, Chemosphere 63 (2006) 285–292.
[85] M.A. Hunne, O.J. Rojas, L.A. Lucia, M. Sain, BioResources 3 (2008) 929–980.
[86] K. Tennakone, I.R.M. Kottegoda, J. Photochem. Photobiol., A 93 (1996) 79–81.
[87] D.S. Wang, J. Zhang, Q. Luo, R. Guo, X.Y. Li, Y. Duan, et al., J. Hazard. Mater. 169 (1–3) (2009) 546–550.
[88] M. Rivero-Huguet, W.D. Marshall, J. Environ. Monit. 11 (2009) 1072–1079.
[89] Y. Matsumura, K. Yoshikata, S. Kunisaki, T. Tsuchido, Appl. Environ. Microbiol. 69 (2003) 4278–4281.
[90] Q.L. Feng, J. Wu, G.Q. Chen, F.Z. Cui, T.N. Kim, J.O. Kim, J. Biomed. Mater. Res. 52 (2000) 662–668.
[91] X. Xu, W.J. Brownlow, S.V. Kyriacou, Q. Wan, J.J. Viola, Biochemistry 43 (2004) 10400–10413.
[92] K.S. Gogoi, P. Gopina, A. Paul, A. Ramesh, S.S. Ghosh, A. Chattopadhyay, Langmuir 22 (2006) 9322–9328.
[93] J.L. Elechiguerra, J.L. Burt, J. Nanobiotechnol. 3 (2005) 6.
[94] S. Pal, Y.K. Tak, J.M. Song, Appl. Environ. Microbiol. 73 (2007) 1712–1720.
[95] A.S. Nair, N.P. Binoy, S. Ramakrishna, T.R.R. Kurup, L.W. Chan, C.H. Goh, et al., ACS Appl. Mater. Interfaces 1 (2009) 2413–2419.
[96] G. Jin, M.P. Prabhakaran, B.P. Nadappuram, G. Singh, D. Kai, S. Ramakrishna, J. Biomater. Sci, Polym. Ed. (2012). Available from: https://doi.org/10.1163/156856211X617399.
[97] T.M. Vogel, C.S. Criddle, P.L. McCarty, Environ. Sci. Technol. 21 (1987) 722.
[98] L.J. Matheson, P.G. Tratnyek, Environ. Sci. Technol. 28 (1994) 2045.
[99] Y. Xu, W. Zhang, Indus. Eng. Chem. Res. 39 (2000) 2238.
[100] S.M. Ponder, J.G. Darab, T.E. Mallouk, Environ. Sci. Technol. 34 (2000) 2564.
[101] J. Cao, W. Zhang, J. Hazard. Mater. 132 (2006) 213.
[102] S.S. Poguberović, D.M. Krčmar, S.P. Maletić, Z. Kónya, D.D.T. Pilipović, D.V. Kerkez, et al., Removal of As(III) and Cr(VI) from aqueous solutions using "green" zero-valent iron nanoparticles produced by oak, mulberry and cherry leaf extracts, Ecol. Eng. 90 (2016) 42–49. Available from: https://doi.org/10.1016/j.ecoleng.2016.01.083.
[103] S. Ding, L. Zhao, Y. Qi, Q. Lv, Preparation and characterization of lecithin-nano Ni/Fe for effective removal of PCB77, J. Nanomater. 2014 (2014) 678489. Available from: https://doi.org/10.1155/2014/678489.
[104] C. Mystrioti, T.D. Xanthopoulou, P. Tsakiridis, N. Papassiopi, A. Xenidis, Comparative evaluation of five plant extracts and juices for nanoiron synthesis and application for hexavalent chromium reduction, Sci. Total. Environ. 539 (2016) 105–113. Available from: https://doi.org/10.1016/j.scitotenv.2015.08.091.
[105] K. Nakata, A. Fujishima, TiO2 photocatalysis: design and applications, J. Photochem. Photobiol. C: Photochem. Rev. 13 (2012) 169–189.
[106] C. Kresge, M. Leonowicz, W. Roth, J. Vartuli, J. Beck, Nature 359 (1992) 710–712.
[107] M. Pelaez, A. de la Cruz, S. Elias, F. Polycarpos, D. Dionysiou, Catal. Today 144 (2009) 19–25.
[108] Y. Liu, J. Li, X. Qiu, C. Burda, *Novel TiO_2 nanocatalysts for wastewater purification: tapping energy from the sun*, Water Sci. Technol. 54 (8) (2006) 47–54. Available from: https://doi.org/10.2166/wst.2006.733.
[109] S. Rengaraj, X.Z. Li, J. Mol. Catal. A: Chem. 243 (2006) 60–67.
[110] M.V. Liga, E.L. Bryant, V.L. Colvin, Q. Li, Water Res. 45 (2011) 535–544.
[111] D.E. Giammar, C.J. Maus, L.Y. Xie, Environ. Eng. Sci. 24 (2007) 85–95.
[112] N. Deedar, A. Irfan, Q. Ishtiaq, J. Environ. Sci. 21 (2009) 402–408.
[113] C. Srisitthiratkul, V. Pongsorrarith, N. Intasanta, Appl. Surf. Sci. 257 (2011) 8850–8856.
[114] H. Choi, M.G. Antoniou, A.A. de la Cruz, E. Stathatos, D.D. Dionysiou, Desalination 202 (2007) 199–206.

[115] M. Cho, H. Chung, W. Choi, J. Yoon, Different inactivation behaviors of MS-2 phage and *Escherichia coli* in TiO photocatalytic disinfection, Appl. Environ. Microbiol. 71 (2005) 270–275. Available from: https://doi.org/10.1128/AEM.71.1.270-275.2005.
[116] L. Zan, W. Fa, T. Peng, Z. Gong, Photocatalysis effect of nanometer TiO and TiO-coated ceramicplate on Hepatitis B virus, J. Photochem. Photobiol. B Biol. 86 (2007) 165–169. Available from: https://doi.org/10.1016/j.jphotobiol.2006.09.002.
[117] X.B. Wang, W.P. Cai, Y.X. Lin, G.Z. Wang, C.H. Liang, J. Mater. Chem. 20 (2010) 8582–8590.
[118] Z. Yuhong, F. Li, H. Fuigui, W. Aiwu, W. Cai, Y. Jinping, et al., Green biosynthesis and characterization of zinc oxide nanoparticles using *Corymbia citriodora* leaf extract and their photocatalytic activity, Green. Chem. Lett. Rev. 8 (2015) 59–63. Available from: https://doi.org/10.1080/17518253.2015.1075069.
[119] C. Parthiban, N. Sundaramurthy, Biosynthesis, characterization of ZnO nanoparticles by using *Pyrus pyrifolia* leaf extract and their photocatalytic activity, Int. J. Innovative Sci. Eng. Technol. 4 (2015) 9710–9718.
[120] L.K. Adams, D.Y. Lyon, P.J.J. Alvarez, Water Res. 40 (2006) 3527–3532.
[121] B. Ramasamy, J. Jeyadharmarajan, P. Chinnaiyan, Environmental Science and Pollution Research, 2021.
[122] S. Recillas, J. Col_on, E. Casals, E. Gonz_alez, V. Puntes, A. S_anchez, et al., 2010, 184(1–3), 425–431.
[123] B.D. Martin, S.A. Parsons, B. Jefferson, Water Sci. Technol. 60 (2009) 2637–2645.
[124] H.L. Lien, W.-X. Zhang, Chemosphere 49 (2002) 371–378.
[125] J.O. Ighalo, P.A. Sagboye, G. Umenweke, (...), S. Ogunniyi, A.G. Adeniyi Environ. Nanotechnol. Monit. Manage. 15 (2021) 100443.

CHAPTER 24

Green and sustainable future and conclusion

Deepankara Vrushabhadas Shastri[1], Shilpa Rajkumar[1], J. John Donald Raj[1], S. Kishore Babu[1] and Kantha D. Arunachalam[2]
[1]Department of Physics and Nanotechnology, SRM Institute of Science and Technology, Kattankulathur, Tamil Nadu, India
[2]Faculty of Sciences, Marwadi University, Rajkot, Gujarat, India

24.1 Introduction

Nanotechnology goods, processes, and applications are predicted to contribute significantly to environmental and climate protection by conserving raw resources, energy, and water, as well as lowering greenhouse gas emissions and hazardous waste. As a result, the usage of nanomaterials is likely to have some environmental benefits and long-term implications [1–6].

Please keep in mind that nanotechnology now plays only a minor part in environmental protection, both in research and in actual applications. In their respective domains, environmental engineering firms pay little attention to nanotechnology.

Increased consumer awareness of environmental protection, along with rising raw material and energy prices, has resulted in a huge variety of products on the market that claim to have specific environmental benefits.

Nanomaterials have unique physical and chemical properties, making them ideal for developing new green goods. Improving the durability of materials against mechanical stress or weathering helps to extend the life of products; nanotechnology-based dirt and waterproof coatings can reduce cleaning work; new insulating materials can improve building energy efficiency; and adding nanoparticles to the material to reduce weight and save energy during transportation are just a few examples.

Nanomaterials are utilized in the chemical industry to increase energy and resource efficiency due to their unique catalytic capabilities. In some cases, nanomaterials can be used to substitute environmentally hazardous substances.

Nanotechnology-optimized energy production and storage goods and processes have high prospects; they are currently in the development stage and are expected to make substantial contributions to climate protection and fixing our energy challenges in the future.

Industrial Applications of Nanoceramics
DOI: https://doi.org/10.1016/B978-0-323-88654-3.00029-9

The primary goal of most "nanoconsumer items" on the market is not environmental conservation. Neither the use of nanosilver in textiles to fight sweat nor the use of carbon nanotubes in golf clubs assist in safeguarding the environment. Manufacturers frequently make such claims, but rarely back them up with evidence [7,8].

Self-cleaning or stain-resistant textile surface coatings are examples of products that are touted as decreasing cleaning efforts and thereby save energy, water, and detergents.

The focus is usually on the long-term potential of where nanotechnology can take us. This, on the other hand, usually indicates untested expectations. To determine the product's true environmental impact (both positive and negative), the complete life cycle must be verified, from raw material production to disposal at the end of the cycle.

In general, the quantity of resources and energy used to manufacture a product is not taken into account when describing environmental benefits.

The goals of green nanotechnology for sustainable future include a variety of objectives. The basic purpose of green technology is to meet social demands in a way that does not deplete or devour the Earth's natural resources. Here are a few examples:

1. Sustainability addresses current needs without jeopardizing future generations' ability to meet their own.
2. Reduction of the number of sources.
3. Changes the manufacturing and consumption patterns to reduce waste and productivity.
4. Creates new and innovative alternatives to fossil fuels and chemical-intensive agriculture systems that have been demonstrated to be harmful to human health and the environment.
5. In energy applications, the most serious issue may be green technology. Alternative fuels, new power generation methods, and energy efficiency are all part of this.
6. Green construction encompasses all areas of construction, from the materials used to the site of the structure.
7. Green chemistry is the development, design, and usage of chemical products and processes that decrease or eliminate the use and production of harmful substances.

24.2 Green nanotechnology and its implications on the environment

Here are some ecologically friendly items, as the name green implies. Green nanotechnology is defined as "the use of twelve principles to reduce and eliminate the use or generation of hazardous substances in the design, production, and application of chemicals and processes," while green chemistry is defined as "the use of twelve principles to reduce and eliminate the use or generation of hazardous substances in the design, production, and application of chemicals and processes" [9–12].

Global climate change and pollution, such as air, soil, and water, are all aspects of environmental sustainability. Even in today's wealthy countries, climate change is a contentious issue. In the fight against climate change, aside from the technical scope, direct proof from research, public awareness, and government must all work together.

Environmental sustainability includes global climate change as well as pollution in the air, land, and water. Climate change is a divisive topic even in today's wealthier countries. Apart from the technological scope, direct evidence from study, public awareness, and government collaboration is also required in the fight against climate change.

Environmental bulk materials in the nanometer range, such as metals, metal oxides/sulfides, polymers, and carbon, exhibit fascinating physical and chemical properties due to their high surface-to-volume ratio. As a result, when compared to a wide number of competitors, multifunctional functions and improved responsiveness/selectivity can be predicted.

The capabilities of environmental nanotechnology can be further enhanced by altering nanomaterials in terms of morphology (particle size and shape), microstructure (pores and surfaces), and composition (heterojunction and doping).

New potential and difficulties for adsorption and/or catalytic performance have arisen as a result of the advent of new distinguishing characteristics such as electrical, thermal, optical, magnetic, and electrochemical properties [11].

24.3 Future challenges

Despite the fact that much research has been done and published on nanoparticles, mature manufacture of the requisite nanomaterials and their safe applications is not always possible. The following are the special issues of environmental nanotechnology, in addition to the generic challenges of nanotechnology.

1. The right mix of activity, selectivity, and consistency: High yield is attributed to high reactivity due to fine surface and microstructure, whether it be adsorption efficiency or catalytic efficiency in environmental applications. For specific environmental challenges, selective adsorption and/or degradation are also necessary. Furthermore, extreme and harmful conditions, which are frequent in environmental applications, can readily degrade the nanocatalyst's composition and/or structure. Stability will be questioned as a fundamental indicator of practical feasibility. Stability is more critical in environmental applications than it is in chemical manufacture and energy because secondary contamination might result from metal leaching from metal-based nanocatalysts and/or newly generated intermediates [13]. Nanomaterials must be designed with activity, selectivity, and stability in mind for precise application control.
2. Mechanistic research: Nanomaterial-powered environmental applications will not only boost performance, but will also initiate diverse processes with distinct mechanisms [8]. The novel mechanism can be deduced from the material's size,

shape, composition, surface properties, or hybridization. The materials' characteristics can also influence competitive adsorption or catalysis of nanoparticles. To optimize environmental nanotechnology, environmental processes and materials science must be elucidated. Extinction tests, in situ spectroscopy, and computational studies are all required, as well as their combination. Chemical engineering work is also necessary for putting theoretical discoveries into practice in the environmental field.

3. Nanocontamination and the nano/biological interface: In most circumstances, possibilities and obstacles occur in tandem, and the rise of nanomaterials is no exception [11]. Nanomaterials and nanotechnology have enabled a variety of biological applications, including detection, diagnosis, and healing, that are impossible to do with bulk materials. At the same time, fate and transportation, human and environmental exposure, and direct injury to the human body should all be considered. Nanopollution is no longer a topic of discussion, but rather a serious threat. Environmental nanotechnology is facing a huge problem in terms of basic knowledge, recognition, and functioning of nano/biological interactions. Much research has been done on the future of nanomaterials and nanotechnology [12,14–26].

24.4 Conclusion

Efforts to monitor and regulate pollution in the environment should be increased. Traditional environmental protection technologies have great prospects thanks to the development of nanomaterials and nanotechnology. As a result, environmental nanotechnology has grown in popularity. Several effective applications have been shown, such as environmental nanocatalysis, selective adsorption of nanoabsorbents, nanomembrane separation, and new detection and monitoring. However, these pressing issues still face significant obstacles. A precise balance must be struck between activity, reactivity, and stability in the rational design and better use of nanomaterials in environmental processes. Nanotechnology and environmental processes should be the subjects of mechanical study. Through the process of unraveling the nano/bio interface, the environmental dangers of nanomaterials must also be thoroughly investigated. Environmental nanotechnology's engagement and efforts will contribute to the future's overall sustainability.

Given that nanotechnology is predicted to become a trillion-dollar industry in the next 10 years, it is critical to examine the influence of each stage of the nanoproduct life cycle on the environment and human health using a life cycle methodology (occupational and final use). Please recycle the product before drawing any conclusions regarding its environmental benefits or shortcomings. Current life cycle evaluation methodologies created for traditional bulk materials, on the other hand, must be evaluated and, if necessary, changed to be acceptable for evaluating nanoparticles. Nano-LCRA and comprehensive environmental assessment are a combination of the two

LCA and risk assessment frameworks. It appears to be very helpful in determining the whole impact of nanoparticles on human health. Using green chemical indicators to analyze the greenness of nanomaterials and nanofabrication processes, as well as applying green chemistry principles to nanofabrication technologies, are some recommended methods to ensure that nanomaterials have an overall "positive" impact. Future pathways for this universal technology include taking a more proactive approach when designing and developing new nanoproducts.

References

[1] Communication from the Commission—Towards a European Strategy for Nanotechnology. 2004.
[2] F. Boccuni, B. Rondinone, C. Petyx, S. Iavicoli, Potential occupational exposure to manufactured nanoparticles in Italy, J. Clean. Prod 16 (8–9) (2008) 949–956. Available from: https://doi.org/10.1016/j.jclepro.2007.04.021.
[3] B.R. Allenby, D. Rejeski, The industrial ecology of emerging, J. Ind. Ecol 12 (3) (2008) 267–269. Available from: https://doi.org/10.1111/j.1530-9290.2008.00051. Available from: http://www3.interscience.wiley.com/cgi-bin/fulltext/121395874/PDFStechnologiesTART.
[4] National Nanotechnology Intiative, Nanotech Facts: What is Nanotechnology, 2010.
[5] World Commission on Environment and Development. Our Common Future (The Brundtland Report), 1987.
[6] D. Vrushabhadas, A.S. Bhaskar, K.D. Arunachalam, A comprehensive study on intrinsic alpha corpuscular radiation damage in monazite crystals using picometer-scale imaging, coupled with SCXRD and Raman spectroscopy, Radiat. Phys. Chem. 212 (2023) 111131. Available from: https://doi.org/10.1016/j.radphyschem.2023.111131.
[7] M.A. Curran, P. Frankl, R. Heijungs, A. Kohler, S.I. Olsen, Nanotechnology and Life Cycle Assesment—A Systems Approach to Nanotechnology and the Environment, 2007.
[8] X. Duan, H. Sun, S. Wang, Metal-free carbocatalysis in advanced oxidation reactions, Acc. Chem. Res 51 (3) (2018) 678–687. Available from: https://doi.org/10.1021/acs.accounts.7b00535, http://pubs.acs.org/journal/achre4.
[9] A. Garner, G.A. Keoleian, Industrial ecology: an introduction; The National Pollution Prevention. 1995.
[10] T. Fleischer, A. Grunwald, Making nanotechnology developments sustainable. A role for technology assessment? J. Clean. Prod. 16 (8–9) (2008) 889–898. Available from: https://doi.org/10.1016/j.jclepro.2007.04.018.
[11] M.F. Hochella, D.W. Mogk, J. Ranville, I.C. Allen, G.W. Luther, L.C. Marr, et al., Natural, incidental, and engineered nanomaterials and their impacts on the Earth system, Science 363 (6434) (2019). Available from: https://doi.org/10.1126/science.aau8299, https://science.sciencemag.org/content/363/6434/eaau8299/tab-pdf.
[12] B.P. Karn, L. Bergeson, Green nanotechnology: straddling promise and uncertainty, Nat. Resour. Environ. 24 (2) (2009) 9–23.
[13] E. Saputra, S. Muhammad, H. Sun, H.M. Ang, M.O. Tadé, S. Wang, Different crystallographic one-dimensional MnO2 nanomaterials and their superior performance in catalytic phenol degradation, Environ. Sci. Technol 47 (11) (2013) 5882–5887. Available from: https://doi.org/10.1021/es400878c.
[14] K. Watlington, Emerging Nanotechnologies for Site Remediation and Wastewater Treatment, US EPA, 2005.
[15] J.A. Rogers, M.G. Lagally, R.G. Nuzzo, Synthesis, assembly and applications of semiconductor nanomembranes, Nature. 477 (7362) (2011) 45–53. Available from: https://doi.org/10.1038/nature10381.

[16] H. Şengül, T.L. Theis, S. Ghosh, Toward sustainable nanoproducts: an overview of nanomanufacturing methods, J. Ind. Ecol 12 (3) (2008) 329–359. Available from: https://doi.org/10.1111/j.1530-9290.2008.00046, http://www3.interscience.wiley.com/cgi-bin/fulltext/121381837/PDFSTART.
[17] B. Tonn, A. Hemrick, F. Conrad, Cognitive representations of the future: survey results, Futures. 38 (7) (2006) 810–829. Available from: https://doi.org/10.1016/j.futures.2005.12.005.
[18] O.M. Koo, I. Rubinstein, H. Onyuksel, Role of nanotechnology in targeted drug delivery and imaging: a concise review, Nanomedicine 1 (3) (2005) 193–212. Available from: https://doi.org/10.1016/j.nano.2005.06.004.
[19] L. Shi, Y. Yin, L.C. Zhang, S. Wang, M. Sillanpää, H. Sun, Design and engineering heterojunctions for the photoelectrochemical monitoring of environmental pollutants: a review, Appl. Catal. B: Environ. 248 (2019) 405–422. Available from: https://doi.org/10.1016/j.apcatb.2019.02.044, http://www.elsevier.com/inca/publications/store/5/2/3/0/6/6/index.htt.
[20] M.M. Khin, A.S. Nair, V.J. Babu, R. Murugan, S. Ramakrishna, A review on nanomaterials for environmental remediation, Energy Environ. Sci. 5 (8) (2012) 8075–8109. Available from: https://doi.org/10.1039/c2ee21818f, http://www.rsc.org/Publishing/Journals/EE/About.asp.
[21] J. Parmar, D. Vilela, K. Villa, J. Wang, S. Sánchez, Micro- and nanomotors as active environmental microcleaners and sensors, J. Am. Chem. Soc 140 (30) (2018) 9317–9331. Available from: https://doi.org/10.1021/jacs.8b05762, http://pubs.acs.org/journal/jacsat.
[22] D.E. Meyer, M.A. Curran, M.A. Gonzalez, An examination of existing data for the industrial manufacture and use of nanocomponents and their role in the life cycle impact of nanoproducts, Environ. Sci. Technol 43 (5) (2009) 1256–1263. Available from: http://pubs.acs.org/doi/pdfplus/10.1021/es8023258.10.1021/es8023258.
[23] J. Theron, J.A. Walker, T.E. Cloete, Nanotechnology and water treatment: applications and emerging opportunities, Crit. Rev. Microbiol. 34 (1) (2008) 43–69. Available from: https://doi.org/10.1080/10408410701710442.
[24] A.K. Sinha, K. Suzuki, M. Takahara, H. Azuma, T. Nonaka, K. Fukumoto, Mesostructured manganese oxide/gold nanoparticle composites for extensive air purification, Angew. Chem. – Int. Ed 46 (16) (2007) 2891–2894. Available from: https://doi.org/10.1002/anie.200605048.
[25] R. Shilpa, R. Mudachathi, Increasing the free spectral range of Fabry-Perot cavities using dissimilar 1D photonic crystals, Mater. Lett. 299 (2021) 130089. Available from: https://doi.org/10.1016/j.matlet.2021.130089.
[26] R. Shilpa, Abhinaya, Yamuna, R. Mudachathi, Thermal and layer thickness dependence of defect and band edge transmissions in 1D PCs based on ZnS/CaF_2 multilayers, IOP Conf. Ser. Mater. Sci. Eng. 1219 (2022) 012015. Available from: http://doi.org/10.1088/1757-899X/1219/1/012015.

Index

Note: Page numbers followed by "*f*" and "*t*" refer to figures and tables, respectively.

F

G

N

T

CPI Antony Rowe
Eastbourne, UK
June 11, 2024